LABORATORY MANUAL SEELEY'S PRINCIPLES OF
ANATOMY & PHYSIOLOGY
SECOND EDITION

ERIC WISE
SANTA BARBARA CITY COLLEGE

LABORATORY MANUAL TO ACCOMPANY SEELEY'S PRINCIPLES OF ANATOMY & PHYSIOLOGY, SECOND EDITION

Published by McGraw-Hill, a business unit of The McGraw-Hill Companies, Inc., 1221 Avenue of the Americas, New York, NY 10020. Copyright © 2012 by The McGraw-Hill Companies, Inc. All rights reserved. Previous edition © 2009. No part of this publication may be reproduced or distributed in any form or by any means, or stored in a database or retrieval system, without the prior written consent of The McGraw-Hill Companies, Inc., including, but not limited to, in any network or other electronic storage or transmission, or broadcast for distance learning.

Some ancillaries, including electronic and print components, may not be available to customers outside the United States.

✪ This book is printed on recycled, acid-free paper containing 10% postconsumer waste.

1 2 3 4 5 6 7 8 9 0 QDB/QDB 1 0 9 8 7 6 5 4 3 2 1

ISBN 978–0–07–735128–1
MHID 0–07–735128–2

Vice President, Editor-in-Chief: *Marty Lange*
Vice President, EDP: *Kimberly Meriwether David*
Senior Director of Development: *Kristine Tibbetts*
Executive Editor: *James F. Connely*
Developmental Editor: *Ashley Zellmer*
Marketing Manager: *Denise M. Massar*
Project Coordinator: *Mary Jane Lampe*
Senior Buyer: *Laura Fuller*
Designer: *Tara McDermott*
Cover Designer: *Elise Lansdon*
Cover Image: © *Getty Images/David Sacks*
Senior Photo Research Coordinator: *John C. Leland*
Compositor: *Aptara, Inc.*
Typeface: *10/12 Minion*
Printer: *Quad/Graphics*

All credits appearing on page or at the end of the book are considered to be an extension of the copyright page.

Photos Courtesy of Eric Wise except:
12.19: © The McGraw-Hill Companies, Inc./Cordi Smith, photographer
TA 13.2: © The McGraw-Hill Companies, Inc./Cordi Smith, photographer
43.12b: © The McGraw-Hill Companies, Inc./Cordi Smith, photographer

Some of the laboratory experiments included in this text may be hazardous if materials are handled improperly or if procedures are conducted incorrectly. Safety precautions are necessary when you are working with chemicals, glass test tubes, hot water baths, sharp instruments, and the like, or for any procedures that generally require caution. Your school may have set regulations regarding safety procedures that your instructor will explain to you. Should you have any problems with materials or procedures, please ask your instructor for help.

www.mhhe.com

CONTENTS

Instructor Preface iv
Student Preface viii
Working in the Lab x

Laboratory Exercises

1. Organs, Systems, and Organization of the Body 1
2. Microscopy 13
3. Cell Structure and Function 23
4. Tissues 41
5. Integumentary System 59
6. Introduction to the Skeletal System 69
7. Appendicular Skeleton 81
8. Axial Skeleton: Vertebrae, Ribs, Sternum, Hyoid 97
9. Axial Skeleton: Skull 111
10. Articulations 129
11. Muscle Physiology 147
12. Introduction to the Study of Muscles and Muscles of the Shoulder and Upper Extremity 159
13. Muscles of the Hip, Thigh, Leg, and Foot 189
14. Muscles of the Head and Neck 209
15. Muscles of the Torso 221
16. Introduction to the Nervous System 233
17. Brain and Cranial Nerves 241
18. Spinal Cord and Somatic Nerves 263
19. Nervous System Physiology: Stimuli and Reflexes 277
20. Introduction to Sensory Receptors 287
21. Taste and Smell 293
22. Eye and Vision 299
23. Ear, Hearing, and Balance 311
24. Endocrine System 325
25. Blood 337
26. Blood Tests and Typing 345
27. Structure of the Heart 353
28. Electrical Conductivity of the Heart 365
29. Functions of the Heart 371
30. Introduction to Blood Vessels and Arteries of the Upper Body 379
31. Arteries of the Lower Body 395
32. Veins and Special Circulations 403
33. Function of Vessels and the Lymphatic System 421
34. Blood Vessels and Blood Pressure 429
35. Structure of the Respiratory System 435
36. Respiratory Function, Breathing, and Respiration 447
37. Physiology of Exercise and Pulmonary Health 459
38. Anatomy of the Digestive System 465
39. Digestive Physiology 487
40. Anatomy of the Urinary System 495
41. Urinalysis 507
42. Male Reproductive System 515
43. Female Reproductive System 525

Appendix A Measurement Conversions A-1
Appendix B Preparation of Materials A-1
Appendix C Lab Reports A-3
Index I-1

INSTRUCTOR PREFACE

Anatomy and physiology can be the crown jewel in our students' education or the bane of their college career. This lab manual was written for the undergraduate student of anatomy and physiology, and it consists of 43 exercises designed to help students learn basic human anatomy and the practical lab applications in physiology.

The diversity of interests in today's anatomy and physiology students is due, in part, to the number of majors that either require or recommend the subject. This lab manual provides a framework for understanding anatomy and physiology for students interested in nursing, radiology, physical or occupational therapy, physical education, dental hygiene, and other allied health majors.

This manual was written to be used with *Seeley's Principles of Anatomy and Physiology*, by Phil Tate. The illustrations are labeled; therefore, students do not need to bring their lecture text to the lab. The lab manual accompanies the lecture text and lecture portion of the course and can be used in either a one-term or a full-year course. The illustrations are outstanding, and the balanced combination of line art and photographs provides effective coverage of material. The amount of lecture material in the manual is limited, so there is little material included that is not part of the lab experience.

Practical lab experience is an invaluable opportunity to reinforce lecture concepts, enrich students' understanding of anatomy and physiology, and allow them to explore new dimensions in the subject area. The educational benefit of reinforcing lecture material with hands-on experiments and acquiring knowledge with a learn-by-doing philosophy makes the anatomy and physiology lab a very special educational environment. Many of us use lab experiences to present conceptually difficult material in physiology and to provide students with different learning styles another avenue for learning.

The 43 exercises in this lab manual provide a comprehensive overview of the human body. Each exercise presents the core elements of the subject matter. As an instructor, you can tailor this manual to match your own vision of the course or use it in its entirety. There are significant differences among anatomy and physiology laboratories, and the advances in physiology equipment, especially computer modules, are numerous and continually evolving. The materials section in each lab is designed for a lab of 24 students and includes the amounts and types of reagents to be used. The labs generally take between 2 and 3 hours to complete.

This lab manual was written for three types of anatomy and physiology courses. For courses that use the cat as the primary dissection animal, cat dissections or mammalian organ dissections follow the material on humans. For courses that use models or charts, numerous cadaver photographs are included so that students can see the representative structures as they exist in the cadaver material. Finally, for courses that use cadavers, this lab manual can be used by studying the human material and omitting the cat dissection sections.

WHAT'S NEW IN THE SECOND EDITION?

Exercise 1
Includes new examples for parietal and visceral

Exercise 2
Entire exercise underwent rewrite for clarity and better flow
Microscope-use directions are now numbered
Included expanded tips on how to use the microscope.
Resolution is now clearly defined

Exercise 3
Revision of objectives for clarity.
Added material to and rewrote the cellular function section.

Exercise 4
Added material on extracellular matrix.
Changed figures for clarity.
Reorganized epithelial table (4.1).

Exercise 5
Added a critical thinking review question.

Exercise 6
Added new objective for bone growth

Exercise 7
Rewrote introduction and added an overview table of the appendicular skeleton (table 7.1).
Added articulation features of the cuneiform bones.

Exercise 8
Rewrote and reorganized the objectives for clarity.
Entire exercise underwent rewrite for clarity and better flow
Reorganization of sternum in figure 8.10 for clarity

Exercise 9
Improved mnemonic for facial and cranial bones.
Reorganized anterior view section of skull.

Exercise 10
Rewrote the introduction for improved flow.
Rewrote sections for clarity.
Addition of activities, such as palpating joints.
Improved graphic for bursa and tendon sheath.

Exercise 11
Updated Ph.I.L.S. simulations to reflect changes in Ph.I.L.S. 3.0

Exercise 12
Consolidated exercise with an Overview of Muscles section.
Reorganized Objectives section to reflect sequence of events in exercise.
Added some critical thinking questions about muscle use in the review section.

Exercise 13
Reorganized objective section.
Minor reorganization of exercise for better flow.
Inclusion of lateral hip rotators.
Addition of activities such as palpation of muscles.
Increased detail in figure 13.1.
Improved photograph of right posterior hip.

Exercise 14
Rewrote exercise for clarity.
Added cat musculature questions to review section.

Exercise 15
Reorganized objectives.
Revised erector spinae muscles into three groups.

Exercise 16
Reordered objectives for flow. Rewrote and reorganized exercise for flow and clarity.

Exericse 17
Rewrote the association areas and primary sensory areas along with other parts of the exercise for clarity.

Exercise 19
Added virtual experiments from Ph.I.L.S. 3.0

Exercise 20
Added graphic on sensory receptors of the skin (figure 20.1c)

Exercise 21
Added and rewrote several sections of the exercise for clarity.

Exercise 22
Made major revisions of the interior of the eye for clarity.
Added additional review questions.

Exercise 23
Modified illustrations for clarity.

Exercise 24
Added sections on receptor sites for hormones, endocrine control of homeostasis, and negative feedback mechanisms.
Added review activity for hormones.

Exercise 25
Improved blood smear photograph in review section.

Exercise 26
Added ELISA section for testing blood.

Exercise 27
Added section on how to locate the coronary arteries.
Revised artwork where appropriate.

Exercise 28
Revised cardiac conduction system.

Exercise 29
Added Ph.I.L.S. virtual experiment on frog heart physiology.

Exercise 31
Revised review questions.

Exercise 32
Rewrote objectives to correlate with exercise.
Added a fetal blood flow question to review section.

Exercise 34
Added question on blood pressure.

Exercise 35
Numbered cat dissection procedures.
Redrew cross section of trachea.

Exercise 36
Expanded and rewrote introduction and objectives.
Modified procedure section, added brief description of Boyle's Law, added an alternative spirometry for schools on limited budget, and expanded.

Exercise 38
Numbered procedures for study of digestive anatomy.

Exercise 40
Revised Introduction and added material to Objectives.
Extensive revision of blood flow through the kidney.
Numbered the procedure of kidney dissection and examination of the kidney, bladder and cat dissection for clarity.
Added paragraph on juxtaglomerular apparatus.
Added additional review questions. Revised overview of kidney photomicrograph (figure 40.7).

Exercise 41
Added additional objective and reorganized objectives to correlate with exercise. Significantly expanded procedure section.
Expanded table 41.1 for ease of completion by student.
Revised review section for depth and breadth.
Modified urine sediment figures (41.2 and 41.3).

Exercise 42
Added additional objectives to correlate with exercise.
Revised epididymis section and spermatic cord section.
Added a review questions on developmental stages in sperm development and pathway of sperm from formation to release.
Added artwork of histology of testis to correlate with photomicrograph (figure 42.4)

Exercise 43
Added additional objective and reorganized objectives to correlate with exercise.
Defined oogenesis.

KEY FEATURES

1. **Dynamic art program.** The figures in this laboratory manual have been carefully rendered to convey realistic, three-dimensional detail.

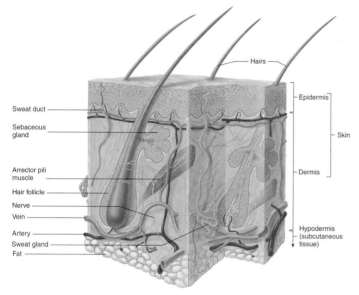

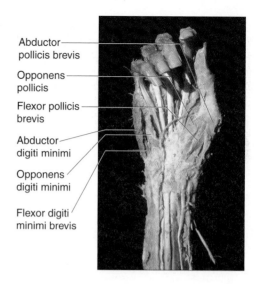

2. **Instructional photographs.** Numerous full-color photomicrographs and dissection images prepare students for what they will encounter in the lab or can supplement discussions when hands-on labs are not available. These labeled photographic references also preserve a record of the lab experience long after it has passed.

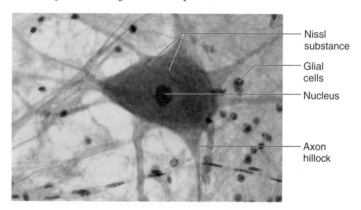

3. **Labels.** Illustrations are labeled for students to learn the names and terminology by looking at real-life examples or models and by referring to the illustrations in the manual.
4. **Focus on the laboratory.** This manual focuses primarily on the material necessary for the laboratory and does not repeat the material presented in the lecture text, with the expectation that students can look up material in the lecture text when necessary.
5. **Integrated use of the cat for dissection specimen.** The cat is used as the dissection animal; however, it is integrated with material on human anatomy, so that animals do not have to be relied upon as dissection specimens if so desired.
6. **Safety.** Safety guidelines appear in the inside front cover for reference. The international symbol for caution (⚠) is used throughout the manual to identify material that the reader should pay close and special attention to when preparing for or performing the laboratory exercise.
7. **Clean up.** At the end of many laboratory exercises, an icon for clean up (✋) reminds the student to clean up the laboratory. Special instructions are given where appropriate.
8. **Data collection.** Collection of data is embedded within each exercise, instead of in a separate table at the back of the manual.
9. **User-friendly format.** Each exercise begins on a right-hand page, and the pages are perforated to allow students to more easily remove the exercises to turn them in and later store them.
10. **Key terms.** Current anatomic terminology is used throughout the laboratory manual. Key terms are boldfaced.

TEACHING AND LEARNING SUPPLEMENTS

In addition to this laboratory manual, an extensive array of supplemental materials is available for use in conjunction with *Seeley's Principles of Anatomy and Physiology*, second edition, by Phil Tate. Instructors can obtain teaching aids to accompany this laboratory manual by visiting www.mhhe.com/tate2

- **Instructor's Manual for the Laboratory Manual** Visit www.mhhe.com/tate2 to view and print the instructor's manual. This helpful preparation guide includes suggestions for coordinating lab exercises with the textbook, set-up instructions and materials lists, and answers to the laboratory review questions at the end of each exercise.
- **Textbook-specific study tools**—The ARIS website features quizzes, interactive learning games, and study tools tailored to coincide with each chapter of *Principles of Anatomy and Physiology*.
- **Course assignments and announcements**—Students of instructors choosing to utilize McGraw-Hill's ARIS tools for course administration will receive a course code to log into their specific course for assignments.
- **Study on the fly content**—Now students with portable media players can study anywhere, anytime.
  - Audio tutorials, developed by topic and correlated to each chapter
  - Audio quizzes
  - Animations

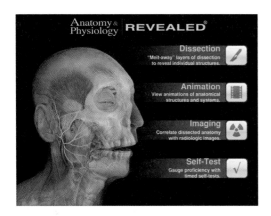

ANATOMY & PHYSIOLOGY REVEALED 3.0

Now covering all body systems, this amazing multimedia tool is designed to help students learn and review human anatomy using cadaver specimens. Detailed cadaver photographs blended with a state-of-the-art layering technique provide a uniquely interactive dissection experience. This easy-to-use program features the following sections:

- **Dissection**—Peel away layers of the human body to reveal structures beneath the surface. Structures can be pinned and labeled, just as in a real dissection lab. Each labeled structure is accompanied by detailed information and an audio pronunciation. Dissection images can be captured and saved.
- **Animation**—Compelling animations demonstrate muscle actions, clarify anatomical relationships, view physiological processes, and explain difficult concepts.
- **Histology**—Labeled light micrographs presented with each body system allow students to study the cellular detail of tissues at their own pace.
- **Imaging**—Labeled X-ray, MRI, and CT images familiarize students with the appearance of key anatomical structures as seen through different medical imaging techniques.
- **Customization**—Instructors can create structure lists for their courses.
- **Self-test**—Challenging exercises let students test their ability to identify anatomical structures in a timed practical exam format or traditional multiple choice. A results page provides analysis of test scores plus links back to all incorrectly identified structures for review.
- **Search**—Users can type in a term to quickly locate any structure in the program. Visit www.aprevealed.com for more information.

Physiology Interactive Lab Simulations (Ph.I.L.S.)

Ph.I.L.S. 3.0 is the perfect way to reinforce key physiology concepts with powerful lab experiments. Created by Dr. Phil Stephens at Villanova University, this program offers *37 laboratory simulations* that may be used to supplement or substitute for wet labs. All 37 labs are self-contained experiments—no lengthy instruction manual required. Users can adjust variables, view outcomes, make predictions, draw conclusions, and print lab reports. This easy-to-use software offers the flexibility to change the parameters of the lab experiment—there is no limit!

Ph.I.L.S. 3.0 offers a *new* interface which makes it even easier to practice lab simulations. *New* post-lab quizzes offer additional interactive assessment opportunities!

"MediaPhys" Tutorial

This physiology study aid offers detailed explanations, high-quality illustrations, and animations to provide a thorough introduction to the world of physiology. MediaPhys is filled with interactive activities and quizzes to help reinforce physiology concepts that are often difficult to understand.

ACKNOWLEDGMENTS

Many people have been involved in the production of this lab manual, and I would like to thank the editorial and marketing teams at McGraw-Hill, including Marty Lange, Michelle Watnick, Denise Massar, James Connely, and Ashley Zellmer. Thanks also go to the production team—Mary Jane Lampe, John Leland, and Tara McDermott—for their input and encouragement.

I would like to dedicate this book to my wife, Ashley.

Please feel free to write me or e-mail me with your comments, suggestions, and criticisms. I value your input and hope that your comments will lead to an even better revision of this laboratory manual.

Eric Wise
Santa Barbara City College
721 Cliff Drive
Santa Barbara, CA 93109
wise@sbcc.edu

LIST OF REVIEWERS

Reynaldo Acurio, *Broward College*
Nahel Awadallah, *Johnston Community College*
Lawrence C. Bird, *Our Lady of the Lake College*
Moges Bizuneh, *Ivy Tech Community College—Central Indiana*
Alan Crandall, *Idaho State University*
Smruti Desai, *Lone Star College Cyfair*
Tracey Emmons, *Sandhills Community College*
Tara Fay, *University of Scranton*
Purti Gadkari, *Wharton County Junior College*
Tammy Gillespie, *Eastern Arizona College*
Gerald Heins, *Northeast Wisconsin Technical College*
Carrie Dawn Hillard, *Northeast Mississippi Community College*
William Huber, *Saint Louis Community College at Forest Park*
Phillip Yuan Pei Jen, *Gordon College*
Steven Kish, *Zane State College*
Natalie Lenard, *Our Lady of the Lake College*
Robert McMullen, *Pikes Peak Community College*
Laurie M. Montgomery, *Community College of Baltimore County*
Amanda Nelson, *University of Wisconsin Green Bay*
Rebecca Roush, *Sandhills Community College*
Joanne Settel, *Baltimore City Community College*
Claudia Stanescu, *University of Arizona*
Terry Thompson, *Wor-Wic Community College*

STUDENT PREFACE

This laboratory manual was written to help you gain experience in the lab as you learn human anatomy and physiology. The 43 exercises explore and explain the structure and function of the human body. You will be asked to study the structure of the body using the materials available in your lab, which may consist of models, charts, mammal study specimens (such as cats), preserved or fresh internal organs of sheep or cows, and possibly cadaver specimens. You may also examine microscopic sections from various organs of the body. You should familiarize yourself with the microscopes in your lab very early so that you can take the best advantage of the information they can provide.

The physiology portion of the course involves experiments that you will perform on yourself or your lab partner. They may also involve the mixing of various chemicals and the study of the functions of live specimens. Because the use of animals in experiments is of concern to many students, a significant attempt has been made to reduce (but, unfortunately, not eliminate) the number of live experiments in this manual. Until there is an effective replacement for live animals, their use will continue to be part of the college physiology lab. Your instructor may have alternatives to live animal experimentation exercises, such as computer simulations. It is important to get the most out of what live specimen experimentation there is. Coming to the lab unprepared and then sacrificing a lab animal while gaining little or no information is an unacceptable waste of life. Use the animals with care. Needless use or inhumane treatment of lab animals is not acceptable or tolerated.

As a student of anatomy and physiology, you will be exposed to new and detailed information. The time it takes to learn the information will involve *more* than just time spent in the lab. You should maximize your time in the lab by reading the assigned lab exercises before you come to class. You will be doing complex experiments, and, if you are not familiar with the procedure, equipment, and time involved, you could ruin the experiment for yourself and/or your lab group. The exercises in physiology are written so that you can fill in the data as you proceed with the experiment. At the end of each exercise are review sheets that your instructor may wish to collect to evaluate the data and conclusions of your experiments. The illustrations are labeled, except on the review pages. All review materials can be used as study guides for lab exams, or they may be handed in to the instructor.

The anatomy exercises are written for cat and human study, although these exercises can be used with or without cats or cadavers. Get *involved* in your lab experience. Don't let your lab partner do all of the dissections or all of the experimentation; likewise, don't insist on doing everything yourself. Share the responsibility and you will learn more.

SAFETY

Safety guidelines appear in the inside front cover for reference. The international symbol for caution (⚠) is used throughout the manual to identify material that you should pay close and special attention to when preparing for or performing laboratory exercises.

CLEAN UP

Special instructions are provided for cleanup at the end of appropriate laboratory exercises and are identified by a unique icon (✋).

HOW TO STUDY FOR THIS COURSE

Some people learn best by concentrating on the visual, some by repeating what they have learned, and others by writing what they know over and over again. In this course, you will have to adapt your learning style to different study methods. You may use one study method to learn the muscles of the body and a completely different method for understanding the function of the nervous system. Some students need only a few hours per week to succeed in this course, whereas others seem to study far longer with a much less satisfactory performance.

You need to come to class. Come to lab on time. The beginning of the lab is when most instructors go over the material and point out what material to omit, what to change, and how to proceed. If you do not attend lab, you do not get the necessary information.

Read the material ahead of time. The subject matter is very visual, and you will find an abundance of illustrations in this manual. Record on your calendar all of the lab quizzes and exams listed on the syllabus provided by your instructor. Budget your time so that you study accordingly.

Work hard! There is absolutely no substitute for hard work to achieve success in a class. Some people do math more easily than others, some people remember things more easily and some people express themselves better. Most students succeed because they work hard at learning the material. It is a rare student who gets a bad grade because of a lack of intelligence. Working at your studies will get you much further than worrying about your studies.

Be *actively* involved with the material and you will learn it better. Outline the material after you study it for a while. Read your notes, go over the material in your mind, and then make the information your own. There are several ways that you can get actively involved.

Draw and doodle a lot. Anatomy is a visual science, and drawing helps. You do not have to be a great illustrator. Visualize the material in the same way you would draw a map to your house for a friend. You do not draw every bush and tree but, rather, create a *schematic* illustration that your friend could use to get the *pertinent* information. As you know, there are differences in maps. Some people need more practice than others, but anyone can do it. The head can be a circle, which can be divided into pieces representing the bones of the skull. Draw and label the illustration after you have studied the material and without the use of your text! Check yourself against the text to see if you really know the material. Correct the illustration with a colored pen so that you highlight the areas you need work on. Go back and do it again until you get it perfect. This does take some time, but not as much as you might think.

Write an outline of the material. Take the mass of information to be learned and go from the general to the specific. Let's use the skeletal system as an example. You may wish to use these categories:

1. Bone composition and general structure
2. Bone formation
3. Parts of the skeleton
   a. Appendicular skeleton
      (1) Pectoral girdle
      (2) Upper extremity
      (3) Pelvic girdle
      (4) Lower extremity
   b. Axial skeleton
      (1) Skull
      (2) Hyoid
      (3) Ribs
      (4) Vertebral column
      (5) Sternum

An outline helps you organize the material in your mind and lets you sort the information into areas of focus. If you do not have an organizational system, then this course is a jumble of terms with no interrelationships. The outline can get more detailed as you progress, so you eventually know that the specific nasal bone is one of the facial bones and the facial bones are skull bones, which are part of the axial skeleton, which belongs to the skeletal system.

Test yourself before the exam or quiz. If you have practiced answering questions about the material you have studied, then you should do better on the real exam. As you go over the material, jot down possible questions to be answered later, after you study. If you compile a list of questions as you review your notes, then you can answer them later to see if you have learned the material well. You can also enlist the help of friends, study partners, or family (if they are willing to do this for you). You can also study alone. Some people make flash cards for the anatomy portion of the course. It is a good idea to do this for the muscle section of the class, but you may be able to get most of the information down by using the preceding technique. Flash cards take time to fill out, so use them carefully.

Use memory devices for complex material. A mnemonic device is a memory phrase that has a relationship to the study material. For example, there are two bones in the wrist right next to one another, the trapezium and the trapezoid. The mnemonic device used by one student was that trapez*ium* rhymes with th*umb* and it is the one under the thumb.

Use your study group as a support group. A good study group is very effective in helping you do your best in class. Work with people who will push you to do your best. If you get discouraged, your study partners can be invaluable support people. A good group can help you improve your test scores, develop study hints, encourage you to do your best, and let you know that you are not the *only* one living, eating, and breathing anatomy and physiology.

Just as a good study group can really help, a bad group can drag you down further than you might go on your own. If you are in a group that constantly complains about the instructor, that the class is too hard, that there is too much work, that the tests are not fair, that you don't really need to know this much anatomy and physiology for your own field of study, and that this isn't medical school, then get yourself out of that group and into one that is excited by the information. Don't listen to people who complain constantly and make up excuses instead of studying. There is a tendency to start believing the complaints, and that begins a cycle of failure. Get out of a bad situation early and get with a group that will move forward.

Do well in the class and you will feel good about the experience. If you set up a study time with a group of people and they spend most of the time talking about parties, sports, or personal problems, then you aren't studying. There is nothing wrong with parties, sports, or discussions about personal problems, but you need to address the task at hand, which is learning anatomy and physiology. Don't feel bad if you must get out of your study group. It is *your* education and, if your partners don't want to study, then they don't really care about your academic well-being. A good study partner is one who pays attention in class, who is prepared ahead of the study time session, and who can explain information that you may have gotten wrong in your notes. You may want to get the phone number of two or three such classmates.

SUPPLEMENTAL MATERIALS

A variety of materials can be purchased separately to supplement this laboratory manual. Please see the **instructor's preface** for a list and description of these items.

TEST-TAKING

Finally, you need to take quizzes and exams in a successful manner. By doing practice tests, you can develop confidence. Do well early in the semester. Study extra hard early (there is no such thing as overstudying). If you fail the first test or quiz, then you must work yourself out of an emotional ditch. Study early and consistently, and then spend the evening before the exam going over the material in a general way and solving those last few problems. Some people do succeed under pressure and cram before exams; however, the information is stored in short-term memory and does not serve them well in their major field. If you study on a routine basis, then you can get up on the morning of a test, have a good breakfast, listen to some encouraging music, maybe review a bit, and be ready for the exam. The morning is a great time to study.

Your instructor is there to help you learn anatomy and physiology, and this laboratory manual was written with you in mind. Relate as much of the material as you can to your own body and keep an optimistic attitude.

Please feel free to write me or e-mail me with your comments, suggestions, and criticisms. I value your input and hope that your comments will lead to an even better revision of this laboratory manual.

Eric Wise
Santa Barbara City College
721 Cliff Drive
Santa Barbara, CA 93109
wise@sbcc.edu

WORKING IN THE LAB

The lab is a busy place, and your first priority in lab is to have a safe laboratory experience. You should read the Laboratory Safety Guidelines on the inside front cover of your lab manual and follow **all** of the safety directions that your instructor provides you. Know where the closest phone is in the event of an emergency, and make sure that you understand any specific emergency procedures for your lab.

Working in the science lab requires you to focus on the procedures and materials at hand. You may work as part of a group in some labs, and it is important that you read your lab material before coming into the lab. Some of the materials you work with may be dangerous, and a thorough prior knowledge of the lab exercise will ensure a safer lab.

Pay attention to the experiment and what is to be done and when. Casual observation and carelessness may lead to incorrect results. Establish a procedure for conducting experiments. If you are working with one or more lab partners, divide the responsibilities before the experiment begins. If you are responsible for a particular portion of the experiment, make sure your lab partners see the results. Make a careful record of the results of your experiment.

Be honest. Fudging data is not tolerated in the scientific community. Record your data as you measure them. If your results do not seem to be what they should, then discuss this with your instructor. Never record data that you think you *should* get; instead, record the *observed* data.

MEASUREMENT IN SCIENCE

Members of the scientific community and people of many nations of the world use the metric system to record quantities such as length, volume, mass (weight), and time. This is because the metric system is based on units of 10, and conversion to higher or lower values is relatively easy when compared with using the U.S. customary system. For example, assume you are working on a bicycle and are using a 1/2-inch wrench. If you need to go up in size, you move to a 9/16-inch, then a 5/8-inch, then an 11/16-inch, or perhaps as large as a 3/4-inch wrench. This requires a bit of computation as you move from one size to the next. On the other hand, if you are using the metric system and a 12-millimeter (mm) wrench is too small, you progressively move to a 13 mm, 14 mm, or 15 mm wrench.

The same idea can be applied to volume or weight. In the case of volume, there are 8 ounces per cup and 128 ounces per gallon. The calculation for the number of ounces in 7 gallons is a little cumbersome (7 gallons × 128 ounces). In the metric system, there are 1000 milliliters in 1 liter, so there are 7000 milliliters in 7 liters. The conversions are much easier. Medical dosages are given frequently in milliliters or cubic centimeters (cc). Under standard conditions, 1 milliliter occupies 1 cubic centimeter, so these values are interchangeable.

You can use the metric system to measure four quantities—length, volume, mass, and time. Examine table 1 and compare the quantity, base unit, and U.S. equivalent.

TABLE 1 Metric System and Equivalents

Quantity	Base Unit	U.S. Equivalent
Length	Meter (m)	1.09 yards (39.4 inches)
Volume	Liter (L)	1.06 quarts
Mass	Gram (g)	.036 ounce (1/454 pound)
Time	Second (s)	Second

If the quantity measured is much larger or smaller than the base unit, then the base unit can be expressed in multiples or fractions of 10. For example, if you had 1000 grams, then you would have a **kilo**gram. If you had one-thousandth of a gram (1/1000 gram), you would have a **milli**gram. Examine table 2 as you answer the following questions:

What is 1/100 gram? _____

What is 1000 seconds? _____

What is 10 meters? _____

What is 1/1,000,000,000 liter? _____

TABLE 2 Decimals of the Metric System

Name	Description	Multiple/Fraction	
Kilo	One thousand times greater	1000	
Deca	Ten times greater	10	
Base unit			
Deci	One-tenth as much	1/10	0.1
Centi	One-hundredth as much	1/100	0.01
Milli	One-thousandth as much	1/1000	0.001
Micro	One-millionth as much	1/1,000,000	0.000001
Nano	One-billionth as much	1/1,000,000,000	0.000000001

The extremes of measurement represent the **range** of the measurements. In the case of height, these are the smallest to the tallest. In terms of weight, the range represents the lightest to the heaviest.

The **mean** is the average for the group. To obtain the mean for a set of data, take the sum of all the individual measurements and divide by the total number of individuals in the group. The **median** is the value in the middle with half of the data sets above and half below this value.

SCIENTIFIC NOTATION

As you can see from table 2, some measurements in science are very small. The amounts of hormones circulating in the blood are very minute, indeed. To provide a shortened notation for numbers that are very large or small, we use scientific notation. A number such as 60,000 is written as 6×10^4. You move the decimal point four places to the left and thus the superscript above the 10 is a 4. Write 6000 in scientific notation: _____. For very small numbers, the superscript is written as a negative number. For example, 0.00006 is written as 6×10^{-5}, as you move the decimal point five places to the right. Convert the following numbers into scientific notation:

4,300,000 _____

0.000034 _____

2200 _____

0.0019 _____

GRAPHING

By graphing data, you can more easily see trends in a sample size. Some graphs are simple line graphs; others are bar graphs or pie charts. One problem in sampling is that you need to have a large enough number to have a valid sample. Let's suppose that you wish to graph the height of everyone in class, and one-half of the basketball team is enrolled in your lab section. This might have a rather unusual effect on your graph (figure 1). On the other

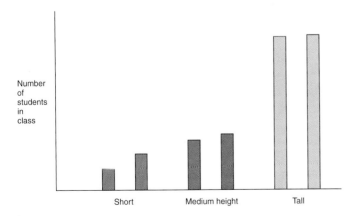

Figure 1

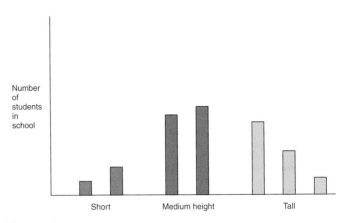

Figure 2

hand, if you can sample your entire school, the effects of the basketball players' sizes in the sample size will be minimized (figure 2).

LAB REPORTS

Your instructor may ask you to write up your results for specific experiments. This process is valuable in that you evaluate your experimental data and form an understanding of the process that comes from the results of your experiment. Writing up scientific experiments generally follows a very specific process, and this is described in Appendix C at the back of the lab manual. You should refer to it before you begin your lab write-ups.

PRACTICAL PROBLEMS

You should make sure that you understand the concepts presented previously by working on the following problems:

1. In terms of base units
   a. What is the base unit of length in the metric system?
   b. What is the base unit of volume in the metric system?
2. How many cubic centimeters are there in 200 mL?
3. Assume a pill has a dosage of 350 mg of medication. How much medication is this in grams?
4. How would you write 0.000345 liters in scientific notation?
5. How many milligrams are there in 4.5 kilograms?
6. How many meters is 250 millimeters?
7. If given a length of 1/10,000 of a meter:
   convert this number into a decimal.
   convert it into scientific notation.
8. Use a word to describe
   a. one-thousandth of a second.
   b. 1000 liters.
   c. one-hundredth of a meter.

Exercise 1
Organs, Systems, and Organization of the Body

INTRODUCTION

Science is the study of natural phenomena, and it follows specific guidelines that make it unique from other disciplines. **Anatomy** is the scientific study of the structure of the human body, and **physiology** is the study of the function of the body. Anatomy and physiology are two closely related fields, and they are well suited for study together.

The study of the human body requires you to understand both how the body or organ is oriented and how it is presented in terms of body regions. In this exercise, you examine the major organ systems of the body, the directional terms, and the levels of organization of the body, from the subatomic level to the whole organism. You also describe the major regions of the body. These topics are discussed in the *Principles of Anatomy and Physiology* text in chapter 1, "The Human Organism."

OBJECTIVES

At the end of this exercise, you should be able to

1. list the levels of structural hierarchy from smallest to largest;
2. list the 11 organ systems of the body;
3. place major organs, such as the heart, lungs, and stomach, in the proper organ system;
4. explain what is meant by *anatomic position;*
5. give directional terms that are equivalent to up, down, front, back, toward the midline, and toward the surface of the body;
6. determine from an illustration whether a section is in the frontal, transverse, or sagittal plane;
7. identify the major body cavities;
8. identify the quadrants and nine regions of the abdomen.

MATERIALS

Models of human torso
Charts of human torso

PROCEDURE
Levels of Organization

The human body can be studied from a number of perspectives. The earliest study involved **gross anatomy,** or cutting up part or all of the body and examining its details. As more sophisticated equipment was developed, other levels of organization became apparent.

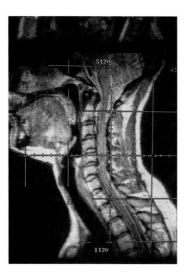

Figure 1.1   MRI of the Neck

Today, the manipulation of atomic nuclei under magnetic fields has led to magnetic resonance imaging (MRI) studies that do not depend on dissection of the body (figure 1.1). The following list shows the levels of organization, along with examples of each:

Level	Examples
Chemical	Oxygen, carbohydrates
Cell	Fibrocytes, red blood cell
Tissue	Nervous, muscular
Organ	Stomach, kidney
Organ system	Digestive system, urinary system
Organism	*Homo sapiens*

Organ Systems

Anatomy can be studied in many ways. **Regional anatomy** is the study of particular areas of the body, such as the head or leg. Most undergraduate college courses in anatomy and physiology (and the format of this lab manual) involve **systemic anatomy,** which is the study of **organ systems,** such as the skeletal system and the nervous system. Although organ systems are studied separately, it is important to realize the intimate connections among the systems. If the heart fails to pump blood as part of the cardiovascular system, then the lungs do not receive blood for oxygenation and the intestines do not transfer nutrients to the blood as fuel. The brain is no longer capable of functioning, and the result is death. From a clinical standpoint, the failure of one system has impacts on many other organ systems.

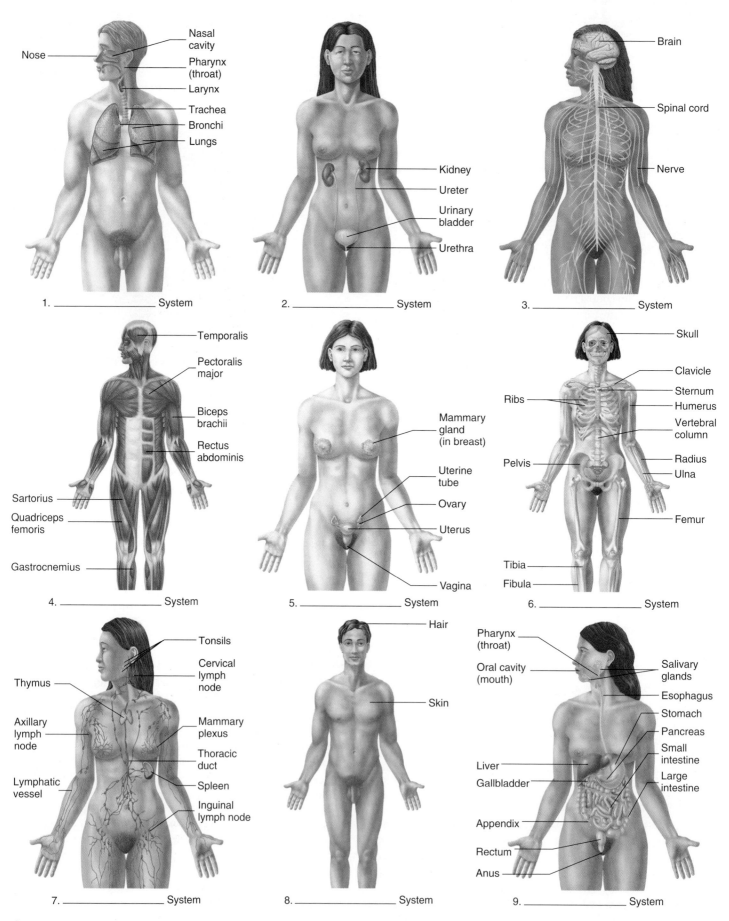

Figure 1.2 Organ Systems of the Human Body. Write the name of the system underneath the figure representing it.

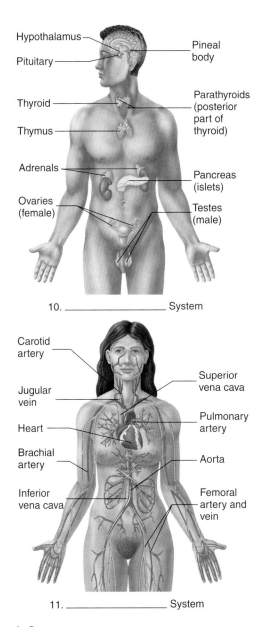

Figure 1.2—*Continued.*

Examine the torso models and charts in the lab and locate various organs. Using figure 1.2, find the following organ systems:

Respiratory   Lymphatic
Urinary   Integumentary
Nervous   Digestive
Muscular   Endocrine
Reproductive   Cardiovascular
Skeletal

The **respiratory system** takes oxygen to the body and releases carbon dioxide. The nose, pharynx, larynx, trachea, and lungs are the organs of the respiratory system. The **lymphatic system** cleanses and returns tissue fluid to the cardiovascular system and assists the body in protecting itself from foreign organisms. The lymphatic system consists of lymph vessels, along with such organs as the thymus, spleen, and tonsils. The **urinary system** rids the body of waste products; it consists of the kidneys, ureters, urinary bladder, and urethra. The **integumentary system** provides the protective covering of the body and is mostly formed by the skin. The **nervous system** is well developed in humans, allowing us to interact and interpret our environment. The brain, spinal cord, and peripheral nerves make up the system. The **digestive system** is responsible for providing nutrition to the tissues. The mouth and salivary glands, along with the esophagus, stomach, intestines, and associated organs, such as the liver, are part of the digestive system. The **muscular system** moves the body and consists of the individual muscles, such as the biceps brachii and the gluteus maximus muscles. In the **endocrine system,** the individual organs produce hormones. Organs such as the hypothalamus, pituitary, thyroid, pancreas, and gonads are endocrine organs. The **reproductive system** is responsible for the maintenance of the species. The sex cells from the male join with the sex cells of the female and produce offspring. The main organs in the system are the ovaries, uterine tubes, uterus, and vagina in females and the testes, ductus deferens, glands producing seminal fluid, and penis in males. The **cardiovascular system** is one primarily of transport. The heart and blood vessels are the organs of this system. The **skeletal system** provides a framework for movement and a mechanism for protecting the body. The individual bones of the body, such as the humerus and femur, are the organs of the system.

A quick way to remember all 11 systems is to remember this phrase "Run Mrs. Lidec." Each letter of the phrase represents the first letter in the name of one of the organ systems.

Anatomic Position

In clinical settings, it is vital to have a proper orientation when dealing with patients. If two physicians are operating on a patient and one tells the other to make an incision to the left, the physician making the cut does not have to ask, "My left or your left?" because the cut is always to the *patient's* left side. When referring to the human body, you will orient the body in the anatomic position. In this position, the body is upright, facing forward, arms and legs straight, palms facing forward, feet flat on the ground, and eyes open (figure 1.3a).

Directional Terms

With the body in the anatomic position, there are specific terms to describe the location of one part with respect to another. Table 1.1 lists the directional terms used for humans.

In quadrupeds (four-footed animals), the directional terms are somewhat different. Note in figure 1.3c that quadrupeds do not have a superior/inferior designation. *Dorsal* refers to the back, and ventral is on the belly side. Anterior (or cephalic) is the front, or head end, of the animal and posterior (or caudal) is the rear, or tail end, of the animal.

There are other terms for location that have unique meanings. For the digestive system, **proximal** refers to regions closer to the mouth, while distal is in reference to regions closer to the anus. **Parietal** is in reference to the body wall when compared to **visceral,** which refers to areas closer to the internal organs. The heart, for example, has a visceral layer closer to the heart proper called the **visceral pericardium,** while it also has a parietal layer farther from the heart called the **parietal pericardium.** Likewise the lungs have

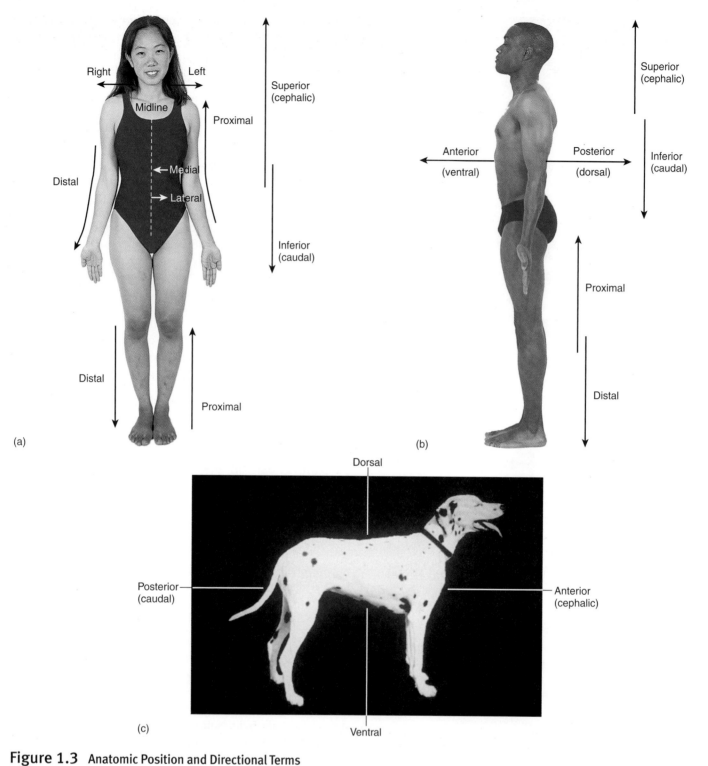

Figure 1.3 Anatomic Position and Directional Terms
(a) For humans, anterior view; (b) for humans, lateral view; (c) for quadrupeds, lateral view.

a **visceral pleura** and a **parietal pleura. Ipsilateral** refers to being on the same side of the body, and **contralateral** refers to being on the other half (left side/right side). The right hand and right arm are ipsilateral, while the ears are contralateral. Directional terms do not change if the body position changes. If you stand on your head, it is still superior to your feet because you reference the body as if it were in anatomic position.

Planes of Sectioning

When you view a picture of an organ that has been cut, it is very important to understand how the cut was made. Just as an apple looks very different when cut crosswise as opposed to lengthwise, so do some organs. Examine figure 1.4 for the following **sectioning planes.** A cut that divides the body or organ into superior and inferior parts is in the **transverse,** or **horizontal, plane.** A cut that

Exercise 1 Organs, Systems, and Organization of the Body

TABLE 1.1 Directional Terms Used for Humans

Term	Meaning	Example
Superior	Above	The nose is superior to the chin.
Inferior	Below	The stomach is inferior to the head.
Medial	Toward the midline	The sternum is medial to the shoulders.
Lateral	Toward the side	The ears are lateral to the nose.
Superficial	Toward the surface	The skin is superficial to the heart.
Deep	Toward the core	The lungs are deep to the ribs.
Ventral (or anterior)	To the front	The toes are ventral/anterior to the heel.
Dorsal (or posterior)	To the back	The spine is dorsal/posterior to the sternum.
Proximal	For extremities, meaning near the trunk	The elbow is proximal to the wrist.
Distal	For extremities, meaning away from the trunk	The toes are distal to the knee.

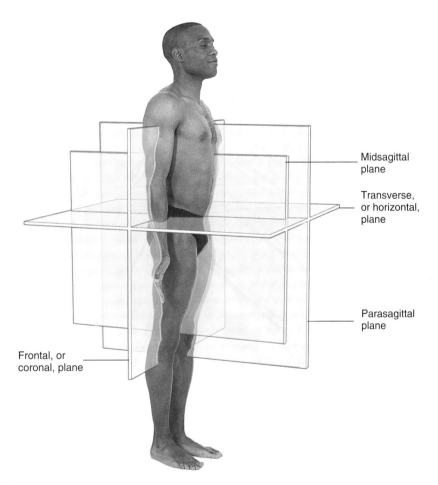

Figure 1.4 Sectioning Planes

divides the body into anterior and posterior portions is in the **frontal,** or **coronal, plane.** A cut that divides the body into left and right portions is in the **sagittal plane.** A cut that divides the body equally into left and right halves is in the **midsagittal** (or **median**), whereas one that divides the body into unequal left and right parts is in the **parasagittal plane.**

Major Body Cavities

A cavity is an enclosed space inside the body. The brain is located in the cranial cavity, the tongue is located in the oral cavity, and the stomach is found in the abdominal cavity. The three largest cavities that do not open to the exterior environment are the **thoracic cavity,** the **abdominal cavity,** and the **pelvic cavity** (figure 1.5). Compare the models and charts in the lab with this figure. The thoracic cavity is located directly superior to the diaphragm and is further divided into the **mediastinum** and the **pleural cavities.** The mediastinum is the region medial to the lungs; it contains, among other things, the heart (in the **pericardial cavity**), esophagus, and trachea. The pleural cavities are lateral to the mediastinum and contain the lungs.

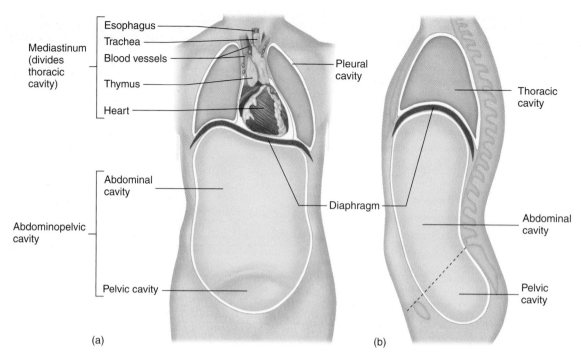

Figure 1.5 Trunk Cavities
(a) Anterior view; (b) lateral view.

Below the diaphragm is the **abdominopelvic cavity,** which can be subdivided into the abdominal cavity and the pelvic cavity. The abdominal cavity contains the stomach, the small intestine, most of the large intestine, and various digestive organs, such as the liver and the pancreas. The pelvic cavity begins at the region of the hips and contains the lower part of the large intestine and some of the reproductive organs (such as the uterus and ovaries) of the female reproductive system. These cavities are lined with **serous membranes.**

Regions of the Body

Examine figure 1.6 for specific areas of the body. You will refer to these areas throughout this lab manual, so a complete study of these regions here is essential. Anatomical regions are listed first, with the commonly used nouns (if appropriate) listed in parentheses. In anatomy, some regions of the body are described differently than you might expect. In anatomical usage, the **arm** is the region between the shoulder and the elbow, and the **leg** is the region between the knee and the ankle.

Cephalic (head)
    Frontal (forehead)
    Orbital (eye)
    Nasal (nose)
    Buccal (cheek)
    Oral (mouth)
    Mental (chin)

Cervical (neck)
Trunk
    Thoracic (chest)
        Pectoral
        Sternal
        Clavicular
        Acromial (shoulder)
    Abdominal (belly)
    Inguinal (groin)
    Genital (pubic)
    Coxal (hip)
Upper extremity
    Axillary (armpit)
    Brachial (arm)
    Cubital (elbow)
    Antebrachial (forearm)
    Carpal (wrist)
    Manual (hand)
    Digital (finger)
Lower extremity
    Femoral (thigh)
    Patellar (kneecap)
    Crural (leg)
    Tarsal (ankle)
    Pedal (foot)
    Digital (toe)

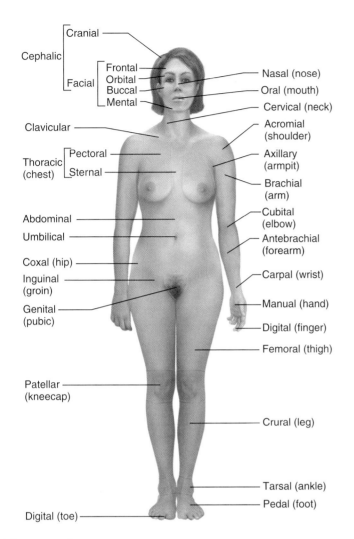
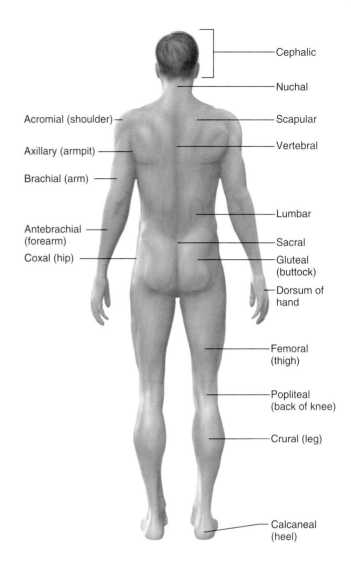

Figure 1.6 Regions of the Body

Find the following locations on your body and provide the appropriate anatomical description for these regions.

Shin _____

Elbow _____

Neck _____

Toes _____

Shoulder _____

Thigh _____

Kneecap _____

Abdominal Regions

The abdomen can be further divided into either four quadrants or nine regions (figure 1.7). In clinical practice, the abdomen is divided into quadrants. In anatomic studies, the nine-region approach is often used. Examine figure 1.7 and locate the following regions:

Four Quadrants Approach
Right-upper quadrant
Left-upper quadrant
Right-lower quadrant
Left-lower quadrant

Nine Regions Approach
Right hypochondriac
Left hypochondriac
Epigastric
Right lumbar
Left lumbar
Umbilical
Hypogastric
Right iliac
Left iliac

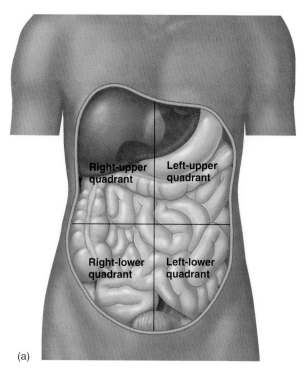 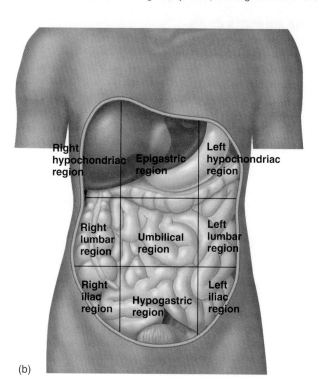

Figure 1.7 Abdominal Regions
(a) Four quadrants; (b) nine regions.

Exercise 1 Review

Organs, Systems, and
Organization of the Body

Name: _____

Lab time/section: _____

Date: _____

1. The scientific discipline that studies the function of the human body is known as _____.

2. Organs are associated into functionally related groups called _____.

3. In terms of reference, the body is placed in what position? _____

4. What specific body cavity lies directly inferior to the diaphragm? _____

5. The specific body cavity that is enclosed by the rib cage is known as the _____.

6. The body cavity surrounded by the hip bones is called the _____.

7. The term *arm* in anatomy refers to the region between the
   a. shoulder and elbow.   b. elbow and wrist.   c. shoulder and wrist.   d. shoulder and hand.

8. The term *leg* in anatomy refers to the region between the
   a. hip and knee.   b. knee and ankle.   c. hip and ankle.   d. ankle and foot.

9. The liver occupies what two regions of the abdomen? _____

Use correct anatomic terminology to describe the following relationships.

10. In terms of up and down, the head is _____ to the toes.

11. In terms of nearness to the trunk, the fingers are _____ to the arm.

12. In terms of nearness to the surface, the brain is _____ to the scalp.

13. In terms of front to back, the nipples are _____ to the shoulder blades.

14. The lungs belong to the _____ system.

15. The liver belongs to the _____ system.

16. The pectoralis major belongs to the _____ system.

17. If you were to sit on a horse's back, you would be on the _____ aspect of the horse.
    a. anterior   b. ventral   c. posterior   d. dorsal

9

18. What is the difference between the abdomen and the abdominal cavity? _____
_____

19. Complete the illustration by correctly placing the following terms:

femoral   crural   pedal   genital   coxal   abdominal   pectoral   sternal
cervical   cephalic   frontal   acromial   axillary   brachial   antebrachial   carpal

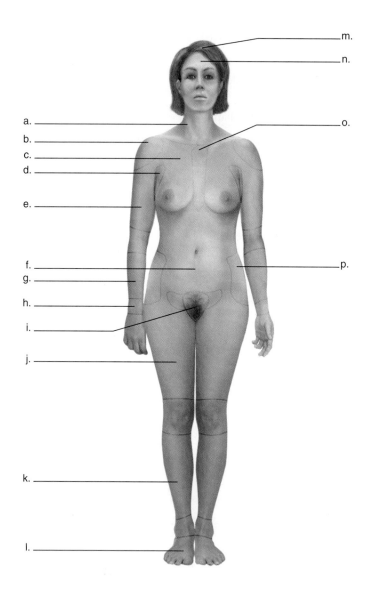

20. In the illustration of the brain, place the appropriate terms next to the planes that represent them.

midsagittal   frontal   transverse

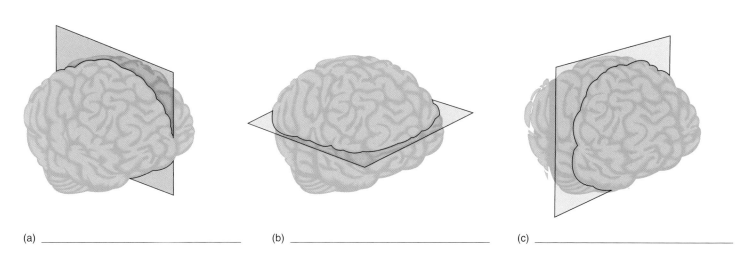

(a) _____   (b) _____   (c) _____

Exercise 2
Microscopy

INTRODUCTION

Originally, the study of anatomy and physiology was based on macroscopic, or gross, observation. This study was limited by the **resolution** of the human eye, which is the ability of the person to distinguish two objects as separate. With the invention and use of the light microscope, much greater detail was observed (microscopes increase the resolution), and thus began the study of cells and tissues. **Light microscopy** involves the use of visible light and glass lenses to illuminate, magnify, and observe a specimen.

The various types of microscopes are discussed in the *Principles of Anatomy and Physiology* text in chapter 3, "Cell Structures and Their Functions." Chapter 3 also has many electron microscope photographs that show details of cellular structure. This exercise covers how to use the compound microscope, how to examine prepared slides under the microscope, and how to make simple slides for study.

OBJECTIVES

At the end of this exercise, you should be able to

1. name the parts of the microscope and their functions presented in this exercise;
2. demonstrate the proper use of the compound microscope;
3. place a microscope slide on the microscope and observe the material, in focus, under all magnifications of the microscope;
4. calculate the total magnification of a microscope based on the lenses used;
5. prepare a wet mount for observation;
6. list the rules for proper microscope use.

MATERIALS

Compound microscope
Prepared slide with the letter *e* (or newsprint and scissors)
Transparent ruler or sections of overhead acetates of rulers
Glass microscope slides
Coverslips
Lens paper
Kimwipes or other cleaning paper
Lens cleaner
Small dropper bottle of water
1% methylene blue solution
Toothpicks
Histological slides of kidney, stomach, or liver
Prepared slides of silk threads

PROCEDURE
Care of the Microscope

There are a few rules concerning microscopes that you should always observe:

1. When carrying the microscope, hold it securely with two hands—one hand under the base and one on the arm.
2. Keep the microscope upright at all times.
3. Keep microscope lenses clean with lens cleaner and softened lens paper. Do *not* use paper towels or your shirt.
4. Use only the fine-focus knob when using the high-power objective lens.
5. Remove slides from the microscope before putting it away.
6. Secure the cord with a rubber band or wrap the cord around the base of the microscope.
7. Store the microscope with the low-power objective lens in place.
8. Put the microscope away in its proper location.

Using the Microscope

1. Examine figure 2.1 and familiarize yourself with the parts of the microscope.
2. Remove the microscope from the storage area. When you carry the microscope, place one hand under the base and the other hand on the arm of the microscope. Never tilt the microscope from an upright position, as lenses or filters may fall and break.
3. Take the microscope to your desk. Unwrap the electrical cord from around the microscope and familiarize yourself with the parts of the microscope. Compare the microscope that you have in lab with the one illustrated in figure 2.1. There may be differences between your microscope and the one in figure 2.1, but you should be able to locate the parts listed in the checklist. Use the checklist to make sure that you find all of the listed parts. Place a check mark next to the appropriate space when you locate each part of the microscope.

Microscope Parts Checklist

____ Base	____ Arm	____ Objective lens
____ Condenser	____ Iris diaphragm lever	____ Body tube
____ Nosepiece	____ Coarse-focus knob	____ Fine-focus knob
____ Stage	____ Ocular (eyepiece) lens	____ Light source

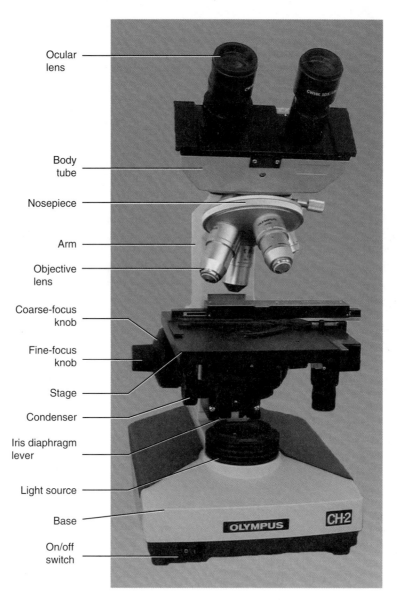

Figure 2.1 Compound Light Microscope

(Courtesy of Olympus America, Inc.)

4. With the microscope in front of you, plug it in, making sure the ocular lens or lenses are facing you. Make sure the cord is not hanging over the counter or in the aisle where someone might trip on the cord or pull the microscope off the counter. If the microscope has an illuminator (light source) dial, make sure that the setting is on the lowest level.
5. The low-power objective lens (typically 4 magnifications or 4×) should be facing down. This lens is often the shortest one and has a red line around the barrel of the lens. If it is not in place, turn the nosepiece until the low-power objective clicks into place. Always use the low-power objective lens when you first look at a microscope slide.
6. Use the knob on the microscope to raise the condenser lens so that it is at its highest level.
7. Obtain a prepared slide with the letter *e* or take a piece of newsprint and, using a pair of scissors, cut out a single letter from the paper. Place the letter on a glass microscope slide and add a drop of water to the piece of paper on the slide.
8. Place a thin coverslip on the slide by touching one edge of the coverslip to the water and lowering it slowly over the newsprint specimen as seen in figure 2.2. If you drop the coverslip on top of the slide, you will probably trap several air bubbles, which may obscure some of your specimen.
9. Locate and turn on the light switch.
10. Place the slide on the microscope stage with the coverslip on the top, the letter centered in the circle on the stage of the microscope, and with the letter upright so you can read it from where you are sitting. Most microscopes have a slide clip that holds the microscope slide in place. Use the stage control knobs to move the specimen so that light is coming through the specimen. This may require that you look at the specimen from the side to see if it is centered in the microscope stage.

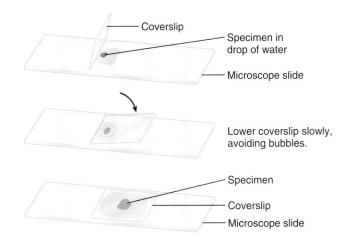

Figure 2.2 Preparation of a Wet Mount

11. Examine the specimen under low power. Use the coarse-focus knob to bring the specimen into focus. This is done by raising the mechanical stage of the microscope completely and then slowly lowering the stage as you look into the ocular lenses. Focusing the microscope requires a little bit of patience. When first looking at microscope slides always use the low-power objective lens.

    The coverslip should be close to the objective lens. Look through the ocular lens and rotate the coarse-focus knob slowly so the objective lens and the slide begin to move away from each other. This should bring the object into focus in the **field of view.** The field of view is the circle that you see as you look into the microscope.

12. Adjust the ocular lenses. Binocular microscopes have one fixed ocular and another that you can adjust. Focus the microscope so that the image in the nonadjustable ocular lens comes into focus. This is usually the right one. Once it is in focus, use the knurled ring on the adjustable ocular to bring that lens into focus for the other eye.

13. Adjust the light. Too much or too little light makes a specimen difficult to see. You can change the light level by either adjusting the light from the rheostat knob on the base or adjusting the iris diaphragm lever in the front of the microscope. The condenser lens will also help with lighting. Generally put the lens close to the specimen.

14. Draw what the specimen looks like in the space provided.

Does the letter appear right side up, or is the image inverted? _____

Is the letter oriented correctly, or is the image flipped horizontally? _____

How much of the letter occupies the field of view? _____

15. After you get the specimen in focus under low power, you should examine the slide under the next higher power. This is done by centering the image that you observe in the field of view and then switching the objective lens to the next higher power.

    The image should be close to being in focus if your microscope is *parfocal*. Do not adjust the height of the mechanical stage as you change the lenses. The next higher-power objective lens should clear the slide. Once you rotate the lens, adjust the focus by using the *fine-focus knob only*. You will probably need to adjust the light by moving either the iris diaphragm lever at the front of the microscope or the rheostat knob to improve the **contrast** of the image. As magnification increases, does the field of view increase or decrease in size?

16. You can look at the specimen under high power by the same procedure by turning the objective lens to the high-power lens. If you cannot focus on the high power or have lost the image that you were looking for, you should return to low power. Make sure the specimen is centered in the field of view and try the process again. If you still cannot find the object under high power, you should ask your instructor to help you.

Microscope Troubleshooting

If you are having a difficult time seeing anything or seeing things in focus, there could be several reasons. Use the troubleshooting tips listed in table 2.1 to correct the problem.

Proper Lighting

On some faintly stained specimens, the material may be difficult to see on low power. One trick is to locate the edge of the coverslip

TABLE 2.1 Troubleshooting Tips

Problem	Solution
Nothing is visible in the lens.	Plug in the microscope.
	Turn the power supply on.
	Rotate the objective lens until you hear it click into place.
	The bulb is burned out; replace the bulb.
You see a dark crescent.	The objective lens is not in proper position; click the lens into place.
All you see is a light circle.	The microscope is out of focus; adjust the coarse-focus knob.
	The light is up too high; turn down the light.
	The iris diaphragm is open too much; close it down.

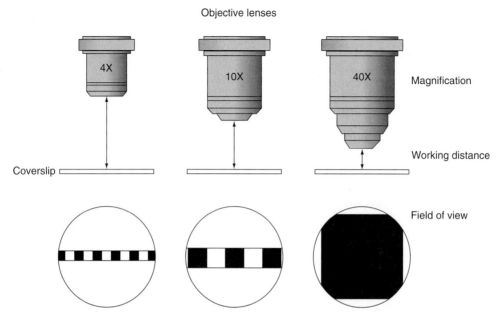

Figure 2.3 Increasing Magnification and Decreasing Field of View
As magnification increases (seen by increasing numbers on the objective lenses), both the working distance (distance between the objective lens and microscope slide) and the field of view decrease.

and turn the coarse-focus knob up and down until the edge is in sharp focus. This lets you know that you are in the approximate focal plane for examining the material on the slide. Move the slide to where the specimen should be, adjust the light, and reexamine it.

Magnification and Field of View

You can determine the size of the object under view if you know the diameter of the field of view. The field of view can be measured directly under low magnification by using a clear ruler. If higher magnifications are used, rulers won't work, and you have to calculate the field of view. There is a relationship between the diameter of the field of view and the magnification used. You can first calculate the total magnification using the following procedure. Look at the barrels of your microscope and determine the magnification of each lens.

Eyepiece (ocular) magnification: _____

Low-power objective magnification: _____

Total magnification (= ocular magnification × objective magnification): _____

Place a transparent section of ruler or a ruled section of acetate on the stage of the microscope. The space between the dark vertical lines is 1 millimeter (mm). Count the number of millimeters at the broadest part of the field of view and enter this number as the diameter of the field of view in the following space.

Diameter of the field of view (mm): _____

You can calculate the length of an object by determining how much of the diameter of the field of view it occupies. Let's say that the diameter of the field of view is 10 mm. If an object takes up one-half of the field of view, then you can estimate its size at 5 mm. If the object takes up only one-third of the field of view, how large is it? Record your answer in the following space.

Object size (mm): _____

As the magnification increases, the field of view decreases proportionally. Thus, if the diameter of the field of view is 10 mm at one magnification and you double the magnification by changing lenses, the field of view is reduced to a diameter of 5 mm. If you switch to a new lens and increase the magnification by 10 times, then the field of view is reduced to one-tenth of the original field of view. Look at figure 2.3 for a representation of this.

Keep the clear ruler or ruled acetate sheet under the microscope and increase the magnification to the next higher power by moving the next larger objective lens into place. Record the total magnification of your microscope with this objective lens.

Total magnification: _____

Examine the ruler under the microscope and record the diameter of the field of view in millimeters.

Diameter of the field of view in millimeters: _____

Has the increase in magnification produced a decrease in the field of view? _____

Is the decrease in the field of view proportional to the magnification? _____

Now calculate the total magnification of the microscope using the high-power objective lens.

Magnification with high-power objective lens: _____

You will not be able to measure the field of view accurately under high power with a ruler or an acetate sheet; however, you should be able to calculate the diameter of the field of view. For example, if the diameter of the field of view is 2.5 mm at 40 power (40×), then it is 0.25 mm at 400× (10 times more magnified yet one-tenth the field of view). Calculate the diameter of the field of view under high power.

Diameter of the field of view under high power: _____

Examination of a Prepared Slide

Examine the prepared slide of silk threads which has three crossed threads. As you focus on the threads under low

power, which color thread is on top? _____

Which color thread is in the middle? _____

Which color thread is on the bottom? _____

Under low power, how many threads or how much of one

thread appears to be in focus? _____

As you switch to the next higher power, how many threads or how much of one thread appears in focus?

_____

The **depth of field** is the thickness of the material that is in focus under a particular magnification. What happens to the depth of field when you increase the magnification in the microscope? _____

Preparation of a Wet Mount

You can make relatively quick and easy observations under the microscope as long as the material is thin enough and small enough. One technique for cell examination is to examine cells from the inside of the oral cavity. Do the following procedure.

1. With a toothpick, gently scrape the inside of your cheek.
2. Smear the cheek material from the toothpick on a clean microscope slide.
3. Place a drop of methylene blue on the smear. Methylene blue is a stain that makes the nucleus more apparent.
4. Place a coverslip on the slide by holding the coverslip at a 45-degree angle and slowly lowering it onto the slide to avoid trapping air bubbles. Examine the specimen under the microscope. The small, oval structures inside the cells are the nuclei. The angular lines around the nuclei are the cell membranes.

5. Draw a single cell in the space provided and label the plasma membrane, nucleus, and cytoplasm.

Your illustration of a cheek cell:

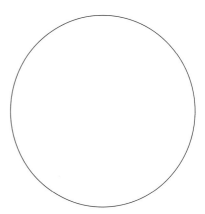

For another observation, remove a hair from your head (preferably one with split ends) and examine it by making a wet mount (figure 2.2). Place the hair in the center of the slide and add a drop of water. Place the coverslip on one edge of the drop and slowly lower it, trying to avoid trapping too many air bubbles in the process. Draw what you see in the following space.

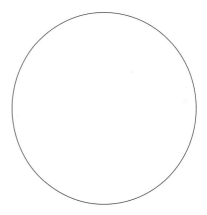

Observation of a Prepared Slide

Examine a prepared slide provided by your instructor. Examine the entire sample using the low-power objective lens. You should scan the entire area, looking for areas you want to observe more closely. Move to the next higher power and adjust the focus using the fine-focus knob. Finally, examine the material with the high-power objective lens and draw what you see in the space provided.

Your illustration of material from a prepared slide:

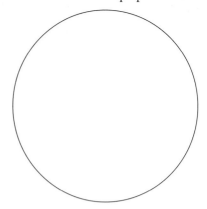

Name of the sample drawn: _____

You should use the iris diaphragm lever to adjust the amount of light that strikes the specimen. Too much or too little light produces significant changes in the observed image.

Oil Immersion Lens

The objective lenses you have used so far are called "dry" lenses. Your lab may be equipped with microscopes that have oil immersion lenses. The techniques for using these lenses are somewhat different from those for dry lenses. Once you have examined the specimen using the high-power dry lens, find the spot you want to examine and center it in the field of view. Add a drop of immersion oil on top of the coverslip and carefully swing the oil immersion lens into place. Use the fine focus only; otherwise you might drive the oil immersion lens through the slide and break it. Once you have examined the slide, swing the lens away and remove the slide, carefully wiping away the immersion oil with a clean piece of lens paper (do not use your shirt or a paper towel; these can scratch the lens). Use only lens paper to clean the oil from the oil immersion lens. Use another paper to remove any remaining oil, if needed.

Cleaning the Microscope

Smudges on the images you view through the microscope may be due to several things. There may be makeup or dirt on the ocular lens (or lenses). There may be dirt, oil, salt, stains, or other material on the objective lenses. To clean a lens, place a small amount of lens-cleaning fluid on a clean sheet of lens paper. Make one circular pass on the lens and throw the paper away. If you continue to clean the lens with the same lens paper, you can grind dirt or dust into it. Use a fresh piece of lens paper and repeat the procedure if further cleaning is needed.

You may want to clean the microscope slide before you examine it. Use a cleaning paper, such as a Kimwipe, to clean oil or dust from the slide.

Finally, dust may have collected inside the microscope over the years or the lenses may be scratched. There is nothing you can do about this, though you may want to bring this to your instructor's attention.

Putting the Microscope Away

When you are finished using the microscope make sure that you do the following:

1. Remove the slide from the stage.
2. Rotate the nosepiece so that the low-power (4×) lens is down.
3. Make sure that the slide clip is centered so the bar is not sticking out from the side of the microscope.
4. Wrap the cord loosely around the microscope or secure the cord with a rubber band.
5. Hold the microscope with two hands, one on the arm and one on the base.
6. Put it away in the correct place.

Exercise 2 Review
Microscopy

Name: _____

Lab time/section: _____

Date: _____

1. If the ocular lens is 10×, what would the total magnification be for the following objective lenses?
    a. 7×_____
    b. 15×_____
    c. 20×_____

2. What is the name of the thin glass plate that is placed on top of a specimen? _____

3. The microscope that you use in lab is a(n)
    a. compound microscope.   b. dissecting microscope.   c. electron microscope.

4. What is the name of the circle you see when you look through the ocular lens of the microscope? _____

5. What is the function of the iris diaphragm of the microscope? _____

_____

_____

6. If the diameter of the field of view is 5.6 mm at 40×, what is the diameter at 80×? _____

7. Place the following labels correctly in the microscope illustration.

   body tube
   objective lens
   stage
   light source
   base
   ocular lens
   arm
   coarse-focus knob

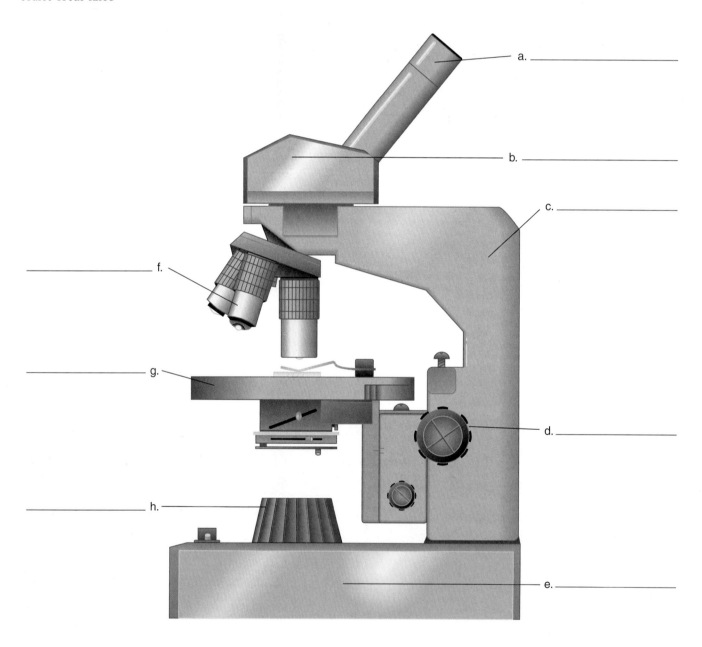

Exercise 2 Microscopy

8. When you switch from a low-power objective lens (for example, 4×) to a higher-power objective lens (for example, 10×), what happens to the working distance between the lens and the coverslip?

___

9. When you change from a low-power objective lens to a higher one, what happens to the field of view? Does it increase or decrease?

___

10. When should you use the low-power lens on the microscope?

___

11. How should you clean the lenses of a microscope?

___

12. What is the proper way to carry the microscope in lab?

___

13. Examine the following "field of view" and determine what the size of the object is.

___

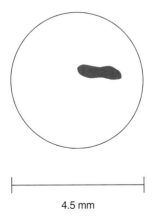

4.5 mm

Exercise 3
Cell Structure and Function

INTRODUCTION

The cell is the structural and functional unit of the human body and is the fundamental unit of living organisms. **Cytology** is the scientific study of cells. Many diseases can be traced to some type of cellular change. Cells perform many functions, some of which are unique to the particular organ where they are found. Generally speaking, however, cells grow, divide, acquire nutrients, release wastes, and respond to local stimuli.

The membrane of the cell is the dynamic interface between the internal, living environment of the cell and the external environment. In humans, most cells are bathed in a liquid medium called extracellular fluid (ECF), which provides nutrients, oxygen, hormones, water, and other materials to the cell. From the interior of the cell, the cell releases ammonia, carbon dioxide, and other metabolic products into this liquid. The cell membrane is important in the exchange of materials between the internal fluid of the cell and the environment surrounding the cell. The exchange of materials between the cell and the ECF maintains the homeostatic balance the cell must have to survive. Even small changes in the concentration of certain materials in the cell might lead to cellular death, so the constant adjustment of water, ions, and other metabolic products is extremely important. In large part, the cell membrane actively regulates what enters and what leaves the cell.

In this exercise, you examine the structure of animal cells, learn how cells of the body divide to make new cells, look at the physical processes that influence cell membrane dynamics, and study the nature of membrane transport. These topics are discussed in chapter 3, "Cell Structures and Their Functions," in the *Principles of Anatomy and Physiology* text.

OBJECTIVES

At the end of this exercise, you should be able to

1. describe the importance of cells in the makeup of the body;
2. list all of the organelles and their functions;
3. describe the three main events of the cell cycle;
4. name the four phases of mitosis and the events occurring in each phase;
5. describe the processes by which substances move across membranes;
6. define the terms *hypertonic, hypotonic,* and *isotonic;*
7. define the terms *diffusion, osmosis,* and *filtration;*
8. describe the movement of water across a selectively permeable membrane;
9. compare and contrast diffusion and osmosis.

MATERIALS

Cell Structure

Models or charts of animal cells
Electron micrographs of cells or a textbook with electron micrographs
Prepared slides of whitefish blastula
Microscopes
Modeling clay (Plasticine)—two colors
Marbles

Brownian Motion

India ink in dropper bottles or small jar of powdered charcoal
Dropper bottle of water
Microscopes
Microscope slides
Coverslips
Vegetable oil or light household oil in dropper bottles
Hot plate

Diffusion Demonstration

Potassium permanganate crystals
100 mL beaker (one per table)
Water

Diffusion

Agar plates (three per table)
0.01 Molar (M) potassium permanganate solution in dropper bottles
0.01 M methylene blue solution in dropper bottles
0.01 M potassium dichromate solution in dropper bottles
Plastic drinking straws
Millimeter ruler
Fine probe, spatula or small forceps
Warming tray

Osmosis Demonstration

Thistle tube or dialysis tubing attached to a glass tube with a string
Dialysis tubing
Rubber band
Dark corn syrup or a 20% sugar solution dyed with food coloring
1% starch solution
Ring stand and clamp
250 mL beaker
Distilled water
Permanent marker

Osmosis Experiment

Four strips of 20 cm-long dialysis tubing (one set of four per table)
Four 200 mL beakers
String or "W" clamp with wooden applicator stick
Scissors
Four solutions (2 L each) of 0%, 5%, 15%, and 30% sucrose, each tinted with a different color of food coloring
1 liter of 15% sucrose solution
Balances
Towels
Pipettes (10 mL)
Pipette pumps

Osmosis and Living Cells

Clean glass microscope slides
Coverslips
5 mL of mammal blood (available at local veterinarian's office)
Distilled water in dropper bottle (one per table)
0.9% saline solution in dropper bottle (one per table)
5% sodium chloride solution in dropper bottle (one per table)
Latex or plastic gloves

Filtration

Filter paper
Funnel
Ring stand with ring clamp
10 mL graduated cylinder
500 mL beaker
Iodine solution in dropper bottles (one per table)
Filtration solution (500 mL of 1% starch, 1% charcoal, and 1% copper sulfate) consists of 5 g each of starch, charcoal, and copper sulfate in one bottle or flask
Stopwatch or clock with second hand

PROCEDURE
Overview of the Cell

There are many types of cells in the body. Some are long and thin, some are spherical, and some are flat. You will examine a representative cell as an example but realize that there is tremendous diversity in cell shapes and functions.

Cells consist of two main parts, the **plasma** (or **cell**) **membrane** and the **cytoplasm.** The plasma membrane is the outer boundary of the cell, and, although it cannot be seen using the light microscope, its location is determined by the difference in color between the cytoplasm and the surrounding liquid on the microscope slide. The cytoplasm is the portion of the cell in which water, dissolved materials, and small cellular **organelles** are found. The fluid in which the organelles are suspended is called the **cytosol.** Small filaments and tubules make up the **cytoskeleton** (also considered part of the cytoplasm). The nucleus directs the cell's activities and stores its genetic information. It is visible with the light microscope and frequently appears as a spherical or an oblong structure. Figure 3.1 shows the plasma membrane and cytoplasm. Locate these on the model or charts in the lab.

Plasma Membrane

The plasma membrane is the "gatekeeper" of the cell. It is selective in what it allows into the cell. The plasma membrane is composed of a **phospholipid bilayer,** proteins, cholesterol, and other molecules. Phospholipids consist of a hydrophilic phosphate group attached to hydrophobic lipid groups (figure 3.2). Interspersed among the phospholipid molecules are **cholesterol molecules,** which provide stability to the membrane or, in larger concentrations, can make the membrane more fluid. Two major types of proteins are also found in the membrane. **Peripheral proteins** are found on the inner or outer surface of the membrane, whereas those passing into the membrane are known as **integral proteins.** Integral proteins may have carbohydrates or other molecules associated with them and frequently serve as cell markers. Some integral proteins function as channels by which specific materials can pass through the membrane. The plasma membrane is important in establishing electrochemical charge differentials, which allow for nerve impulse conduction and muscle contraction. It is also a selectively permeable membrane that provides an entrance or exit to the cell for some materials while excluding other material from entering or exiting the cell's interior. Proteins may anchor one cell to another, provide a place for metabolic reactions to take place, act as cell markers that identify a particular cell, or act as membrane receptors or channels.

Cytoplasm

Most of the inside of the cell is cytoplasm, which consists of the inner fluid portion of the cell, known as the cytosol; the inner framework of the cell, called the cytoskeleton; and the small, specialized units of the cell, called organelles. The cytosol is composed of water with dissolved materials, such as sugars, ions, proteins, and amino acids.

Figure 3.3 illustrates the cytoskeleton (not visible under the light microscope), which consists of microtubules, microfilaments, and intermediate filaments, all of which provide shape to the cell, a place to anchor organelles, and resistance to gravitational and other forces acting on the cell. **Microtubules** are made of the protein tubulin and are approximately 25 nanometers (nm) in diameter. **Intermediate filaments** are composed of fibrous proteins and are approximately 10 nm in diameter. **Microfilaments** are made of **actin** and are approximately 8 nm in diameter.

Organelles

The term *organelle* literally means small organ. Each organelle serves a particular function in the cell. There are two types of organelles—membranous and nonmembranous. **Mitochondria** are rod-shaped organelles with a double membrane. The major function of the mitochondrion (plural *mitochondria*) is to convert the stored chemical energy in food molecules to stored chemical energy in molecules of adenosine triphosphate (ATP). **Ribosomes** are the smallest of the organelles (about 25 nm in diameter) and are nonmembranous. Their function is to produce proteins. The **endoplasmic reticulum** is an organelle composed of a network of enclosed channels. There are two types of endoplasmic reticulum: **rough endoplasmic reticulum (RER),** which has attached ribosomes, and **smooth endoplasmic reticulum (SER),** which does

Exercise 3 Cell Structure and Function

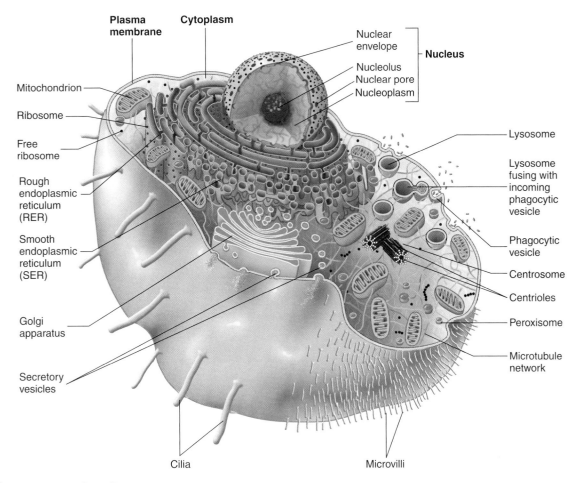

Figure 3.1 Overview of a Cell

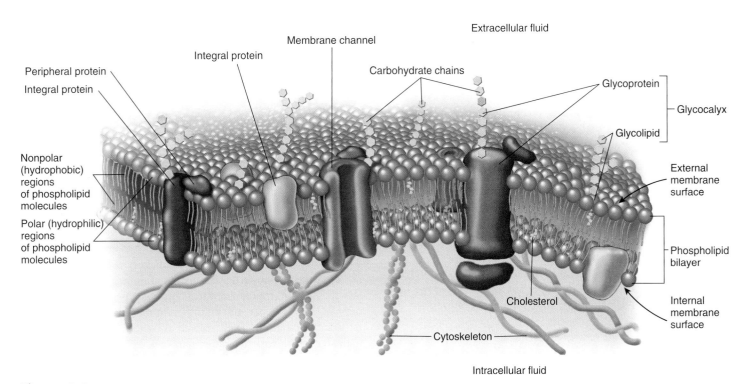

Figure 3.2 Plasma Membrane

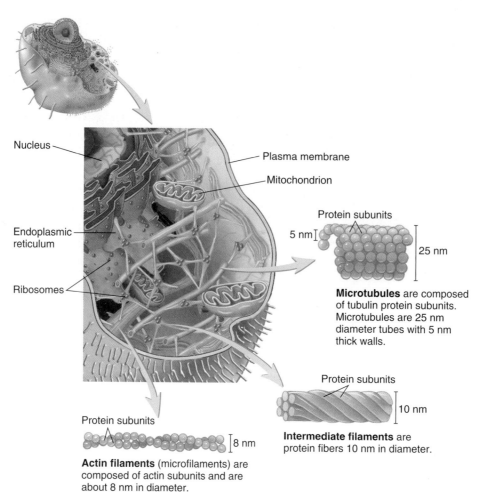

Figure 3.3 Cytoskeleton

not have ribosomes on its surface. The rough endoplasmic reticulum is associated with the nucleus and produces proteins for transport and use outside the cell. The smooth endoplasmic reticulum is often a distal extension of the rough endoplasmic reticulum. It produces lipid compounds (including phospholipids and steroids) and detoxifies material. The **Golgi apparatus** receives material from the endoplasmic reticulum and other parts of the cytoplasm and serves as an assembly and packaging organelle. It has **cisternae,** which are flattened, membranous sacs, and it forms vesicles to transport the molecules it assembles. The nucleus has two major functions: One is to house the genetic information of the cell, and the other is to control the various tasks of the cell. These two functions are carried out by DNA (deoxyribonucleic acid), which combines with proteins to form a material called chromatin in the nucleus. The nucleus is bounded by a **nuclear envelope,** which is a double envelope. The nuclear envelope has holes in it called nuclear pores, which allow the movement of materials into or out of the nucleus.

One or more structures known as the nucleoli (sing. *nucleolus*) are inside the nucleus. The nucleoli consist of portions of chromosomes and thus contain DNA and protein. They make rRNA, which forms ribosomes, the protein-producing organelles in the cytoplasm of the cell. Examine a model or chart in lab and find the nucleus, the nuclear envelope, and the nucleolus. Look at figure 3.1 and table 3.1 for a brief description of organelles and their functions. Your text has more detailed descriptions of these organelles, and you should refer to it for more information.

Vesicles

Vesicles, fluid-filled sacs inside the cell, digest subcellular material, transport material out of the cell, and carry on enzymatic activities. Vesicles protect the integrity of the plasma membrane. If a substance were to be removed from the cell by simply opening a hole in the plasma membrane, the cell would probably burst. The vesicle fuses with the plasma membrane ejecting the larger molecules without disrupting the plasma membrane. Two specialized vesicles in the cytoplasm are lysosomes and peroxisomes. Lysosomes digest material with enzymes in a process known as phagocytosis. Therefore, they are sometimes known as phagocytic vesicles. Peroxisomes use enzymes to convert potentially toxic hydrogen peroxide to water and oxygen. Hydrogen peroxide is formed in cells by the metabolism of fatty acids and amino acids, as well as by the interaction of water with an unstable form of oxygen known as oxygen free radicals.

Other Cellular Components

Additional structures occur in the cell that do not fall neatly into the category of plasma membrane or cytoplasm. Cilia and flagella

TABLE 3.1 Organelles and Their Functions in Cells

Organelle	Structure	Function
Mitochondrion	Double membrane, rod-shaped with cristae	ATP production, fatty acid oxidation
Ribosome	No membrane, rRNA subunits	Protein production
Rough endoplasmic reticulum	Single membrane, enclosed channels with ribosomes	Protein production for export
Smooth endoplasmic reticulum	Single membrane, enclosed channels without ribosomes	Lipid and steroid synthesis, detoxification
Golgi apparatus	Single membrane, cisternae as flattened sacs forming vesicles	Assembly of macromolecules, transport from cell for secretion
Lysosome	Single membrane, contains hydrolytic enzymes	Digestion of material
Peroxisome	Single membrane, contains peroxidase	Conversion of H_2O_2 to $H_2O + O_2$
Nucleus	Double membrane with nuclear pores, chromatin, and nucleoli	Storage of genetic information, regulation of cellular activity

are two such structures. These structures extend from the body of the cell and consist of microtubules covered by the plasma membrane. Most cells do not have cilia, and flagella are found only in the sperm cells of humans. Cilia are found on cells that are involved in movement, such as the movement of mucus along the free edge of the respiratory passage or in the uterine tubes. Cilia and flagella have a similar structure, but cilia are shorter than flagella.

Centrosomes are also unique cellular structures. They are typically found close to the nucleus, contain centrioles, and are involved in microtubule formation. They also are involved in the formation of a structure known as the spindle apparatus, which is involved in cellular division.

Finally, microvilli are small extensions of the surface of some cells that are involved in the absorption of material (such as the cells of the digestive or urinary systems).

The Cell Cycle

One of the great wonders of science is the mechanism by which a single cell, the result of the fusion of egg and sperm, develops into a complex, multicellular organism, such as a human. Various estimates put the total number of cells in the human body in the trillions. All of these cells came from the first cell, or zygote. In this part of the lab exercise, you examine the mechanism by which this occurs.

Most cells produce more cells by a process known as the cell cycle. The cell cycle can be divided into three events: **interphase, mitosis,** and **cytokinesis.** These events are illustrated in figure 3.4. Note that, for most of the time in the average life of a cell, it is in interphase.

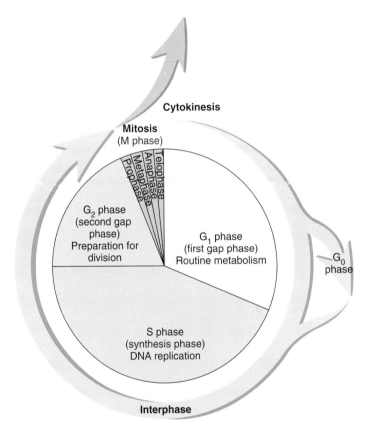

Figure 3.4 Cell Cycle

Interphase

Interphase is the time when a cell undergoes growth and duplication of DNA in preparation for the next cell division. If a cell is not going to divide any further (as with brain cells and some muscle cells), then interphase is regarded as the time when a cell carries out normal cellular function.

Interphase has three separate phases known as the G_1 phase, S phase, and G_2 phase. In the G_1 phase (G stands for gap), cells are growing in size and producing organelles. In the S phase (S stands for synthesis), the DNA of the cell is duplicated. The double helix of the DNA molecule unzips and two new, identical DNA molecules are produced. In the final phase of interphase, the G_2 phase, the cell continues to grow and prepares for the process of mitosis.

Some cells, such as muscle cells and brain cells, do not undergo further division and are said to be in the G_0 (G zero) phase.

Mitosis

Mitosis is a continuous event divided into four phases. Mitosis is nuclear division, and it involves the division of genetic information to produce two identical nuclei. In order for mitosis to occur, the chromatin in the nucleus of the cell must condense into compact units called chromosomes. A chromosome consists of two chromatids held at the center by a centromere. Examine figure 3.5 for the structure of a chromosome. You should also note the structure of chromosomes as you study the cells undergoing mitosis.

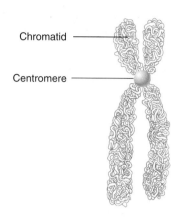

Figure 3.5 Structure of an Isolated Chromosome

The four phases of mitosis—prophase, metaphase, anaphase, and telophase—are described next. Refer to figure 3.6 as you read the descriptions.

Prophase

Cells in interphase have a distinct nuclear envelope, and the genetic information is dispersed in the nucleus as chromatin. The first indication that a cell is undergoing mitosis is the condensation of chromatin into chromosomes. Each chromosome consists of two elongated arms known as chromatids, which are connected to each other by a centromere.

In addition to the thickening of the chromosomes, the nucleolus disappears and the nuclear envelope begins to disassemble. In order for the chromosomes to separate and move apart, the nuclear envelope must be absent. During prophase, the mitotic apparatus becomes apparent. The mitotic apparatus consists of spindle fibers, which attach to the chromosomes at regions of the centromere known as the kinetochores. Two asters are points of radiating astral fibers at each end (pole) of the cell. In the center of the astral fibers are two small structures known as centrioles.

Metaphase

In this phase, the chromosomes align between the poles of the cell in a region known as the metaphase plate.

Anaphase

In anaphase, the chromatids separate at the centromere, and each chromatid is known as a daughter chromosome. The spindle fibers, which were attached to the kinetochores during prophase, pull the daughter chromosomes toward opposite poles of the cell. The centromere region moves first, and the arms of the chromosomes follow.

Telophase

Once the daughter chromosomes reach the poles, telophase begins. The daughter chromosomes begin to unwind into chromatin, the nucleolus reappears, and the nuclear envelope begins to re-form. The mitotic apparatus disassembles, thus terminating mitosis.

Cytokinesis

The splitting of the cell's cytoplasm into two parts is known as cytokinesis. Although cytokinesis is a distinct process, it frequently begins during late anaphase or early telophase. In late anaphase, as the chromosomes are moving to the poles, the plasma membrane begins to constrict at a region known as the cleavage furrow. This begins the process of dividing the cytoplasm (figure 3.7) as the cell splits into two separate daughter cells. The cytoplasm and the organelles are effectively divided into two parts.

Examine a slide of whitefish blastula and look for the various phases of the cell cycle in those cells. Most of the cells that you see are in a particular part of the cell cycle. What is this phase, and why are most of the cells in this phase? _____
_____
_____

Compare the slide with figure 3.6. Draw representative cells in each phase of mitosis in the following space.

Exercise 3 Cell Structure and Function

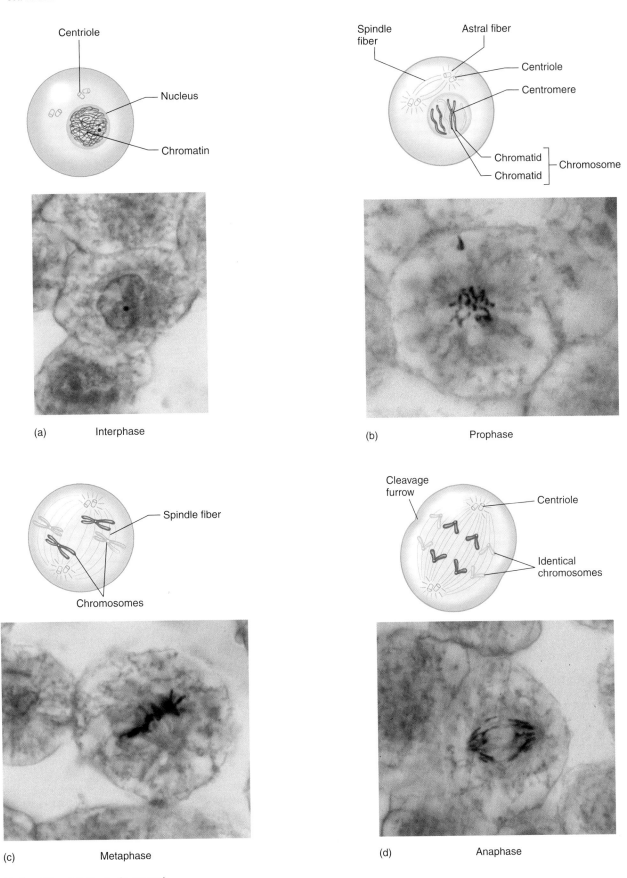

Figure 3.6 The Cell Cycle (1,000×)
Compare the stages of each part of the cell cycle.

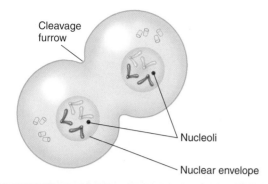

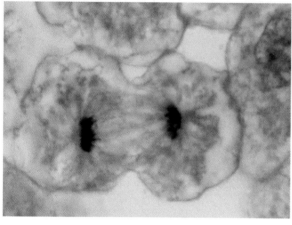

(e) Telophase

Figure 3.6—*Continued.*

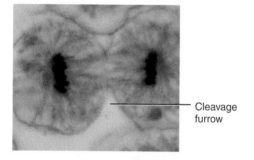

Figure 3.7 Cytokinesis (1,000×)

TABLE 3.2	Major Events of Mitosis
Prophase	Chromatin condenses to form chromosomes.
	Nuclear envelope disappears.
	Spindle apparatus forms.
	Nucleolus disappears.
Metaphase	Chromosomes align on the metaphase plate.
Anaphase	Chromosomes split and daughter chromosomes migrate to poles; cytokinesis often begins.
Telophase	Chromosomes reach poles; nuclear envelope re-forms.
	Chromosomes unwind to chromatin; cytokinesis divides the cytoplasm.
	Nucleolus reappears.

Mitosis Activity

Review the phases of mitosis in table 3.2. With two colors of modeling clay, make chromosomes. You should have a long chromosome and a short chromosome of each color, for a total of four chromosomes. Each chromosome should have two chromatids, and the chromosomes should be joined by a marble, which represents the centromere. Draw a large circle on a sheet of paper to represent a cell. Manipulate the clay chromosomes to show how mitosis occurs. After you have done this, describe the process of mitosis in your own words on a piece of paper. List each stage and what happens in that stage.

Cellular Function

Not only do cells duplicate themselves, but they also interact with the environment around them. Some of these interactions are linked to physical processes that occur independently of living organisms (such as diffusion) and some are due to the interface of the living cell membrane and the material inside the cell or outside the cell. In this regard, plasma membranes have many functions. They may be barriers that keep material inside the cell or prevent material from coming into the cell. Plasma membranes can transport material inside the cell or outside the cell.

Small molecules and ions travel across the cell membrane by **diffusion,** which is defined as the movement of particles from regions of higher concentration to regions of lower concentration. Some molecules or ions require a gate or channel in order to move across the plasma membrane. They move by diffusion across the membrane, but this movement is facilitated by gates so this is known as **facilitated diffusion.** Water moves across a membrane by a process known as **osmosis.**

If molecules or ions move against the diffusion gradient (for example from regions of low concentration to regions of high concentration) then the cell must expend energy. This movement whereby cells use energy to move material is known as **active transport.**

If molecules are too large to pass through the membrane by gates or channels, the cell can employ a different process. Material enters the cell by a process known as **endocytosis.** Materials exit a cell by a process known as **exocytosis.** Both of these involve the encapsulation of material with a phospholipid bilayer. In the case of endocytosis, material from outside the cell is engulfed as the plasma membrane forms a pouch, and the material is brought into the cell. In exocytosis, secretory vesicles engulf material, then the vesicle fuses with the plasma membrane, ejecting the contents to the outside of the cell.

This portion of the lab is done most efficiently if the timing of the experiments overlaps. While you are waiting for the results of one experiment, begin another.

Brownian Motion

All matter has **kinetic energy** unless it is at absolute zero (−273°C). Kinetic energy is the energy of motion, and it is the driving force of the movement of atoms and molecules. Atoms and molecules are too small to be seen, even with the use of a light microscope, yet their movement can be inferred as they hit large particles that are visible under the microscope. Items such as ink particles or powdered charcoal can be seen vibrating or jiggling, and we interpret this as the collective collisions of many molecules striking a larger structure and causing it to move. This is named **Brownian motion,** after botanist Robert Brown, who first described this in the nineteenth century. You will observe Brownian motion by examining particles in the following activity:

1. Place a drop of India ink or a small amount of powdered charcoal on a clean glass microscope slide and add a drop of water.
2. Carefully set a coverslip on the slide and gently place a thin line of vegetable oil or household oil around the edge of the coverslip to prevent the slide from drying out.
3. Examine the slide under high power and describe the movement of the particles in the space provided.

   Description of the movement of particles: _____
   _____

4. Remove the slide from the microscope and place it on a warm surface (such as a hot plate on the low-temperature setting) for a few seconds until the slide becomes warm.
5. Quickly return the slide to the microscope and note the movement of the particles, compared with the initial observation.
6. Describe any difference in the space provided. _____
   _____

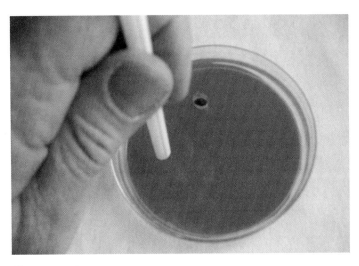

Figure 3.8 Placement of the Wells in Agar

4. Leave the beaker undisturbed, but note the changes that occur during 1 hour.
5. Record your observations in the space provided. While you are waiting, continue with the other experiments.

   Description of potassium permanganate movement:
   _____
   _____

Diffusion

Particles in gases and solutions move by kinetic energy, the energy of motion. Random collisions in gases or solutions spread particles from regions of high concentration to regions of low concentration. This is diffusion. If a bottle of perfume is poured into a dish in a room, the perfume molecules diffuse from the dish into the air of the room. The concentration of particles is higher in the dish and lower in the air; the molecules move down the **concentration gradient.** If the particles become uniformly dispersed, then the system has reached **equilibrium.**

If dye particles are placed in water, the same effect happens as the particles begin to spread out in the water. Water is the liquid into which material dissolves, so water is called a **solvent.** The material that becomes dissolved in water is the **solute,** and the combination of solvent and solute is the **solution.** Water is one of many kinds of solvents, yet it is the solvent vital to the life of the cell. Water is a polar molecule that dissolves ions and carries sugars, amino acids, and other materials to and from cells. Diffusion, driven by kinetic energy, is a process equally important to the cell. An example of the essential nature of diffusion is the movement of oxygen into the blood vessels of the lungs. Oxygen in the air is at a higher concentration than in the blood of the lung capillaries; consequently, oxygen moves from the air to the blood.

1. You can demonstrate diffusion by placing a crystal of potassium permanganate in a 100 mL beaker of water. Do this as a group at each table.
2. Fill a 100 mL beaker almost to the top with tap water.
3. Place the beaker on your table and drop a small crystal of potassium permanganate into it.

Many factors can affect the rate of diffusion, including changes in temperature, changes in concentration of the solute, size or weight of the solute particles, and interactions between the solute and the solvent. In the following two experiments, you will examine the effects of the weight of the particle and of temperature on the diffusion rate.

Diffusion Rates and Particle Weight

You may want to do this experiment as a group of three or four students. Agar, a liquid that forms a gel, can be used as a medium in which to measure diffusion rates of materials. Agar consists of water and algal polysaccharides. The liquid nature of agar is such that diffusion occurs in the gel at a slow rate.

1. Using a plastic drinking straw and a petri dish filled with agar, gently make three stabs into the dish so that they are approximately equidistant from one another (figure 3.8). Do not twist the straw, or you will break the agar and leave a crack into which fluid will run.
2. Remove the small plug of agar with a fine probe if needed so a well is left in the agar.
3. Into one of the wells, place a drop of 0.01 M potassium permanganate solution (molecular weight 158).
4. Into another well, place an equal amount of 0.01 M methylene blue solution (molecular weight 320).
5. In the last well, place the same amount of 0.01 M potassium dichromate solution (molecular weight 294). Potassium permanganate is a purple solution, methylene blue is blue, and potassium dichromate is yellow.

Leave the petri dish on your desk and make observations of the diffusion rate every 20 minutes for 80 minutes. Record the diameter of the diffusion in chart 3.1 and continue with the following experiments.

CHART 3.1	Diffusion Rate and Particle Weight Diameter of Diffusion (in mm)			
	20 min	40 min	60 min	80 min
Potassium permanganate				
Methylene blue				
Potassium dichromate				

Caution
These dyes stain clothes, skin, and lab notebooks!

Effects of Temperature on Diffusion Rates

1. Prepare two more petri dishes in the same way, except this time take one of the petri dishes from a refrigerator and another from a warming tray (such as an electric warming tray on the lowest setting).
2. Make three wells in the cold petri dish and place three drops of the respective dyes as you did previously.
3. This time, however, place the petri dish back in the refrigerator and examine after 80 minutes. You only need to record the diameter of diffusion at the end of the 80 minutes.
4. Follow the same procedure for the petri dish on the warming tray.
5. Record your results in chart 3.2.

CHART 3.2	Diameter of Diffusion (in mm)	
	80 min (Cold)	80 min (Warm)
Potassium permanganate		
Methylene blue		
Potassium dichromate		

Compare the diffusion rates of the cold and warm temperatures with the dish left at room temperature for 80 minutes. How does an increase or a decrease in temperature affect the diffusion rate? _____

_____

_____

What can you say about temperature and the kinetic energy of the system? _____

_____

Osmosis

In the preceding diffusion experiments, there was no barrier to the movement of the particles. Plasma membranes are barriers to certain molecules, while they allow other molecules to pass through. This type of membrane is called a **selectively permeable membrane.** Water, some alcohols, oxygen, and carbon dioxide move easily across the plasma membrane, whereas larger molecules, such as proteins or charged particles (ions), are prevented from crossing the membrane.

The movement of water across a selectively permeable membrane from regions of higher water concentration (for example, more pure water) to lower water concentration (for example, less pure water) is known as **osmosis.** Osmosis is a particular kind of diffusion, and the process can be viewed from the perspective of the solvent or the perspective of the solute.

The Solvent Perspective of Osmosis

In diffusion, material moves from higher concentrations to lower concentrations. The same can be seen with osmosis. If you have two solutions separated by a selectively permeable membrane, one a 10% sugar solution (or 90% water) and one a 5% sugar solution (or 95% water), then the water will move from higher water concentration (95% water) to lower water concentration (90% water). The greater the difference between the two solutions, the greater the concentration gradient, which, in this case, is known as the **osmotic potential.**

The Solute Perspective of Osmosis

A 10% sugar solution has more solutes than a solution of 5% sugar. The 10% sugar solution is said to be **hypertonic** to the 5% sugar solution. The 5% sugar solution is said to be **hypotonic** to the 10% sugar solution. If these two solutions are separated by a selectively permeable membrane, then water flows *from* the hypotonic solution *to* the hypertonic solution. If enough water flows across the membrane and the two solutions reach the same concentration of sugar, then *equilibrium* is established and the net movement of water stops. If solutions have the same concentration of solutes, they are said to be **isotonic** to one another.

Demonstration of Osmosis

Observe osmosis with a thistle tube osmometer or with a dialysis bag filled with dark corn syrup or colored sugar solution, attached to a glass tube with a string. The thistle tube osmometer consists of a hollow bell attached to a long tube. The tube is filled with molasses or sugar solution, and the large opening of the tube is covered with dialysis tubing and secured with a rubber band. The osmometer is placed in a beaker of water and clamped to a ring stand (figure 3.9). The level of the corn syrup or sugar solution is indicated with a mark from a permanent marker.

Examine the setup throughout the lab period. You may find that the liquid in the tube eventually stops rising. This occurs when the gravitational pressure equals the force exerted by the process of osmosis. The amount of force required to balance, or equilibrate, osmosis is the **osmotic pressure.**

Exercise 3 Cell Structure and Function

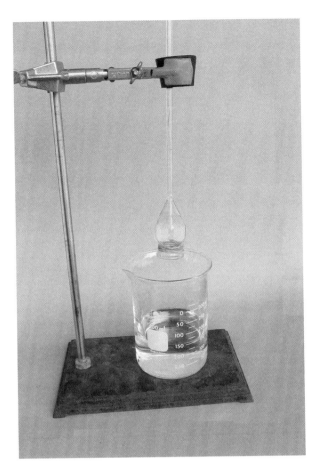

Figure 3.9 Osmosis Demonstration

Examine an osmometer (or prepare one if your lab instructor indicates for you to do so) that contains a starch solution.

Does the liquid move up the tube? _____

Why or why not? _____

_____

What does this say about the osmotic activity of starch?

_____

_____

Osmosis and the Concentration Gradient

The osmotic potential varies, depending on the concentration gradient between the two solutions. In this experiment, you determine the effects of various concentrations of a sucrose solution on the rate of osmosis.

1. Fill four dialysis bags that have been soaking in water with 10 mL of a 15% sucrose solution. You can do this by placing a pipette pump (or bulb) on the end of a 10 mL pipette, drawing liquid to reach the 10 mL mark, and filling the dialysis bag.
2. Secure the bottom end of the bag prior to filling it with sucrose solution from a pipette and pipette pump, and then tie or clamp both ends of the bag (figure 3.10).

(a)

(b)

Figure 3.10 Dialysis Bags
(a) Filling; (b) tying.

3. Rinse each bag in distilled water and blot with a towel.
4. Weigh each bag to the nearest tenth of a gram and record the weights in chart 3.3. This weight is the **initial weight** of the bag.

CHART 3.3 Initial Weight of Dialysis Bags

Bag 1 _____ grams

Bag 2 _____ grams

Bag 3 _____ grams

Bag 4 _____ grams

5. Place bag 1 in a beaker filled about two-thirds to the top with a 0% sugar solution. Make sure the bag is covered with the solution and leave it there for 20 minutes.
6. Place bag 2 in a 5% sugar solution and leave it for 20 minutes.
7. Place bag 3 in a 15% sugar solution and leave it for 20 minutes.
8. Place bag 4 in a 30% sugar solution and leave it for 20 minutes.
9. Remove the bags from the beakers after 20 minutes, and blot and weigh each bag.
10. Remember to place each bag back in its proper solution! Record the weight of the bags each 20 minutes for a total of 80 minutes in chart 3.4.

CHART 3.4 Weight of Bags (in g) for Each Time Period

Bag	20 min	40 min	60 min	80 min
1	_____	_____	_____	_____
2	_____	_____	_____	_____
3	_____	_____	_____	_____
4	_____	_____	_____	_____

Calculate the change of weight of each bag from the initial weight for each of the time periods. For example, let's assume a bag initially weighed 20.5 g and the recorded weights are as follows:

20 min	40 min	60 min	80 min
23.5 g	24.2 g	25.0 g	25.6 g

The change in weight is as follows:

3.0 g	3.7 g	4.5 g	5.1 g

Graph the change in weight for each bag of your experiment in chart 3.5 and connect the points with a line. Indicate which line represents which solution.

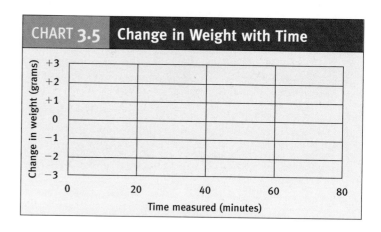

CHART 3.5 Change in Weight with Time

Which of the bags (if any) gained weight? _____

Which of the bags (if any) lost weight? _____

Determine the osmotic relationship (hypertonic, hypotonic, isotonic) of the bag to the solution in the beaker. Record your answer in chart 3.6.

CHART 3.6 Osmotic Relationship

Bag 1: The bag solution is _____ to the beaker solution.

Bag 2: The bag solution is _____ to the beaker solution.

Bag 3: The bag solution is _____ to the beaker solution.

Bag 4: The bag solution is _____ to the beaker solution.

Does the change in weight correlate to what you know about osmosis? If so, how does it correlate?

_____

_____

_____

Caution
Make sure you wear protective gloves while conducting the next experiment to avoid any potential transmission of disease.

Osmosis and Cell Membranes

The importance of isotonic solutions can be demonstrated by the following procedure:

1. Place a drop of fresh mammalian blood on a slide. Make sure you wear protective gloves while conducting this experiment.
2. To this slide, add a drop of physiological saline (0.9%) and place a coverslip on the slide.
3. Observe the cells on high power under the microscope and note their shape.

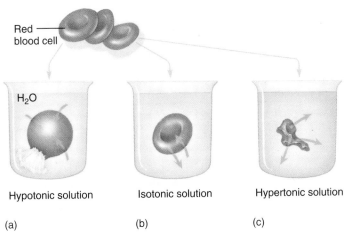

Figure 3.11 Stages of Red Blood Cells
(a) Inflated; (b) normal; (c) crenated.

Figure 3.12 Filtration Apparatus

4. Describe this in the space provided. _____
5. Place another drop of blood on a new slide.
6. To this slide, add a drop of 5% sodium chloride (NaCl) solution.
7. Place a coverslip on the slide and record your observations.
8. Examine the slide for at least a few minutes or until a change of shape becomes obvious.

   Description of the shape of the cells in 5% NaCl: _____

9. Finally, with a third slide, repeat the previous procedure but add a drop of distilled water instead of the sodium chloride. Immediately observe this slide and then continue to look at it for a few minutes. Record your observations.

   Description of the shape of cells in distilled water: _____

When red blood cells lose water, they undergo a process known as **crenation.** The loss of water causes the membrane to look wrinkled.

   Did any of the cells show crenation? _____
   Which solution might produce this? _____

When water moves into a cell at a rapid rate, the cell becomes inflated and sometimes bursts in a process known as **hemolysis.**

   Did any of the cells undergo hemolysis? _____
   Which solution might produce this effect? _____

Examine figure 3.11 for the various effects of solutions on red blood cells.

Filtration

The process of filtration is very important in certain cells of the body. It occurs as the pressure of a fluid forces particles through a filtering membrane. Filtration is a major component of kidney function. Hydrostatic pressure from the blood forces urea, ions, sugars, and other materials from the blood through the membranes of the kidney cells. In this way, small particles in the blood are forced into kidney tubules, whereas larger molecules, such as proteins, remain in the blood. In this experiment, you learn the basic principles of filtration, such as the **selectivity** of the filtration membrane and the **filtration rate** of a system.

1. Fold a piece of filter paper in half and then fold it in half again, so that it forms a cone (figure 3.12).
2. Place the cone in a funnel mounted on a ring stand over a beaker.
3. Into this cone filter, add the filtration solution, which is a mixture of copper sulfate, powdered charcoal, and starch in water.
4. Fill the funnel to near the brim and let the filtrate (the material passing through the filter) collect in the beaker.
5. When the funnel is approximately half full, place a 10 mL graduated cylinder under it and record the time it takes to fill the cylinder to the 2 mL mark.
6. Divide the time by 2 and calculate the filtration rate expressed as mL/minute.
7. Record this value.

   Number of seconds to produce 2 mL: _____

   Filtration rate: _____ mL/min

You should be able to determine what material passes through the membrane by the following method. Remove the funnel from the beaker. If the solution in the beaker has a blue cast to it, then copper sulfate passed through the membrane. If black particles are found in the filtrate, then charcoal passed through the membrane. If you add a few drops of iodine to the beaker and the solution turns black, then starch passed through the membrane. Iodine reacts with starch and produces a blue-black color. Record what material passed through the membrane in chart 3.7.

CHART 3.7 Filtration

Substance	Yes	No
Copper sulfate	_____	_____
Activated charcoal	_____	_____
Uncooked starch	_____	_____

The force that drives filtration in the funnel is the force of gravity on the liquid. In the kidney, the force that drives filtration is blood pressure.

Exercise 3 Review
Cell Structure and Function

Name: _____
Lab time/section: _____
Date: _____

1. Cells in the body have a fluid surrounding them. What is the name of this fluid? _____

2. The cytoplasm has a liquid portion. What is it called? _____

3. Name the major parts of the cell. _____
   _____
   _____

4. What structure surrounding the cytoplasm is composed mostly of a phospholipid bilayer? _____

5. Which organelle is responsible for ATP production? _____

6. Which organelle makes protein for use outside the cell? _____

7. Which organelle in the cell produces lipids? _____

8. The DNA that controls most of the function of the cell is found in what organelle? _____

9. What cellular structure is responsible for ribosome production? _____

10. What are the three parts of the cell cycle? _____
    _____
    _____
    _____

11. Of the three parts of the cell cycle, in which one is DNA duplicated? _____

12. In what part of the cell cycle do chromosomes first split apart? _____

13. What part of the cell cycle involves the division of the cytoplasm? _____

14. Describe the four phases of nuclear (mitotic) division and what occurs during those phases.
    _____
    _____
    _____

15. Name the cellular structures seen in the illustration using the terms provided.

    ribosomes   smooth endoplasmic reticulum   plasma membrane
    mitochondrion   vesicle   nucleus
    Golgi apparatus   cytoplasm   nucleolus
    rough endoplasmic reticulum   cilia

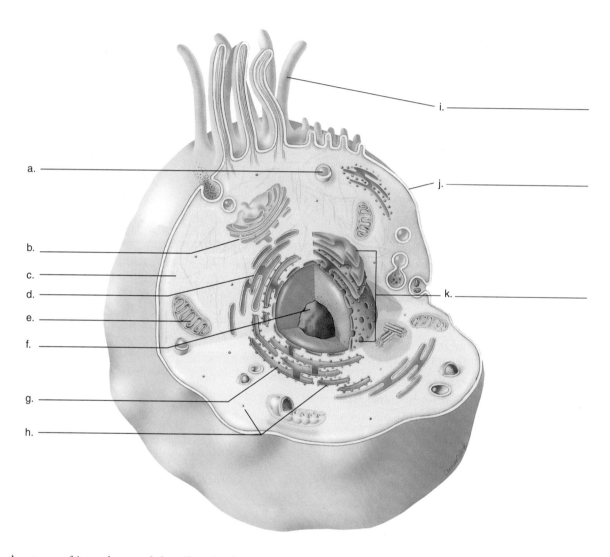

16. Name the stages of interphase and describe what happens during those phases.

    _____

    _____

    _____

17. Particles that are visible in the light microscope and move erratically by molecular collision represent what process? _____

    _____

18. Particles in a gas or liquid moving from a region of higher concentration to a region of lower concentration represent what

    process? _____

Exercise 3 Cell Structure and Function

19. Name the phases of the cell cycle as illustrated. Use the terms provided.

    interphase   prophase   metaphase   telophase   anaphase

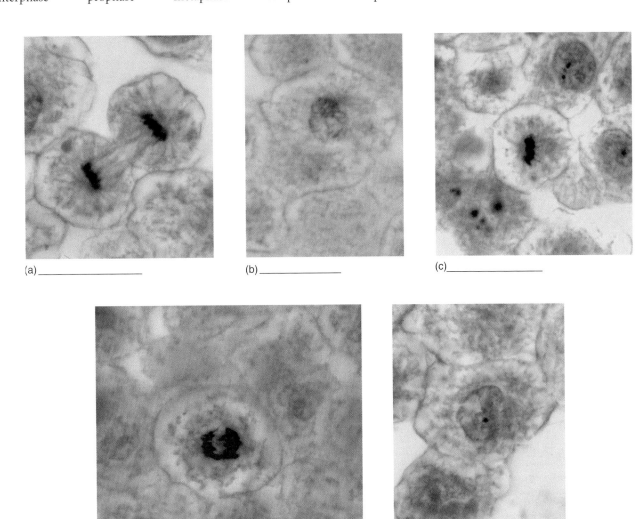

(a) _____

(b) _____

(c) _____

(d) _____

(e) _____

20. What is a concentration gradient? _____

_____

_____

21. Define *solute*. _____

_____

_____

22. When particles are evenly dispersed or distributed in solution, what condition is the solution said to be in? _____

_____

_____

23. Given two different molecules, one of 230 molecular weight and one of 415 molecular weight, which one would diffuse the farthest if they were both allowed to diffuse for the same length of time? Why? _____

_____

24. What process occurs as water moves from regions of higher water concentration to regions of lower water concentration across a semipermeable membrane? _____

25. A solution containing more solute is _____ (hypertonic/hypotonic/isotonic) to a solution containing less solute.

26. A solution with 5% sugar is _____ (isotonic/hypertonic/hypotonic) to a 3% sugar solution.

27. If the two solutions in question 26 were separated by a selectively permeable membrane, which solution would lose water?

_____

28. The pressure needed to stop osmosis is called the _____.

29. Define *hemolysis*. _____

_____

30. Is selectivity needed for filtration to occur? Why or why not? _____

_____

31. What would happen to the filtration rate if you were to apply pressure to the filtration system? _____

_____

What might happen to the filtration membrane if the water pressure were too high? _____

_____

Why might this be of concern to people who have both kidney disease and high blood pressure? _____

_____

Exercise 4
Tissues

INTRODUCTION

The study of tissues, called **histology,** is the study of anatomy at the microscopic level. The term *tissue* in histology refers to cells and extracellular material that have a particular function. There are four main tissue types in the human body: **epithelial tissue, connective tissue, muscular tissue,** and **nervous tissue.** Two or more tissues that form an enclosed structure make up an organ of the body. The heart is an organ that is composed primarily of muscular tissue but it also contains epithelial tissue and connective tissue. The study of histology is very important because many organic dysfunctions of the human body are diagnosed at the tissue level. Surgical specimens are routinely sent to pathology labs so that accurate assessment of the health of the tissue, and consequently the health of the person, can be made.

Tissues are discussed in chapter 4, "Tissues, Glands, and Membranes," in the *Principles of Anatomy and Physiology* text.

In this exercise, you examine numerous slides of organ tissue and begin an introduction to histology. In later exercises, you revisit histology as you examine various organ systems.

OBJECTIVES

At the end of this exercise, you should be able to

1. recognize the various types of epithelium;
2. associate a particular tissue type with an organ, such as kidney or bone;
3. examine a slide under the microscope or a picture of a tissue and name the tissue represented;
4. distinguish between cartilage and other connective tissues;
5. list the three parts of a neuron;
6. describe the muscle cell types according to location and structure.

MATERIALS

Microscope
Colored pencils

Epithelial Tissue Slides

Simple squamous epithelium
Simple cuboidal epithelium
Simple columnar epithelium
Stratified squamous epithelium
Transitional epithelium
Pseudostratified columnar epithelium

Muscular Tissue Slides

Skeletal muscle
Cardiac muscle
Smooth muscle
All three muscle types

Nervous Tissue Slides

Spinal cord smear

Connective Tissue Slides

Dense regular connective tissue
Dense irregular connective tissue
Elastic connective tissue
Reticular connective tissue
Loose (areolar) connective tissue
Adipose tissue
Hyaline cartilage
Fibrocartilage
Elastic cartilage
Ground bone
Cancellous bone
Bone marrow
Blood

PROCEDURE

Before you begin this exercise, you should be thoroughly familiar with the microscopes in your lab. If you are not familiar with using a microscope review Exercise 2. As you examine various tissues, look for distinguishing features that will identify each tissue. It is a good idea to examine more than one slide of a particular tissue, so that you can see a range of samples of that tissue. Often, the material you see in the lab is from a slice of an organ and, as such, includes more than one tissue type. For example, a sample of cartilage taken from the trachea contains epithelial tissue, fat, and other connective tissues in addition to the cartilage you want to study. If you are looking for smooth muscle from the digestive tract, you may also find both epithelial tissue and connective tissue in the slide. Use the figures in this exercise to help you locate tissues on the prepared slides. Examine each slide by holding the slide up to the light and visually locating the sample. Then put the slide on the microscope and examine it on low power, scanning around the slide. Move to progressively higher magnifications after you have identified the tissue. If you cannot identify the tissue after some searching, ask your lab partner or your instructor for help.

EXTRACELLULAR MATRIX

Humans are composed not only of cells but also of nonliving material known as the **extracellular matrix (ECM).** The ECM supports cells, anchors cells, separates tissues from one another, regulates communication between cells, and assists in wound healing. The extracellular matrix consists of fibers, binding proteins, and ground substance.

TABLE 4.1	Epithelial Classification
Cell Type	Layer
Simple squamous epithelium	Simple
Simple cuboidal epithelium	Simple
Simple columnar epithelium	Simple
Pseudostratified columnar epithelium (modified simple epithelium)	Simple
Stratified squamous epithelium	Stratified
Transitional epithelium (modified stratified epithelium)	Stratified

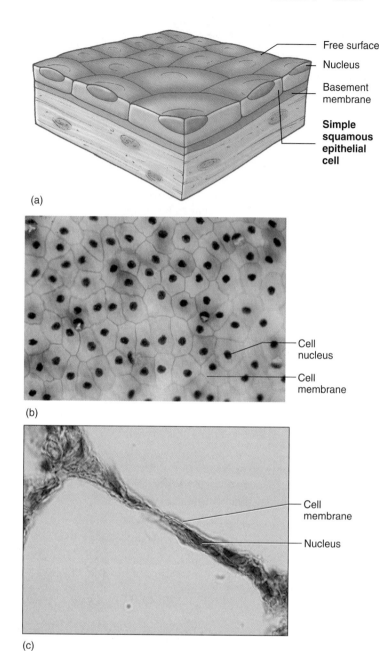

Figure 4.1 Simple Squamous Epithelium
(a) Diagram; (b) photomicrograph of mesothelium, surface view (400×); (c) photomicrograph of lung tissue, side view (1000×).

EPITHELIAL TISSUE

Epithelial tissue is a highly cellular tissue, covering or lining parts of the body (such as the skin on the outside or lining the digestive tract on the inside) or found in glandular tissue, such as the sweat glands or the pancreas. In most cases, epithelial tissue adheres to the underlying layers by way of a **basement membrane,** which is a noncellular adhesive layer. In a region such as the skin, the epidermis is made of epithelial tissue, and the basement membrane connects the epidermis to the underlying dermis. Epithelial tissue is classified according to the shape of the cells and the number of layers present. The cell shapes are **squamous** (flattened), **cuboidal,** and **columnar.** The arrangement of layers is **simple** (cells in a single layer), or **stratified** (cells stacked in more than one layer). When you look at epithelial tissue under the microscope, examine the edge of the sample because epithelial tissue is frequently found as a lining. Epithelial tissue is listed by type in the following discussion. Examine table 4.1 for an overview of epithelium.

Simple Epithelia

Simple epithelium is only one cell layer thick. The cells are located on the basement membrane, which may adhere to connective tissue, muscle, or even other epithelial tissue. Each epithelium is further classified according to shape.

Simple Squamous Epithelium

This epithelial type is composed of thin, flat cells that lie on the basement membrane like floor tiles. From a side view, these cells look flat, and their resemblance to floor tiles is apparent. Examine a prepared slide of simple squamous epithelium, which is seen in surface view or side view. Simple squamous epithelium is in the air sacs of lungs; it also lines blood vessels and is called **endothelium.** It can be found as the surface layer of many membranes, where it is called **mesothelium.** It provides a slippery surface (as found on the inside of blood vessels), allows filtration, and permits diffusion (as in the lungs). Compare your slide with the photograph in figure 4.1. Draw what you see under the microscope in the space provided. Note whether your slide shows the cell as a surface or side view.

Illustration of simple squamous epithelium:

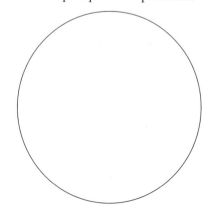

Exercise 4 Tissues

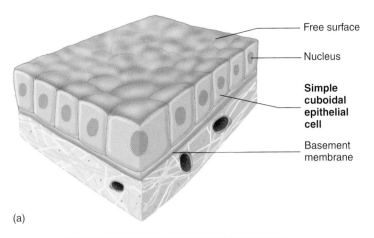

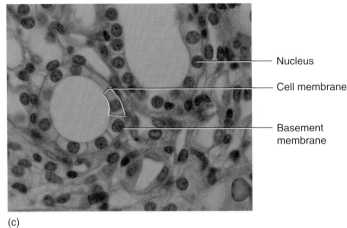

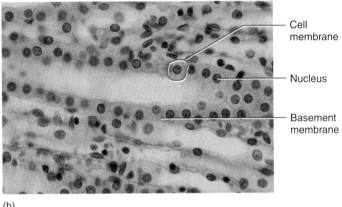

(b)

Figure 4.2 Simple Cuboidal Epithelium
(a) Diagram; (b) photomicrograph of kidney, long section (400×); (c) photomicrograph of kidney, cross section (400×).

Simple Cuboidal Epithelium

Simple cuboidal epithelium consists of cubelike or wedge-shaped cells that are mostly uniform in size. These cells form many of the major glands and glandular organs of the body and are the major cell type in the kidneys. Simple cuboidal epithelium lines tubules such as sweat ducts. It is often involved in the secretion of fluids (sweat, oil) or in reabsorption (as in the kidneys). Examine a prepared slide of simple cuboidal epithelium and compare it with figure 4.2. Draw what you see under the microscope in the space provided and label the nucleus of the cell and the basement membrane.

Illustration of simple cuboidal epithelium:

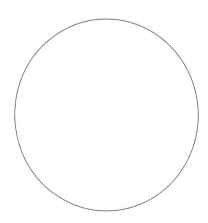

Simple Columnar Epithelium

This epithelium resembles tall columns anchored at one end to the basement membrane. The nuclei are frequently in a row. Nonciliated simple columnar epithelium lines the inner portion of the digestive tract and provides an absorptive area for digested food. Mucus-secreting **goblet cells** are often found with it. Ciliated simple columnar epithelium lines the uterine tubes. Examine a prepared slide of simple columnar epithelium and compare it with figure 4.3. Draw a representation of what you see in the space provided.

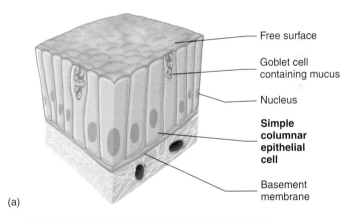

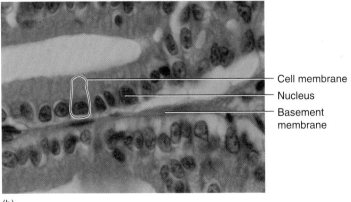

Figure 4.3 Simple Columnar Epithelium
(a) Diagram; (b) photomicrograph of stomach (400×).

Illustration of simple columnar epithelium:

Pseudostratified Columnar Epithelium

Pseudostratified columnar epithelium is an unusual type of epithelium. This tissue may first look as if it occurs in a few layers, but all of the cells rest on the basement membrane. The nuclei are at different levels in the tissue, making it look at first glance like it is stratified. Pseudostratified columnar epithelium lines some portions of the respiratory passages, where it protects the lungs by trapping dust particles in a mucous sheet and moving the particles away from them. Examine a prepared slide of pseudostratified columnar epithelium and compare it with figure 4.4. Draw a representative sample of what you see in the space provided.

Illustration of pseudostratified columnar epithelium:

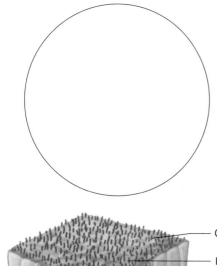

Stratified Epithelia

Stratified epithelium is so named because many epithelial cells occur in layers on a basement membrane. The term *stratified* comes from the word *strata* and refers to the layering of these cells. You will examine two common types belonging to this group.

Stratified Squamous Epithelium

Stratified squamous epithelium is a tissue that covers the outermost layer of skin around us. It also lines the vaginal canal and mouth. The multiple layers of this tissue protect the underlying tissue from mechanical abrasion. This epithelium may have cuboidal-shaped cells at the basement layer, but it derives its name from the cell shape at the free surface. Stratified squamous epithelium comes in two distinct types, keratinized and nonkeratinized (keratin is a tough protein that hardens cells in the outer layer of the skin). Examine a prepared slide of stratified squamous epithelium and locate the basement membrane. Compare your slide with figure 4.5 and draw what you see under the microscope in the space provided. Locate the basement membrane and count the layers of cells between it and the surface of this cell type.

Illustration of stratified squamous epithelium:

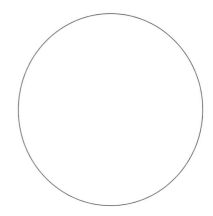

Record the numbers of layers of cells here: _____

Transitional Epithelium

Transitional epithelium is unusual in that it has some remarkable stretching capabilities. Transitional epithelium lines the ureter,

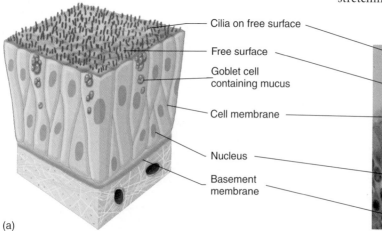

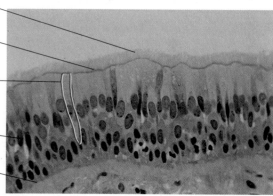

(b)

Figure 4.4 Pseudostratified Ciliated Columnar Epithelium
(a) Diagram; (b) photomicrograph of trachea (400×).

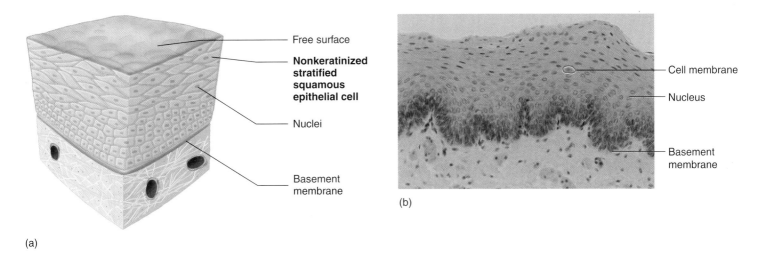

Figure 4.5 Stratified Squamous Epithelium
(a) Diagram; (b) photomicrograph of esophagus (400×).

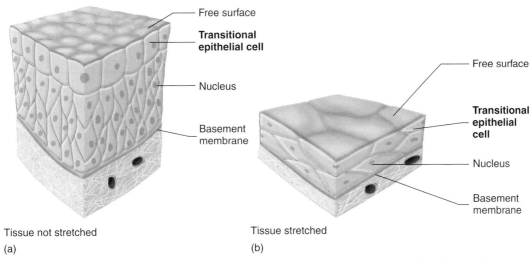

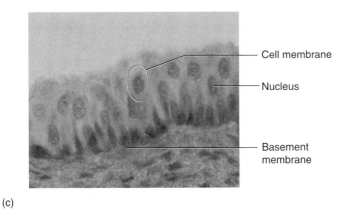

Figure 4.6 Transitional Epithelium
(a) Diagram, relaxed; (b) diagram, stretched; (c) photomicrograph of bladder, relaxed (400×).

urinary bladder, and proximal urethra and allows these organs to expand as urine collects within them. Examine a prepared slide of transitional epithelium. Look at figure 4.6 and draw the cell type in the space provided.

Illustration of transitional epithelium:

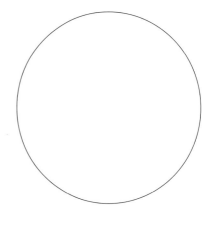

Cell Connections

Cells are held together in many ways. Epithelial cells are held to other structures by the basement membrane, as discussed previously. Cells are often held together at specific regions known as **desmosomes. Tight junctions** hold the cells even more closely together. **Gap junctions** allow communication between cells. These are illustrated in figure 4.7.

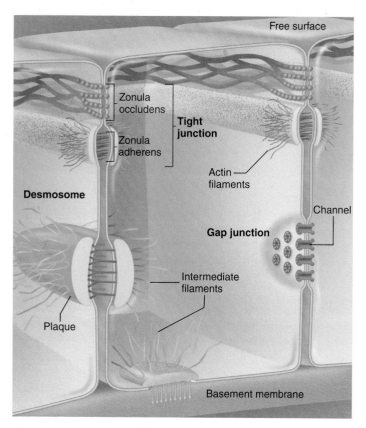

Figure 4.7 Cell Connections

MUSCULAR TISSUE

Like epithelial tissue, muscular tissue is a cellular tissue with the tissue having mostly cells and little matrix. These cells are contractile and shorten in length due to the sliding of protein filaments across one another. There are three types of muscle tissue in the body: skeletal, cardiac, and smooth muscle.

Muscle Types

Skeletal Muscle

The cells of skeletal muscle are called **fibers.** When you refer to the muscles of your body, you are referring to organs made mostly of skeletal muscle. Skeletal muscle is sometimes known as striated muscle because it has obvious **striations** (they look like stripes) in the fiber. Skeletal muscle is **voluntary** because you have conscious control over this type of muscle in your body. The individual muscle cells have many nuclei and are thus called **multinucleate,** or **syncytial.** The nuclei are elongated and occur on the periphery of the cell. Examine a prepared slide of skeletal muscle under high power. Compare this slide with figure 4.8. Draw a representation of the muscle in the space provided. How many widths of muscle cells fit across the diameter of your microscope when viewed at high power?

Illustration of skeletal muscle:

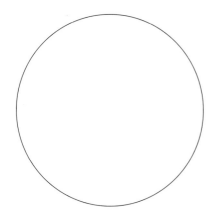

How many muscle fiber widths do you count? _____
By counting the cell widths, you can further distinguish among skeletal, cardiac, and smooth muscle.

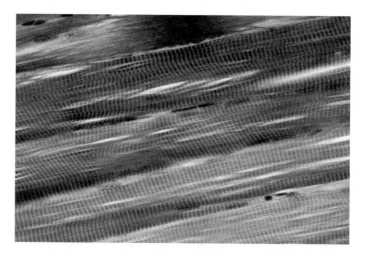

Figure 4.8 Skeletal Muscle (400×)

Cardiac Muscle

Cardiac muscle is found only in the heart, and it is the main tissue making up that organ. Cardiac muscle is similar to skeletal muscle in that it is striated, but the striations are much less obvious, and the individual cell diameter is less than those of skeletal muscle cells. Cardiac muscle cells are called **myocytes.** Another difference between cardiac muscle and skeletal muscle is that cardiac muscle is branched. Cardiac muscle contracts without conscious thought and is therefore called **involuntary muscle.**

The cells are mostly uninucleate, with the nucleus centrally located and oval in shape. Cardiac muscle cells are joined together by **intercalated disks,** which facilitate the transmission of the electrical impulses in the heart. As one muscle cell receives an impulse, it sends it on to the next cell. Look at a prepared slide of cardiac muscle and compare it with figure 4.9, noting the striations, intercalated disks, and nuclei of the cells. Count the number of widths of cardiac muscle fibers across the diameter of the microscope under high power. Draw the cells in the space provided.

Illustration of cardiac muscle:

Record the number of cell widths that occupy the diameter of your field of view at high power: _____

muscle is relaxed. When the muscle is contracted, the nucleus appears corkscrew-shaped. Examine a prepared slide of smooth muscle and note the narrow diameter of the fibers. How many fiber widths do you count across the diameter of the field of view of your microscope under high power? Compare your slide with figure 4.10 and draw an illustration of smooth muscle in the space provided.

Illustration of smooth muscle:

Number of cell widths under high power: _____

Examine a prepared slide of all three muscle types, if available. Work with your lab partner and identify each muscle cell. Fill in chart 4.1, comparing the muscle types and their characteristics.

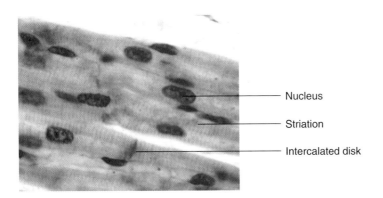

Figure 4.9 Cardiac Muscle (1000×)

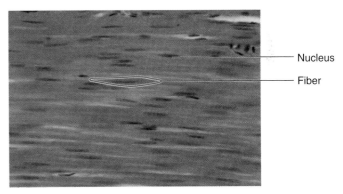

Figure 4.10 Smooth Muscle
Photomicrograph of intestine (400×).

Smooth Muscle

Smooth muscle is **nonstriated** in that the fibers do not have stripes perpendicular to the length of the fiber. Smooth muscle, like cardiac muscle, is **involuntary** and is found in the digestive tract, where it propels food along by a process known as peristalsis. Smooth muscle is also found in abundance in the uterus and is the driving force behind uterine contractions during labor. The cells of smooth muscle are spindle-shaped and uninucleate, and the nucleus is centrally located and elongated in shape when the

NERVOUS TISSUE

Nervous tissue, like epithelial and muscular tissues, is also a cellular tissue. Nervous tissue is found in the brain, spinal cord, ganglia, and peripheral nerves of the body. The conductive cell of this tissue is the **neuron,** which receives and transmits electrochemical impulses. A neuron is a specialized cell with three

CHART 4.1	Muscle Characteristics		
Muscle Type	Number of Nuclei	Location	Voluntary/Involuntary
	Many		
	One		
	Mostly one		

major regions: the **dendrites,** the **nerve cell body** (or **soma**), and the **axon.** Examine figure 4.11 for the regions of a typical neuron.

Examine the neurons of a smear of the spinal cord. Look for purple star-shaped structures, which are the nerve cell bodies. Compare them with figure 4.12.

Special support cells of the nervous system are called **glial cells,** or **neuroglia.** These cells do not conduct impulses, but they guide developing neurons to synapses, remove some neurotransmitters, and perform other functions. You will study glial cells in greater detail in Exercise 20.

CONNECTIVE TISSUE

The tissues that you have seen so far, epithelial, muscular, and nervous, are composed mostly of cells with very little extracellular matrix. This is not the case with connective tissue. There is usually more extracellular matrix than cells in connective tissue. This **matrix** usually consists of fibers and fluid, gel, or solid ground substance. Because of the abundance of matrix, connective tissue is classified by *specific tissue.*

Connective tissues have diverse appearances and function, and specific tissues do not seem to have much in common with one another; however, all arise from an embryonic tissue known as **mesenchyme.** Connective tissue is described in more detail next and is outlined in table 4.2.

Loose Connective Tissue

Loose connective tissue (or **areolar connective tissue**) is a complex collection of fibers and cells that has a distinctive look. Loose connective tissue is found as a wrapping around organs, as sheets of tissue between muscles, and in many areas of the body where two different tissues meet (such as the boundary layer between fat and muscle). Loose connective tissue consists of large, pink-stained **collagenous fibers;** smaller, dark **elastic** and **reticular fibers** (fine collagenous fibers); and a collection of cells, including **fibroblasts,**

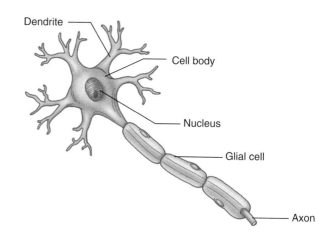

Figure 4.11 Regions of a Typical Neuron

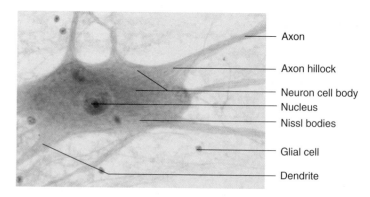

Figure 4.12 Spinal Cord Smear (400×)

fibrocytes, mast cells, and **macrophages.** Examine a prepared slide of loose connective tissue and compare it with figure 4.13. Make a drawing of it in the space provided.

Exercise 4 Tissues

TABLE 4.2 Connective Tissue

Loose connective tissue
Dense regular collagenous connective tissue
Dense irregular collagenous connective tissue
Dense regular elastic connective tissue
Dense irregular elastic connective tissue
Adipose tissue
Reticular connective tissue
Bone marrow
    Yellow bone marrow
    Red bone marrow
Cartilage
    Hyaline cartilage
    Fibrocartilage
    Elastic cartilage
Bone
Blood

connective tissue, or run in many directions, in which case it is known as **dense irregular collagenous connective tissue.** Dense regular collagenous connective tissue is found in tendons and ligaments. Dense irregular collagenous connective tissue is found in the deep layers of the skin and the white of the eye. The fibers in these tissues are made of a protein called **collagen. Fibrocytes** and **fibroblasts** may be found in the tissue, in addition to the collagenous fibers. Fibroblasts secrete fibers and then mature into fibrocytes. Examine a slide of dense connective tissue and compare it with figure 4.14. Draw a section of dense regular or dense irregular collagenous connective tissue in the space provided.

Illustration of dense connective tissue:

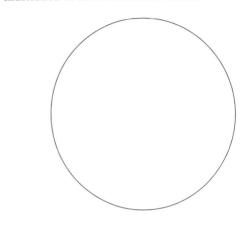

Illustration of loose connective tissue:

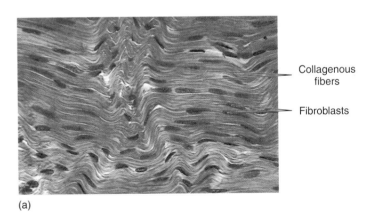

(a)

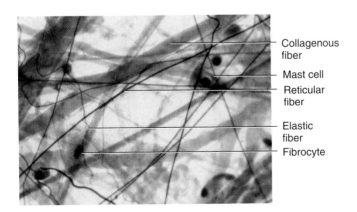

Figure 4.13 Loose (Areolar) Connective Tissue (400×)

Dense Connective Tissue

This tissue is composed of **collagenous fibers** that may either be parallel, in which case it is known as **dense regular collagenous**

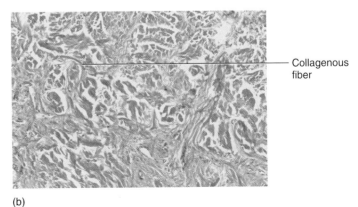

(b)

Figure 4.14 Dense Connective Tissue

(a) Dense regular collagenous connective tissue (400×); (b) dense irregular collagenous connective tissue (100×).

Dense Elastic Connective Tissue

This tissue has cellular components of fibroblasts and fibrocytes. The fibroblasts actively secrete fibers and subsequently develop into fibrocytes. The fibers in this tissue are made of the protein **elastin** and are called **elastic fibers.** Elastic tissue in the walls of arteries is called **dense irregular elastic connective tissue;** it can be identified as dark, squiggly lines in the artery walls. Elastic tissue that forms the vocal cords is known as **dense regular elastic connective tissue.** Examine figure 4.15 as you look at a prepared slide of elastic tissue. Then draw the specific tissue in the space provided.

Illustration of elastic connective tissue:

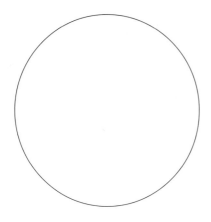

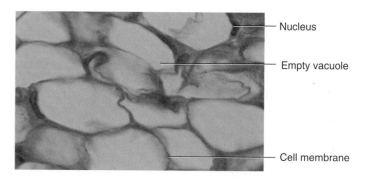

Figure 4.16 Adipose Tissue (400×)

Illustration of adipose tissue:

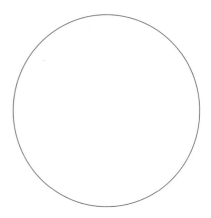

Adipose Tissue

Adipose tissue is unusual as a connective tissue in that it is highly cellular. The cells that make up adipose tissue are called **adipocytes** (which is a technical way to say *fat cells*). Adipocytes store lipids in large vacuoles, whereas the nucleus and cytoplasm remain on the outer part of the cell near the plasma membrane. In prepared slides of adipose tissue, the fat typically has been dissolved in the preparation process. You should locate the large, empty cells with the nuclei on the edges of some of the cells. Compare what you see in the microscope with figure 4.16 and make a drawing in the space provided. You may need to close down the iris diaphragm or reduce the amount of light shining on the specimen to best see this tissue.

Reticular Connective Tissue

Reticular connective tissue also has fibroblasts and fibrocytes, but the tissue has **reticular fibers.** These fibers are made of fine strands of collagen. Reticular connective tissue is found in soft internal organs, such as the liver, spleen, and lymph nodes. Reticular connective tissue provides an internal framework for these organs. If your slide is stained with a silver stain, then the reticular fibers appear black. If your slide is stained with Masson stain, then the reticular fibers appear blue. In either case, look for fibers that appear branched among small, round cells. The other cells in the prepared slide are the cells of the organ (such as spleen, lymph node, or tonsil) that the reticular connective tissue is holding together. Examine a prepared slide of reticular tissue and compare it with figure 4.17. Draw a sample of what you see in the space provided.

Illustration of reticular connective tissue:

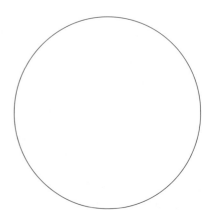

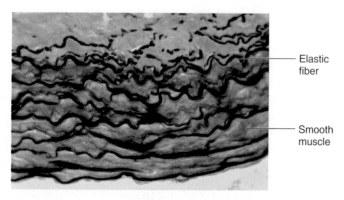

Figure 4.15 Dense Irregular Elastic Connective Tissue (400×)

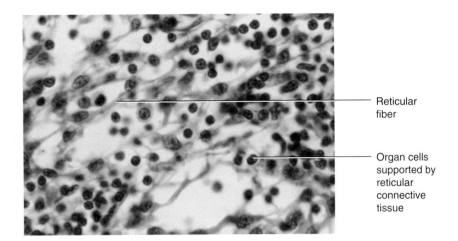

Figure 4.17 Reticular Connective Tissue (400×)

Bone Marrow

Bone marrow is a special connective tissue that contains many cells. **Yellow bone marrow** contains mostly adipose tissue and is found in the shafts of the long bones of adults. Another type of marrow is known as **red bone marrow,** which, in addition to adipocytes, contains precursor cells of red blood cells and white blood cells. These are known as stem cells. Examine a prepared slide of red bone marrow and compare it with figure 4.18. Draw a sample of what you see in the space provided.

Illustration of red bone marrow:

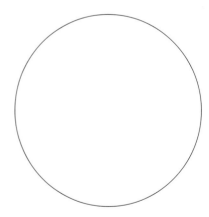

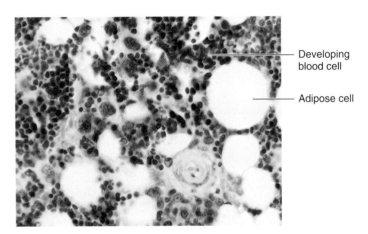

Figure 4.18 Bone Marrow
Photomicrograph of red bone marrow (400×).

Cartilage (Cartilage and Perichondrium)

Cartilage is a type of connective tissue in which the matrix is composed of a pliable material that allows for some degree of movement. The matrix of cartilage is made of glucosamine and chondroitin sulfate, and it forms a semisolid gel in which fibers and cells are found. In prepared slides, the cells sometimes come out of the matrix and leave a cavity. This cavity is known as a **lacuna** and is diagnostic of cartilage tissue. The **perichondrium** is dense irregular connective tissue on the surface of cartilage. There are three types of cartilage in the human body, and they vary by the number and type of protein fibers they contain. These three types are **hyaline cartilage, fibrocartilage,** and **elastic cartilage.**

Hyaline Cartilage

The most common cartilage in the body is **hyaline cartilage,** which is found at the apex of the nose, covering the ends of many long bones at joints, between the ribs and the sternum, in the respiratory passages, and in many other locations. Hyaline cartilage is clear and glassy in fresh tissue but will probably appear light pink in prepared stained slides. The cells that occur in hyaline cartilage are called **chondrocytes,** and the fibers are **collagenous fibers.** Find the chondrocytes in a prepared slide of hyaline cartilage. The fibers will not appear distinct because they blend into the color of the matrix. The chondrocytes of hyaline cartilage frequently occur in pairs. Some people say that they look like pairs of eyes staring back at you as you look at them in the microscope.

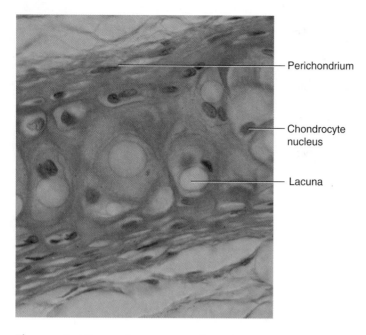

Figure 4.19 Hyaline Cartilage (400×)

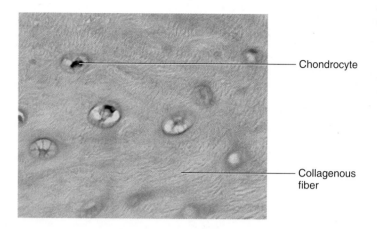

Figure 4.20 Fibrocartilage (400×)

Illustration of fibrocartilage:

Look around the edge of the sample of hyaline cartilage and locate the perichondrium. Compare the material under the microscope with figure 4.19. Draw what you see in the space provided.

Illustration of hyaline cartilage:

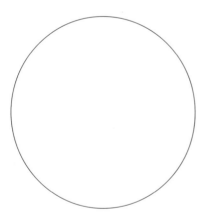

Fibrocartilage

This tissue is similar to hyaline cartilage in that it has chondrocytes and collagenous fibers, but it differs from hyaline cartilage in that it has many more collagenous fibers. These fibers are visible in the prepared slides. Fibrocartilage is found in areas where more stress is placed on the cartilage, such as in the intervertebral disks, the symphysis pubis, and the menisci of each knee. Examine a prepared slide of fibrocartilage and locate the collagenous fibers and chondrocytes. Compare the slide with figure 4.20 and make a drawing of fibrocartilage in the space provided.

Elastic Cartilage

Elastic cartilage is unique in that it contains elastic fibers, which give the tissue its flexible nature. Elastic cartilage is found in the external ear and in a part of the larynx called the epiglottis. If the prepared slide is stained with a silver stain, then the elastic fibers appear black. If the slide is not stained in this way, then the fibers may be hard to distinguish. Examine a prepared slide of elastic tissue and look for the chondrocytes and elastic fibers. Compare your slide with figure 4.21 and make a drawing of elastic cartilage in the space provided.

Illustration of elastic cartilage:

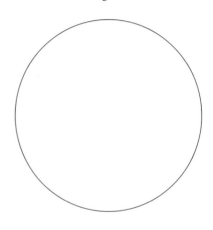

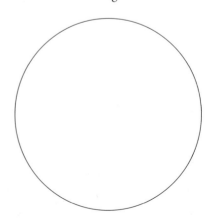

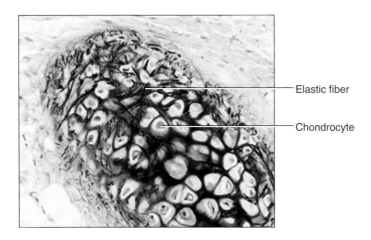

Figure 4.21 Elastic Cartilage (400×)

Bone

Bone, or osseous tissue, is a type of connective tissue in which the matrix is made of a mineral component called **hydroxyapatite crystal,** or **bone salt.** Collagenous fibers occur in the tissue along with osteocytes, or bone cells. Two types of bone are common in the body. **Compact bone** is the more common type; it makes up the dense material in a long section of a bone. **Cancellous (spongy) bone** is found in the end regions of long bones and has plates of bone interspersed with bone marrow. Look at a prepared slide of ground bone and compare it with figure 4.22. Locate the large, circular regions called **osteons** (which resemble a cross section of a tree trunk with the growth rings), the **central canal,** and the **lacunae** where **osteocytes** occur in the tissue. Osteons are the long, thin units of bone with the space in the middle of them called the central canals. A more detailed examination of bone appears in Exercise 8. After you study the slide, draw a sample of it in the space provided.

Illustration of ground bone:

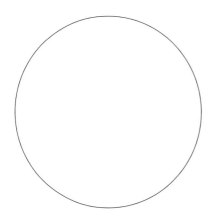

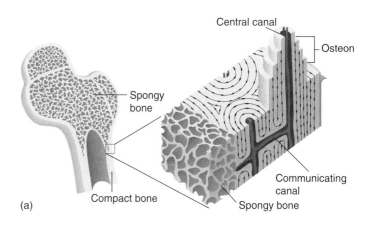

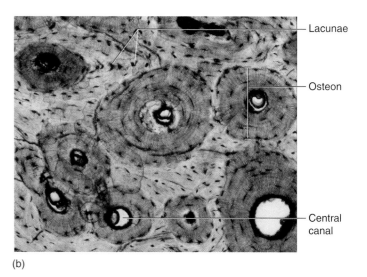

Figure 4.22 Ground Bone
(a) Diagram of ground bone; (b) photomicrograph (100×).

platelets occur in blood. Examine a prepared slide of blood and look for the numerous pink disks in the sample. These are the erythrocytes. You may find larger cells with lobed, purple nuclei. These are the leukocytes. The details of blood are studied in Exercise 25. Examine the slide under high power and compare it with figure 4.23. Make a drawing of blood in the space provided.

Illustration of blood:

Blood

Blood is a connective tissue in which the matrix is fluid and no fibers are visible. The matrix of the blood is called **plasma,** and the cells of the blood are either **red blood cells (erythrocytes)** or **white blood cells (leukocytes).** Some cellular fragments called

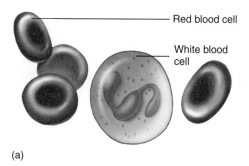 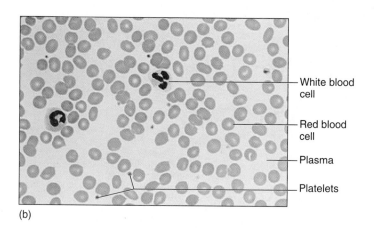

Figure 4.23 Blood
(a) Diagram; (b) photomicrograph (400×).

Study Hint

Examine as many different images of tissues as you can. Drawing the material observed in the lab is a good memory tool. Try to make associations between the abstract image of tissue and something common. For example, you can think of bone as looking like a series of tree rings. Adipose tissue may look like angular balloons. If you have difficulty distinguishing between two specimens, try to find a characteristic that separates the two. For example, smooth muscle and dense regular collagenous connective tissue may look similar in certain stained preparations. You will note that smooth muscle has nuclei inside the fibers, whereas dense connective tissue has nuclei of the fibrocytes between adjacent fibers. Review the types of connective tissue in table 4.2.

Exercise 4 Review

Tissues

Name: _____

Lab time/section: _____

Date: _____

1. List the four main tissues of the body. _____
   _____

2. What is the name of the noncellular layer that attaches epithelial tissue to lower layers? _____
   _____

3. How many layers are in simple epithelium? _____

4. Name the three general cell shapes of epithelial tissue. _____
   _____

5. A single, flattened layer of epithelial cells represents what cell type? _____

6. What functional problem could occur if the digestive tract were lined with stratified squamous epithelium instead of simple columnar epithelium? _____

7. A multiple layer of flattened epithelial cells represents what cell type? _____

8. Can you make the hairs on your arm "stand on end?" Based on your results, what type of muscle is responsible for doing this?
   _____

9. What cell type lines the inside of the urinary bladder? _____

10. What muscle cell types have the following features?

    a. striated and voluntary _____

    b. striated and involuntary _____

    c. nonstriated and involuntary _____

11. What muscle cell type has intercalated disks? _____

12. The intestine is lined with what kind of muscle? _____

13. The heart is composed primarily of what kind of muscle? _____

14. The muscles of the arm are composed primarily of what cell type? _____

15. What cell type is responsible for the transmission of electrochemical impulses? _____

16. What cell type is the conductive cell of the nervous system? _____

17. What is the matrix in blood called? _____

18. Name the three types of fibers in loose connective tissue. _____
    _____

19. What type of connective tissue is springlike and found in the middle walls of arteries? _____

20. What is the cell type in adipose tissue? _____

21. Name the outer connective tissue layer around cartilage. _____

22. What kinds of fibers are in fibrocartilage? _____

23. Describe the functional difference between fibrocartilage and hyaline cartilage. _____
    _____
    _____

24. What is another name for calcium salts in bone? _____

25. In a cross section of a bone, you can usually see two types of bone tissue. What are these called? _____
    _____

26. How do you distinguish generally between epithelial tissue and connective tissue? _____
    _____
    _____

Exercise 4 Tissues

27. Label the following photomicrographs by tissue type and by using the terms provided.
   a. collagenous fibers, elastic fibers

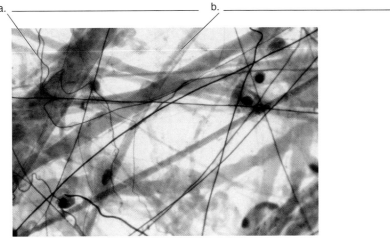

a. _____ b. _____

Tissue Type (400×) _____

b. cilia, basement membrane, nucleus

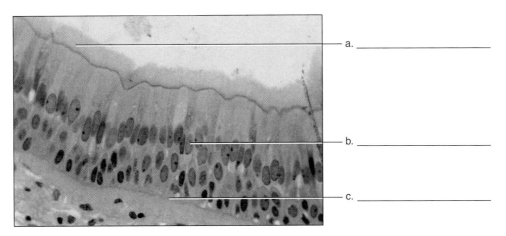

a. _____
b. _____
c. _____

Tissue Type (400×) _____

c. intercalated disks, striations, nucleus

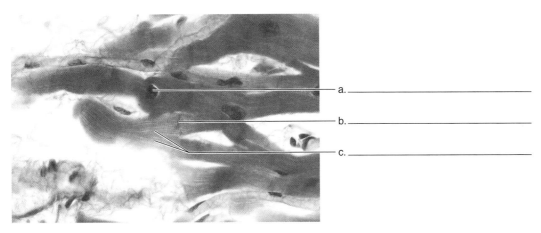

a. _____
b. _____
c. _____

Tissue Type (100×) _____

d. nerve cell body, glial cells

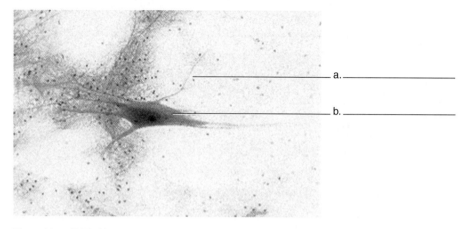

a. _____

b. _____

Tissue Type (100×) _____

Exercise 5
Integumentary System

INTRODUCTION

The integumentary system consists of the skin and associated structures, such as hair, nails, and various glands. The integument consists primarily of a cutaneous membrane composed of an epidermis and a dermis. Below these two layers is an underlying layer called the hypodermis, which anchors the integument to deeper structures. The skin is the largest organ of the body and is vital in terms of water and temperature regulation and protection against microorganisms. These topics are discussed in chapter 5, "Integumentary System," in the *Principles of Anatomy and Physiology* text. In this exercise, you examine the structure of skin, hair, and nails and name the layers of the epidermis and dermis.

OBJECTIVES

At the end of this exercise, you should be able to

1. list the two layers of the integument;
2. list all of the layers of the epidermis from superficial to deep;
3. describe the structure and function of sweat glands and sebaceous glands;
4. draw a hair in a follicle in longitudinal section;
5. discuss the protective nature of melanin.

MATERIALS

Models and charts of the integumentary system
Microscopes
Prepared microscope slides of
    Thick skin (with Pacinian corpuscles)
    Hair follicles
    Pigmented and nonpigmented skin

PROCEDURE
Overview of the Integument

Examine models or charts of the integumentary system and locate the two major regions of the **integument,** the **epidermis** and the **dermis.** The epidermis is a cellular layer and can be distinguished as the most superficial layer of the integument. The dermis is the deeper layer. Also locate the **hypodermis,** not a layer of the integument but a structure anchoring the integument to underlying bone or muscle. Compare the models in the lab with figures 5.1 and 5.2.

Locate the **hair follicles** in the dermis, **sebaceous (oil) glands,** and **sudoriferous (sweat) glands.** The sweat glands are connected to the surface by **sweat ducts.**

Microscopic Examination of the Integument

Examine a prepared slide of thick skin under low power. Locate the hypodermis, dermis, and epidermis. The hypodermis can be distinguished by the presence of significant amounts of adipose tissue found there, whereas the dermis consists primarily of collagenous fibers. The epidermis is the most superficial layer and can be determined by the epithelial cells of that layer. You can look for cells that have obvious nuclei.

Dermis

The dermis is responsible for the structural integrity of the integumentary system. It is composed primarily of collagenous and other fibers, along with blood vessels, nerves, sensory receptors, hair follicles, and glands. **Blood vessels** are located in the dermis, taking nutrients to both the dermis and the epidermis. The blood vessels do not enter the epidermis, but the nutrients and oxygen diffuse from the dermis into the cells of the epidermis. Blood vessels are also important because they release heat to the external environment. As the internal temperature of the body rises, blood flows into the capillaries of the dermis, allowing heat to radiate from the skin. In cold weather, precapillary sphincters close, decreasing the amount of blood flowing to the dermis, keeping heat in the body core.

There are many **nerves** present in the dermis. Sensory information, such as light touch, changes in temperature, and the feeling of pain, is transmitted from receptors in the dermis to the central nervous system. The sensory structures of the skin are covered in Exercise 20.

The dermis consists of two major regions, the deeper **reticular layer** and the superficial **papillary layer.** The reticular layer is the thickest layer of the dermis. It is primarily composed of irregularly arranged collagenous fibers, along with some elastic and reticular fibers. The collagenous fibers provide strength and flexibility. The elastic fibers, as their name implies, provide elasticity to the skin. The papillary layer is found between the deeper reticular layer and the superficial epidermis. It is named for the dermal **papillae,** which are bumps that interdigitate with the epidermis, providing good adhesion between the two layers. It prevents side slippage of the epidermis over the dermis. It contains more elastic fibers than the reticular layer. Examine a slide of thick skin and locate the two layers of the dermis, as seen in figure 5.2.

Epidermis

The epidermis is composed of stratified squamous epithelium with the superficial layer infused with the protein keratin. Keratin is a hardening material that toughens the skin and protects the lower layers of the epidermis from abrasion. Keratin also protects the skin from ultraviolet radiation.

In slides stained with hematoxylin and eosin (HE) stain, the dermis is often colored pink, and the superficial region with multicolored layers is the epidermis. The colors may vary in the slides used in your lab. Compare these layers in the prepared slide with figure 5.3.

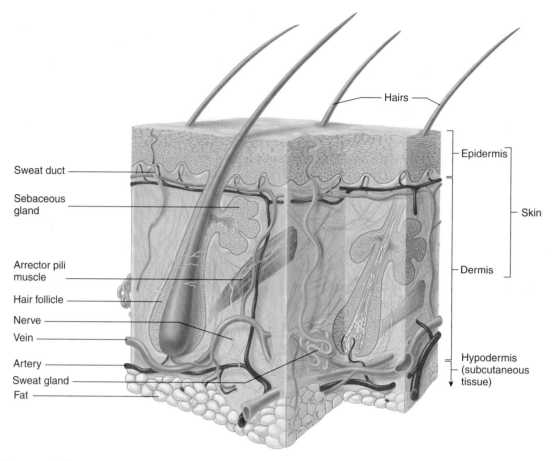

Figure 5.1 Diagram of the Integument

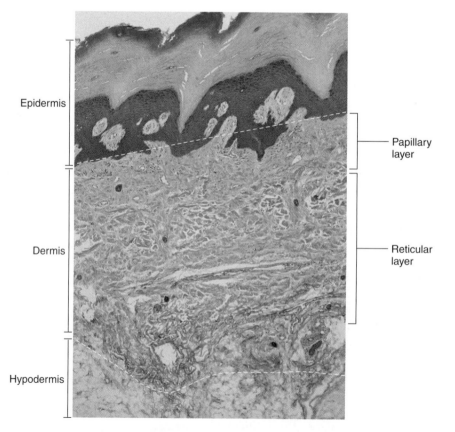

Figure 5.2 Photomicrograph of the Integument (40×)

Strata of the Epidermis

In thick skin, the deepest layer of the epidermis is known as the **stratum basale.** This single layer of cells is attached to the dermis by the **basement membrane,** a noncellular layer. Cells in the stratum basale divide repeatedly and push up toward the superficial layers as the cells increase in age. This process is illustrated in figure 5.4. In the stratum basale are specialized cells known as **melanocytes.** These cells produce the brown pigment **melanin.** Melanin protects the stratum basale from the damaging effects of ultraviolet radiation. The number of melanocytes is relatively constant among people of all skin colors, but the amount of melanin produced by the melanocytes is variable. People of darker pigmentation produce more melanin, whereas people of lighter pigmentation have melanocytes that produce less melanin. People of lighter pigmentation are more susceptible to the effects of ultraviolet radiation and skin cancer.

The layer containing many cells superficial to the stratum basale is the **stratum spinosum.** In prepared slides, the cells appear star-shaped because the cell membrane pulls away from the other cells, except where the cells are joined at microscopic junctions called desmosomes. The stratum basale and stratum spinosum are often classified as the **stratum germinativum.**

A layer of granular cells is present in the next layer, the **stratum granulosum.** The cells have purple-staining keratohyalin granules in their cytoplasm. The granules are precursors of keratin that is found in the outermost layer of the epidermis.

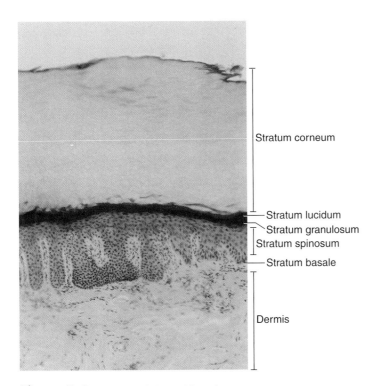

Figure 5.3 Strata of the Epidermis
Photomicrograph (100×).

5. In the stratum corneum, the dead cells have a hard protein envelope, contain keratin, and are surrounded by lipids.

4. In the stratum lucidum, the cells are dead and contain dispersed keratohyalin.

3. In the stratum granulosum, keratohyalin granules accumulate and a hard protein envelope forms beneath the plasma membrane; lamellar bodies release lipids; cells die.

2. In the stratum spinosum, keratin fibers and lamellar bodies accumulate.

1. In the stratum basale, cells divide by mitosis and some of the newly formed cells become the cells of the more superficial strata.

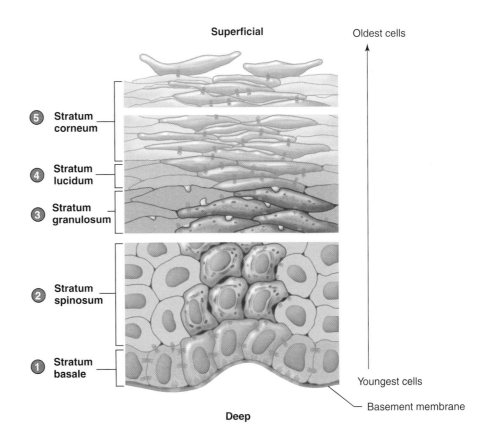

Figure 5.4 Layers of the Epidermis and the Process of Maturation

Superficial to this layer is the **stratum lucidum,** so named (*lucid,* light) because this layer appears translucent in fresh specimens. This layer generally appears clear or pink in prepared slides stained with HE. It is found only in the palms of the hands and soles of the feet.

The most superficial layer of the integument is the **stratum corneum,** which is a tough layer consisting of dead, flattened cells that form the surface. Cells of the stratum corneum have been toughened by complete **keratinization;** thus, the epidermis is said to be made of **keratinized stratified squamous epithelium.** Examine a prepared slide of thick skin and locate the various strata of the epidermis, as illustrated in figure 5.3.

Cells in the epidermis go through three major phases. In the stratum basale, the cells are involved in rapid cell division. Superficial to this layer, cells undergo a process of producing precursor molecules to keratin. The final phase is the completion of keratinization, which leads to the death of the epithelial cells. The most superficial cells of the epidermis are tough. They protect the skin from microorganisms and desiccation and eventually slough off. The epidermis renews itself about every 6 weeks.

Integumentary Glands

A significant number of glands occur in the dermis and hypodermis (figure 5.5*a*). These include **sweat glands, lactiferous (milk) glands, sebaceous (oil) glands,** and **ceruminous (earwax) glands.** The sweat glands are composed of simple cuboidal epithelium and come in two different types. **Merocrine sweat glands,** the most common type, are sensitive to temperature, and produce normal body perspiration. Perspiration reduces the temperature of the body by **evaporative cooling. Apocrine sweat glands** secrete water and a higher concentration of organic acids than merocrine glands. The organic acids, along with bacterial action, produce the characteristic pungent body odor. These glands are concentrated in the region of the axilla and groin and are associated with hair follicles. Examine a prepared slide of skin and look for clusters of cuboidal epithelial cells with purple nuclei in the dermis or hypodermis. These are the sweat glands illustrated in figure 5.5*b*. Note that the glands are tubules, and when these are cut in section on your slide you are mostly seeing the tubules cut in cross section.

Sebaceous glands are typically associated with hair follicles and are larger than sweat glands. You may not see a hair follicle with the sebaceous gland if the section was made through the gland and not through the gland and follicle. These glands secrete an oily material called **sebum,** which lubricates the hair and causes the skin to repel water. Examine a slide of hair follicles and compare the slide with figure 5.5*c*.

Hair

Hair is considered an accessory structure of the integumentary system; it consists of the **shaft,** the portion that erupts from the skin surface; the **root,** the deeper part which is enclosed by the **follicle** (a pocket of epidermis that folds into the dermis); and the **hair bulb,** which is the actively growing portion of the hair. At the center of the bulb is a small structure known as the **papilla,** a region with blood vessels and nerves that reach the hair

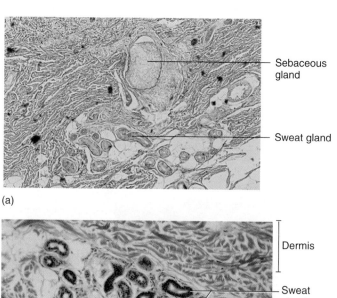

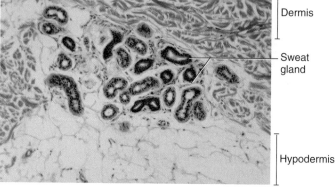

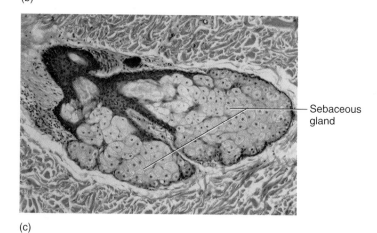

Figure 5.5 Integumentary Glands
(a) Overview of glands; (b) sweat glands (100×); (c) sebaceous glands (100×).

bulb. Look at the prepared slide of hair, locate these structures, and compare them with figures 5.6 and 5.7.

There are are two types of hair. **Determinate hair** grows to a specific length and then stops. Determinate hair is found in many places in the body, including the axilla, groin, arms, legs, eyelashes, and eyebrows. **Indeterminate hair** has continuous growth. It is found on the scalp and in facial hair in males. Examine models and charts in the lab and compare them with figure 5.6.

The hair root is enclosed by a follicle, which is a pocket of epidermis in the dermis; thus, hair is considered an epidermal derivative. The follicle is composed of an outer dermal layer of connective tissue and an inner epithelial layer. These form the **root sheath** of

Exercise 5 Integumentary System

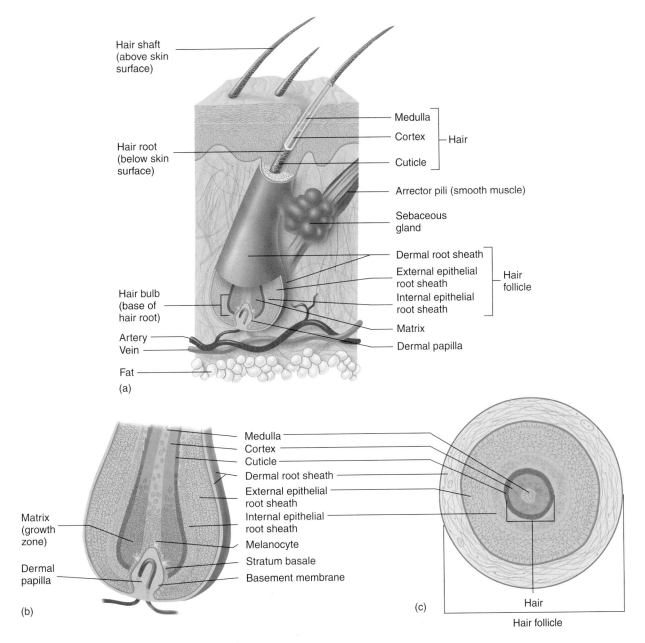

Figure 5.6 Diagram of Hair in a Follicle
(a) Overview of follicle in skin; (b) detail of hair bulb; (c) cross section of hair in follicle.

the hair. The **follicle** encloses the hair in the same way a bud vase encloses the stem of a rose.

Arrector Pili

When the hair "stands on end," it does so by the contraction of **arrector pili muscles.** An arrector pili muscle is a cluster of parallel smooth muscle fibers that connects the hair follicle to the upper regions of the dermis. In response to cold conditions, the arrector pili muscles contract, producing goose bumps. The arrector pili also contract when a person is suddenly frightened. Find an arrector pili muscle in a prepared slide and compare it with figures 5.6 and 5.7. If you have trouble finding an arrector pili muscle, look for slender slips of smooth muscle that angle from the follicle to the superficial regions of the dermis. You may have to look at more than one slide to find them.

Cross Section of Hair

The central portion of the hair is known as the **medulla,** which is enclosed by an outer **cortex.** The cortex may contain a number of pigments, which give the hair its particular color. The brown and black pigments are due to varying types and amounts of melanin. Iron-containing melanin pigments produce red hair. Gray hair is the result of a lack of pigment or the presence of air in the hair root and shaft. The layer superficial to the cortex is the **cuticle** of the hair; it looks rough in microscopic sections. Figure 5.6 shows a cross section of hair.

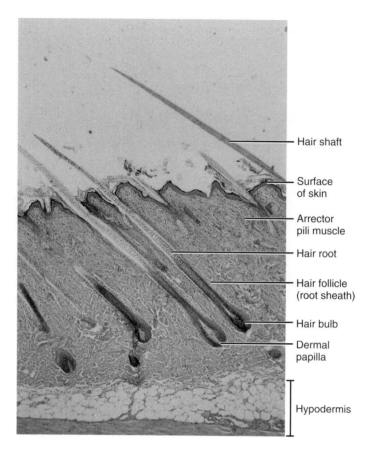

Figure 5.7 Photomicrograph of Hair in a Follicle (40×)

Nails

Nails, the keratinized cells of the stratum corneum of the skin, are located on the distal ends of the fingers and toes. The nail itself is called the **nail body;** as it grows, the **free edge** of the nail is the part you clip. The **cuticle** of the nail is the **eponychium,** and the **nail root** is underneath the eponychium and is the area that generates the nail body. The **nail bed** is the layer deep to the **nail body,** whereas the **lunula** is the small, white crescent at the base of the nail. The **hyponychium** is the region under the free edge of the nail. Examine figure 5.8 and compare your fingernails with the diagram. On each side of the nail is the **nail groove,** a depression between the body of the nail and the skin of the fingers. The ridge above the groove is the **nail fold.**

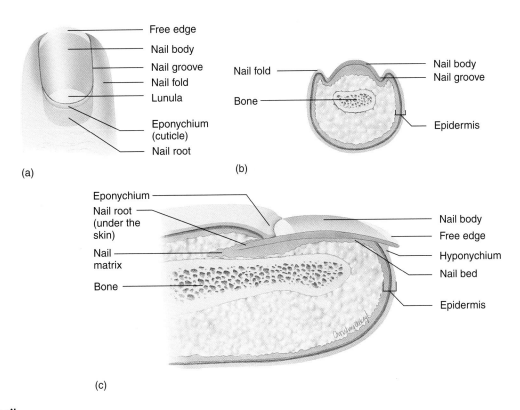

Figure 5.8 Fingernail
(a) Dorsal view; (b) cross section; (c) longitudinal section.

Exercise 5 Review
Integumentary System

Name: _____
Lab time/section: _____
Date: _____

1. The main organ of the integumentary system is the skin. Name three structures that are associated structures in the integumentary system. _____

2. Which one of the three layers (epidermis, dermis, hypodermis) is not part of the integument? _____

3. The dermis has two main layers. Which one of these is the more superficial? _____

4. What is the most common connective tissue fiber in the dermis? _____

5. The release of heat from the body by blood vessels occurs in what main layer of the integument? _____

6. Which stratum is the deepest layer of the epidermis? _____

7. Which stratum of the epidermis is the most superficial? _____

8. Approximately how long does it take for the epidermis to renew itself? _____

9. What cell type produces a pigment that darkens the skin? _____

10. What material makes the epidermis tough? _____

11. What integumentary gland secretes oil? _____

12. Which kind of sweat gland (merocrine, apocrine) is involved in evaporative cooling? _____

13. What part of the hair is found on the outside of the skin? _____

14. What does an arrector pili muscle do? _____

15. The outermost portion of the hair is known as the
    a. cuticle.   b. lunula.   c. root.   d. medulla.

16. The hair of the axilla is considered determinate/indeterminate hair (circle the correct answer).

17. Electrolysis is the process of hair removal by using electric current. Explain how this might destroy the process of hair growth in relation to the hair bulb. _____

18. Since hair color is determined by pigment in the cortex, and the hair shaft is dead, explain the fallacy of a person's hair turning white overnight. _____

_____

19. Since the hair shaft is dead, explain the fallacy of "feeding" the hair, as claimed by some cosmetics firms. _____

_____

20. Two common blisters of the integument are watery blisters, filled with clear fluid, and blood blisters, filled with blood. Based on your knowledge of the blood supply to the integument, describe what layers might be damaged in these two blisters. _____

_____

21. Label the following illustration using the terms provided.

    hair root   hair shaft
    hair follicle   sweat gland
    arrector pili   sebaceous gland
    epidermis   stratum corneum
    dermis
    hypodermis

Exercise 8
Axial Skeleton: Vertebrae, Ribs, Sternum, Hyoid

INTRODUCTION

The axial skeleton consists of 80 bones, including the ribs, hyoid, bones of the skull, vertebral column, and sternum. In this exercise, you examine all parts of the axial skeleton except for the skull, which will be covered separately in Exercise 9. The **vertebral column** consists of 26 bones. The hyoid is a small, floating bone between the floor of the mouth and the upper anterior neck; it assists the tongue in swallowing and the larynx in speech.

The thoracic cage consists of the 12 pairs of ribs and the sternum, which both protect the lungs and heart yet provide for flexibility during breathing. These topics are discussed in the *Principles of Anatomy and Physiology* text in chapter 7, "Anatomy of Bones and Joints."

OBJECTIVES

At the end of this exercise, you should be able to

1. find a specific vertebra (such as T6) on an articulated vertebral column;
2. name selected vertebrae on an articulated vertebral column;
3. name the bony features of an individual vertebra and determine the region of the vertebral column to which it belongs;
4. describe the features that determine cervical, thoracic, lumbar, sacral, and coccygeal vertebrae and identify the atlas and the axis;
5. describe the normal and abnormal spinal curvatures;
6. distinguish between the different types of ribs, the markings of the individual ribs, and whether they come from the left or right side of the body;
7. identify the bony features of the sternum;
8. locate the major features of the hyoid bone.

MATERIALS

Articulated skeleton or plastic casts of a skeleton
Disarticulated skeleton or plastic casts of bones
Charts of the skeletal system
Wooden applicator sticks or plastic drinking straws cut on a bias for pointer tips
Foam pads of various sizes (to protect bone from hard countertops)
Cardboard box or piece of wood about 1 foot square and 3 inches deep

PROCEDURE

Review the bones of the axial skeleton in the lab and in Exercise 6 (figure 6.1). In this exercise, you study the bones of the axial skeleton by examining both the disarticulated bones and the articulated skeleton. Hold each bone up to the skeleton to see how it is positioned in relation to the other bones of the body. When examining bones in the lab, do not use your pen or pencil to locate a structure. These leave marks on the bones. Carefully use a wooden applicator stick or plastic straw to point out structures. Your instructor may want you to place real bone material on foam pads to cushion the bone from the tabletop.

Vertebrae

Overview of the Vertebrae

There are five regions of the vertebral column composed of cervical, thoracic, and lumbar vertebrae and vertebrae that make up the sacrum and the coccyx. The **vertebral column** in humans is much different than in other mammals because we are the only habitually bipedal mammal. In humans, the vertebrae increase in size from the cervical to the sacral vertebrae. This is due to the increase in weight on the lower vertebrae. The diameter of the vertebral canal increases from the sacral region to the cervical region. This is due to the greater number of neural fibers in the spinal cord as sensory information is transmitted to the brain and motor information is transmitted from the brain. Between the vertebrae are fibrocartilaginous pads known as **intervertebral disks.**

Spinal Curvatures

The spine has four curvatures, which alternate from superior to inferior. The **cervical curvature** is convex (bowed forward) when seen from anatomic position. The **thoracic curvature** is concave, whereas the **lumbar curvature** is convex and the **sacral (pelvic) curvature** is concave. These allow for balance when standing upright. Locate the curvatures in figure 8.1. In addition to the spinal curvatures just described, there are abnormal spinal curvatures. **Scoliosis** is a lateral curvature. **Kyphosis** is an exaggerated thoracic curvature, and **lordosis** is an exaggerated lumber curvature.

Obtain a vertebra and examine it for the features represented in a typical vertebra (figure 8.2). Place the vertebra in front of you so that you can see through the large hole known as the **vertebral foramen,** where the spinal cord is located. Note the large **body** of the vertebra, which supports the weight of the vertebral column and is in contact with the intervertebral disks. Note the **vertebral,** or **neural, arch,** which consists of two **pedicles** (ped´i-kls) and two **laminae** (lam´i-nay). The pedicles are the parts of the arch that extend from the body of the vertebra to the two lateral projections called the **transverse processes.** Each lamina is a broad, flat structure between the transverse process and the dorsal **spinous process.** If you rotate the vertebra as illustrated in figure 8.2, you should be able to see the vertebral body in lateral view with the **superior articular process** and the **superior articular facet** visible. The superior articular facet is a smooth face that articulates with the vertebra above. There is also an **inferior articular process** and an **inferior articular facet** that articulates with the vertebra below. The inferior articular facet of a

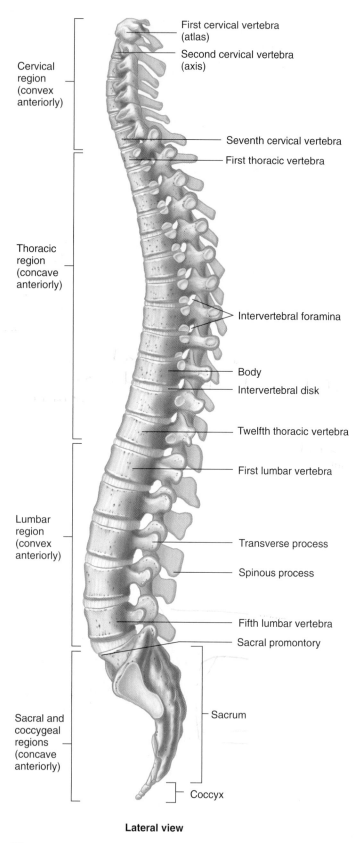

Figure 8.1 Vertebral Column, Lateral View

superior vertebra joins with the superior articular facet of the next inferior vertebra. If you put two adjacent vertebrae together and look at the lateral aspect, you can see the **intervertebral foramen,** a hole that allows the spinal nerve to exit from the vertebral column.

Cervical Vertebrae

There are seven **cervical vertebrae,** and these can be distinguished from all other vertebrae in that each vertebra has three foramina. The **vertebral foramen** is the largest opening, and the two **transverse foramina** house the vertebral arteries and vertebral veins. The transverse foramina are specific to the cervical vertebrae. Cervical vertebrae 2–6 have **bifid** (bī′fid) **spinous processes** (*bifid* = split in two), and the bodies of the cervical vertebrae are less massive than those of the inferior vertebrae. Examine the isolated cervical vertebrae in the lab and compare them with figure 8.3.

The first cervical vertebra (C1) is known as the **atlas** and is the only cervical vertebra without a body. The atlas carries the weight of the head and was named after the Greek mythological figure Atlas, a giant who carried the weight of the world on his shoulders. The atlas joins with the head and provides for range of motion, as when you move your head to indicate "yes." The second cervical vertebra (C2) is the **axis,** and it has a unique process called the **dens,** or **odontoid** (ō-don′toyd) **process,** which runs superiorly through the atlas. The odontoid process allows the atlas to rotate on the axis and provides for the range of motion in your head when you rotate your head to indicate "no." The seventh cervical vertebra (C7) is known as the **vertebra prominens.** It has a spinous process that projects sharply in a posterior direction and can be palpated (felt) as a significant bump at the posterior base of the neck. Look at an atlas, an axis, and a vertebra prominens in the lab and compare them with figure 8.4.

Thoracic Vertebrae

There are 12 thoracic vertebrae. The **thoracic vertebrae** are distinguished from all other vertebrae by their markings on the posterolateral surface of the body of the vertebrae, which are attachment points for the ribs. Ribs 1, 11, and 12 attach to just 1 vertebra, in which case the marking on the vertebral body is known as a **rib facet.** In some sections of the thoracic region, the head of a rib is attached to 2 vertebrae. In this case, each vertebra has a smooth attachment point known as a **demifacet.** The head of the rib spans both of the vertebrae and articulates with the demifacet of the superior and inferior vertebrae.

The thoracic vertebrae also have longer spinous processes than the cervical vertebrae, and the spinous processes of the thoracic vertebrae tend to point inferiorly. Examine figure 8.5 and note the characteristics of the vertebrae as seen in the lab.

Lumbar Vertebrae

There are five lumbar vertebrae. The **lumbar vertebrae** are distinguished from the other vertebrae by having neither transverse foramina nor rib facets. The spinous processes of the lumbar vertebrae tend to be more horizontal than the thoracic vertebrae, and the bodies of the lumbar vertebrae are larger, since they carry more weight than the thoracic vertebrae, yet the twelfth thoracic and the first lumbar vertebrae are remarkably

Exercise 8 Axial Skeleton: Vertebrae, Ribs, Sternum, Hyoid

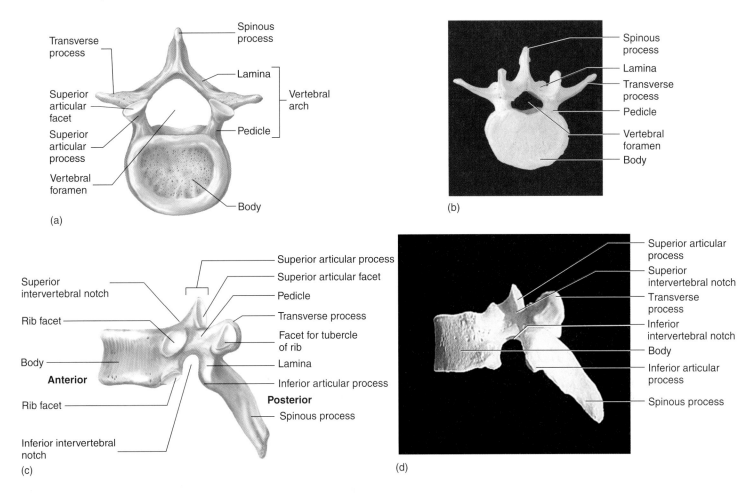

Figure 8.2 Features of a Typical Vertebra
(a) Diagram, superior view; (b) photograph, superior view; (c) diagram, lateral view; (d) photograph, lateral view.

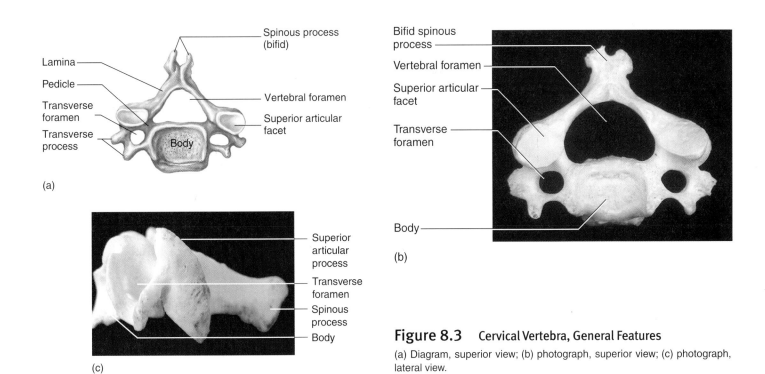

Figure 8.3 Cervical Vertebra, General Features
(a) Diagram, superior view; (b) photograph, superior view; (c) photograph, lateral view.

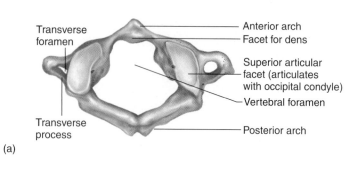
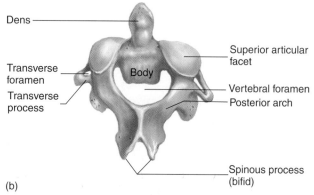
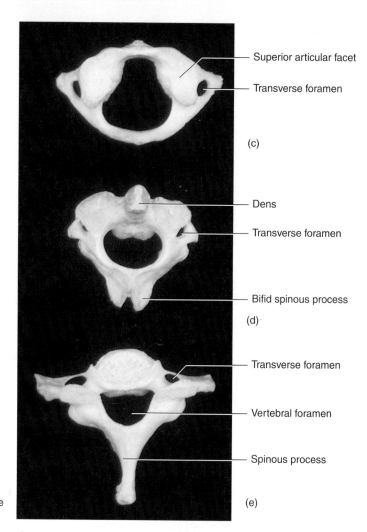

Figure 8.4 Atlas, Axis, and Vertebra Prominens
(a) Atlas, superior view; (b) axis, supero-posterior view. Photographs of three vertebrae, superior view; (c) atlas; (d) axis; (e) vertebra prominens.

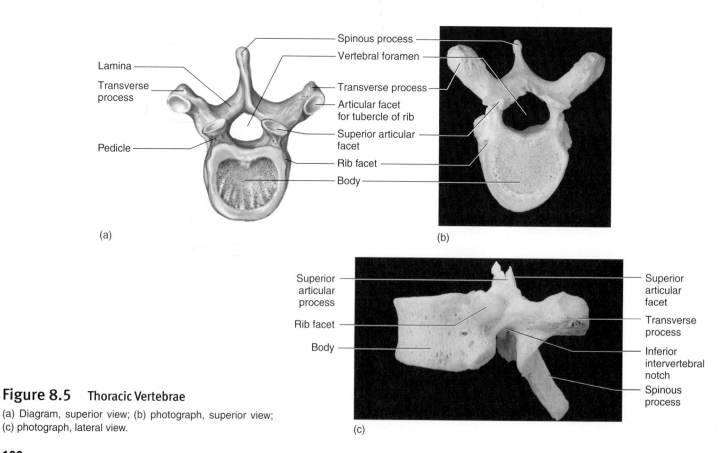

Figure 8.5 Thoracic Vertebrae
(a) Diagram, superior view; (b) photograph, superior view; (c) photograph, lateral view.

similar. The last thoracic vertebra has rib facets on it, and the first lumbar does not. Look at the lumbar vertebrae in the lab and compare them with figure 8.6.

Sacrum

The **sacrum** (sā′krum) is a large, wedge-shaped bone composed of five fused vertebrae. The lines of fusion are called **transverse lines,** and these can be seen on both the anterior and posterior sides. Notice how the sacrum is shaped like a shallow, triangular bowl. If you place the sacrum in front of you, as you would a cereal bowl, the shallow depression is the **anterior surface.** The two rows of holes you see are the **anterior sacral foramina.** On the posterior surface are the **posterior sacral foramina.** Compare a sacrum in the lab with figure 8.7.

The **sacral promontory** is a rim on the anterior superior part of the sacrum, and the **alae** (ále, **ala** singular) are two expanded regions of the sacrum lateral to the promontory. The roughened areas lateral to the alae are the **auricular surfaces** of the sacrum; each joins with the ilium to form the **sacroiliac joint.** As additional force is applied to the sacrum, it wedges itself tighter into the ilium. If you examine the posterior surface of the sacrum, you will notice the **posterior sacral foramina,** the **median** and **lateral sacral crests,** and the superior and inferior openings of the **sacral canal.** The **sacral hiatus** is the inferior opening of the sacrum. As with the other vertebrae, the **superior articular processes** and the **superior articular facets** join with the next most superior vertebra (the fifth lumbar vertebra). These features can be seen in figure 8.7.

The sacrum in humans is wedge-shaped; this allows the upper body to carry more weight than if the sacrum were box-shaped, as it is in many quadrupeds. Hold a block of wood or a box between your hands, as illustrated in figure 8.8a. Have your lab partner push down on the box and see how much force is required for the box to slip through your hands. This would be similar to the forces acting on a rectangular sacrum. Turn the box or block of wood so that it forms a wedge in your hands, as illustrated in figure 8.8b. Have your lab partner push down on the box. Is it easier or harder to dislodge the "sacrum" in this way?

Coccyx

The **coccyx** (kok′siks) is the terminal portion of the vertebral column, and it usually consists of four fused vertebrae. The coccyx may be fused with the sacrum in some older females or in individuals who have fallen backward and landed on a hard surface. Examine the coccyx in the lab and compare it with figure 8.7.

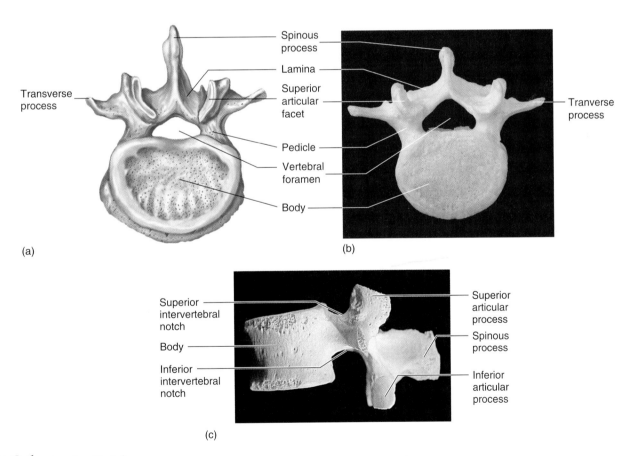

Figure 8.6 Lumbar Vertebrae
(a) Diagram, superior view; (b) photograph, superior view; (c) photograph, lateral view.

Ribs

There are 12 pairs of **ribs** in the human; along with the sternum, they make up the **thoracic cage.** Each rib has a **head** that articulates with the body of one or more vertebrae. On the head is a **facet,** which is the site of articulation with the vertebral body. The **neck** is a constricted region near the head of the rib. Locate this on figure 8.9. The process near the neck on ribs 1–10 is the **tubercle** of the rib. This tubercle articulates with the transverse process of the vertebra. The **angle** of the rib (**costal angle**) is the part of a rib that makes a sharp bend. On an articulated skeleton, it is located about the same distance from the midline of the body as the inferior angle of the scapula. Some ribs have a truncated **sternal end** that attaches to a costal cartilage prior to joining the sternum. The superior edge of the rib is more blunt than the inferior edge. The depression that runs along the inferior side of each rib is known as the **costal groove.** With the costal groove in an inferior position and the blunt sternal end toward the midline, determine whether you are looking at a left or right rib.

The first 7 pairs of ribs are **true,** or **vertebrosternal, ribs.** A true rib is one that attaches to the sternum by its own cartilage. Ribs 8 through 12 are **false ribs** because they do not attach to the sternum by their own cartilage. Ribs 8 to 10 attach to the sternum by way of the cartilage of rib 7 and are known as **vertebrochondral ribs.** Ribs 11 and 12 do not attach to the sternum at all and are known as **floating,** or **vertebral, ribs** (a specific type of false rib). Examine the articulated skeleton in the lab and compare it with figure 8.10.

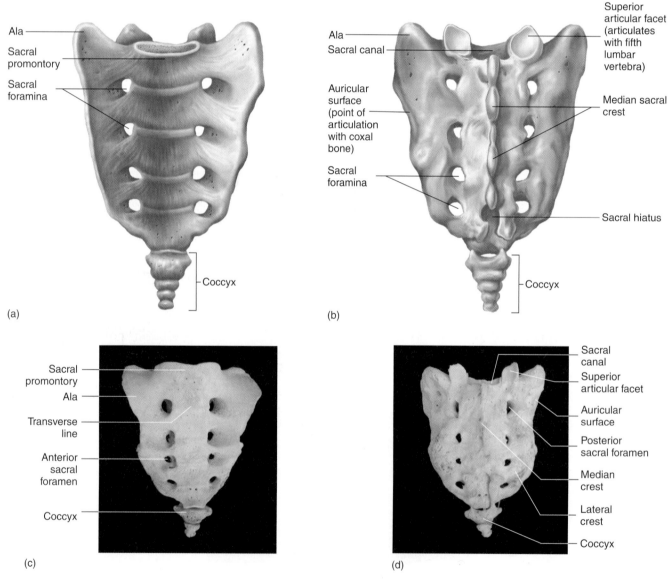

Figure 8.7 **Sacrum and Coccyx**
Diagram (a) anterior view; (b) posterior view; photograph (c) anterior view; (d) posterior view.

Exercise 8 Axial Skeleton: Vertebrae, Ribs, Sternum, Hyoid

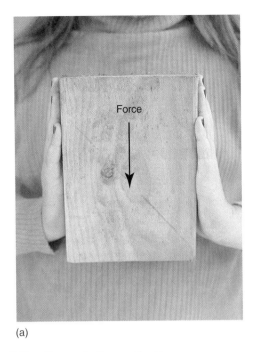
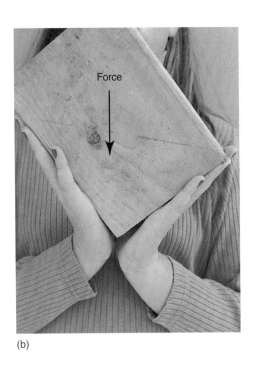

Figure 8.8 Forces Acting on the Sacrum
(a) Rectangular sacrum; (b) wedge-shaped sacrum.

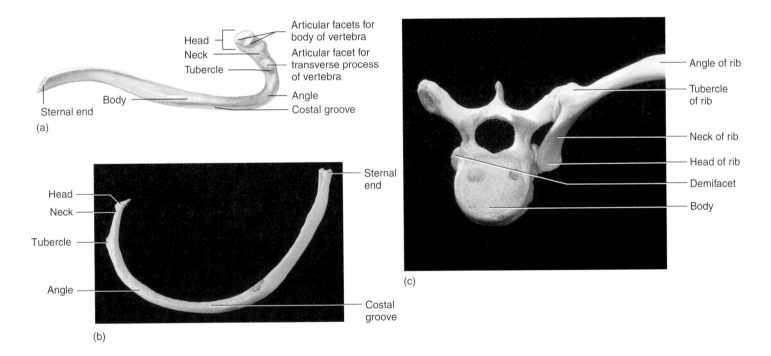

Figure 8.9 Rib
(a) Diagram, anterior view; (b) photograph, inferior view; (c) photograph of rib and vertebra, superior view.

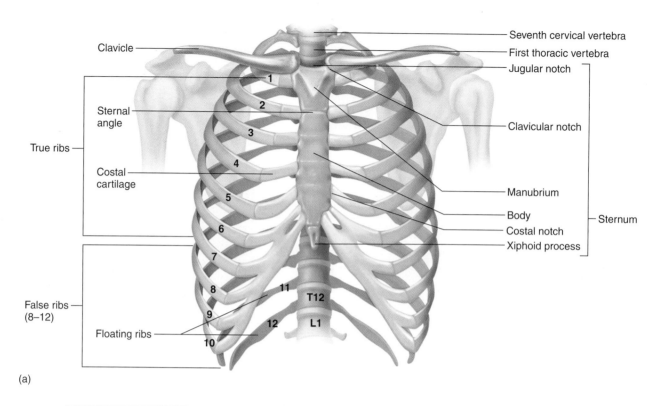

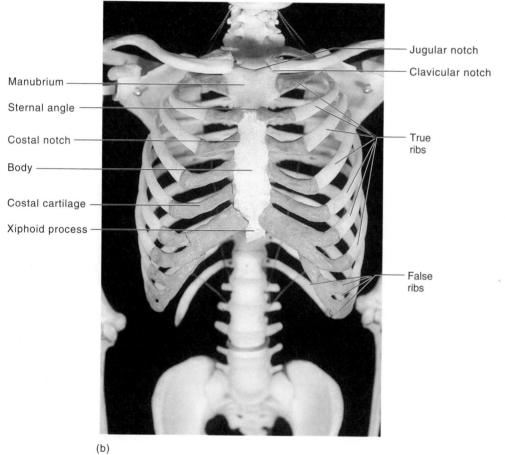

Figure 8.10 Thoracic Cage
(a) Diagram; (b) photograph.

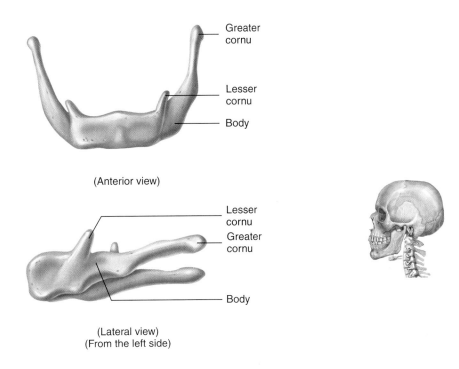

Figure 8.11 Hyoid Bone

Sternum

The **sternum** is composed of three fused bones. The superior segment is the **manubrium,** and the depression at the top of the manubrium is the **jugular,** or **suprasternal, notch.** The lateral indentations on the manubrium are sites of articulation with the clavicles known as **clavicular notches.** The main portion of the sternum is the **body,** and between the body and the manubrium is the **sternal angle.** This is a landmark for finding the second rib when using a stethoscope to listen to heart sounds. Locate the **costal notches** on the body of the sternum. These are where the cartilages of the ribs attach.

The narrow, bladelike part that is the most inferior segment of the sternum is the **xiphoid** (zī'foyd) **process.** Care must be taken when performing CPR (cardiopulmonary resuscitation) so that pressure is applied to the body of the sternum and not to the xiphoid process. If the force were applied to the xiphoid process, it could be fractured and driven into the liver. Examine the structures of the sternum on lab specimens and in figure 8.10.

Hyoid

The **hyoid** is a floating bone (it has no direct bony attachments) at the junction of the floor of the mouth and the neck. The hyoid is anchored by muscles from the anterior, posterior, and inferior directions and aids tongue movement and swallowing. Locate the central **body** of the hyoid, the **greater cornua** (kor´nū-ah; sing. **cornu**), and the **lesser cornua** on the material in the lab and in figure 8.11.

Exercise 8 Review

Axial Skeleton: Vertebrae, Ribs, Sternum, Hyoid

Name: _____

Lab time/section: _____

Date: _____

1. Label the following illustration using the terms provided.

   body   spinous process
   lamina   transverse process
   pedicle   vertebral arch

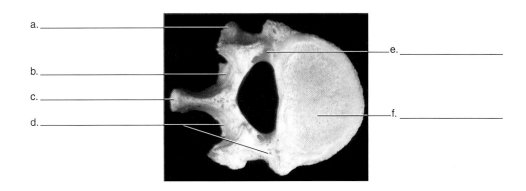

a. _____
b. _____
c. _____
d. _____
e. _____
f. _____

2. What are the names of the fibrocartilage pads between adjacent bodies of the vertebrae? _____

3. What structures make up the vertebral arch? _____

4. The superior articular process of a vertebra articulates (joins) with what specific structure? _____

5. Which spinal curvature is the most superior one? _____

6. On what specific vertebra would you find the odontoid process? _____

7. Which vertebra has no body? _____

8. What two features do cervical vertebrae have that no other vertebrae have? _____

9. Match the cervical vertebrae with their numeric descriptions.

   axis   C1
   vertebra prominens C2
   atlas   C7

107

10. The vertebra in the following illustration comes from what part of the spinal column? _____

    Why? _____

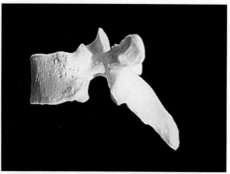

11. How many lumbar vertebrae are in the human body? _____

12. Distinguish between the posterior sacral foramina and the sacral canal. _____
    _____

13. Name the horizontal lines that result from the fusion of the sacral vertebrae. _____

14. What is the joint between the sacrum and the hip bones called? _____

15. Match the region, or vertebrae from the region, with the features or structures found there.

    Region/Vertebrae   Features/Structures
    _____ cervical   a. ala
    _____ thoracic   b. vertebrae with the largest vertebral bodies
    _____ lumbar   c. vertebrae with rib facets
    _____ sacral   d. typically, four fused vertebrae
    _____ coccyx   e. transverse foramina

16. Name the major features of the hyoid bone. _____
    _____

17. Does the hyoid bone have any solid bony attachments? _____

18. A rib that attaches to the sternum by the cartilage of rib 7 has what name? _____

19. Is the angle of the rib on the anterior or posterior side of the body? _____

20. Which ribs (by number) are the floating ribs? _____

21. What part of a rib articulates with the transverse process of a vertebra? _____

22. Label the following illustration using the terms provided.

angle   costal groove   head   tubercle   neck

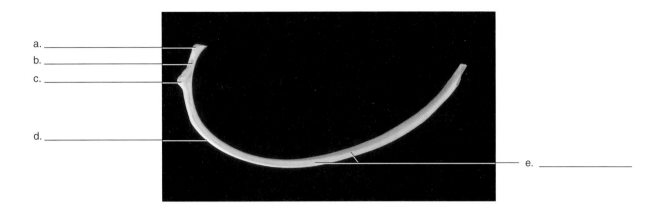

a. _____
b. _____
c. _____
d. _____
e. _____

23. The most inferior portion of the sternum is the
    a. body.   b. manubrium.   c. angle.   d. xiphoid.

24. The superior portion of the sternum is called the _____.

Exercise 9
Axial Skeleton: Skull

INTRODUCTION

The bones of the skull can be divided into three groups: bones of the braincase, bones of the face, and bones of the middle ear. The skull is the most complex region of the skeletal system and not only houses the brain but also contains a significant number of sense organs. In this exercise, you learn the details of the skull, particularly those of the braincase and the face. These topics are discussed in the *Principles of Anatomy and Physiology* text in chapter 7, "Anatomy of Bones and Joints." You will study the bones of the middle ear in this lab manual in Exercise 23. Be careful if you handle real bone material, as bone is fragile and can easily be broken.

OBJECTIVES

At the end of this exercise, you should be able to

1. list all of the bones of the braincase and the major features of those bones;
2. list all the bones of the face and the major features of those bones;
3. name all the bones that occur singly or in pairs in the skull;
4. locate the major foramina of the skull;
5. find the specific bony markings of representative bones;
6. name the major sutures of the skull;
7. list all the fontanels of the fetal skull.

MATERIALS

Disarticulated skull, if available
Articulated skulls with the calvaria cut
Fetal skulls
Foam pads to cushion skulls from the desktop
Wooden applicator sticks or plastic straws for pointer tips

PROCEDURE

General Considerations

As you locate the various bones on a skull in the lab, make sure you do not poke pencils, pens, or fingers into the delicate regions of the eyes or nasal cavity. Be very careful handling skulls. Do not break the delicate structures in these regions.

Overview of Bones of the Braincase

Familiarize yourself with the bones of the skull in the **braincase** and the **face**. The skull bones in the braincase are listed next with a number, in parentheses, after the name of the bone, indicating whether the bone occurs singly or in pairs. You can remember the bones of the cranium with the mnemonic "of pets." Each letter represents a cranial bone (*o* for occipital, *f* for frontal, etc.). Take a skull back to your lab table and locate the bones listed in figure 9.1.

Bones of the Braincase

Frontal (1)	Parietal (2)
Occipital (1)	Temporal (2)
Sphenoid (1)	Ethmoid (1)

Overview of Face

The bones of the face are listed next with (1) indicating a single bone and (2) indicating bones that are paired. Locate these bones on a skull in the lab, as shown in figures 9.1 and 9.2.

Bones of the Face

Maxilla (2)	Nasal (2)
Mandible (1)	Palatine (2)
Vomer (1)	Zygomatic (2)
Lacrimal (2)	Inferior nasal concha (plural *conchae*) (2)

The facial bones can be remembered by use of this mnemonic device: "Monkeys in New Zealand like peaches very much." In the mnemonic device, all of the bones except the last two (vomer and mandible) are paired bones.

Once you have found the major bones of the skull, use the following descriptions and illustrations to find the specific structures. The skull is described from a series of views, and you should locate the anatomical features listed for those views.

Anterior View

The large bone that makes up the forehead is the **frontal bone.** It is a single bone that makes up the superior portion of the **orbits** (eye sockets) and has two ridges above the eyes called **supraorbital margins** (eyebrows). There are holes in each of these ridges, called **supraorbital foramina,** where nerves and arteries reach the face. Most of what you can see inferior to the frontal bone are facial bones. The bony part of the nose is composed of paired **nasal bones,** which join with the cartilage that forms the tip of the nose. Lateral to the nasal bones are the upper portions of the maxillae. The maxillae form the inferior, medial portion of the orbit, extend lateral and inferior to the nose, and hold the upper teeth. The two maxillae have sockets called **alveoli** (ăl′vē-oh-lī) (sing. **alveolus**), which contain the teeth and extensions of bone between each pair of sockets called **alveolar processes.** The **infraorbital foramen,** a

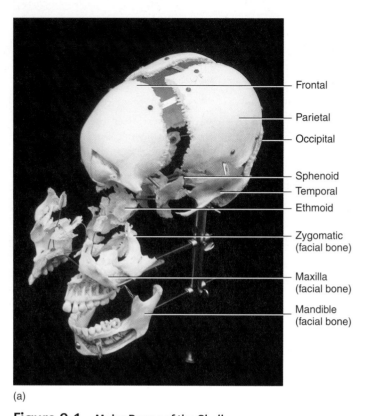

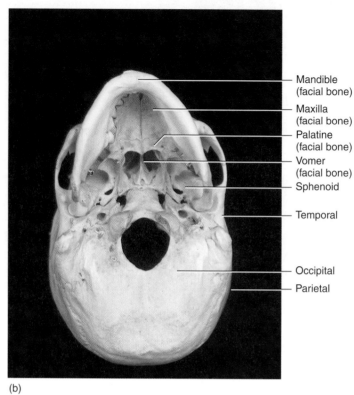

Figure 9.1 Major Bones of the Skull
(a) Disarticulated skull, 3/4 view; (b) inferior view.

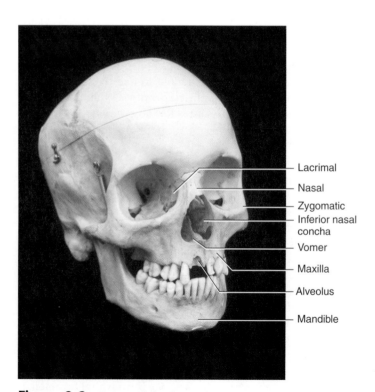

Figure 9.2 Bones of the Face

small hole below the eye in the maxilla, is a passageway for nerves and blood vessels. Posterior to the maxilla are the thin **lacrimal** (lăk′rĭh-mul) **bones,** which contain the **naso-lacrimal canal,** a tube that drains tears from the eye to the nose. Locate these bones in figure 9.3. Posterior to the lacrimal bones is the **ethmoid** (ĕth′moyd) **bone,** which is very delicate and frequently broken on skulls that have been mishandled. Posterior to the ethmoid is the **sphenoid** (sfē′noyd) **bone,** which forms the posterior wall of the orbit and contains not only the **optic canal** (a passageway for the optic nerve) but also the **superior orbital fissure** and the **inferior orbital fissure.** Note the bones on the lateral side of the orbit. These are the **zygomatic bones.**

The most inferior bone of the face is the **mandible,** commonly known as the lower jaw. The mandible also has alveoli and alveolar processes and holds the lower teeth (figure 9.4). The mandible begins as two bones in utero and fuses at the midline of the chin. This fusion results in the **mandibular symphysis** (sĭhm′ fah-sĭs). Lateral to the mandibular symphysis are the **mental foramina,** which conduct nerves and blood vessels to the tissue anterior to the jaw.

Superior View

Locate the major suture lines of the skull from the superior view, as seen in figure 9.5. The frontal bone is separated from the pair of parietal bones by the **coronal suture.** The parietal bones are separated from one another by the **sagittal suture.** The parietal bones are separated from the occipital bone by the **lambdoid** (lamb′doyd) **suture** (named after the Greek letter lambda, which is

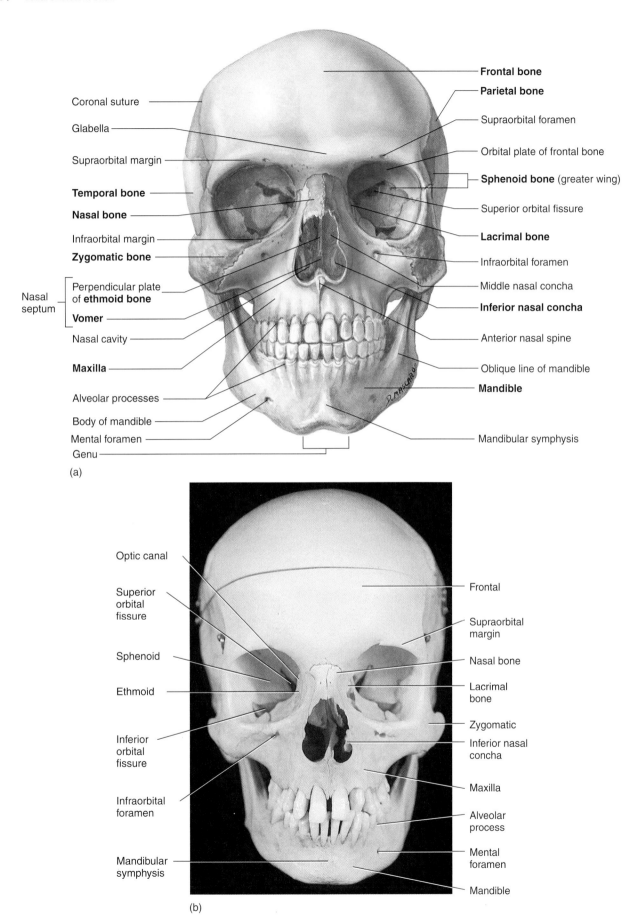

Figure 9.3 Skull, Anterior View

(a) Diagram; (b) photograph.

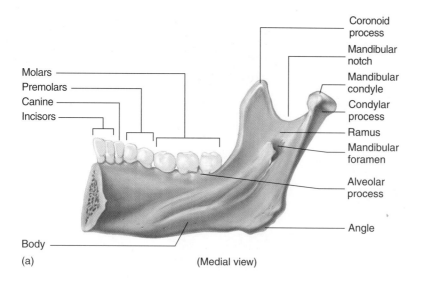

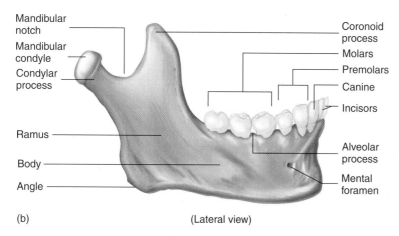

Right half of the **mandible**

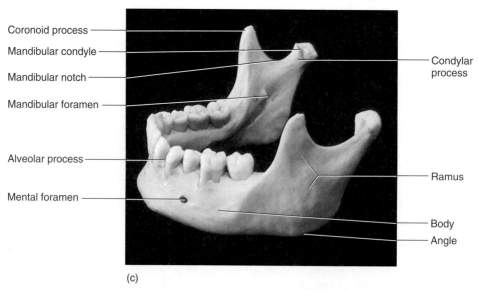

Figure 9.4 Mandible

Diagram (a) medial view; (b) lateral view. (c) Photograph, lateral view.

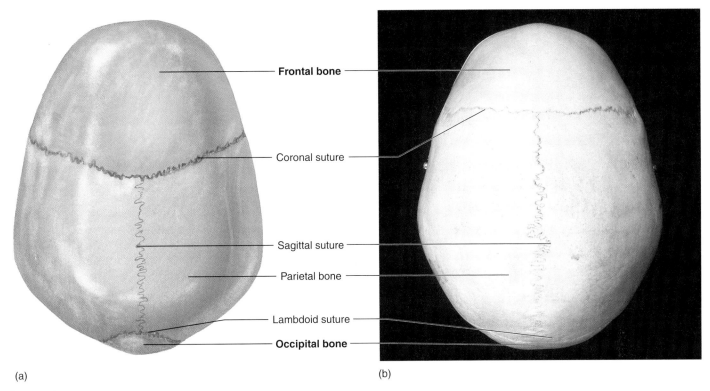

Figure 9.5 Skull, Superior View
(a) Diagram; (b) photograph.

Y-shaped). There may be small bones between the occipital bone and the parietal bones (or between other skull bones), and these are known as **sutural,** or **wormian, bones.**

Lateral View

The coronal and lambdoid sutures also may be seen from the lateral view as well as the **squamous suture,** which separates the **temporal bone** from the **parietal bone.** Locate the bones of the braincase from a lateral view. These are the **frontal, parietal, occipital, temporal, sphenoid,** and **ethmoid bones.** You may be able to see a bump at the back of the occipital bone from this angle. This is known as the **external occipital protuberance.** The hole on the side of the head where the ear attaches is the **external auditory (acoustic) meatus.** The large process posterior and inferior to this opening is the **mastoid process.** A long, thin spine medial to the mastoid process is the **styloid** (stī´loyd) **process,** which has muscles that connect it to the hyoid bone, tongue, and larynx. The sphenoid bone can also be seen from this view just anterior to the temporal bone. A bony process is found lateral to and forming a thin bridge of bone superficial to the sphenoid bone. This is the **zygomatic** (zī-goh-mă´tik) **arch,** which makes up the upper part of the cheek and is commonly known as the cheekbone. The zygomatic arch is composed of the **zygomatic process** of the temporal bone and the **temporal process** of the zygomatic bone. The zygomatic bone forms the lateral wall of the orbit. On the inner wall of the orbit is the ethmoid bone, the lacrimal bone, the maxilla, and the nasal bone. Examine these structures in figure 9.6.

The mandible has a **condylar** (kon´dih-lar) **process** with a terminal **mandibular condyle,** articulating with the temporal bone; a **coronoid process,** which lies medial to the zygomatic arch; and a **mandibular notch,** a depression between the condylar process and coronoid process. The vertical section of the mandible is the **mandibular ramus** (*ramus* = branch), and the horizontal portion of the mandible is the **body.** The **angle** of the mandible is at the posterior, inferior part of the bone at the junction of the body and the ramus. On the inside of each ramus of the mandible is the **mandibular foramen,** which is a conduit for an artery, a vein, and a nerve. The parts of the mandible can be identified in figures 9.4 and 9.6.

Inferior View

Place the skull in front of you with the mandible removed (figure 9.7). The largest hole in the skull, the **foramen magnum,** should be close to you. The foramen magnum is in the occipital bone and is where the brain joins the spinal cord. Lateral to the foramen magnum are the **occipital condyles,** which are processes that articulate with the first cervical vertebra. A small bump at the posterior part of the occipital bone is the external occipital protuberance, an attachment site for muscles. At the junction of the occipital bone and the temporal bone is the **jugular foramen,** a hole through which the internal jugular vein passes. If you carefully insert a pipe cleaner or wooden applicator stick into the jugular foramen and turn the skull over (with the top of the cranium removed), you will see that the jugular foramen is in the posterior part of the skull.

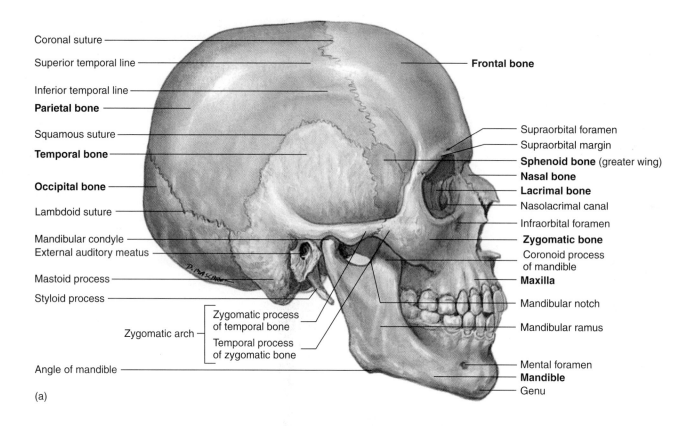

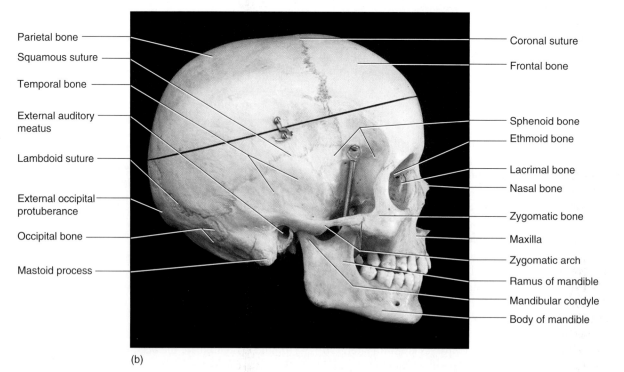

Figure 9.6 Skull, Lateral View
(a) Diagram; (b) photograph.

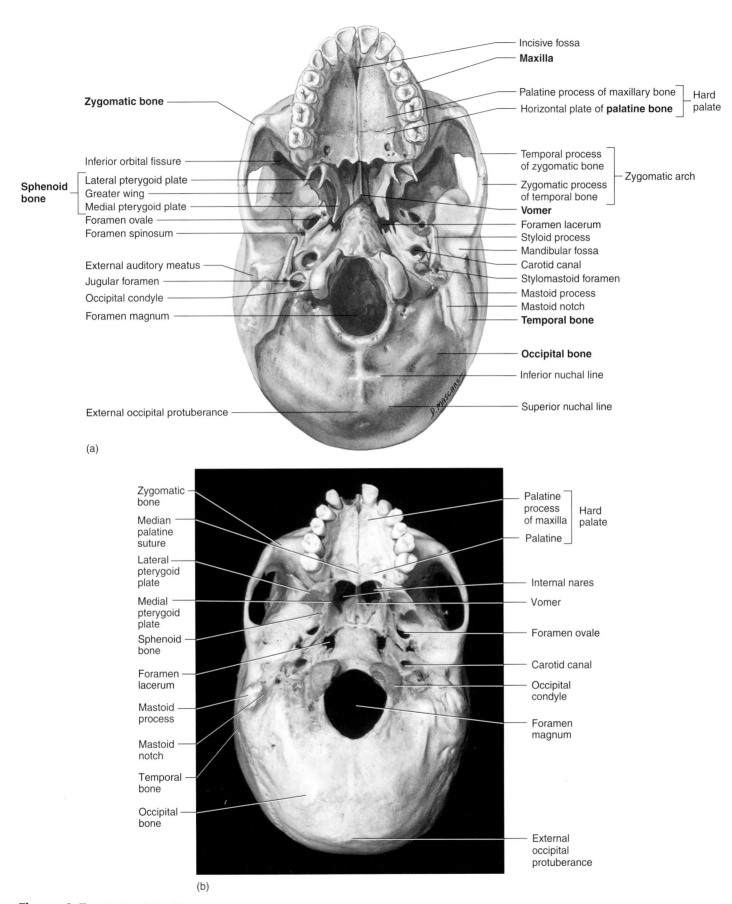

Figure 9.7 Skull, Inferior View
(a) Diagram; (b) photograph.

You can see the mastoid process of the temporal bone and a depression just medial to the process known as the **mastoid notch**. Find the **styloid processes** in your specimen, although they may be hard to locate since they often get broken in lab specimens. The styloid process is an attachment point for muscles that move the tongue, larynx, and hyoid. Medial to the styloid process is the **carotid canal**, through which passes the internal carotid artery, which takes blood to the brain. If you carefully insert a pipe cleaner into the carotid canal of a real skull, you will notice that the canal bends at about a 90-degree angle; if you turn the skull over, you will note that the opening occurs in the middle of the skull. You cannot do this with most plastic casts of skulls because the canal is filled in. At the junction of the temporal bone and the sphenoid bone is the **foramen lacerum** (lah-ăr´um), which joins with the carotid canal as it enters the skull. The temporal bone also has a **mandibular fossa**, which is the articulation site of the mandible. The zygomatic process of the temporal bone can also be seen from this view.

From the inferior view, the sphenoid bone can be seen as a bone that runs from one side of the skull to the other. The most lateral parts of the sphenoid are **greater wings**. Two pairs of flattened processes can also be seen in this view, the **lateral pterygoid** (tear´ih-goyd) **plates** and the **medial pterygoid plates** (*pterygoid* = winglike). These are attachments for muscles that extend from the sphenoid to the mandible. Just posterior to the pterygoid plates is the **foramen ovale** (ō-val´-eh), which conducts one of the branches of the trigeminal nerve to the mandible.

In the midline of the skull and sometimes looking like a part of the sphenoid is the **vomer**. This is a single bone of the face that forms part of the **nasal septum**. The two large holes on each side of the vomer are the **internal nares**. Connected to the vomer and forming part of the **hard palate** is the **palatine bone**. The palatine bones are L-shaped bones with a horizontal plate and a vertical plate. The horizontal plates normally join at the **median palatine suture**. If this suture does not fuse completely at birth, an individual has a cleft palate. Also part of the hard palate are horizontal shelves of the maxillae. These are the **palatine processes of the maxillae**.

The major openings of the skull are presented in table 9.1. Locate the openings and note the structures that pass through these holes.

Interior of the Cranial Vault

With the top of the skull removed, you can see that the **cranial cavity** is divided into three major regions. These are the **anterior cranial fossa**, a depression anterior to the lesser wings of the sphenoid; the **middle cranial fossae**, which lie between the **lesser wings of the sphenoid** and the petrous portion of the temporal bone; and the **posterior cranial fossa**, which is posterior to the petrous portion of the temporal bone. Examine figure 9.8 for a view of the interior of the cranium.

Beginning with the anterior cranial fossa, you should find the centrally located **ethmoid bone**. A sharp ridge known as the **crista galli** projects from the main portion of this bone. The small, horizontal plate of bone with numerous holes lateral to the crista galli is the **cribriform** (krĭ´ bri-form) **plate**. The holes in this plate transmit nerves that carry the sense of smell from the nose to the brain. If the skull was cut close to the orbit, you can see the **frontal sinus**, a hollow space in the anterior portion of the frontal bone.

The dividing line between the anterior and middle cranial fossae is the sphenoid bone. Locate the lesser wings of the sphenoid and the **sella turcica** (sell´ah-tur´sik-ah) just posterior to it. The sella turcica has a small depression called the **hypophyseal** (hī-poh´fah-seal) **fossa**, in which the pituitary gland sits. The posterior raised part of the sella turcica is known as the **dorsum sellae** (sell´ ā). The **greater wings of the sphenoid** are more inferior than the lesser wings, and each one contains the **foramen rotundum**, which takes a branch of the trigeminal nerve to the maxilla. The **foramen ovale** can be seen from this view as well.

Behind the sphenoid bone is the temporal bone, which has a flattened lateral section known as the **squamous portion** and a heavier mass of bone known as the petrous portion. The **petrous portion** divides the middle and posterior cranial fossae. The petrous portion also has a hole in the posterior surface, which is the **internal auditory meatus**. This is a passageway for the nerves that come from the inner ear.

The posterior cranial fossa is located dorsal to the petrous portion of the temporal bone. It contains the foramen magnum and the **jugular foramina**. Most of this fossa is formed by the occipital bone.

Midsagittal Section of the Skull

Be extremely careful with the specimen, since many of the internal structures are fragile. Use figure 9.9 as a guide. Locate the **nasal septum**, which is composed of the **vomer**, the **perpendicular**

TABLE 9.1 Openings of the Skull

Opening	Function or Structure in Opening
Carotid canal	Internal carotid artery
External acoustic meatus	Opening for sound transmission
Foramen lacerum	Cartilage
Foramen magnum	Spinal cord, vertebral arteries
Foramen ovale	Mandibular branch of trigeminal nerve
Foramen rotundum	Maxillary branch of trigeminal nerve
Foramen spinosum	Meningeal blood vessels
Inferior orbital fissure	Maxillary branch of trigeminal nerve
Infraorbital foramen	Infraorbital nerve and artery for the face
Internal acoustic meatus	Vestibulocochlear nerve and facial nerve
Jugular foramen	Internal jugular vein, vagus, and other nerves
Mandibular foramen	Mandibular branch of trigeminal nerve
Mental foramen	Mental nerve and blood vessels
Optic canal	Optic nerve
Stylomastoid	Facial nerve exits skull
Superior orbital fissure	Nerves to the eye and face
Supraorbital foramen	Supraorbital nerve and artery for the face

Exercise 9 Axial Skeleton: Skull

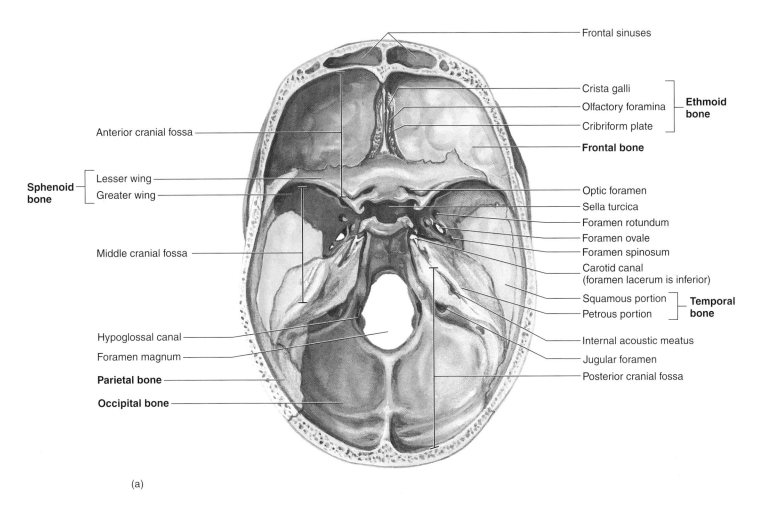

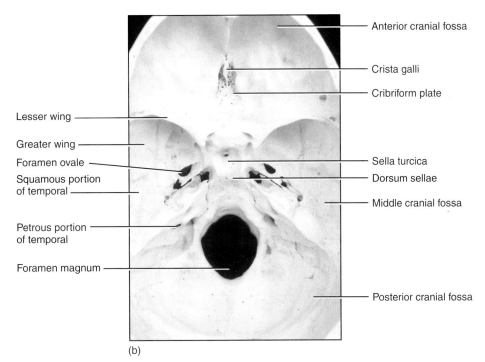

Figure 9.8 Interior of the Cranium
(a) Diagram; (b) photograph.

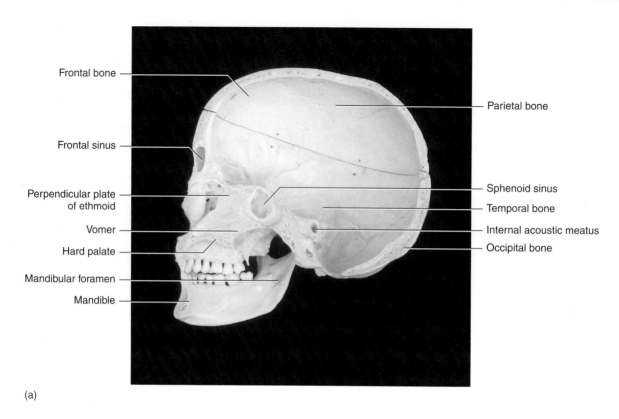

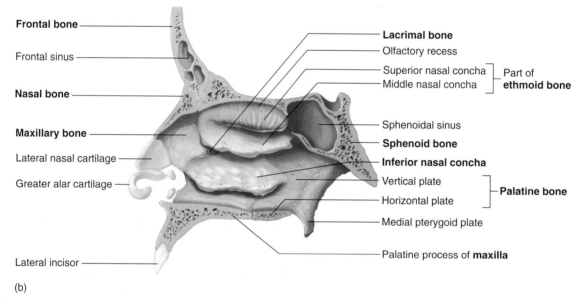

Figure 9.9 Skull, Midsagittal Section
(a) Photograph; (b) diagram of nasal cavity.

plate of the ethmoid bone, and the **nasal cartilage** (absent in skull preparations). If the nasal septum is removed, you should be able to see the **superior nasal concha** (konk′ah) and the **middle nasal concha** of the ethmoid bone. Below these is the **inferior nasal concha,** which is a separate bone. Look for the junction between the palatine bone and the palatine process of the maxilla. These two bony plates make up the hard palate. If the mandible is present, locate the **mandibular foramen,** which is on the inner aspect of the mandible and transmits branches of the trigeminal nerve and blood vessels to the mandible.

Sinuses

There are many sinuses and air cells in the skull. These sinuses provide shape to the skull while decreasing its weight, and some add resonance to the voice. The **paranasal sinuses** are located around the region of the nose and are named for the bones in which they are found. They include the **frontal sinus,** the **maxillary sinus,** the **ethmoidal labyrinth (sinuses),** and the **sphenoidal sinus.** These sinuses may fill with fluid when a person has a cold and harbor bacteria in secondary infections. Locate the sinuses in skulls in the lab and compare them with figure 9.10.

Fetal Skull

The development of the skull is problematic in humans because we are large-brained, bipedal mammals. Our hips are narrow, which provides for efficient locomotion, yet our skulls are large. It is difficult for a narrow-hipped mammal to give birth to a large-headed infant. One adaptation that we have to this is to have significant brain growth after birth. In humans, the adult brain is more than three times the size of the newborn's, and the skull must grow to accommodate this increase. One adapation for this increase after birth is the presence of fontanels (soft spots) in the skull that allow for both the passage of the skull through the birth canal and the growth of the skull during childhood. The fetal skull is discussed in Exercise 10 and can be seen in figure 10.3. Examine fetal skulls, or charts and models of skulls in lab, and look for the fontanels.

Select Individual Bones of the Skull

Ethmoid

The **ethmoid bone** is located in the middle of the skull. The **perpendicular plate** of the ethmoid can be seen in midsagittal view or from the anterior view through the external nares. The **orbital plate** is the part of the ethmoid that lines the medial wall of the orbit. The **middle nasal conchae** can be seen from the nasal cavity, but the **superior nasal conchae** are best seen by looking at an inferior view of the skull through the internal nares or at a midsagittal view with the nasal septum removed. Examine isolated ethmoid bones in the lab and find the **crista galli, olfactory foramina, cribriform plate,** and other structures, as shown in figure 9.11.

Sphenoid

The **sphenoid bone** is seen in figure 9.12. Examine an isolated sphenoid bone in the lab and locate the **greater wings,** the **lesser wings,** the **medial** and **lateral pterygoid plates,** the **sella turcica,** the **hypophyseal fossa,** the **dorsum sellae,** and other features, as seen in figure 9.12.

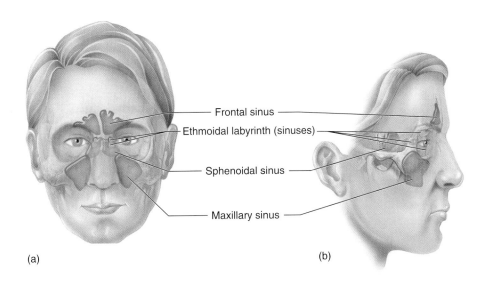

Figure 9.10 Paranasal Sinuses of the Skull
(a) Anterior view; (b) lateral view.

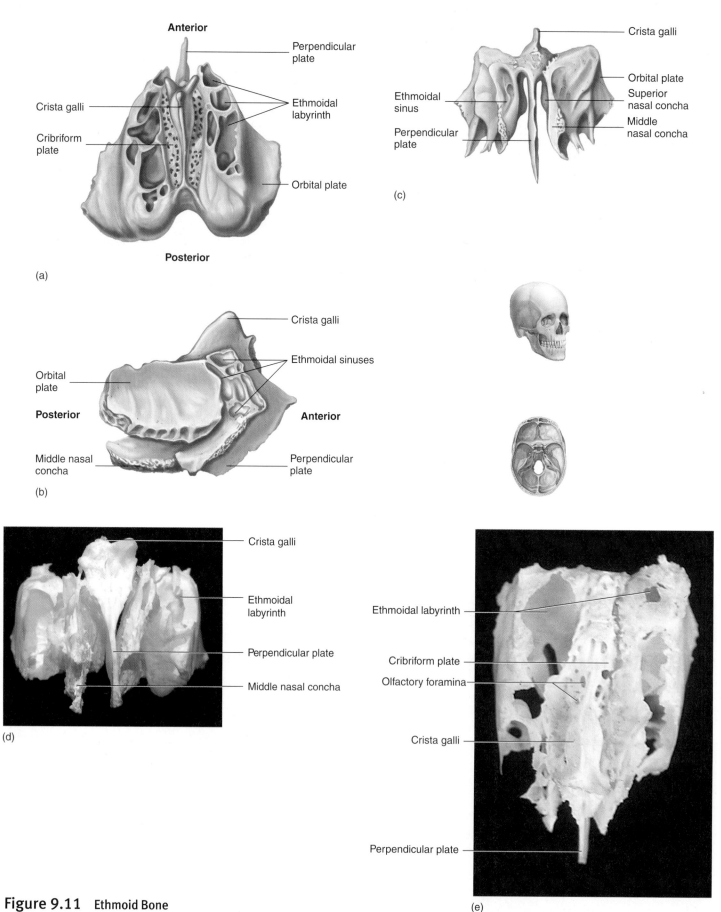

Figure 9.11 Ethmoid Bone
Diagram (a) superior view; (b) lateral view; (c) anterior view. Photograph (d) anterior view; (e) superior view.

Exercise 9   Axial Skeleton: Skull

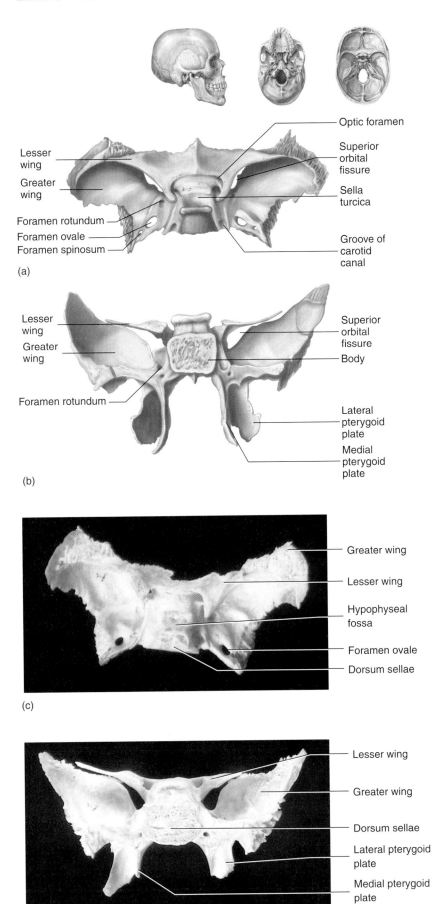

Figure 9.12 Sphenoid Bone

Diagram (a) superior view; (b) posterior view. Photograph (c) superior view; (d) posterior view.

Temporal

The temporal bone is a paired cranial bone. The **squamous portion** is the lateral part of the bone; it forms part of the braincase. There is also a medial part of the temporal called the **petrous portion.** The petrous portion contains the ear ossicles and the opening of the **internal acoustic meatus,** as seen in the medial view of the temporal bone in figure 9.13. Examine an isolated temporal bone in lab and locate these features, as well as the **zygomatic process,** which articulates with the zygomatic bone, and the **mastoid process,** which can be palpated (felt) as a bump posterior to the ear.

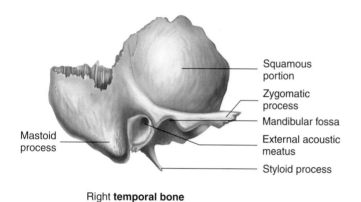

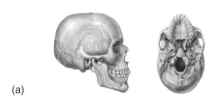

(a)

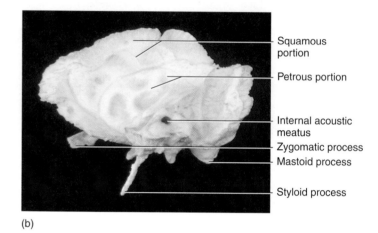

(b)

Figure 9.13 Temporal Bone
(a) Diagram, lateral view. (b) Photograph, medial view.

Exercise 9 Review

Axial Skeleton: Skull

Name: _____

Lab time/section: _____

Date: _____

1. The eyebrows are directly superficial to what bone? _____

2. What is the common name for the zygomatic bone? _____

3. What is the name of the process posterior to the earlobe? _____

4. The hard palate is made up of what bones? _____

5. What are the names of the major paranasal sinuses? _____
_____

6. The mandible fits into what part of the temporal bone to form the jaw joint? _____

7. What bone is found just posterior to the ethmoid bone in the orbit? _____

8. The sella turcica is found in what bone? _____

9. What is the name of the bone that makes up most of the temple (the region superior and anterior to the ears)? _____
_____

10. What are the names of the bones that surround the opening of the nose? _____
_____

11. The upper teeth are held by what bones? _____

12. In what bone would you find the foramen magnum? _____

13. What are the two bony structures that make up the nasal septum? _____

14. The sagittal suture separates the _____ from the _____.
    a. sphenoid, ethmoid
    b. left parietal, right parietal
    c. frontal, parietal
    d. parietals, occipital

15. Which bone is *not* located in the orbit?
    a. maxilla　　b. zygomatic　　c. ethmoid　　d. sphenoid　　e. temporal

16. Which bone is *not* a paired bone of the skull?
    a. zygomatic　　b. temporal　　c. lacrimal　　d. vomer

17. Label the following illustration using the terms provided.

    median palatine suture　　carotid canal
    occipital bone　　zygomatic arch
    vomer　　palatine
    foramen magnum　　lateral pterygoid plate
    palatine process of maxilla　　mastoid process
    zygomatic　　maxilla
    occipital condyle　　temporal
    internal nares　　foramen ovale
    sphenoid

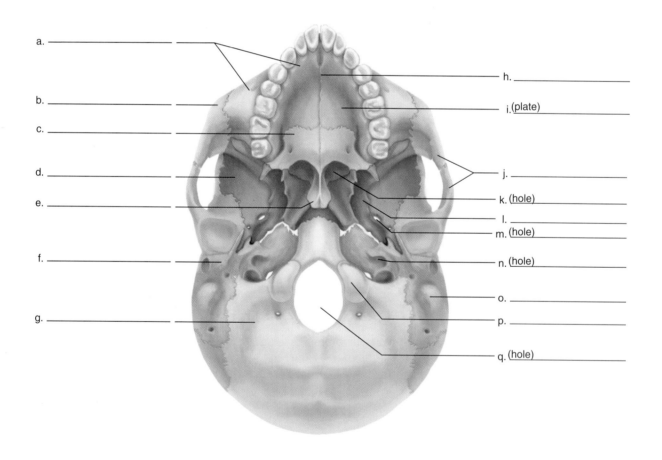

a. _____
b. _____
c. _____
d. _____
e. _____
f. _____
g. _____
h. _____
i. (plate) _____
j. _____
k. (hole) _____
l. _____
m. (hole) _____
n. (hole) _____
o. _____
p. _____
q. (hole) _____

Exercise 9 Axial Skeleton: Skull

18. Label the following illustration using the terms provided.

    ramus   mandibular notch
    mental foramen   mandibular condyle
    mandibular foramen   body
    coronoid process   angle
    alveolar process

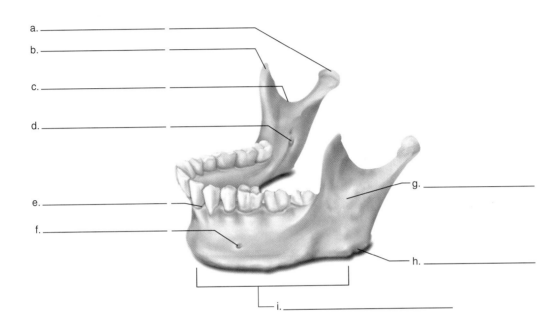

a. _____
b. _____
c. _____
d. _____
e. _____
f. _____
g. _____
h. _____
i. _____

19. Based on what you know about the maxillary sinus, why would a significant impact to the maxilla create a more difficult situation for healing than would the fracture of a long bone? _____

_____

20. Label the following illustration using the terms provided.

frontal sinus   lesser wing of sphenoid   greater wing of sphenoid
foramen magnum   anterior cranial fossa   crista galli
sella turcica   cribriform plate   petrous portion of temporal
foramen ovale

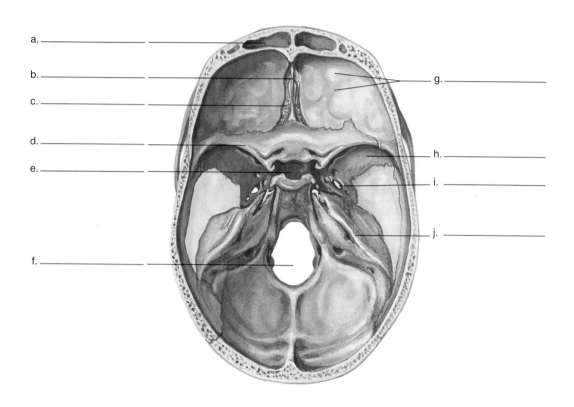

a. _____
b. _____
c. _____
d. _____
e. _____
f. _____
g. _____
h. _____
i. _____
j. _____

Exercise 10
Articulations

INTRODUCTION

The study of articulations is the study of the joints between bones and is known as **arthrology. Articulations** are the areas of interplay between the skeletal system and the muscular system. In this exercise, you study the joints according to their composition, by their movement, and how movable they are. You also study specific joints of the body and actions at joints.

Joints can be classified according to their physical composition, or they can be classified according to their range of movement. In terms of composition, joints are classified as fibrous, cartilaginous, or synovial. In terms of range of movement, joints are immovable, semimovable, or freely movable. As you study the joints in the lab, use the joints in your body as reference and mimic the action of a specific articulation. These topics are discussed in the *Principles of Anatomy and Physiology* text in chapter 7, "Anatomy of Bones and Joints."

OBJECTIVES

At the end of this exercise, you should be able to

1. distinguish among fibrous, cartilaginous, and synovial joints;
2. list all of the fontanels of the fetal skull;
3. discuss the nature of a synovial joint;
4. locate the fibrous capsule, synovial membrane, synovial fluid, and articular cartilage in a dissected synovial joint;
5. list five types of synovial joints;
6. explain the structure of the knee, hip, jaw, and shoulder;
7. describe or illustrate by example all of the actions at joints.

MATERIALS

Mammal joint with intact synovial capsule
Dissection tray with scalpel or razor blades, blunt probe, and protective gloves
Waste container
Model or chart of joints, including those of the shoulder, elbow, knee, hip, and jaw
Articulated skeleton
Fetal skulls

PROCEDURE
Types of Joints

The three major groups of joints classified according to composition are fibrous, cartilaginous, and synovial joints. Fibrous joints are typically composed of connective tissue fibers between two bones. They permit little movement. Cartilaginous joints consist of cartilage between two bones and are generally more movable than fibrous joints, although some cartilaginous joints may have no movement at all. Synovial joints have the most complex structure, including a joint capsule, an inner membrane, and synovial fluid, and they are the most movable of the joints (table 10.1).

Fibrous Joints

Fibrous joints connect one bone to another with collagenous fibers. If the fibers are close together, the joint does not allow for much, if any, movement between the bones. One example of this type of joint is called a **suture,** and it occurs between adjacent bones in the cranium. In these joints, the bones of the cranium are tightly bound together by dense, fibrous connective tissue. As a person approaches the age of 35 or so, some of the sutures of the skull begin to fuse from the region closest to the brain toward the superficial surface of the skull. This fusion leads to the complete union of two bones and is then called a **synostosis** (sin-os-tō´sis). The two frontal bones in the fetal skeleton fuse together as a synostosis and form a single frontal bone. Examine a skull in the lab and locate the sutures.

Another type of fibrous joint is a **gomphosis** (gom-fō´sis), which is represented by teeth in the sockets of the maxilla and the mandible. This type of joint is a peg in a socket. A gomphosis connects the bone of the jaw to the tooth by fibrous connective tissue called **periodontal ligaments.** Examine a jaw or complete skull in the lab and locate a gomphosis. Compare your specimen with figure 10.1.

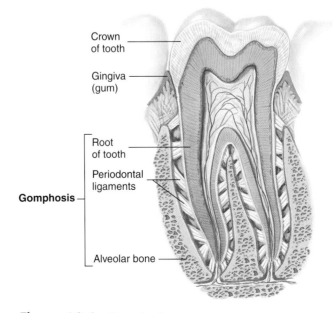

Figure 10.1 Gomphosis

TABLE 10.1 Classification of Joints

	Examples
Fibrous Joints—Joints Held Together by Collagenous Fibers	
Suture	Frontal and parietal bones
Gomphosis	Teeth and mandible
Syndesmosis	Distal tibia and distal fibula
Cartilaginous Joints—Joints Held Together by Cartilage	
Synchondrosis	Epiphyseal plate, humeral head and shaft
	Costal cartilages, first rib and sternum
Symphysis	Joint between two vertebral bodies
Synovial Joints—Joints Enclosed by Synovial Capsule	
Gliding	Between carpal bones
Hinge	Humerus and ulna
Pivot	Atlas and axis
Ellipsoid	Radius and scaphoid
Saddle	Trapezium and first metacarpal
Ball-and-socket	Acetabulum and femur

Another type of fibrous joint is called a **syndesmosis** (sin´dez-mō´sis). The fibrous connective tissue is longer in these joints than in sutures or gomphoses. An example of a syndesmosis is the connection between the radius and ulna or the distal tibia and fibula. Examine an articulated skeleton in the lab and compare it with figure 10.2.

Fontanels

Between the fibrous joints in a developing fetus or newborn are large, membranous regions known as fontanels. The fontanels are the soft spots of an infant's skull. There are four sets of fontanels, and some of these permit the passage of the skull through the birth canal by enabling the bones of the cranium to slide over one another. After birth, the fontanels allow for further expansion of the skull. The **frontal (anterior) fontanel** is between the frontal bone and the parietal bones. The **occipital (posterior) fontanel** is between the occipital bone and the parietal bones. The **sphenoidal (anterolateral) fontanels** are paired structures on each side of the skull and are located superior to the sphenoid bone. The **mastoid (posterolateral) fontanels** are also paired structures posterior to the temporal bone. Most fontanels fuse before 1 year of age, although the frontal fontanel may fuse as late as age 2. Locate these structures in figure 10.3 and on the material available in the lab.

Cartilaginous Joints

If bones are held together by cartilage, the articulation is known as a **cartilaginous joint.** As in fibrous joints, if the cartilage is thin between bones, the joint is immovable. Cartilaginous joints are known as **synchondroses** (sĭn´ kon-dro-sees; sing. **synchondrosis**). If the cartilage is long, there is more movement in the joint. In one case, the cartilaginous joint is a phase in the development of the skeleton. This is an immovable joint called the **epiphyseal** (eh´pih-fĭs-ē-ul) **plate,** which is illustrated in figure 10.4a. This joint eventually fuses to form a single bone.

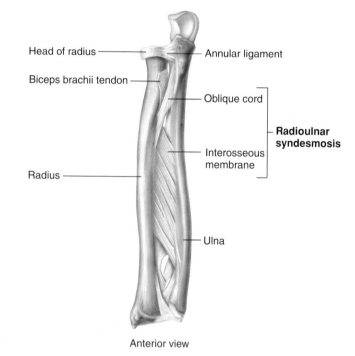

Figure 10.2 Syndesmosis

Anterior view

Another example of a synchondrosis is the articulation between the first rib and the sternum by the **costal cartilage.** In this joint, hyaline cartilage binds the rib to the sternum yet lets it move somewhat (figure 10.4b).

A **symphysis** (sim´fi-sis) is a **fibrocartilaginous** pad between the pubic bones (for example, the pubic symphysis) or in the intervertebral disks. These joints also allow for some movement. The physical stress on a symphysis is greater than that on the costal cartilages, and the joint reflects this in its fibrocartilage composition. Fibrocartilage endures much greater stress than hyaline cartilage. Locate a symphysis in lab and compare it with figure 10.5.

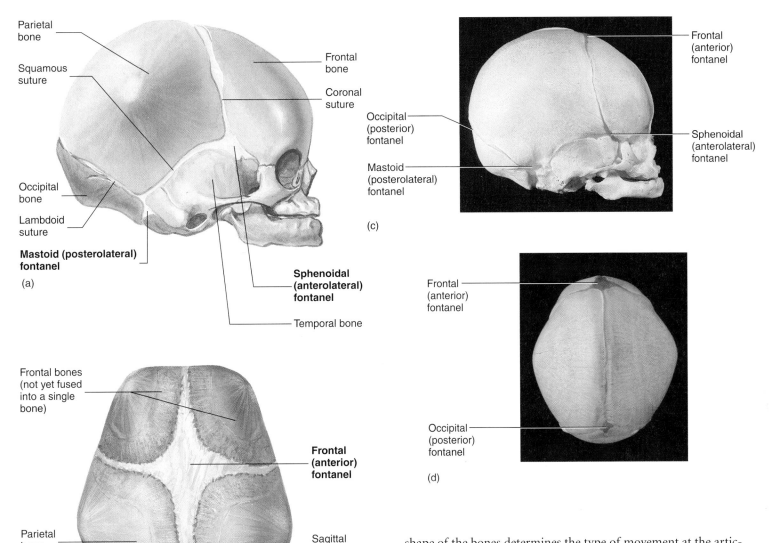

Figure 10.3 Fontanels of the Fetal Skull
Diagram (a) lateral view; (b) superior view. Photograph (c) lateral view; (d) superior view.

Synovial Joints

Joints that allow for extensive movement are called **synovial joints.** The outer part of the synovial joint is the **joint capsule,** which is made of an outer **fibrous capsule** and an inner **synovial membrane.** The synovial membrane secretes **synovial fluid,** a lubricating liquid that reduces friction inside the joint. The space inside the joint is called the **synovial cavity,** and each bone of the joint ends in a hyaline cartilage cap called the **articular cartilage.** There are other structures in synovial joints that are characteristic of specific joints, and these are discussed later in this exercise. The shape of the bones determines the type of movement at the articulation. In general, the more movable a joint is, the less stable it is. The stability of a joint depends on the number and types of ligaments, tendons, and muscles and on the way the bones fit together. Compare the features of the joint in figure 10.6 with models or charts in the lab.

Dissection of a Synovial Joint

To understand the structure of a synovial joint, it is beneficial to examine a fresh or recently thawed mammal joint. Wear protective gloves as you dissect the joint, and wash your hands thoroughly with soap and water after the dissection. Place the joint in front of you on a dissecting tray and cut into the joint capsule with a scalpel or razor blade.

Note the tough, white material that surrounds the joint. This is the joint capsule, and it may be fused with **ligaments** that bind the bones of the joint together. Notice the synovial fluid, which is a slippery substance that provides a slick feel to the inside of the capsule and reduces friction between the bones. Once you cut into the joint, examine the articular cartilage on the ends of the bones. *Carefully* cut into this cartilage with a scalpel or razor blade, and notice how the material chips away from the bone. When you have finished with the dissection, make sure to rinse off the dissection equipment, dispose of the joint in the appropriate animal waste container, and wash your hands.

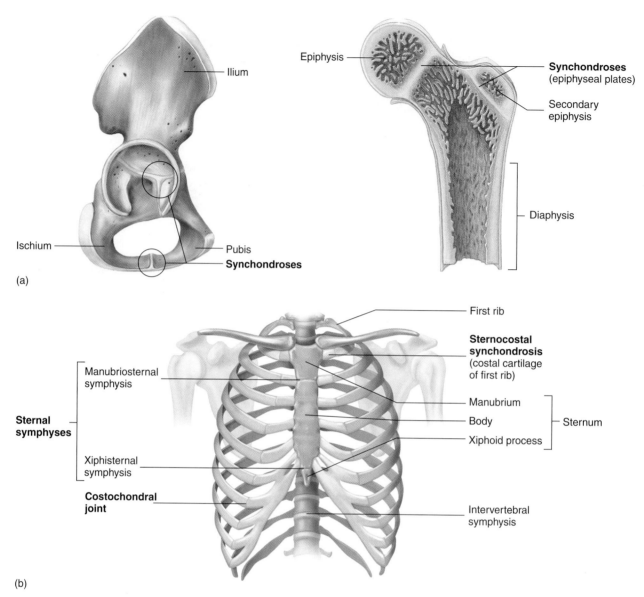

Figure 10.4 Synchondroses
(a) Hip and femur with immovable synarthroses; (b) thorax with movable synchondrosis.

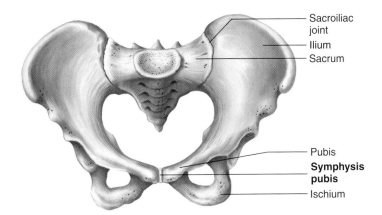

Figure 10.5 Symphysis

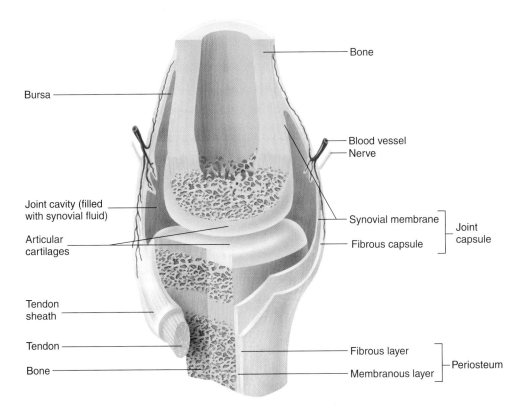

Figure 10.6 Synovial Joint Structure

Modified Synovial Structures

Bursae and tendon sheaths are modified synovial structures. **Bursae** (sing. **bursa**) are small synovial sacs between tendons and bones or other structures. The bursae cushion the tendons as they pass over the other structures. **Tendon sheaths** are modified synovial structures, such as those encircling the tendons that pass through the palm of the hand. Tendon sheaths lubricate the tendons as they slide past one another. Examine figure 10.7 for these structures.

Joints Classified by Movement

Articulations are immovable, semimovable, or freely movable. Immovable joints are known as **synarthrotic joints.** The bones are tightly bound by fibers or hyaline cartilage. Semimovable joints are known as **amphiarthrotic joints,** and they may be fibrous or cartilaginous. Freely movable joints are known as **diarthrotic joints,** and they are always synovial joints. Specific types of synovial joints are discussed next.

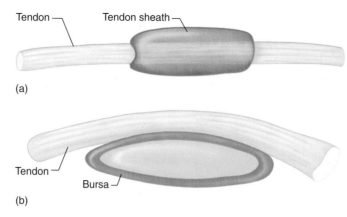

Figure 10.7 Modified Synovial Structures—Tendon Sheaths and Bursae

Synovial Joints Classified by Movement

The synovial joints are classified according to the type of movement they allow between the articulating bones. The joints are listed here in the general order of least movable to most movable. Examine the joints on an articulated skeleton in lab and compare them with figure 10.8.

1. **Plane,** or **gliding, joints** allow for movement between two flat surfaces, such as between the superior and inferior facets of adjacent vertebrae or intertarsal or intercarpal joints.
2. **Hinge joints** allow for angular movement, such as in the elbow, in the knee, or between the phalanges of the fingers. You can increase or decrease the angle of the two bones with this joint.
3. **Pivot joints** allow for rotational movement between two bones, such as in the movement of the atlas and the axis when moving the head to indicate "no." They also occur at the proximal radius and ulna.
4. **Ellipsoid,** or **condyloid, joints** allow significant movement in two planes, such as at the base of the phalanges. Ellipsoid joints consist of a convex surface paired with a concave surface. The junction between the radius and scaphoid bone is a good example of an ellipsoid joint. Others are located between the atlas and occipital bone or the metacarpals and phalanges of digits 2 through 5.
5. **Saddle joints** have two concave surfaces that articulate with one another. An example of a saddle joint is that between the trapezium and the first metacarpal of the thumb. This provides for more movement in the thumb than the ellipsoid joint of the wrist.
6. **Ball-and-socket joints** consist of a spherical head in a round concavity, such as in the shoulder and the hip. There is extensive movement in these joints, yet they are inherently less stable due to the freedom of movement that they afford.

Uniaxial joints move in only one plane. Hinge, pivot, and gliding joints are uniaxial joints. Biaxial joints move in two planes. Ellipsoid and saddle joints are examples of biaxial joints. Multiaxial joints move in many planes. Ball-and-socket joints are multiaxial joints.

Specific Joints of the Body

There are several types of specific joints in the body; some are primarily hinge joints, such as the jaw, elbow, and knee, while others are ball-and-socket joints, such as the shoulder and hip joints.

Temporomandibular Joint

The **temporomandibular joint (TMJ)** is the only diarthrotic joint of the skull. The **articular disk** is a pad of fibrocartilage that provides a cushion between the mandibular condyle and the temporal bone. This joint is both a hinge and a gliding joint. Numerous ligaments strengthen this joint (figure 10.9). Examine a skull with the mandible attached for the nature of the temporomandibular joint. Place your fingers on the temporomandibular joint and palpate (feel) it as you move your jaw.

Shoulder Joint

The **shoulder (humeral) joint** is known as the **glenohumeral (humeroscapular) joint** because the glenoid fossa articulates with the head of the humerus. The shallow glenoid fossa is deepened by the **glenoid labrum,** a cartilaginous ring that surrounds the cavity. Numerous bursae, ligaments, a tough joint capsule, and the **rotator cuff muscles** also stabilize the joint. Compare figure 10.10 with a model or an actual joint in the lab.

Elbow Joint

The **elbow (humeroulnar** and **humeroradial) joint** is a complex joint having both hinge and pivot characteristics. The ulna locks into the humerus tightly but the **annular ligament** that wraps around the radius is commonly torn from the radial head when the forearm is abruptly pulled (figure 10.11). Palpate the annular ligament as you rotate your forearm.

Hip Joint

The **hip joint** is known as the **acetabulofemoral joint.** As with the shoulder joint, the **acetabular labrum** deepens the hip socket. Numerous ligaments, including the iliofemoral, pubofemoral, and

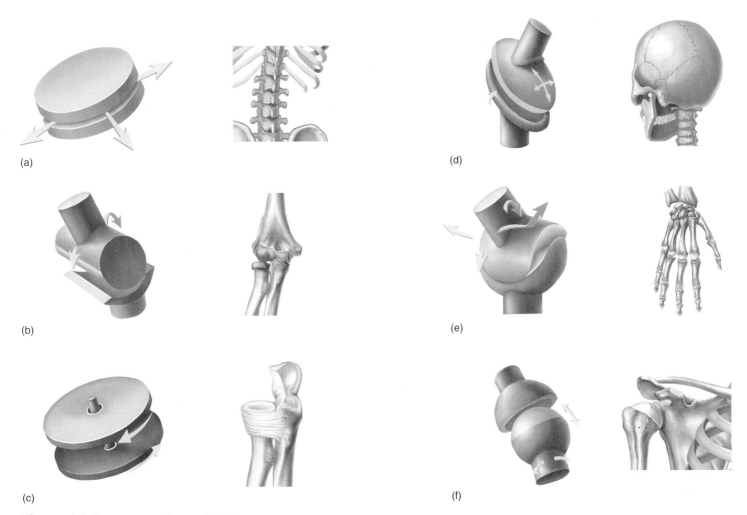

Figure 10.8 Classes of Synovial Joints
(a) Plane, or gliding, joint; (b) hinge joint; (c) pivot joint; (d) ellipsoid, or condyloid, joint; (e) saddle joint; (f) ball-and-socket joint.

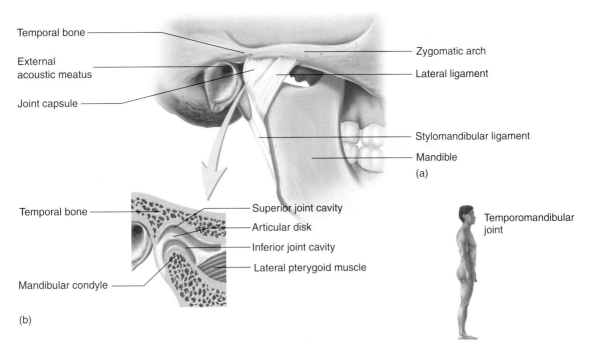

Figure 10.9 Temporomandibular Joint
(a) Lateral view; (b) sagittal section.

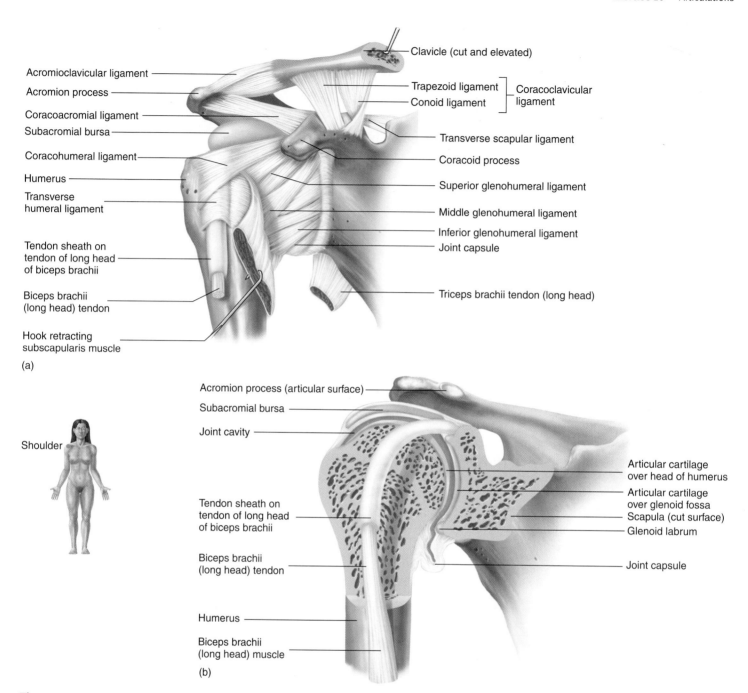

Figure 10.10 Glenohumeral Joint
(a) Anterior view; (b) frontal section.

ischiofemoral, bind the femur to the hip bone. The **ligamentum teres** is a band of dense connective tissue that attaches the acetabulum to the **fovea capitis** of the femur. Compare the models, charts, or specimens in the lab with figure 10.12.

Knee Joint

The **tibiofemoral joint,** also known as the **knee joint,** is the largest, most complex joint of the body. It consists of several major ligaments, including the **tibial (medial) collateral ligament,** the **fibular (lateral) collateral ligament,** the **anterior cruciate ligament,** and the **posterior cruciate ligament.** Another important structure is the **patellar (quadriceps femoris) tendon,** which runs from the quadriceps femoris muscle to the patella. The **patellar ligament** connects the patella and the tibial tuberosity. The last major structures of the tibiofemoral joint are the **medial** and **lateral menisci,** wedge-shaped pads of fibrocartilage that provide a cushion between the femur and tibia. The knee is primarily a hinge joint with a little lateral movement allowed. Palpate the fibular and tibial collateral ligaments and the patellar tendon. Compare specimens in the lab with figure 10.13.

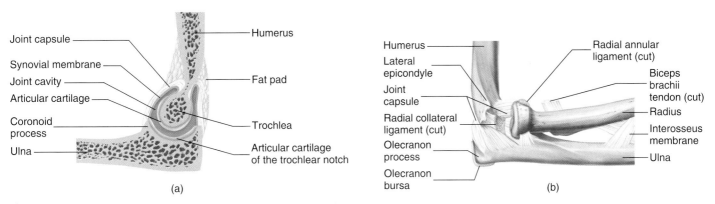

Figure 10.11 Right Elbow Joint
(a) Median section showing humerus and ulna; (b) lateral view with ligaments removed.

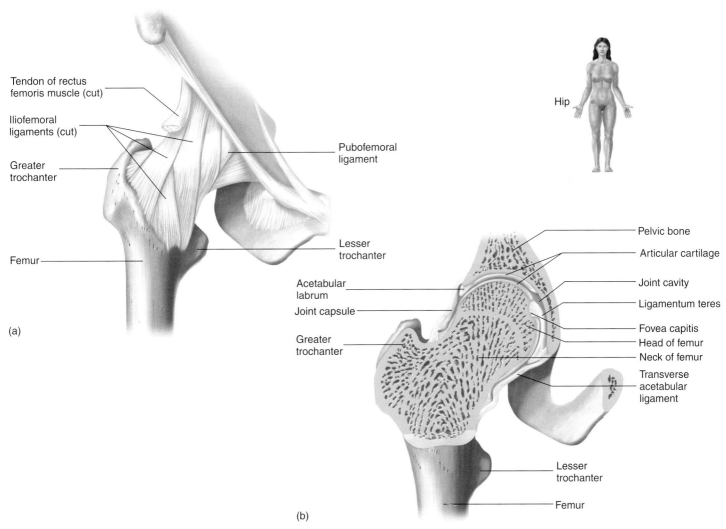

Figure 10.12 Acetabulofemoral Joint
(a) Anterior view; (b) frontal section.

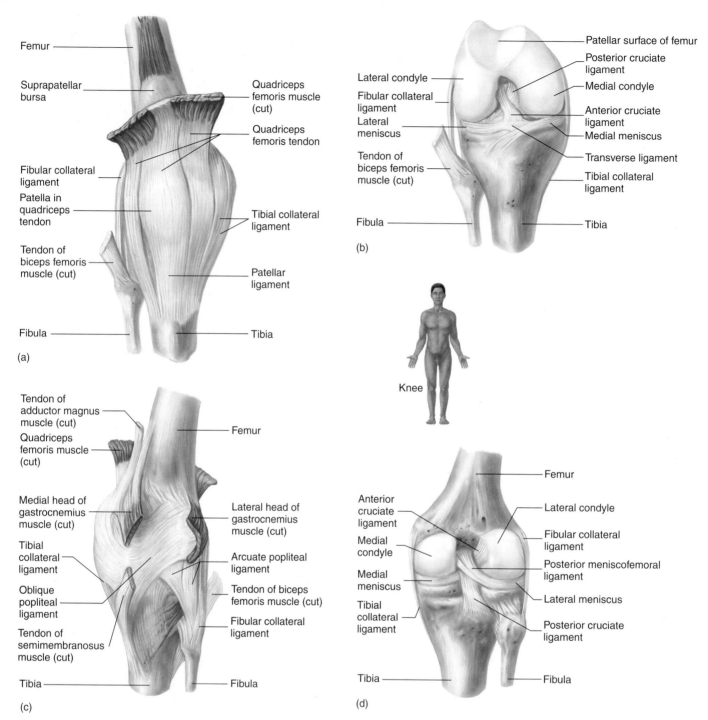

Figure 10.13 Tibiofemoral Joint
(a) Anterior surface view; (b) anterior deep view; (c) posterior surface view; (d) posterior deep view; (e) lateral section.

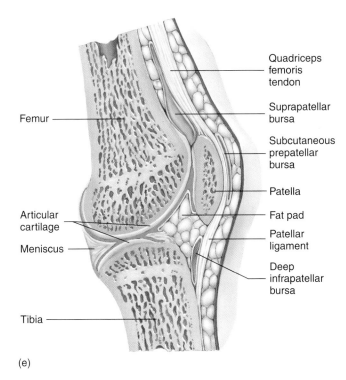

Figure 10.13—*Continued.*

Actions at Joints

Many kinds of movement occur at joints. These movements, controlled by muscles, are called actions. Actions can decrease a joint angle, increase a joint angle, and cause rotation at a joint, among other movements. The specific types of actions are as follows:

Flexion is a decrease in the joint angle from anatomic position. If you bend your elbow, you are flexing your forearm. Flexion of the thigh is in the anterior direction, yet flexion of the leg is in the *posterior* direction. Bending forward at the waist is flexion of the vertebral column. Looking at your toes is flexion of the head. Examine figure 10.14 for examples of flexion.

Extension is a return to anatomic position of a part of the body that was flexed. If you are looking at your toes and lift your head back to anatomic position, you are extending your head. If you straighten your knee after it is bent (flexed), you are extending the leg. Examine figure 10.14 for examples of extension. Extension of the part of the body beyond anatomic position is known as **hyperextension.** When you are about to roll a bowling ball and your arm reaches the very back of the arc, you are hyperextending your arm.

Abduction is movement of the limbs in the coronal plane away from the body (*abduct* = to take away). Abduction is taking away a part of the body in a lateral direction. This can be seen in figure 10.15.

Adduction is the return of the part of the body to anatomic position after abduction. In doing "jumping jacks," you are abducting and adducting in series. Think of adduction as "adding" a limb back to the body, as seen in figure 10.15.

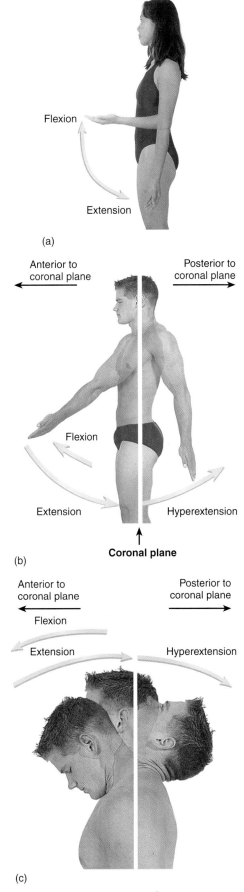

Figure 10.14 Flexion and Extension
Flexion and extension of the (a) forearm; (b) arm; (c) head; (d) trunk; (e) leg.

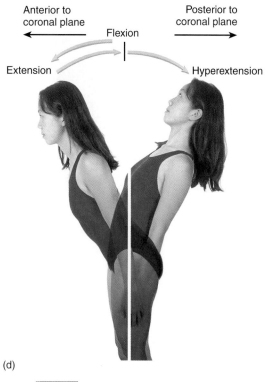

(d)

Figure 10.15 Abduction and Adduction of the Arm

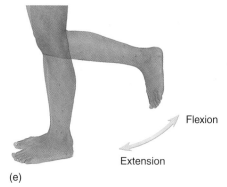

(e)

Figure 10.14—*Continued.*

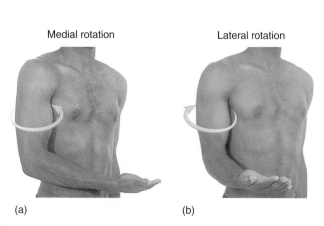

(a)   (b)

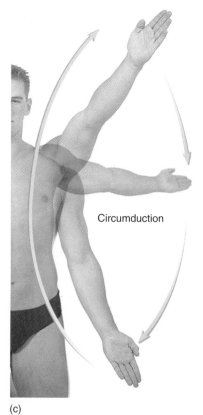

(c)

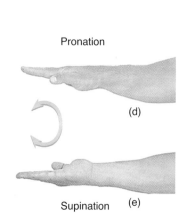

(d)

(e)

Figure 10.16 Rotation of Joints

(a–b) Rotation of arm; (c) conical movement of arm; (d–e) movement of hand.

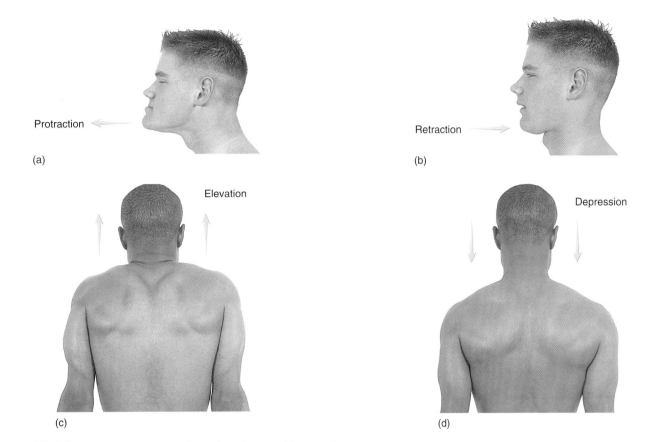

Figure 10.17 Protraction, Retraction, Elevation, and Depression
(a) Protraction of the mandible; (b) retraction of the mandible; (c) elevation of the scapulae; (d) depression of the scapulae.

Rotation is the circular movement of a part of the body. **Lateral rotation** moves the anterior surface of the limb toward the lateral side of the body. **Medial rotation** turns the anterior surface of the limb toward the midline.

Supination is lateral rotation of the hand. The hands are supinated when the body is in anatomic position.

Pronation is medial rotation of the hands. When you turn your palms posteriorly from anatomic position, you are pronating your hands. Rotate various joints of the body and compare them with figure 10.16.

Circumduction is the movement of a muscle in a conical shape, with the point of the cone being proximal. Examine figure 10.16c for an example of circumduction.

Protraction is a horizontal movement in the anterior direction, as in jutting the chin forward.

Retraction is the reverse of protraction. The jaw that moves from anterior to posterior is retracted. These two actions are illustrated in figure 10.17.

Elevation means to move in a superior direction. Elevation of the shoulders occurs when you shrug your shoulders.

Depression is the opposite of elevation; it is movement in the inferior direction. Elevation and depression are seen in figure 10.17.

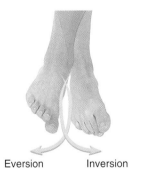

Figure 10.18 Inversion (Supination) and Eversion (Pronation) of the Foot

Inversion describes movement of the feet—turning the soles of the feet medially so they face each other.

Eversion means turning the soles of the feet laterally. Inversion and eversion can be seen in figure 10.18.

Fixing a muscle prevents motion in either direction. This is done with opposing muscles contracting simultaneously. The muscle that has the main force on a joint is called the **prime mover.** Muscles that assist with the prime mover are **synergists,** whereas those that oppose the muscle are **antagonists.**

Exercise 10 Review

Articulations

Name: _____

Lab time/section: _____

Date: _____

1. The study of articulations, or joints, is called _____.

2. What kind of joint (based on joint composition or structure) is one in which the bones are held together by collagenous fibers? _____

3. When two bones fuse into a single bone, this union is called a(n) _____.

4. The teeth are held in the jaw by what specific kind of joint? _____

5. What kind of joint is slightly movable and is held together by fibrous connective tissue? _____

6. Name all the fontanels in the fetal skull. _____

7. Which one of the fontanels is the most dorsal? _____

8. In terms of their structural composition, bones that are held together by cartilage (cartilaginous joints) are also known as _____ joints.

9. The epiphyseal plate is a cartilaginous joint. It is also called a(n) _____.

10. Label the following illustration using the terms provided.
    synovial membrane
    articular cartilage
    fibrous capsule
    joint cavity
    periosteum

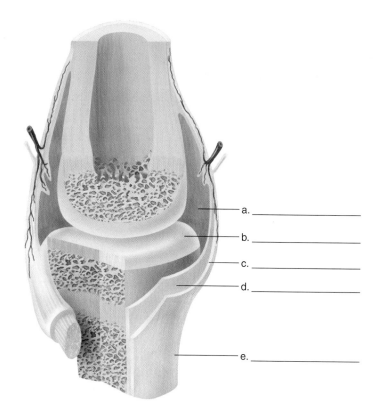

a. _____
b. _____
c. _____
d. _____
e. _____

11. What is the name of a joint that is held together by a joint capsule? _____

12. Synovial fluid is secreted by what structure? _____

13. A skull suture is what kind of joint, in terms of movement? _____

14. Rank the following joints in terms of least movable to most movable, with 1 being the least movable and 5 being the most movable.
    _____ gliding
    _____ saddle
    _____ suture
    _____ syndesmosis
    _____ ball-and-socket

15. Which one of the following joints has the greatest range of movement?
    a. gomphosis   b. suture   c. synchondrosis   d. hinge

Exercise 10 Articulations

16. In which of these joints would you find a meniscus?
    a. cartilaginous   b. fibrous   c. synovial

17. What is the function of the meniscus in the knee? _____

18. Match the joint in the left column with the type of joint in the right column.
    _____ hip   a. hinge
    _____ radiocarpal   b. ball-and-socket
    _____ tibiofemoral   c. plane
    _____ intercarpal   d. ellipsoid

19. What is the function of the labrum in the shoulder joint? _____

20. The joint between the trapezium and the first metacarpal is what kind of joint? _____

21. A class of joint with great movement is a(n) _____.

22. The joint between the femur and the tibia is what type of joint? _____

23. What kind of joint is located at the wrist (between the radius and the scaphoid bones)? _____

24. Label the following illustration using the terms provided.
    anterior cruciate ligament
    meniscus
    fibular collateral ligament
    tibial collateral ligament

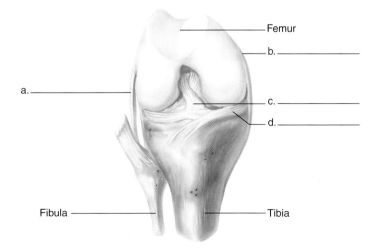

Exercise 11
Muscle Physiology

INTRODUCTION

Muscles have several functional characteristics. Muscles show **contractility** when they are stimulated. The contraction of the muscle can bring about various functions, such as walking, eating, talking, and breathing. Muscles also have **extensibility.** They can be stretched when opposing muscles contract. Muscles have **elasticity** in that they recoil when stretched, and muscles have **excitability** because they respond to stimuli. These topics are covered in the *Principles of Anatomy and Physiology* text in chapter 8, "Histology and Physiology of Muscles."

In this exercise, you examine the nature of muscle contractility and excitability. Skeletal muscles contract due to stimulation by nerves controlling them. Normally, a series of nerve impulses begins to stimulate a muscle and continues with increasing strength, producing the uniform muscle contraction. When these multiple nerve impulses diminish, the muscle relaxes. In this exercise, you explore the nature of skeletal muscle contraction as initiated by external electrical stimulation and apply this information to the functions of the skeletal muscle in your body.

Two events are fundamental to an understanding of muscle physiology. One is an electrochemical event that occurs in muscle membranes, and the other is the physical contraction of the muscle itself. The normal contraction of skeletal muscle occurs when an electrochemical **nerve impulse** travels down an axon and reaches the **synapse** between the nerve and the muscle. This synapse between the corresponding neuron and muscle fiber is known as the **neuromuscular junction.** This junction is seen in figure 11.1. **Acetylcholine (ACh)** is released by the terminal regions of the neuron and stimulates an electrochemical impulse, which travels along the length of the muscle fiber.

When the impulse reaches the **transverse,** or **T, tubules** of the muscle, **calcium** is released from the sarcoplasmic reticulum, and the **actin** and **myosin** myofilaments join together, producing a power stroke in the muscle cell. As a muscle fiber is **depolarized,** the entire fiber contracts maximally. This characteristic represents the **all-or-none principle** of muscle fibers. A single skeletal muscle cell is composed of many myofibrils, each of which consists of discrete units called **sarcomeres,** which are connected to each other at a region known as the **Z disk.** Skeletal muscle is striated because of the regular arrangement of myofilaments. When a muscle contracts, the actin myofilaments and myosin myofilaments slide over one another, causing the muscle to shorten. This is the **sliding filament model,** and it is described in more detail in your textbook.

One neuron stimulates many muscle fibers. This complex of a neuron and the associated muscle cells is known as a **motor unit,** which is illustrated in figure 11.2. When slight contractions occur, few muscle fibers contract because only a few motor units are activated. When a muscle contracts more forcefully, more motor units are stimulated and they subsequently contract.

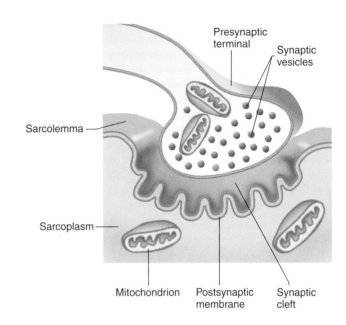

Figure 11.1 Neuromuscular Junction

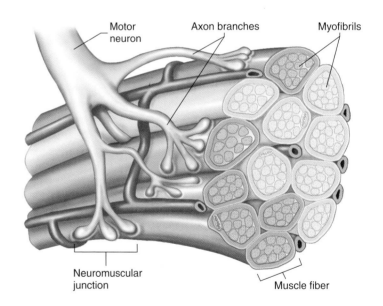

Figure 11.2 Motor Unit

Entire muscles in the body do not contract with an all-or-none response but, rather, exhibit a **graded response** in which muscles gradually increase from slight to more forceful contractions. An increase in the force of contraction is known as **summation.** Summation can occur by increasing the frequency of impulses to an individual fiber or by increasing the stimuli to more and more muscle fibers.

One way to increase the overall contractile strength of the entire muscle is to stimulate more muscle fibers (each of which contracts completely) in that muscle. The gradual increase of

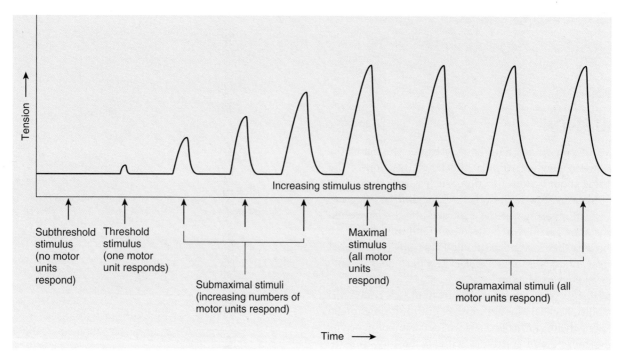

Figure 11.3 Multiple Motor Unit Summation

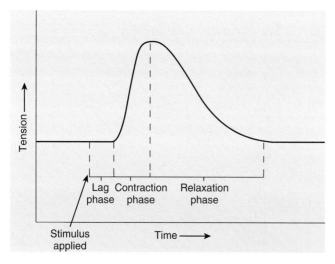

Figure 11.4 Three Phases of a Muscle Twitch

contraction strength of a muscle by adding progressive numbers of fibers is known as **multiple motor unit summation.**

The impulse that causes a muscle to contract must exceed a **threshold** value before any contraction can occur. A **subthreshold stimulus** is one that will not elicit a response in the muscle. Once the muscle contracts, the addition of progressively stronger stimuli results in stronger muscle contractions. The stimulation where all of the muscle fibers are forcefully contracting is known as the **maximal stimulus.** Stimulation beyond maximal stimulus produces no greater contraction strength. This is known as **supramaximal stimulus.** Examine figure 11.3 for the subthreshold, threshold, maximal, and supramaximal stimulus of a muscle in multiple motor unit summation.

When a muscle fiber is stimulated with a single, quick electrical impulse, the fiber undergoes a contraction known as a **twitch.** In a twitch, the fiber has three phases: a lag phase, a contraction phase, and a relaxation phase. After the initial impulse, known as a **stimulus,** the muscle fiber undergoes a **lag** or **latent phase** (figure 11.4). This is the time when the electrical impulse travels across the muscle cell membrane and calcium ions are released from the sarcoplasmic reticulum. No contraction occurs at this time. After this short period, due to the time it takes calcium to attach to troponin, the fiber goes through the **contraction phase** as the myofilaments slide across one another and the muscle fiber shortens. Finally, there is a **relaxation phase** in which the fiber returns to a resting state.

A stimulus above threshold level that occurs soon after an initial stimulus does not cause the muscle fiber to contract. This demonstrates the **refractory period,** which is the time when a stimulus, delivered after a previous stimulus, produces no contraction. If a stimulus is applied shortly after the refractory period, a muscle contracts, and the contraction strength is more pronounced. This effect is called **wave summation** (figure 11.5).

If rapid, repeated stimuli are sent to a muscle, then the muscle produces a series of contractions called **incomplete tetanus** (figure 11.6). If the frequency (number of pulses per second) of the stimuli increases, the contractions fuse in a smooth contraction of the muscle known as **complete tetanus.** Numerous, sequential stimulations of a muscle produce **multiple wave summation** in the muscle. Multiple wave summation results in the smooth, continuous muscle contractions that normally occur in the body. Complete tetanus occurs even in fast muscular movements, such as the flicking of a finger. Examine figure 11.6 for recordings of incomplete tetanus and complete tetanus.

In this exercise, you examine the muscular contraction in a simulated frog or a real frog and discover what occurs as voltage is increased or the frequency is changed. Due to the decline in some frog populations (such as leopard and grass frogs in North America),

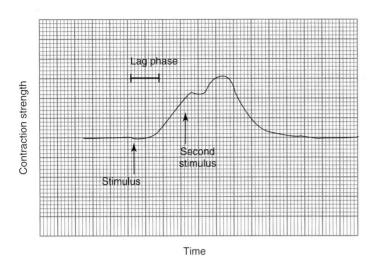

Figure 11.5 Wave Summation

simulated activities may be preferable to the use of live frogs. Bullfrogs are not only plentiful but are a pest species in many areas and may be used without significant impact on their population levels.

OBJECTIVES

At the end of this exercise, you should be able to

1. demonstrate the procedure to determine threshold stimulus;
2. differentiate between tetanus and treppe;
3. describe incomplete tetanus and complete tetanus;
4. demonstrate maximum recruitment with lab equipment and a frog;
5. compare the lab experiments on frogs with muscular contractions in humans;
6. describe the process of muscle recruitment and fatigue in humans.

MATERIALS

Virtual Experiment Lab

Ph.I.L.S. CD* (available through McGraw-Hill)
Compatible computer

Live Frog Lab

Frog (one per experiment)
Frog Ringer's solution in dropper bottles (150 mL per experiment)
Nylon thread
Duograph, physiograph
Myograph transducer
Stimulator and cables
Scissors
Clean, live animal dissection pan
Sharp pithing probes
Glass hooks
Scalpel
BSLSTM stimulator
Force transducer assembly—SS12LA
Pin electrodes ELSTM2
HDW100A tension adjuster
Data acquisition unit MP30
BSL *PRO* template file: FrogMuscle.gtl
Ring stand
Anchoring board
Goggles
Latex gloves
Forceps
Dissection pins
Animal disposal container
Multiple weights (5, 10, 15 lb)

*The Ph.I.L.S. Virtual Physiology Lab CD may be purchased by calling 800-262-4729, or contacting your campus bookstore. To purchase the online version, go to www.mhhe.com/phils ‹file://www.mhhe.com/phils›

PROCEDURE

Simulated Frog Muscle Experiments—Physiology Interactive Lab Simulations

Load the Physiology Interactive Lab Simulations (Ph.I.L.S.) disk. You will need at a minimum either a Windows XP, Vista or Pentium III computer or a Macintosh 10.2, G3 processor and, at least, Adobe Flash 8.x Player to run the program.

For all of the virtual muscle experiments you should follow the standard procedure outlined next.

1. Click on the number of the exercise that you want to complete. These are under the **Skeletal Muscle Function** section on the CD and are listed as:

    4. Stimulus—Dependent Force Generation
    5. Length–Tension Relationship
    6. Principles of Summation and Tetanus
    7. EMG and Twitch Amplitude

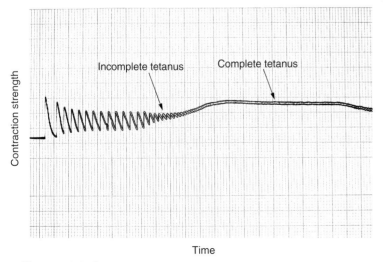

Figure 11.6 Incomplete and Complete Tetanus

After you have selected your choice you should:

2. Read the objectives and introduction to the lab simulation. You can click on highlighted terms to see pictures of the material in question.

3. Take the pre-lab quiz. If you get a question wrong, the program will let you know and provide you with the correct answer.

4. Click on the **Wet Lab** tab and read the material. The highlighted terms open up videos of an actual procedure.

5. Click on **Continue,** which opens up the Laboratory Exercise.

6. Turn the power on the **Data Acquisition Unit.** The green light lets you know it is on.

7. Connect all the plugs to their respective locations. You must drag the tip of the plug to the middle of the adapter in order for it to connect.

8. Adjusting the voltage, frequency, or tension can be done by clicking on the up or down arrow repeatedly or clicking on the arrow and holding it until the desired value is obtained.

9. Take the post-lab quiz to determine your understanding of the lab.

The information for the specific lab exercises on the CD follows.

Select "4. Stimulus—Dependent Force Generation" on the CD.

In this laboratory exercise you examine the impact that an increase in voltage stimulus has on the strength of contraction. You should see a picture that shows a virtual muscle, electrodes to stimulate the muscle, and an input that records the "pull" of the muscle after it has been stimulated. The red and blue wires deliver electric impulses (shocks) to the frog muscle and the black wire records how much force the muscle exerts. The directions ask you to select at least 1 volt to get a response. After you select 1 volt, click on the shock button. You should see an upward curve.

To record the amplitude you need to click on the highest point of the curve on the graph and then somewhere on the baseline, the area where the curve flattens out. You need to select these two points in order to have the data recorded in the journal. Click on the **Journal** entry icon. You should see the data points automatically plotted on the graph. Click on the **X** in the upper right corner of the graph table and you will return to the experiment.

Reduce the voltage to 0. Hit the shock button again. Enter the material in the journal as previously described. Do this repeatedly until you fill in the entire journal from 0 to 1.6 volts. Notice that at low voltages the force line stays flat. This is because muscles have a threshold of contraction below which no contraction occurs. Record the threshold voltage (when you first see a contraction) for the muscle in the following space.

Threshold voltage: _____

What happens to the muscle contraction strength as you increase the voltage?

Record your response in the following space. _____

Although it was not demonstrated in this exercise, there is a point at which the increase in voltage does not lead to an increase in contraction strength. This is known as **maximum recruitment** and it is the voltage at which all the muscle fibers are contracting.

Phases of Muscle Contraction

In this part of the exercise you will measure the lag phase, the contraction phase, and the relaxation phase. Notice how the blue stimulus marker precedes the actual contraction of the muscle. This is the **lag** or **latent phase,** the time between when the stimulus was delivered and when the muscle started to contract. The lag phase is the time after the muscle is stimulated when calcium diffuses from the transverse tubules to the myofibrils. Click on the red line of the graph (at the point when the muscle was stimulated) and then at the point of the red line where you see a spike in voltage. This is when the muscle started to contract. Enter the value in the following space.

Lag phase: _____

The left side of the curve is steeper than the right side of the curve. The left side of the curve is known as the **contraction phase.** How long does this phase take? It is measured by clicking on the graph where the line moves sharply up from the baseline and then clicking at the top of the curve. If the program will not allow you to enter the data, click **Erase** and redo the experiment. Select the appropriate areas and click on them. Enter the time in the following space.

Contraction phase: _____

Click at the top of the curve and then when the red line reaches the baseline and record the **relaxation phase.** Enter the time of the relaxation phase in the following space.

Relaxation phase: _____

Select "5. Length–Tension Relationship" on the CD.

In this laboratory exercise you examine the strength of contraction based on the initial length of the contracting muscle. Follow the directions as previously outlined and in the directions in the program. The muscle can be stretched prior to stimulation. Notice how the muscle is attached to a horizontal bar. On the left side of this bar are two arrows. Click on the upper one to increase the tension on the muscle.

Move the arrow up so that you record the increase in length in 0.5 mm increments. This takes about two clicks of the upper arrow. What occurs to the strength of contraction as you stretch the muscle from 26 mm to 30 mm? _____
Where do you predict the greatest overlap of actin and myosin myofilaments to be? Why? _____
_____

Select "6. Principles of Summation and Tetanus" on the CD.

In this exercise you adjust the **frequency** (number per minute) of the stimulus. Most muscles do not contract by a single twitch. Neurons typically send many impulses that repeatedly stimulate the muscle, producing a smooth contraction. As in the previous experiment, you will need to read the material presented and answer the questions before doing the experiment. Once you get to the experimental stage, turn on the power button to the Data Acquisition Unit, drag the appropriate cables to their inputs, and set the voltage between 1.6 and 2 volts. The voltage remains the same in all of the trials. Under low frequency (few stimulations per unit time) you will see that the muscle contractions all have the same height and return to the baseline. When you increase the frequency, the muscle contraction tracing leaves the baseline, and you have **summation.** You can get a visual demonstration of what this looks like by clicking on the highlighted term *summation.*

You should record the time interval for summation, in milliseconds, in the following space.

Summation: _____

As you decrease the interval between stimuli, notice how the peaks begin to merge. If you increase the frequency more, then the contractions on the graph occur above the normal contraction peaks, but you still see individual contractions. This is **incomplete tetanus.** Record the time interval in milliseconds for incomplete tetanus in the following space.

Incomplete tetanus: _____

When the two peaks merge into a straight line it represents continuous muscle contraction or **complete tetanus.** Record the time interval in milliseconds for complete tetanus in the following space.

Complete tetanus: _____

Select "7. EMG and Twitch Amplitude" on the CD.

The EMG records the electrical activity of contracting muscles. You will record two events. One is the electrical activity of contracting muscle and the other is the force generated by the muscles of the arm and forearm squeezing on a rubber bulb. As opposed to the earlier exercises where you shock the muscle, in this exercise you are picking up recording information only. In this exercise, in addition to placing the plugs in their respective areas, you must also attach the recording electrodes to the proper location on the subject's right arm.

You need to record each data set separately for the lab exercise. Once you select the highest point on the blue line for pressure you should return to the baseline and select a point on the baseline and click there. You must select the **Journal icon** and click on it to enter the data. You must also close the journal box by clicking on the **X** in the right upper corner before you record the EMG data.

Select the high point on the EMG and click on it. Click on the lowest point before entering the material in your journal.

What is the relationship between the force of contraction and the amount of electrical activity in the muscle? Record your answer in the following space. _____

Live Frog Experimentation

You may do this experiment in small groups, or your instructor may elect to do a demonstration for the class. If you are doing this experiment as a group, read the entire exercise first and then follow the directions. The directions are first for physiograph recordings. A section on using **Biopac** follows for labs equipped with this hardware. Preparing the frog is the same whether you use a physiograph or the Biopac setup, and it is described next.

Frog Preparation

If the frog has not been pithed, follow the directions under number 1. If the frog has been pithed, begin at number 2.

1. The most humane way to conduct frog muscle experiments is to pith the frog quickly by inserting a sharp probe into the braincase.
    a. To do this, firmly grasp the frog and bend the head over with your index finger, as illustrated in figure 11.7.
    b. Insert the probe into the braincase and twirl it around in a conical manner, destroying the brain. This is known as a **single pith.** The frog will be killed at this point yet still have reflexes in the lower limbs.
    c. To stop the reflexes, insert the sharp probe into the vertebral canal and run it toward the caudal end of the frog. This is known as a **double pith.** Be careful not to thrust the probe into your hand in your attempt to locate the vertebral canal.
2. Once the frog is pithed, carefully snip its skin above the hip joint and peel it back to the foot (figure 11.8).
    a. Separate the posterior muscles of the thigh and locate the **sciatic nerve,** which appears as a thin, white, glossy thread that runs along the lateral aspect of the femur. You may have to use a scalpel and tease the muscles away from the sciatic nerve.

(a)

(b)

Figure 11.7 Pithing a Frog
(a) Single pith—probe inserted into braincase; (b) double pith—probe inserted into vertebral canal.

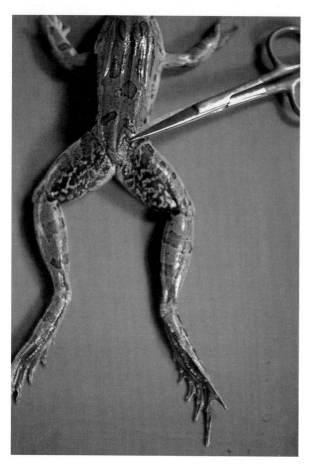

Figure 11.8 Preparation of the Frog, Skin Removal
Cut skin at this point and remove skin from thigh and leg.

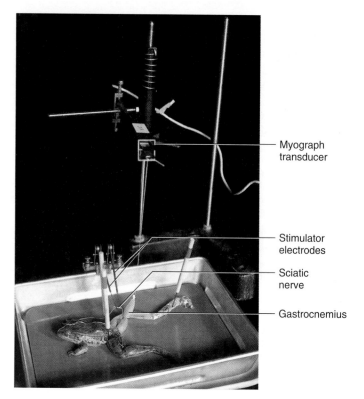

Figure 11.9 Frog Hookup to Recording Apparatus

b. Using a glass hook, carefully lift the sciatic nerve away from the thigh muscles, keeping it moist with frog Ringer's solution (figure 11.9).
c. **Ligate** the nerve by tying a thread around the proximal end of the nerve, near the sacrum, and cut the nerve above the ligature.
d. Separate the thigh muscles from the femur and remove them, leaving the femur exposed. Be careful not to cut or damage the sciatic nerve as you do this.
e. Locate the **gastrocnemius muscle** of the frog and tie the tendon of the muscle with thread.
f. Cut the calcaneal tendon distal to the ligature and lift the gastrocnemius away from the other muscles and from the tibiofibula (a fused bone in frogs).
g. Cut the muscles and the tibiofibula just distal to the knee, so that you have the femur, the sciatic nerve, the gastrocnemius, and the knee joint intact (figure 11.9).
h. Anchor the femur to a board or mount it on a tray and lay the sciatic nerve on the gastrocnemius muscle.
i. Keep the muscle and the nerve moist during the entire experiment. Do not tug on the nerve but gently lay it on the gastrocnemius muscle.
j. Attach the thread tied to the calcaneal tendon to the end of a myograph transducer leaf (figure 11.9). This should be connected to a recording device, such as a physiograph, duograph, kymograph, or physiology computer. If you lightly tap on the leaf of the muscle transducer, you should see a response in your recording apparatus.

There are three critical areas of concern in this lab: the preparation of the muscle, the stimulation of the muscle, and the recording of the response. Failure in any of these three areas will cause poor results or no results at all. The first of these areas is the preparation of the muscle, which has already been outlined. The second and third areas are discussed next.

Physiograph Setup

Stimulator Setup
The preparation of the stimulator first involves determining how many stimuli you deliver in a particular time. Most stimulators can deliver repeating stimuli or single pulses. These are measured as the frequency of the stimulus, and a good frequency to start with is two pulses per second. Another important factor is the duration of the stimulus. This is how long the stimulus is delivered to the sciatic nerve. A duration of 2 milliseconds (ms) usually produces good results.

Determination of Threshold Stimulus
1. Turn the voltage to zero on the stimulator.
2. Set the stimulator frequency to two pulses per second.
3. Adjust the impulse duration to 2 milliseconds.
4. Set the switch to repeat on the stimulator.
5. Lay the sciatic nerve on the stimulator electrodes and slowly increase the voltage until you see the contraction of the muscle. If you have reached 5 to 8 volts and you still have no

response, turn the voltage down, shut the stimulator off, and recheck your connections and settings. Once you see the muscle contract, then the lowest voltage that produces a response is the threshold stimulus.

6. Record this value in the space provided.

   Threshold stimulus: _____

Recording Apparatus Setup
Your lab may be equipped with one or more different physiologic recorders. Follow your instructor's directions to set up the apparatus if you are to do the experiment in groups, or pay close attention if your instructor demonstrates the experiment. Pay particular attention to the settings of the apparatus.

Multiple Motor Unit Summation (Maximum Recruitment)
The **maximum recruitment** is the lowest voltage stimulus at which *all* of the muscle fibers are stimulated.

1. To demonstrate this, set the duration of the pulse to 2 milliseconds, keep the frequency at two pulses per second, and set the stimulus on repeat.
2. Begin the recording and increase the voltage until you see a response (**threshold**) in the muscle. Continue to increase the voltage slowly, and you should see the contraction force increase as indicated by an increase in the height of the tracing.
3. As you continue to increase the voltage slowly, the contraction tracings will not get any higher. The minimum voltage it takes to produce the maximum height is known as the maximum recruitment voltage.
4. Record the maximum recruitment voltage in the space provided.

   Maximum recruitment voltage: _____

Treppe
1. Set the voltage reading at the maximum recruitment voltage and the frequency at one pulse per second.
2. As you stimulate the muscle, notice that there is a stepwise increase in the peaks of the contractions as the muscle contracts over a period of time. This increase in contraction strength is known as **treppe** (trep′eh). Treppe is thought to occur due to the increased availability of calcium in the muscle fibers and is a possible rationale for "warming up" before exercising. Compare your recording with figure 11.10.

Phases of Muscle Contraction
1. If you increase the chart speed to 50 mm per second, you can obtain tracings of the muscle where the **lag phase, contraction phase,** and **relaxation phase** can be seen. If you can record the time when you stimulate the muscle, then you can calculate the lag phase of muscle contraction.
2. Make a tracing of the contraction and compare it with figure 11.11. If the chart speed is 50 mm per second in your tracing, what is the length of the lag phase? This assumes that you have a mark indicating the time of stimulation.

   Lag phase: _____

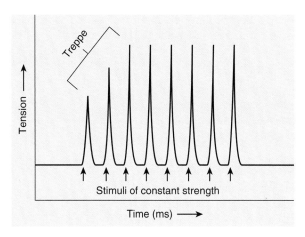

Figure 11.10 Treppe

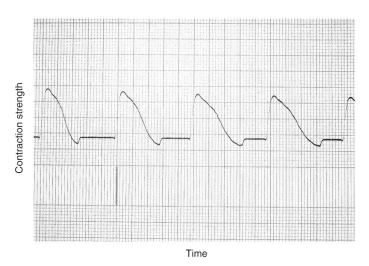

Figure 11.11 Phases of Muscle Contractions in a Single Twitch Contraction

How long is the contraction phase? Record the length of time.

Contraction phase: _____

How long is the relaxation phase? Record the length of time.

Relaxation phase: _____

Fatigue in Frog Muscle
Fatigue in muscles can be due to a decrease in oxygen to the muscle, a decrease in ATP, an increase in metabolic by-products (such as lactic acid and carbon dioxide), or a combination of all three.

You should set up the stimulator to the maximum recruitment voltage and increase the tension on the frog muscle. Stimulate the muscle at 20 pulses per second. The muscle should contract and reach a peak and then the tracing should decrease over time. How long does it take for the muscle to fatigue? Record the time in the space provided.

Length of time for muscle to fatigue: _____

Make sure that all members of your lab group see the frog demonstration and the recordings made from the experiment. Clean your dissection equipment and place the remains of the frog into the appropriate container designated by your instructor.

Biopac Setup for MP30

1. Plug in MP30 to wall outlet.
2. Connect MP30 to computer.
3. Plug SS12LA into channel 2.
4. Turn on the MP30.

Preparation of Stimulator

1. Plug stimulator into wall outlet.
2. Attach Trigger to Analog Out of MP30.
3. Attach ELSTM2 Needle electrodes.
4. Set the voltage range to 10 volts but turn the voltage knob to zero.

Computer Setup

1. Start the BSL *PRO* software. You should see an untitled window.
2. Select the MP30 menu and select Show Stimulator.
3. Go to the File menu and choose **Graph Template (*GTL)** > **File Name: FrogMuscle.gtl.**
4. Do NOT close the stimulator window.
5. Select 0–50 grams force range.
6. Select the Setup Channels from the menu and choose the wrench icon from the Channel menu and locate the scaling button. Make sure the stimulator is set to 0 volt (with the knob turned counterclockwise) and the pulse light NOT blinking. Click on the Cal 1 button.
7. Enter the Cal 2 value, which equals Cal 1 Input Value + 50.
8. Click OK on the scaling window.
9. Go to the Channel 2 window and locate the wrench icon and the scaling window.
10. With only the S hook on the tension adjuster, click Cal 1.
11. Attach a 50-gram weight to the adjuster and click Cal 2.
12. Click OK and exit the scaling window.
13. Attach the tension adjuster to the ring stand with the holes facing down. The adjuster should be set so that when the frog is attached the string will be vertical. Attach a small S hook to the adjuster.

Frog Preparation

1. Loop and tie the thread to the calcaneal tendon. Cut the tendon from the calcaneus.
2. Pin the frog down to a board. Make sure that the thigh and leg are anchored. Moisten the muscle with frog Ringer's solution.
3. Cut gently along the length of the thigh muscles and expose the sciatic nerve.
4. Gently dissect the nerve using a glass hook. Do not damage the nerve or tug on it.
5. Loop the thread to the transducer hook and adjust the tension so the thread is not slack. The thread should be vertical as it attaches to the hook.
6. Remove the covers from the electrode tips (remember to replace them at the end of the experiment) and place them underneath the sciatic nerve. Make sure that the electrode tips do not touch each other.

Running the Experiment

1. Click the Start button.
2. Adjust the level to 0.1 volt.
3. Click the on/off button in the stimulator window.
4. Increase the voltage by 0.2 volt until you see a response in the CH 2 window.
5. When you get a response, this is the threshold voltage.

   Record the value here: _____.

6. Continue to increase the voltage until you get no further increase in contraction. This is the maximum recruitment.

   Record the value here: _____.

7. Reduce the voltage and stop the recording.

Summation, Tetanus, and Fatigue

1. Go to the stimulator window and select "continuous pulses."
2. Adjust the pulses to 1 Hz with the scroll bar.
3. Adjust the voltage to what you obtained for maximum recruitment.
4. Start the recording and increase the frequency slowly by 1–2 Hz every second until the muscle fatigues.
5. Turn off the stimulator and stop the recording.
6. Save your data by selecting the **File menu.** Choose **Save As** and select the file type: **BSL Pro files (*.ACQ) File name: (your name).** Choose **Save.**

Human Muscle Physiology

In this part of the exercise, you study the effects of muscle recruitment and fatigue. These experiments are best done with the use of a device that can measure electrical activity, such as a physiograph or computer and hardware unit, such as the Biopac MP30 and the EMG1 student module.

Multiple Motor Unit Summation

1. Attach the appropriate recording electrodes to the forearm (a positive, negative, and ground) and connect the electrodes to the physiograph or hardware unit.
2. You will first need to establish a baseline recording by adjusting the sensitivity of the equipment to register the maximum grip strength that you can produce. To do this, you should grip your hand tightly while recording the activity and make sure that you have adequate displacement displayed. Ask your instructor for specific directions for the equipment in your lab.
3. Once you have produced a reasonable recording for baseline data, begin the recording by slightly gripping your hand. Pause for a second and then produce a slightly stronger grip. Pause once more for a second and then close your hand as tightly as possible. Stop the recording.

Exercise 11 Muscle Physiology

4. Examine the height of the displacement of the physiograph recording or the electronic recording produced. As more and more muscle fibers are recruited, the increase in the electrical activity produces a stronger signal.

   What do you predict to be the difference between your favored arm (for example, the right arm in a right-handed person) and your weak arm? _____

   _____

   Produce a recording of your *maximum* grip strength in your other arm and compare the two recordings. _____

   _____

Muscle Fatigue

You can examine the effects of muscle fatigue by using the same setup as described previously; instead of using grip strength, examine the nature of muscles when they contract maximally for long periods of time. Fatigue results from a decrease in ATP, oxygen, or food available to the muscle. Hold a 5- or 10-pound weight for a moment to establish your baseline recording. Once you have produced an adequate recording, begin the experiment as follows.

1. Hold a weight (5, 10, or 15 lb) with your forearm held at a 90-degree angle to your body.
2. Begin the recording and wait for the muscle to begin to fatigue. This is seen when the forearm begins to drop somewhat. Stronger individuals should hold heavier weights.
3. When the weight can no longer be held and the forearm begins to fall significantly, stop the recording.

   How does the strength of the signal compare at the beginning of the recording with that at the end of the recording?

   _____

   Does the stability of the recording stay the same, or is there variation in the electrical activity? _____

   _____

   How long does it take for the muscle to fatigue? _____

If you measured fatigue in the frog, compare your results with the frog in terms of time for fatigue to begin.

Exercise 11 Review
Muscle Physiology

Name: _____
Lab time/section: _____
Date: _____

1. What role do nerves play in skeletal muscle contraction? _____

2. What chemical crosses the synapse, causing a muscle to contract? _____

3. Where is calcium released to cause muscle contraction? _____

4. What are the two types of myofilaments found in muscle cells that cause muscle contraction? _____
   _____

5. Does a single muscle fiber or an entire muscle contract by an all-or-none response? _____

6. Name the first phase after a stimulus in a muscle contraction. _____

7. Define *subthreshold stimulus*. _____

8. What happens to the strength of contraction during wave summation? _____
   _____

9. Describe *tetanus*. _____
   _____

10. What is maximum recruitment? _____
    _____
    _____

11. How do tetanus and twitch, as demonstrated in the lab, correlate to human muscle contraction? Which one is more reflective of human muscle response? Why? _____
    _____

12. In the following illustration, which one is the contraction phase, the lag period, and the relaxation phase?

   a. _____

   b. _____

   c. _____

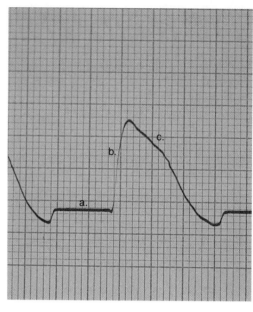

13. If you did the live frog lab, what was the threshold stimulus you obtained in lab? _____

14. If you did the live frog lab, what was the voltage at which you first got maximum recruitment? _____

15. Explain a muscle spasm in terms of recruitment of muscle fibers. _____

16. How does your favored arm compare with your weak arm in terms of electrical activity? _____

17. How does the muscle respond to weight when it is first contracting versus when it is fatigued? _____

Exercise 12
Introduction to the Study of Muscles and Muscles of the Shoulder and Upper Extremity

INTRODUCTION

The next four exercises focus on skeletal muscles. There are over 600 skeletal muscles in the human body. The muscles are grouped according to their location, with most muscles occurring as pairs. In the following exercises, you learn about some of the major muscles of the body. The muscles presented in the following labs are covered in the *Principles of Anatomy and Physiology* text in chapter 9, "Gross Anatomy and Functions of Skeletal Muscles." The origin, insertion, action, and innervation are listed for about 100 muscles in this lab manual. Your instructor may wish to customize the list, so that you learn specific muscles or specific things about each muscle. The study of muscles can be both fun and challenging. Begin your study of muscles early and do not wait to learn them until the night before the lab exam.

Some students find the study of human musculature to be a daunting task, whereas others recall the muscle section of the course as their favorite part of the study of human anatomy and physiology. The satisfaction of studying muscles can be increased if you set obtainable goals. This is best accomplished by choosing small numbers of muscles to study at a sitting and using flash cards as study aids. It is vital that you know the bones and bony markings before studying muscles. Look over Exercise 7 if you need to review the skeleton. You may also want to make a quick sketch of the individual muscles to help you visualize them.

Skeletal muscle is voluntary muscle in that it contracts when consciously stimulated by specific nerves. In this exercise, you are introduced to concepts that apply to all of the muscles, and you then study a specific region, learning the major muscles of the shoulder and upper limb. Some of these muscles originate on the scapula and move or stabilize the humerus. Others originate on the humerus and insert on the forearm.

The muscles of the forearm and hand are very important, particularly in terms of rehabilitating limbs after surgery to relieve carpal tunnel syndrome or after trauma to the limb. You should learn the origins and insertions as you learn the muscles of the forearm, since many of these muscles look quite similar. By knowing the origins and insertions of a muscle, you will not easily mistake it for another muscle. You can compare the musculature of the human with that of the cat.

Overview of Muscles

A thorough study of muscles involves knowing not only the name of the muscle but also its origin, insertion, action, and innervation. The muscles are listed in this exercise by name, followed by their origin. The **origin** is the attachment point of the muscle that does not move during muscular contraction. The muscles also are listed with their **insertion,** which is the attachment that moves during contraction. Each muscle has an **action,** which represents the effects the muscle has on a part of the body (such as *flexion* of the arm). Finally, the **innervation** of a muscle is the nerve attachment that controls it. If the motor portion of a nerve is severed, the muscle innervated by that nerve cannot function.

You should also review the muscle nomenclature and actions at joints as described in Exercise 10. There are two views to muscle action. One is that muscles act on an appendage, such as flexion of the hand. Another view is that muscles act on a joint, such as flexion of the wrist. These can be seen as equivalent. Muscles can have many actions at a joint. In addition to moving a joint, muscles can also fix a joint.

Origins and Insertions

Examine an articulated skeleton as you study the muscles. Visualize the muscle as it attaches to the origin and the insertion. As you study each muscle, locate it on your body and try to determine which part is the stable origin and which part is the moving insertion.

Actions

To fully understand muscles, you must know their actions at joints and associate the actions with the body in anatomic position. When you are learning the action of a muscle, imagine a piece of string tied between the origin and the insertion of the muscle. Think of how the bones move as you pull the insertion closer to the origin. Mimic the action of the muscle as you learn it. When you perform the action of a particular muscle, you should be able to feel that muscle tighten.

Muscle Nomenclature

Another aid to learning muscles is to understand how they are named. Muscles are named by a number of criteria:

Location: Some muscles are named by their location, such as the **tibialis** anterior, the **frontalis,** or the **temporalis.**

Size: Muscles such as the adductor **magnus** are named for their size. The gluteus **minimus** is the smallest of the gluteal muscles.
Shape: The **deltoid** muscle is shaped like a delta, or triangle.
Orientation: The **transversus** abdominis muscle has fibers that are oriented horizontally, or in a transverse direction.
Origin and insertion: The **infraspinatus** muscle originates on the infraspinous fossa of the scapula, whereas the flexor **hallucis** longus muscle inserts on the hallux, or big toe.
Number of heads: The **triceps** brachii muscle has three heads.
Action: The **extensor** digitorum is a muscle that extends the fingers.
Length: Muscles with the term **longus** or **brevis** are named for the overall length of the muscle.

OBJECTIVES

At the end of this exercise, you should be able to

1. locate the muscles of the shoulder and upper extremity on a torso model, chart, cadaver (if available), or cat (if available);
2. list the origin, insertion, and action of each muscle presented;
3. describe what nerve controls each muscle;
4. name all the muscles that have an action on a joint, such as all of the muscles that laterally rotate the arm or muscles that flex the hand;
5. distinguish between the cat musculature and the human musculature and compare them for similarities and differences;
6. reproduce the actions of a muscle by moving the appropriate limbs to demonstrate the actions.

MATERIALS

Human torso model and upper extremity model
Human muscle charts
Articulated skeleton
Cadaver (if available)
Cat

Materials for cat dissection:
  Dissection trays
  Scalpel and two or three extra blades
  Gloves (household latex gloves work well for repeated use)
  Blunt (Mall) probe
  String and tags
  Pins
  Forceps and sharp scissors
First aid kit in lab or prep area
Sharps container
Animal waste disposal container
Cat wetting solution

PROCEDURE

Overview

Most people are familiar with the larger skeletal muscles, such as the pectoralis major, gluteus maximus, deltoid, and trapezius. You should examine the charts and models in lab and locate as many of the most common muscles of the body as you can on the charts or models and in figure 12.1. You should refer to this figure in the subsequent exercises if you need an overview of the muscles.

Human Muscles

Examination of Muscles

Look at the charts, models, and cadaver (if available) in the lab and locate the muscles presented in this section. Review the bones of the hand as well as the bones of the arm and forearm in Exercise 7 to locate precisely the bony origins and insertions. Refer to the following descriptions as you examine the muscles. The specific details of the muscles are provided in table 12.1.

Muscles of the Shoulder

Some muscles of the shoulder act on the scapula. The **trapezius** is a diamond-shaped muscle that originates in the midline of the vertebral column and head and inserts laterally. If the scapula is fixed, the head moves, yet, if the vertebral column and head are fixed, the trapezius has an action on the scapula. Figure 12.2 illustrates the trapezius.

Exercise 12 Introduction to the Study of Muscles and Muscles of the Shoulder and Upper Extremity

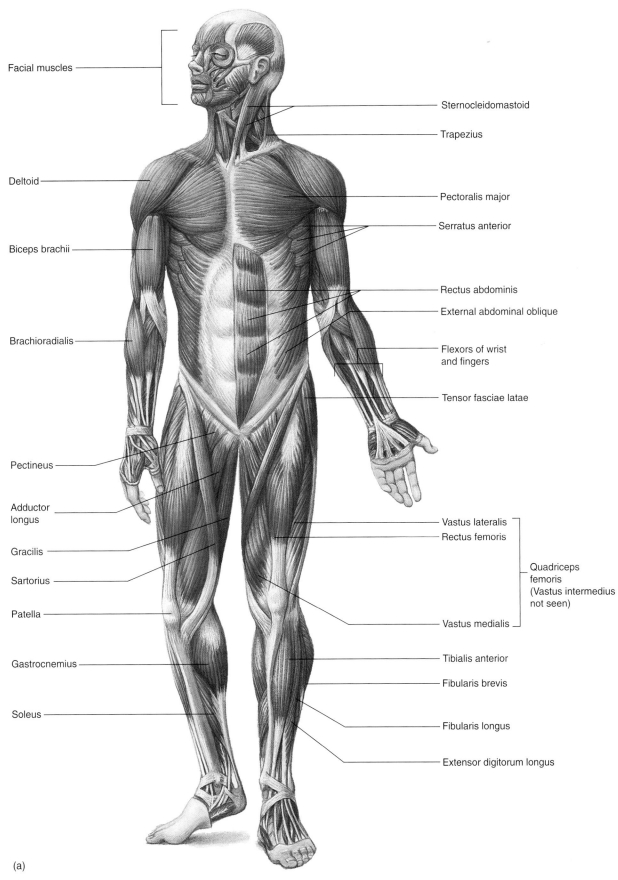

Figure 12.1 Overview of the Muscles of the Body
(a) Anterior view; (b) posterior view.

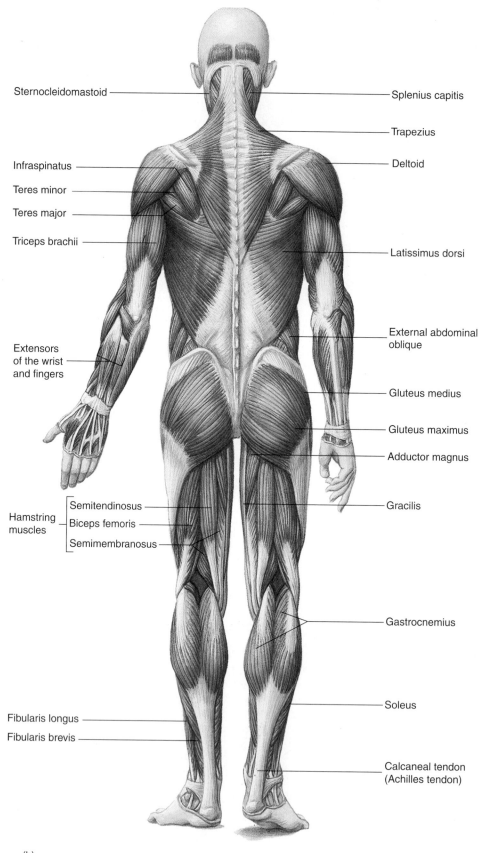

(b)

Figure 12.1—*Continued.*

TABLE 12.1 Muscles of the Shoulder and Upper Extremity

Posterior Muscles Acting on the Shoulder and Arm

Name	Origin	Insertion	Action	Innervation
Trapezius	Occipital protuberance, nuchal ligament, spinous processes of C7–T12	Clavicle, acromion process, and spine of scapula	Extends head, rotates and adducts scapula, fixes scapula	Accessory (XI) nerve
Latissimus dorsi	T7–12, L1–5, sacrum, iliac crest, ribs 10–12	Intertubercular groove of humerus	Extends, adducts, medially rotates arm, depresses shoulder	Thoracodorsal nerve
Deltoid	Clavicle, acromion process, spine of scapula	Deltoid tuberosity of humerus	Abducts, flexes, extends arm, medially and laterally rotates arm	Axillary nerve
Rhomboideus major	Spines of T1–4	Inferior, medial border of scapula	Adducts scapula	Dorsal scapular nerve
Rhomboideus minor	Nuchal ligament, spines of C6–7	Medial border of scapula at scapular spine	Adducts scapula	Dorsal scapular nerve
Levator scapulae	C1–4	Superior angle of scapula	Elevates scapula, abducts and rotates neck	Spinal nerves C3–5 and dorsal scapular nerve
Supraspinatus*	Supraspinous fossa	Greater tubercle of humerus	Abducts arm (stabilizes shoulder joint)	Suprascapular nerve
Infraspinatus*	Infraspinous fossa	Greater tubercle of humerus	Extends and laterally rotates arm (stabilizes shoulder joint)	Suprascapular nerve
Subscapularis*	Subscapular fossa	Lesser tubercle of humerus	Extends and medially rotates arm (stabilizes shoulder joint)	Subscapular nerve
Teres minor	Lateral border of scapula	Greater tubercle of humerus	Extends, laterally rotates, and adducts arm, stabilizes shoulder joint	Axillary nerve
Teres major	Lateral border of scapula	Crest of lesser tubercle of humerus	Extends, adducts, and medially rotates arm	Subscapular nerve
Triceps brachii	Infraglenoid tuberosity of scapula, lateral and posterior surface of humerus	Olecranon process of ulna	Extends and adducts arm, extends forearm	Radial nerve

Anterior Muscles Acting on the Shoulder and Arm

Name	Origin	Insertion	Action	Innervation
Pectoralis major	Clavicle, sternum, cartilages of ribs 1–7	Crest of greater tubercle of humerus	Flexes, adducts, and medially rotates arm	Anterior thoracic nerves
Pectoralis minor	Ribs 3–5	Coracoid process of scapula	Depresses glenoid cavity, raises ribs 3–5	Anterior thoracic nerves
Biceps brachii	Supraglenoid tubercle, coracoid process of scapula	Radial tuberosity	Flexes arm, flexes forearm, supinates hand	Musculocutaneous nerve
Coracobrachialis	Coracoid process of scapula	Midmedial shaft of humerus	Flexes and adducts arm	Musculocutaneous nerve

Anterior Muscles Acting on the Forearm

Name	Origin	Insertion	Action	Innervation
Brachialis	Anterior, distal surface of humerus	Ulnar tuberosity	Flexes forearm	Musculocutaneous and radial nerves
Brachioradialis	Lateral supracondylar ridge of humerus	Styloid process of radius	Flexes forearm	Radial nerve
Supinator	Lateral epicondyle of humerus, anterior ulna	Proximal radius	Supinates hand	Radial nerve
Pronator teres	Medial epicondyle of humerus, coronoid process of ulna	Lateral, middle shaft of radius	Pronates hand, flexes forearm	Median nerve
Pronator quadratus	Distal part of anterior ulna	Distal radius	Pronates hand	Median nerve (anterior interosseus branch)

*Musculotendinous cuff muscles

TABLE 12.1 Muscles of the Shoulder and Upper Extremity (continued)

Anterior Muscles Acting on the Forearm and Hand

Name	Origin	Insertion	Action	Innervation
Palmaris longus	Medial epicondyle of humerus	Palmar fascia	Flexes hand	Median nerve
Flexor carpi radialis	Medial epicondyle of humerus	Metacarpals 2 and 3	Flexes and abducts hand	Median nerve
Flexor carpi ulnaris	Medial epicondyle of humerus and ulna	Pisiform, hamate, metacarpal 5	Flexes and adducts hand	Ulnar nerve
Flexor digitorum superficialis	Medial epicondyle of humerus, coronoid process of ulna, proximal radius	Middle phalanges of digits 2–5	Flexes proximal and middle phalanges 2–5, flexes hand	Median nerve
Flexor digitorum profundus	Ulna, interosseus membrane	Distal phalanges of digits 2–5	Flexes phalanges and flexes hand	Median and ulnar nerves
Flexor pollicis longus	Anterior portion of radius and interosseus membrane	Distal phalanx of pollex (thumb)	Flexes thumb and hand	Median nerve
Flexor pollicis brevis	Trapezium	Proximal phalanx of digit 1	Flexes thumb	Median and ulnar nerves
Abductor pollicis brevis	Scaphoid, trapezium	Proximal phalanx of digit 1	Abducts thumb	Median nerve
Opponens pollicis	Trapezium	Metacarpal 1	Opposes thumb	Median nerve
Abductor digiti minimi	Pisiform	Proximal phalanx of digit 5	Abducts digit 5	Ulnar nerve
Flexor digiti minimi brevis	Hamulus of hamate	Proximal phalanx of digit 5	Flexes digit 5	Ulnar nerve
Opponens digiti minimi	Hamulus of hamate	Metacarpal 5	Opposes digit 5	Ulnar nerve

Posterior Muscles Acting on the Forearm and Hand

Name	Origin	Insertion	Action	Innervation
Extensor carpi radialis longus	Lateral supracondylar ridge of humerus	Metacarpal 2	Extends and abducts hand	Radial nerve
Extensor carpi radialis brevis	Lateral epicondyle of humerus	Metacarpal 3	Extends and abducts hand	Radial nerve
Extensor carpi ulnaris	Lateral epicondyle of humerus, proximal ulna	Metacarpal 5	Extends and adducts hand	Radial nerve
Extensor digitorum	Lateral epicondyle of humerus	Middle and distal phalanges of digits 2–5	Extends phalanges 2–5, extends hand	Radial nerve
Abductor pollicis longus	Posterior radius and ulna, interosseus membrane	Metacarpal 1	Abducts and extends thumb, abducts hand	Radial nerve
Extensor pollicis longus and brevis	Posterior radius and ulna, interosseus membrane	Proximal and distal phalanges of pollex (thumb)	Extends thumb	Radial nerve

The **rhomboideus** muscles are deep to the trapezius, so reflection of the trapezius is necessary to see them. These muscles originate on the vertebral column and insert on the medial border of the scapula. As they contract, they adduct the scapulae (pull them to the midline). The **rhomboideus major** is a larger and more inferior muscle than the **rhomboideus minor.** Compare the material in lab with figure 12.2.

The **levator scapulae** is named for what it does, elevate the scapula. The levator scapulae originates on the lateral side of the neck and inserts on the upper scapula. If the neck is fixed, the levator scapulae raises the scapulae, as in shrugging. If the scapula is fixed, the muscle rotates or abducts the neck.

The **deltoid** is located on top of the shoulder and abducts the arm. Locate the deltoid on material in the lab and compare it with figure 12.3. It is a fan-shaped muscle whose distal attachment partially covers the insertion of the **pectoralis major.** The pectoralis major has fibers that run horizontally across the chest region, and it is a superficial muscle of the chest. Deep to the pectoralis major muscle is the **pectoralis minor** muscle, whose fibers run in a more vertical direction, as seen in figure 12.4.

Another muscle of the posterior surface is the **latissimus dorsi,** which arises from the vertebrae by way of a broad, flat **lumbodorsal fascia.** The latissimus dorsi originates on the back, but the insertion is on the anterior aspect of the humerus. It is a

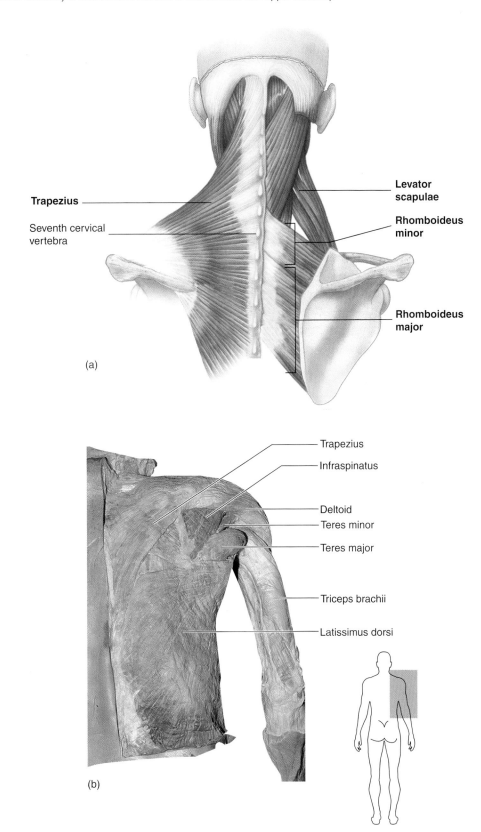

Figure 12.2 Trapezius and Associated Muscles
(a) Diagram; (b) photograph.

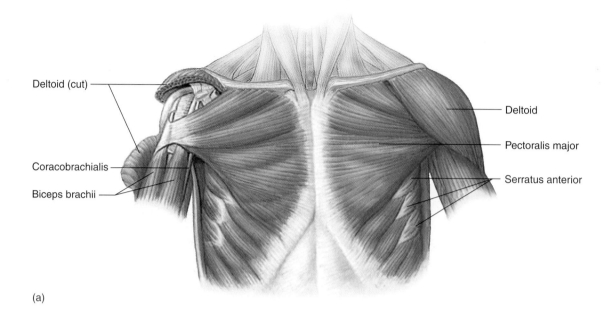

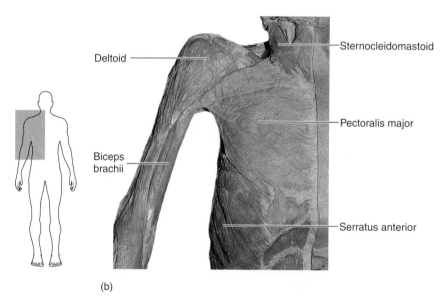

Figure 12.3 Superficial Muscles of the Anterior Chest
(a) Diagram; (b) photograph.

powerful extensor of the arm and is known commonly as the swimmer's muscle (figures 12.2 and 12.5).

Deep to the trapezius and the deltoid are muscles that originate on the scapula proper. The **supraspinatus** is named for its origin on the supraspinous fossa, and it is a synergist to the deltoid. The **infraspinatus** is a muscle originating on the infraspinous fossa. It runs laterally and posteriorly to the humerus, and it laterally rotates the arm. The **subscapularis** is named for its origin on the subscapular fossa. The insertion of the subscapularis is on the anterior surface of the humerus. When it contracts, it medially rotates the arm. The subscapularis originates on the anterior surface of the scapula, between the scapula and the ribs, and thus cannot be seen in a posterior view. Locate these muscles in the lab and compare them with figure 12.5.

The scapula gives rise to two other muscles, the teres minor and the teres major. The **teres minor** appears like a slip of the infraspinatus, whereas the **teres major** crosses to the medial side of the humerus in a way similar to the latissimus dorsi (figure 12.5). The **rotator (musculotendinous) cuff** muscles stabilize the shoulder joint by holding the head of the humerus in the glenoid fossa.

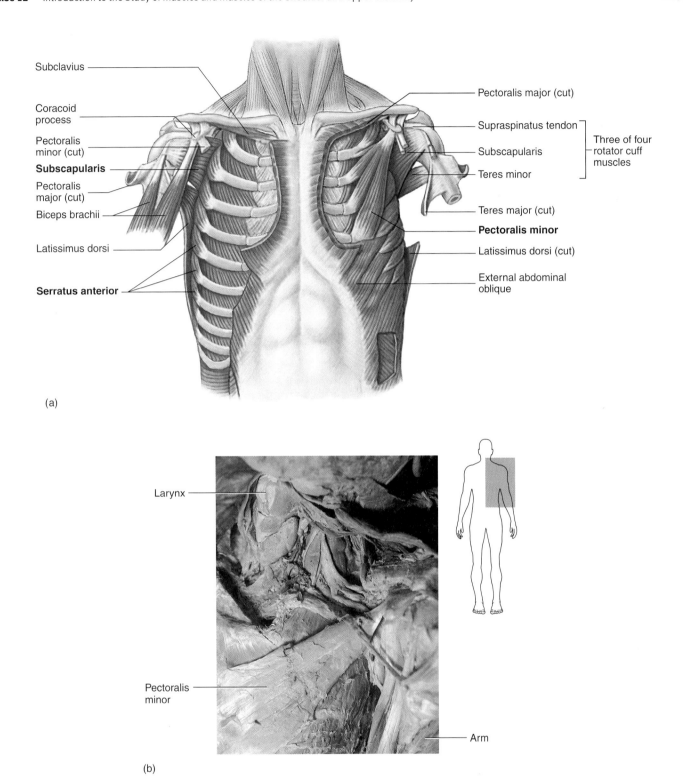

Figure 12.4 Deep Muscles of the Anterior Chest
(a) Diagram; (b) photograph.

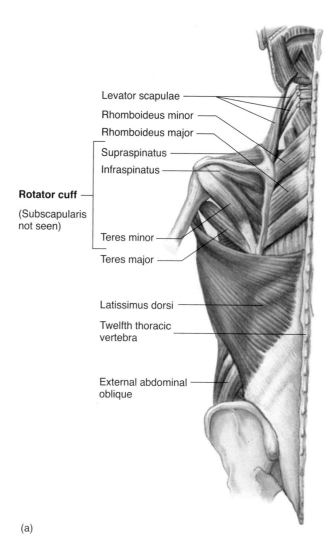
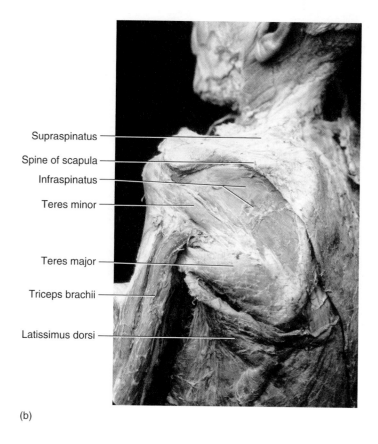
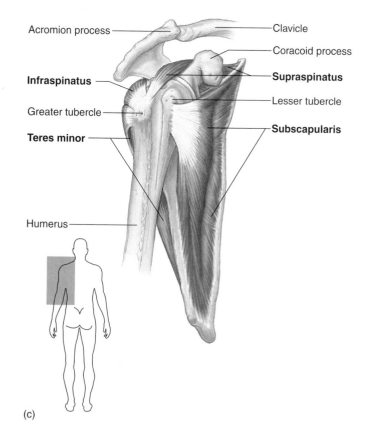

Figure 12.5 Scapular Muscles
(a) Diagram of left posterior view; (b) photograph of left posterior view; (c) diagram of right lateral view.

Muscles composing the cuff are the supraspinatus, infraspinatus, subscapularis, and teres minor. A rotator cuff injury occurs if there is damage to any of these muscles.

The **biceps brachii** muscle is a two-headed muscle of the arm. It neither originates nor inserts on the humerus, yet it has an action on the arm. Flexion of the forearm occurs primarily from the contraction of this muscle. The biceps brachii has a long tendon that originates on the scapula and runs between the intertubercular groove of the humerus and a short tendon that originates on the coracoid process. The insertion of the biceps brachii is on the radius. The **triceps brachii** is the only major muscle on the posterior surface of the humerus. The triceps brachii extends the arm and forearm and thus is an antagonist to the biceps brachii (figure 12.6).

Alongside the short head of the biceps brachii is a small slip of muscle known as the **coracobrachialis.** This muscle is a short, diagonal muscle of the arm. Underneath the biceps brachii on the anterior surface of the arm is the **brachialis** muscle. The brachialis crosses the anterior surface of the elbow joint and flexes the forearm. The **brachioradialis** is a distal muscle of the arm and is the most lateral muscle of the forearm. Examine figures 12.6 and 12.7 for these muscles and compare them with the descriptions in table 12.1.

Muscles of the Forearm and Hand

The muscles of the forearm, due to their similar appearance, provide greater challenges than do the muscles of the shoulder and arm. As you study these muscles, it is important that you locate them and determine their origin and insertion. The origins and insertions will help you determine if you are looking at the correct muscle. The muscles of the forearm and hand are generally named for their action (**pronate** the hand or **flex** the digits) or for their insertion (**carpi** for inserting on carpals or metacarpals, **digitorum** for fingers, and **pollicis** for thumb). In a few instances, they are named for the shape of the muscle (**teres** for round or **quadratus** for square).

Notice that much of the muscle mass for the fingers is located on the forearm. The fingers move by a force applied from a distance, analogous to a puppet controlled by puppet strings. If the muscle mass for the hand muscles were located on the hands proper, then the hands would look like softballs. By having the muscle mass in the forearm, an efficiency of form is evident that allows for a powerful grip yet precise movements of the fingers.

Muscles That Supinate and Pronate the Hand

The first group of muscles in this study includes those muscles that insert on the radius. The **supinator** is a muscle that originates on the arm and forearm and supinates the hand. It wraps around the radius and is the deepest, proximal muscle of the forearm. Examine this muscle in the lab and compare it with figure 12.7.

The **pronator teres** is a round muscle that pronates the hand. It originates on the medial side of the arm and forearm and inserts on the lateral side of the radius. The pronator teres is a bit different from the other superficial forearm muscles in that it runs at an oblique angle on the forearm, whereas the other muscles run parallel along the length of the forearm.

The **pronator quadratus** is a square muscle deep to the other forearm muscles on the distal part of the radius and the ulna. It pronates the hand. Compare the two pronator muscles in the lab with figures 12.7 and 12.8.

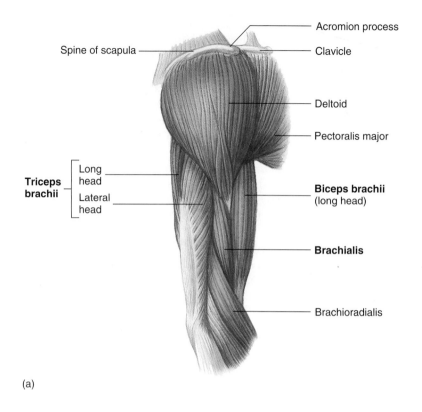

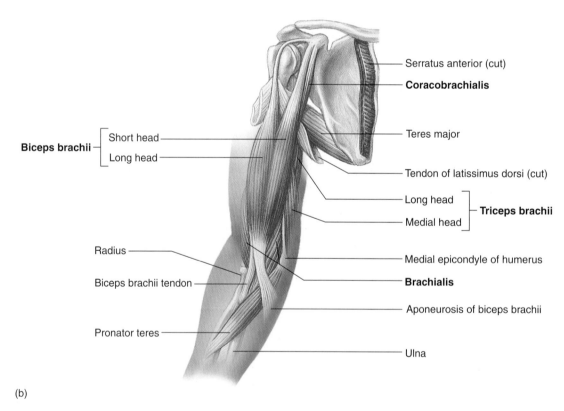

Figure 12.6 Muscles of the Right Arm
Diagram (a) lateral view; (b) anterior view. Photograph (c) lateral view; (d) posterior view.

Exercise 12 Introduction to the Study of Muscles and Muscles of the Shoulder and Upper Extremity

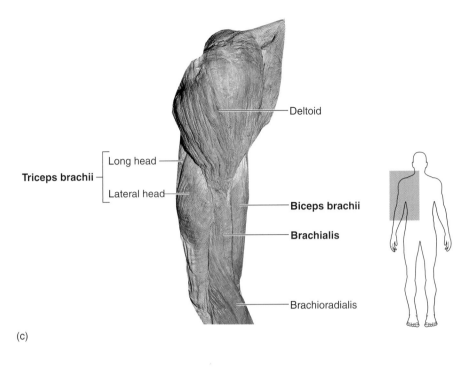

(c)

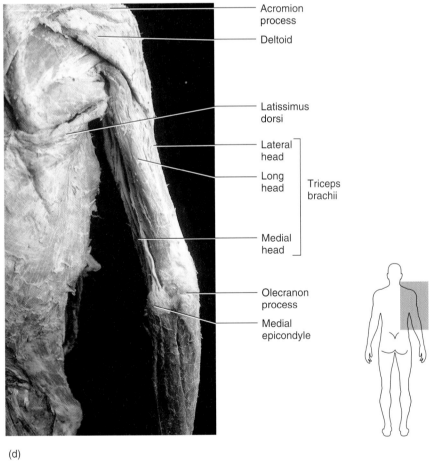

(d)

Figure 12.6—*Continued*.

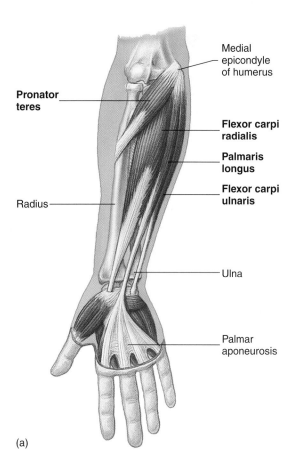
(a)

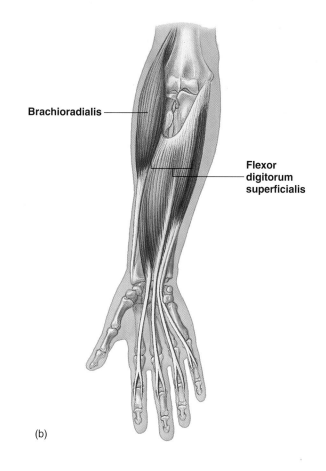
(b)

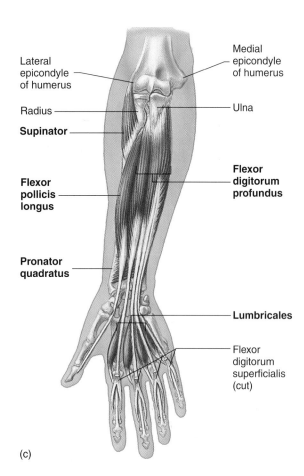
(c)

Figure 12.7 Muscles of the Forearm, Anterior View
(a) Superficial group; (b) muscles of middle depth; (c) deep muscles.

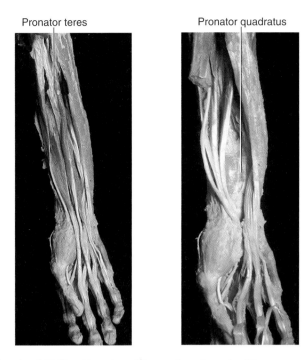

Figure 12.8 Muscles of Pronation, Anterior View

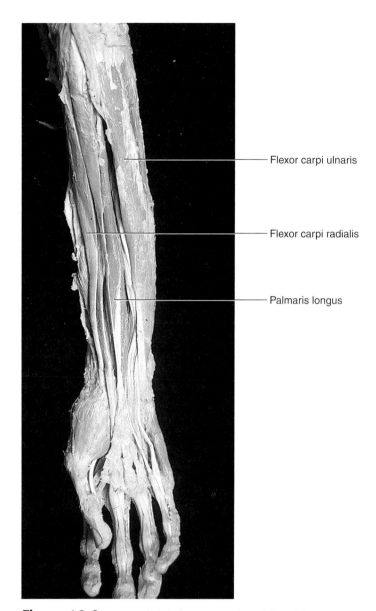

Figure 12.9 Superficial Flexor Muscles of the Right Forearm and Hand, Anterior View

Flexor Muscles

The next muscles to be studied are grouped by their insertion on the hand. The superficial **palmaris longus** muscle is absent in about 10% of the population. It is centrally located in the middle, anterior forearm and inserts into a broad, flat tendon known as the **palmar aponeurosis.** This aponeurosis has no bony attachment but, rather, attaches to the fascia of the underlying muscles.

The **flexor carpi radialis** muscle inserts on the metacarpals on the radial side of the hand. The pulse of the radial artery is often taken at the wrist just lateral to the tendon of the flexor carpi radialis muscle. Most of the flexor muscles of the hand originate from the medial epicondyle of the humerus, as seen with the flexor carpi radialis. Tendons of this muscle run underneath a connective tissue band known as the **flexor retinaculum,** which anchors the tendons to the wrist and prevents the tendons from pulling away from the wrist when the hand is flexed.

The **flexor carpi ulnaris** muscle also has an origin on the medial epicondyle of the humerus and inserts on the carpals and a metacarpal of the ulnar side of the hand. The flexor carpi ulnaris is a medial muscle of the forearm. Examine these muscles in the lab and in figures 12.7 and 12.9.

The **flexor digitorum superficialis** is a superficial flexor muscle of the digits. It is not the most superficial muscle of the forearm but is under the palmaris longus, the flexor carpi radialis, and flexor carpi ulnaris. The flexor digitorum superficialis is superficial to the deep digit flexor, which is discussed next. As with the other flexor muscles, the flexor digitorum superficialis has an origin on the medial epicondyle of the humerus. The insertion tendons form a V on the middle phalanges of digits 2 through 5.

The **flexor digitorum profundus** is a deep muscle of the digits. It is an exception to the other hand flexors in that it does not originate on the medial epicondyle of the humerus but on the ulna and the membrane between the radius and the ulna (interosseus membrane). It is a deep muscle of the forearm that runs underneath the flexor digitorum superficialis. At the insertion point, the tendons of the flexor digitorum profundus actually run through the split tendons of the flexor digitorum superficialis and extend to the distal phalanges of digits 2 through 5.

The **flexor pollicis longus** originates on the radius and the interosseus membrane and runs along the lateral forearm to insert on the anterior of the thumb. Flexion of the thumb is the curling of the thumb as if you were ready to flip a coin. Examine these muscles and compare them with figures 12.7 and 12.10.

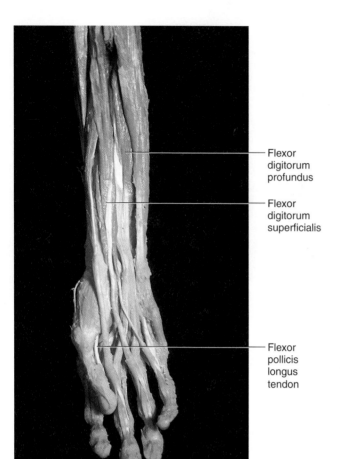

Figure 12.10 Deep Flexor Muscles of the Right Forearm and Hand

- Flexor digitorum profundus
- Flexor digitorum superficialis
- Flexor pollicis longus tendon

Intrinsic Muscles of the Hand

There are many muscles that originate on the hand and have an action on the hand. Some of these arise on the pad of muscles at the base of the thumb known as the **thenar eminence.** These muscles are named for their action on the thumb, including the **flexor pollicis brevis,** the **abductor pollicis brevis,** and the **opponens** (up-pō´-nenz) **pollicis.** The pad of tissue at the base of digit 5 is known as the **hypothenar eminence,** and the muscles that arise there are the **abductor digiti minimi,** the **flexor digiti minimi brevis,** and the **opponens digiti minimi.** Find these muscles on models or cadavers in lab and compare them with figure 12.11.

Extensor Muscles

As a general rule, most of the extensors originate on the lateral epicondyle or lateral supracondylar ridge of the humerus. The extensor muscle tendons are held to the posterior surface of the wrist by a connective tissue band known as the **extensor retinaculum.** The **extensor carpi radialis longus** muscle originates on the humerus and inserts on the second metacarpal (on the radial side) of the hand. The **extensor carpi radialis brevis** is deep to the longus, and it inserts on the dorsum of the third metacarpal. Both of these muscles extend and abduct the hand. Examine these muscles in figure 12.12.

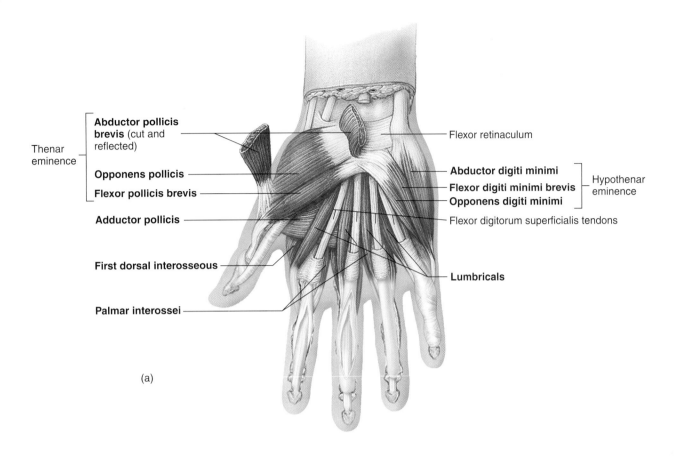

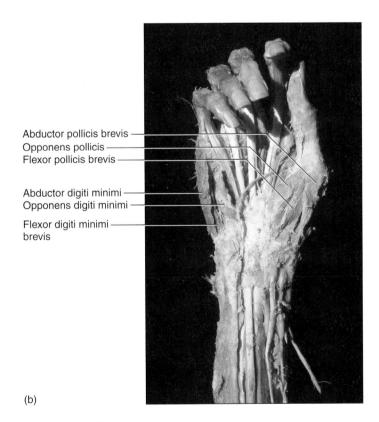

Figure 12.11 Intrinsic Muscles of the Right Hand
(a) Diagram; (b) photograph.

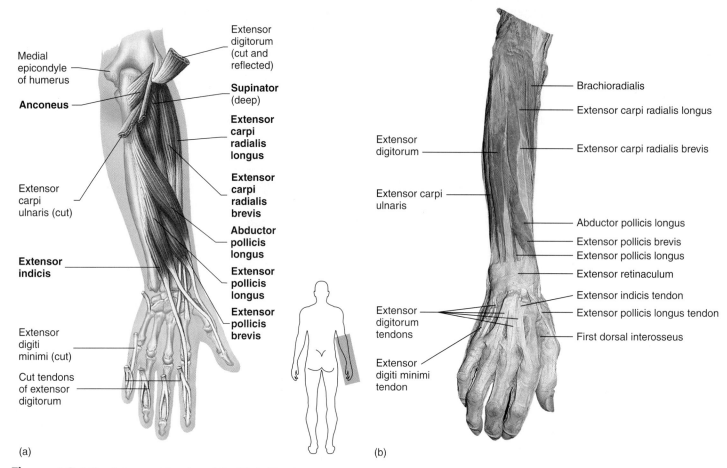

Figure 12.12 Extensor Muscles of the Right Hand
(a) Diagram; (b) photograph.

The **extensor carpi ulnaris** originates on the lateral epicondyle and inserts on the fifth metacarpal (on the ulnar side) of the hand. As it contracts, the hand is extended and adducted. The **extensor digitorum** is a singular muscle on the back of the hand (remember, there are two flexor digitorum muscles). As an extensor muscle, it originates on the lateral epicondyle of the humerus. It inserts on the middle and distal phalanges of the second through fifth digits. The tendons of this muscle can be seen on the dorsal surface of the hand and in figure 12.12.

The **abductor pollicis longus** muscle inserts on the metacarpal of the thumb. By pulling your thumb away from the index finger, in an anterior direction, you are abducting the thumb. There are two extensor pollicis muscles, the **extensor pollicis longus** and the **extensor pollicis brevis.** Extension of the thumb is done when you flip a coin. At the end of the flip, the thumb is extended. The tendons of the extensor pollicis muscles and the abductor pollicis muscles form a depression in the form of a triangle at the base of the thumb. This depression is known as the anatomical snuff box. These muscles can be seen in figures 12.12 and 12.13.

Cat Dissection

The objective in using a cat for the exercise on musculature is to provide a study specimen for dissection and application to the human system. Differences occur between cat musculature and human musculature, but you should focus on similar structures in order to gain an appreciation of human musculature. Read all of the material in the exercise before beginning the dissection. Dissection is a skill in which you try to separate the overlying structures from those underneath while keeping intact as much material as you can.

Dissection Concerns

Wear an apron or an old overshirt as you dissect. Dissection instruments are sharp, and you must be careful when using scalpels. Do not cut *down* into the specimen but, rather, lift structures gently and try to make incisions so the scalpel blade cuts *laterally*. Cut *away* from yourself and your lab partners. *If you do cut yourself, notify the instructor immediately!* Wash the cut with antimicrobial soap and seek medical advice.

The laboratory should be well ventilated; if your eyes burn and you develop a headache, get some fresh air for a moment. Dissecting without having your face directly over the specimen may help. Occasionally, students have allergic reactions to formaldehyde. This usually consists of feeling that your breathing is becoming restricted. If this happens, notify your instructor.

At the end of the exercise, remove the scapel blade and wash your dissection equipment with soap and water. Place all used blades and other sharp material in the **sharps container** in lab.

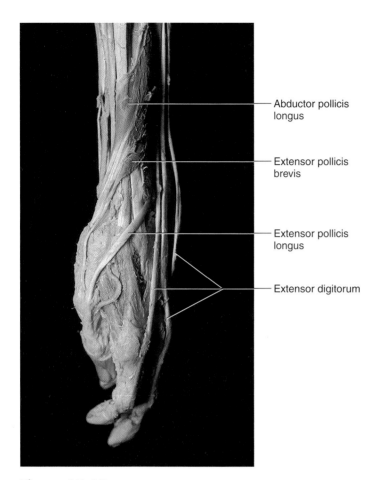

Figure 12.13 Posterior Muscles of the Thumb

Labels: Abductor pollicis longus; Extensor pollicis brevis; Extensor pollicis longus; Extensor digitorum

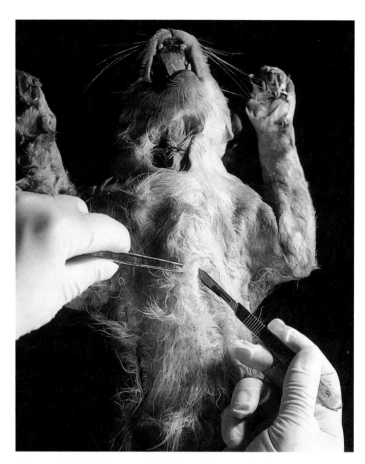

Figure 12.14 Lateral Cutting with a Scalpel

Remove all of the excess animal material on the dissection trays and put it in the appropriate animal waste container. *Do not dump excess animal material in the lab sinks!* Wash the dissection trays and place them in the appropriate area to dry.

Cat Care

Keep your cat in a plastic bag. It is recommended that you retain the skin of the cat as a protective wrapping when you are finished with the day's dissection or that you take a small towel (or an old T-shirt) and wrap the specimen in the cloth, soaking it in a cat wetting solution before you place it back in the bag. Usually one lab period is required to skin the cat and prepare the specimen for further study.

External Features

Take a dissection tray and a cat specimen to your table. Remove the cat from the plastic bag and place it on the tray. You may have excess fluid in the plastic bag, and this should be disposed of properly as directed by your instructor. Place the cat on its back and determine whether you have a male specimen or a female specimen. Both sexes have multiple **teats** (nipples), yet the males have a **scrotum** near the base of the tail with a **prepuce** (penile foreskin) ventral to the scrotum. Females have a **urogenital opening** anterior to the anus without a scrotum and prepuce. Once you have identified the sex of your specimen, compare yours with others in the class, so that you can identify the sexes externally. If you have difficulty determining the sex, ask your instructor for help.

Examine the cat and notice the **vibrissae,** or whiskers, in the facial region. Other variations from humans are the presence of **claws** and **friction pads** on the extremities and a **tail.** Review the planes of the body, as illustrated in Exercise 1, before you begin the dissection. The terms used in that exercise will be of great importance in the dissection procedures.

Removal of the Skin

Begin your dissection by lifting the skin in the pectoral region with a forceps and making a small cut with a scalpel or sharp scissors in the midline. Work a blunt probe gently into the cut, so that you free the skin from the underlying fascia and muscle somewhat. Be careful—the muscles are close to the skin. Insert your scalpel, blade side up, and make small incisions in the skin, cutting away from the underlying muscle, as illustrated in figure 12.14.

Make a cut that runs up the midline of the sternal region until you reach the neck. Likewise, cut posteriorly until you reach

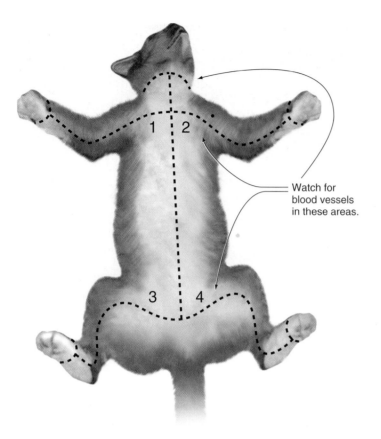

Figure 12.15 Removal of the Skin of the Cat
After cutting down the midline, incisions should be made into the limbs (follow the numbers) and the skin gently removed from underlying structures.

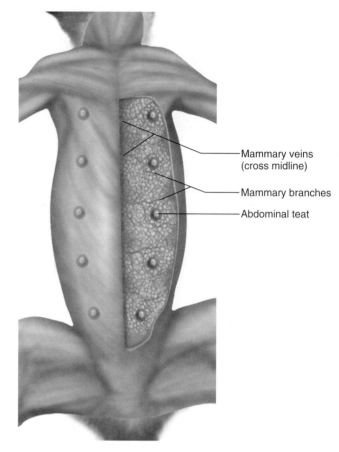

Figure 12.16 Mammary Glands of the Cat

an area craniad to the genital region. Leave the genitals intact and carefully make an incision that runs perpendicular to your first cut. Likewise, make a perpendicular incision along the upper thoracic region (figure 12.15). Continue your caudal incision along the medial aspect of the thigh. Watch out for superficial blood vessels in this area (especially the great saphenous vein), and stop when you reach the knee. Cut the skin around the knee and begin removing it as a layer from the dorsal side of the cat. Cut the skin from the base of the tail and remove the skin from the back.

Once you reach the shoulders, turn the cat back to the ventral side and remove the skin on the medial side of the arm until you reach the elbow. Stop at this level and remove the skin from the lateral aspect of the arm, working back toward the shoulder. Be careful not to damage the superficial veins on the lateral side of the forearm. Continue your removal of the skin from the ventral body region. If your cat is a female, locate the mammary glands, which are elongated, beige, lobular tissue on each side of the midline on the ventral side (figure 12.16).

Make an incision on the ventral side of the neck along the midline. Be careful—there are many blood vessels along the neck. Do not cut through these blood vessels. Continue up into the face and look for beige lumps of glandular tissue in the region of the mandible. These are the **salivary glands.** Cut the skin carefully from the face and remove it from the head by cutting around the ears (figure 12.17).

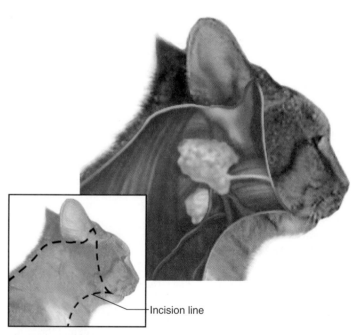

Figure 12.17 Removal of Skin from the Head of a Cat

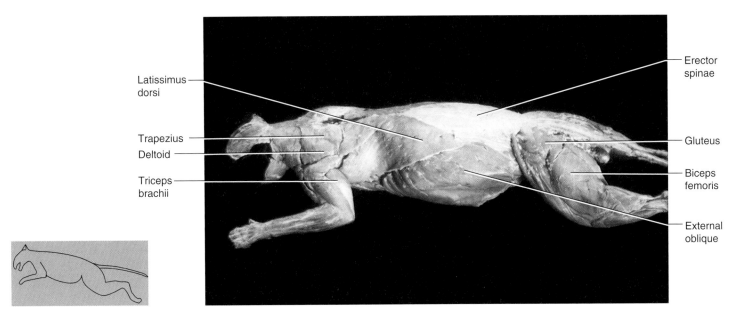

Figure 12.18 Major Muscles of the Cat

You should be able to remove all the skin from the cat at this time. Examine the undersurface of the skin and note the superficial muscles that cause the skin to move. These are the **cutaneous maximus** and the **platysma.**

Return to the cat and remove as much fat as you can. **Subcutaneous fat** is variable from cat to cat, and your specimen may have little or it may have significant amounts. The muscle also is covered by fascia, a connective tissue wrapping. Remove the fascia from the muscle so that the fiber direction is apparent.

Once you have removed the skin from your cat and removed the superficial fascia, you should identify the major muscles of the cat. Look for the large latissimus dorsi muscle of the back and the external abdominal oblique muscle. You should also find the deltoids and triceps brachii muscles of the shoulder region and the gluteus and biceps femoris muscles of the hip and thigh region. Compare your cat with figure 12.18.

Individual Muscles of the Cat

Begin your **dissection** of the muscles by understanding that the term *dissect* means to separate. When you isolate one muscle from another, locate the tendons of that muscle. The **tendon** is the attachment point of the muscle to a bone. Broad, flat tendons are known as **aponeuroses.**

If you have not already removed the skin from the forelimb of the cat, you should do so now. This can be accomplished by making a longitudinal incision along the length of the forelimb. Be very careful not to cut the tendons or the blood vessels or nerves. Pull the skin off as if you were removing a pair of knee socks.

As you get to the tips of the digits, cut the skin from the digits. Pay particular attention and keep the tendons intact. It is a good procedure to dissect only one side of the cat at a time. If you make an error on one side, you will have the other side to dissect.

As you make the dissection in the cat, you can leave many of the muscles of the forelimb intact. Deeper muscles can generally be seen by moving the more superficial muscles off to the side.

Once you locate the muscle, tug gently on it to locate its **origin** and **insertion.** If you pull too hard, you may rip the muscle. The outer wrapping of the muscle is the **fascia,** and it should be removed to find the fiber direction of the muscle. The main part of the muscle is the **belly.** You may need to cut a muscle to locate deeper muscles. This is done by **transecting** the muscle, which is to cut the muscle into two sections perpendicular to the fiber direction. After transecting the muscle, you may want to **reflect** it, or pull it toward its attachment site. When separating two muscles, you may find a cottonlike material between them. This is loose connective tissue that forms part of the fascia.

Pectoral Muscles of the Cat

Four major muscles are seen in a superficial view of the pectoral region. These are the pectoantebrachialis, pectoralis major, pectoralis minor, and xiphihumeralis. The **pectoantebrachialis** has no corresponding muscle in the human. It originates on the sternum and inserts on the forelimb. Transect and reflect the pectoantebrachialis to see the pectoralis major. The **pectoralis major** originates on the sternum and inserts on the upper humerus. The **pectoralis minor** is a large muscle in cats and inserts on the upper humerus. The **xiphihumeralis** is another cat muscle that has no corresponding human muscle; it originates on the sternum and inserts on the proximal humerus along with the pectoralis major and minor. Locate these muscles on your cat and in figure 12.19.

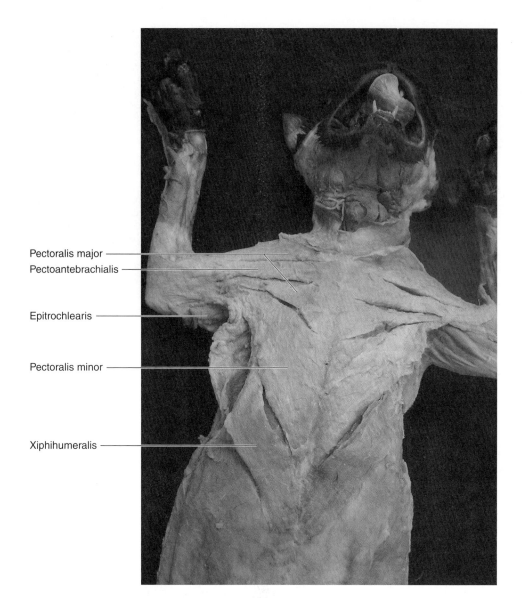

Figure 12.19 Muscles of the Pectoral Region of the Cat

Shoulder Muscles

In humans, there is a single trapezius on each side of the body. In cats, each trapezius consists of three muscles. Examine the cat from the dorsal side and locate the **clavotrapezius,** the **acromiotrapezius,** and the **spinotrapezius.** All of these muscles originate on the vertebral column, with the clavotrapezius also originating on the occipital bone. The clavotrapezius inserts on the clavicle, the acromiotrapezius on the acromion process of the scapula, and the spinotrapezius on the spine of the scapula. Compare these muscles with figure 12.20.

Two other muscles of the region are the **latissimus dorsi** and the **levator scapulae ventralis.** The **levator scapulae** is deep to the trapezius. You will need to cut a part of the trapezius known as the clavotrapezius in order to see the levator scapulae. In cats, there is an additional muscle called the levator scapulae ventralis, which inserts on the scapular spine. Examine figure 12.20 for the levator scapulae muscles. These two muscles are similar to those in the human. Locate these muscles on the cat and compare them with figure 12.20.

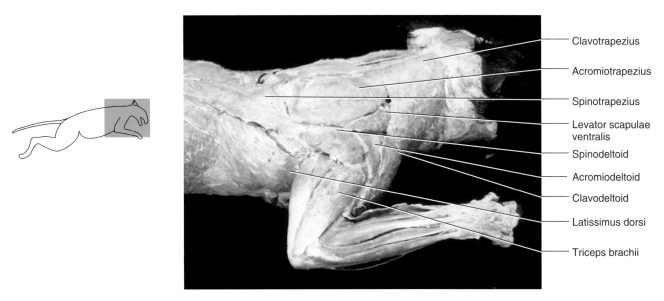

Figure 12.20 Superficial Muscles of the Right Shoulder of the Cat

In humans, there is a single deltoid muscle, but in cats the deltoid consists of three muscles, as does the trapezius. Like the trapezius, the deltoid muscles are named for their bony attachments. The **clavodeltoid** (clavobrachialis) originates on the clavicle, the **acromiodeltoid** on the acromion process, and the **spinodeltoid** on the spine of the scapula. Insertions of these muscles are on the arm or forelimb. You will need to dissect the trapezius carefully to see the **rhomboideus** muscles. These muscles originate on the vertebral column and insert on the scapula. Locate these muscles on the cat and compare them with figures 12.20 and 12.21.

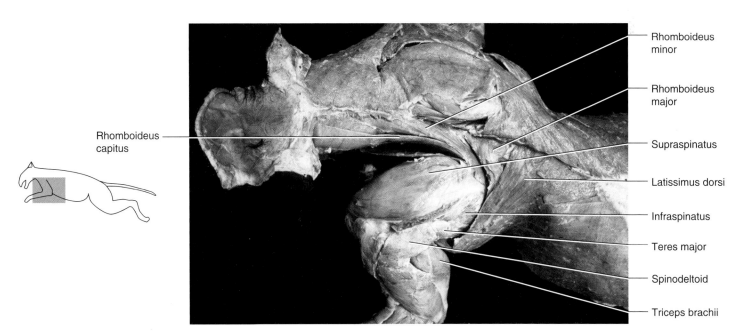

Figure 12.21 Deep Muscles of the Left Scapula of the Cat

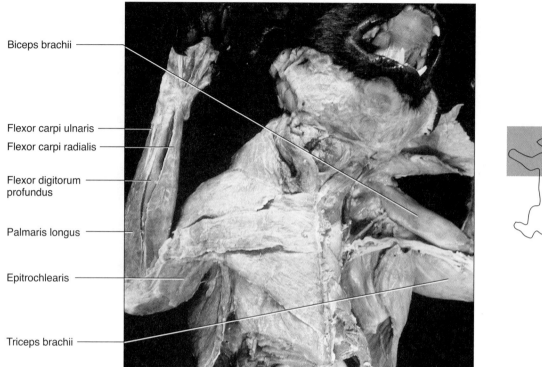

Figure 12.22 Muscles of the Proximal Forelimb of the Cat, Ventral View

Scapular Muscles That Act on the Humerus

The deep muscles of the scapula can be seen by reflecting the overlying muscles. The **supraspinatus, infraspinatus, subscapularis, teres major,** and **teres minor** are roughly equivalent to the same muscles in the human. Locate the supraspinatus, infraspinatus, and teres major and find them in figure 12.21.

Forelimb Muscles

The muscles that have an action on the forelimb of the cat typically either flex or extend the forelimb as their primary action. The **epitrochlearis** does not have a corresponding muscle in humans (figure 12.22). The epitrochlearis is on the medial side of the humerus, and it extends the forelimb. The **biceps brachii** is also a medial muscle, and it flexes the forelimb. The **triceps brachii, anconeus, brachioradialis,** and **brachialis** are lateral or posterior muscles. The triceps brachii and the anconeus extend the forelimb, whereas the brachialis flexes the forelimb. Find the lateral muscles, using figures 12.22 and 12.23 as a guide.

Superficial Muscles on the Medial Aspect of the Forelimb

Most of the muscles of the forelimb of the cat run parallel to the radius and ulna. An exception to this is the **pronator teres,** which is a small slip of muscle that runs obliquely down the forelimb. It pronates the wrist. The **palmaris longus** is a broad, flat muscle, superficially located on the forelimb with insertions into the digits. In humans, the palmaris longus terminates at the palmar fascia.

The **flexor carpi radialis** is a thin muscle inserting on the second and third metacarpals. It is named for its action, its insertion, and its location. The **flexor carpi ulnaris** has an origin on the humerus and ulna and an insertion on the medial metacarpals and carpals. The **flexor digitorum superficialis (flexor digitorum sublimis)** is located in the forelimb as a middle-level muscle, underneath the palmaris longus. It originates on fascia of other forearm muscles, as opposed to originating on bone. Examine the superficial muscles of the medial forelimb and compare them with figure 12.22.

Deep Muscles on the Medial Aspect of the Forelimb

Underneath the upper layer of muscles are many muscles with varied actions. The **supinator** is a deep muscle that runs diagonally from the lateral epicondyle of the humerus to the proximal radius. It is the deepest of the proximal forelimb muscles. The **flexor digitorum profundus** is an extensive muscle that inserts on the first through fifth digits. It replaces the flexor pollicis longus for the thumb flexion, since this muscle is absent in cats. The **pronator quadratus** is a square muscle located between the radius and ulna deep to the flexor digitorum profundus. Find the deep muscles in the cat and compare them with figure 12.24.

Muscles on the Lateral Aspect of the Forelimb

The lateral muscles of the forelimb are the extensor group; the **extensor carpi radialis longus** muscle is deep to the brachioradialis and inserts on the second metacarpal. The **extensor carpi radialis brevis** is underneath the extensor carpi radialis longus and inserts on the third metacarpal.

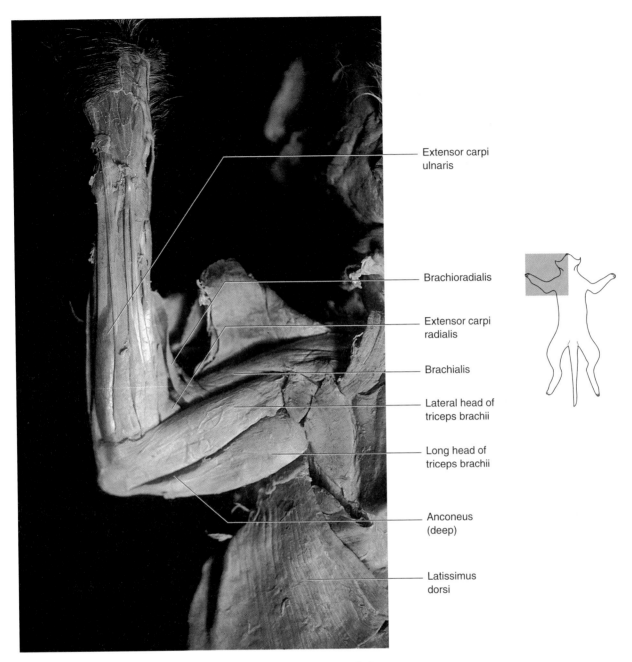

Figure 12.23 Muscles of the Left Proximal Forelimb of the Cat, Dorsal View

Locate the **extensor carpi ulnaris,** which is next to the **extensor digitorum lateralis,** inserting on the fifth metacarpal. The **extensor digitorum communis** can be located on the lateral aspect of the forelimb. It is a broad muscle, inserting by tendons on the second through fifth digits. Locate these muscles on the cat and in figure 12.25. The **extensor digitorum lateralis** is specific to the cat; it inserts with the tendons of the extensor digitorum communis to the digits. The **extensor pollicis brevis** is a well-developed muscle in the cat, whereas the abductor pollicis longus is absent in cats.

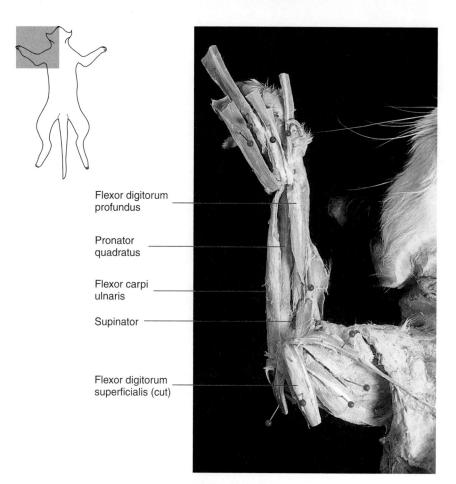

Figure 12.24 Deep Muscles of the Right Medial Forelimb of the Cat

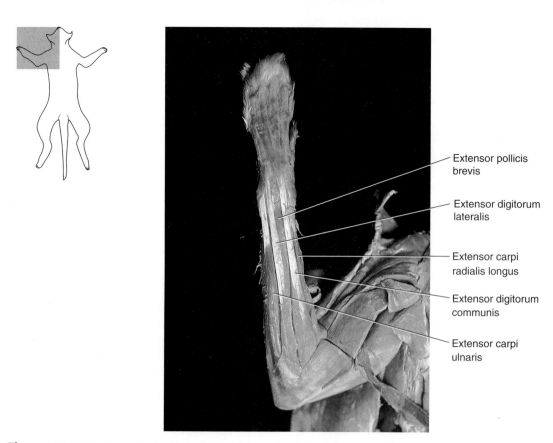

Figure 12.25 Lateral Muscles of the Left Forelimb of the Cat

Exercise 12 Review

Introduction to the Study of Muscles
and Muscles of the Shoulder
and Upper Extremity

Name: _____
Lab time/section: _____
Date: _____

Answer the following questions in terms of human muscles.

1. What is the action of the deltoid muscle? _____

2. Name the origin of the supraspinatus muscle. _____

3. What is the insertion of the pectoralis minor? _____

4. What is the origin of the levator scapulae? _____

5. What is the insertion of the latissimus dorsi? _____

6. Does the biceps brachii muscle originate or insert on the humerus? _____

7. What is the origin of the trapezius? _____

8. The pectoralis major has what action? _____

9. What is the insertion of the subscapularis? _____

10. What is the action of the triceps brachii? _____

11. What is the origin of the brachialis? _____

12. Name all the muscles that flex the arm. _____

13. Which muscles are antagonists of the triceps brachii? _____

14. Label the muscles in the following illustration using the terms provided.

   brachialis   biceps brachii   pectoralis major
   triceps brachii   deltoid   brachioradialis

   a. _____
   b. _____
   c. _____
   d. _____
   e. _____
   f. _____

15. What is the origin of the flexor carpi ulnaris? _____

16. What is an antagonist to the supinator muscle? _____

17. Where does the flexor digitorum superficialis insert? _____

18. What is the insertion of the extensor carpi ulnaris muscle? _____

19. What is the action of the flexor carpi radialis muscles? _____

20. What muscle extends both the hand and the phalanges? _____

21. What muscles flex the hand? _____

22. What muscles extend the thumb? _____

23. Where are the extensor carpi muscles located, on the anterior or posterior side of the forearm? _____

Exercise 12 Introduction to the Study of Muscles and Muscles of the Shoulder and Upper Extremity   **187**

24. Label the muscles or structures in the following illustration using the terms provided.

    flexor carpi ulnaris   pronator teres   palmar aponeurosis
    flexor carpi radialis   palmaris longus

    a. _____
    b. _____
    c. _____
    d. _____

    e. _____

25. What muscles do you use when you throw a baseball? Mimic the action slowly and try to feel what muscles contract.
    _____

26. What muscles do you use when you do a pushup? _____

27. What muscles do you use when you use a hammer to hit a nail? _____

Answer the following questions in terms of cat muscles.

28. What are the three muscles that correspond to the single trapezius in humans? _____
    _____

29. What is the action of the epitrochlearis? _____

30. Which muscle does not correspond to a human muscle? Circle one.

   a. deltoid

   b. pectoralis minor

   c. xiphihumeralis

   d. latissimus dorsi

31. How does the deltoid of the cat differ from the deltoid of the human? _____

32. Which muscle is more developed in cats, the flexor digitorum superficialis or the flexor digitorum profundus? _____
_____

Exercise 13
Muscles of the Hip, Thigh, Leg, and Foot

INTRODUCTION

The muscles of the hip, thigh, leg, and foot are primarily muscles of locomotion. They can be grouped generally into muscles that have an action on the thigh, the leg, or the foot. These muscles are discussed in the *Principles of Anatomy and Physiology* text in chapter 9, "Gross Anatomy and Functions of Skeletal Muscles."

The size and shape of the muscles of the hip and thigh in humans are different from those of other mammals in that humans are bipedal. This two-legged walking habit is much different from a quadruped's locomotion in terms of balance and forward movement. When a quadruped walks, typically two legs remain on the ground at a time. When a biped walks, the inherent instability of standing on one leg brings into focus the importance of stabilizing the support limb and maintaining the center of balance. Think about what happens to the mechanics of movement and the center of balance when you lift a leg to walk or run.

The muscles of the leg originate typically from the femur, tibia, or fibula and insert on the bones of the foot. Anterior and posterior actions on the thigh and leg are flexion or extension. Those for the foot are described as **dorsiflexion** (decreasing the angle between the shin and dorsum or top of the foot) or **plantar flexion** (an action that is seen as you stand on your toes to reach something high). Review the other actions of the leg and foot in Exercise 10 in this lab manual. Some of the muscles of the feet are similar to those of the hand, such as the extensor digitorum muscles.

OBJECTIVES

At the end of this exercise, you should be able to

1. locate the muscles of the hip, thigh, leg, and foot on a model, chart, cadaver (if available), or cat;
2. list the origin, insertion, and action of each muscle presented;
3. describe what nerve controls each muscle;
4. name all the muscles that have a specific action at a joint, such as all of the muscles that flex the thigh or those that dorsiflex the foot.

MATERIALS

Human torso model
Human leg models
Human muscle charts
Articulated skeleton
Cadaver (if available)
Cat
Materials for cat dissection
    Dissection trays
    Scalpel with two or three extra blades
    Gloves (household latex gloves work well for repeated use)
    Blunt (Mall) probe
    String and tags
    Pins
    Forceps and sharp scissors
    First aid kit in lab or prep area
    Sharps container
    Animal waste disposal container

PROCEDURE

Review the muscle nomenclature and the action of muscles as outlined in Exercises 10 and 12. Examine muscle models or charts in the lab and locate the muscles described in the following section and in table 13.1. Correlate the shape of the muscle with the name and study the models or charts. You may want to look at an articulated skeleton as you review the origins and insertions of the muscles, so that you can better see the muscle attachment points. Once you have learned the origin and the insertion, it should be easier to understand the action of the muscle. This is done, in part, by imagining how the bony attachments of the origin and insertion would come together if the muscle pulled them closer to one another. If you have a cadaver, pay particular attention to the origins and the insertions of the muscles of the hip, leg, and foot.

Human Muscles

Muscles of the Hip and Thigh

The **iliopsoas** muscle is a major flexor of the thigh. It originates inside the abdominopelvic cavity and crosses over the pelvic brim to insert on the posterior aspect of the femur. The iliopsoas is actually two muscles, the **iliacus** and **psoas major.** A companion muscle is the **psoas minor.** This exercise treats these muscles as one.

The **sartorius** (*sartorius* = tailor) is a slender muscle that originates in a superior, lateral position and inserts in an inferior, medial location. It allows you to sit cross-legged and is thus named the "tailor's muscle." The sartorius is the most superficial muscle on the anterior thigh. The most superficial muscle of the medial aspect of the thigh is the gracilis. The **gracilis** is a thin muscle that adducts the thigh. Locate these muscles in the lab and in figure 13.1.

Adductor muscles are located in the medial aspect of the thigh. The gracilis is an adductor muscle, as are the pectineus and

TABLE 13.1 Muscles of the Hip, Thigh, Leg, and Foot

Name	Origin	Insertion	Action	Innervation
Muscles of the Medial Thigh				
Iliopsoas	T12, L1–5, and iliac fossa	Lesser trochanter of femur	Flexes thigh and lumbar vertebrae	Femoral nerve, spinal nerves L1–3
Sartorius	Anterior superior iliac spine	Proximal, medial tibia	Flexes and laterally rotates thigh, flexes leg	Femoral nerve
Gracilis	Pubis, near symphysis	Proximal, medial tibia	Adducts thigh, flexes leg	Obturator nerve
Pectineus	Pubic crest	Pectineal line of femur	Adducts thigh	Femoral and obturator nerves
Adductor brevis	Pubis	Linea aspera at proximal portion of femur	Adducts, flexes, and laterally rotates thigh	Obturator nerve
Adductor longus	Pubis	Linea aspera at middle portion of femur	Adducts, flexes, and laterally rotates thigh	Obturator nerve
Adductor magnus	Pubis, ischium	Linea aspera, adductor tubercle of distal femur	Adducts, extends, and laterally rotates thigh	Obturator and tibial nerves
Muscles of the Anterior Thigh				
Quadriceps Femoris Muscles				
Rectus femoris	Anterior inferior iliac spine	Tibial tuberosity by the patellar ligament	Flexes thigh, extends leg	Femoral nerve
Vastus lateralis	Greater trochanter of femur, linea aspera of femur	Tibial tuberosity by the patellar ligament	Extends leg	Femoral nerve
Vastus intermedius	Proximal, anterior femur	Tibial tuberosity by the patellar ligament	Extends leg	Femoral nerve
Vastus medialis	Linea aspera of femur, on medial side	Tibial tuberosity by the patellar ligament	Extends leg	Femoral nerve
Muscles of the Posterior and Lateral Thigh				
Tensor fasciae latae	Anterior superior iliac spine	Iliotibial band of fascia lata to lateral condyle of tibia	Flexes, abducts and medially rotates thigh	Superior gluteal nerve
Gluteus maximus	Outer iliac blade, iliac crest, sacrum, coccyx	Gluteal tuberosity of femur, iliotibial band of fascia lata	Extends, abducts, and laterally rotates thigh; braces knee	Inferior gluteal nerve
Gluteus medius	Outer iliac blade	Greater trochanter of femur	Abducts and medially rotates thigh	Superior gluteal nerve
Gluteus minimus	Outer iliac blade	Greater trochanter of femur	Abducts and medially rotates thigh	Superior gluteal nerve
Hamstrings				
Biceps femoris	Ischial tuberosity, linea aspera of femur	Head of fibula, lateral condyle of tibia	Extends thigh, flexes leg	Tibial and fibular nerves
Semitendinosus	Ischial tuberosity	Proximal, medial tibia	Extends thigh, flexes leg	Tibial nerve
Semimembranosus	Ischial tuberosity	Medial condyle of tibia, collateral ligament	Extends thigh, flexes leg	Tibial nerve
Muscles of the Anterior Leg				
Tibialis anterior	Tibia, interosseous membrane	Metatarsal 1 and medial cuneiform	Dorsiflexes and inverts foot	Deep fibular nerve
Extensor digitorum longus	Lateral condyle of tibia, shaft of fibula	Middle and distal phalanges of digits 2–5	Extends toes 2–5, dorsiflexes and everts foot	Deep fibular nerve
Extensor hallucis longus	Anterior shaft of fibula, interosseous membrane	Distal phalanx of hallux (big toe)	Extends hallux, dorsiflexes foot, inverts foot	Deep fibular nerve
Muscles of the Anterior Leg				
Fibularis longus	Head and shaft of fibula, lateral condyle of tibia	Metatarsal 1, medial cuneiform	Plantar flexes and everts foot	Superficial fibular nerve
Fibularis brevis	Shaft of fibula	Metatarsal 5	Plantar flexes and everts foot	Superficial fibular nerve
Fibularis tertius	Fibula, interosseous membrane	Dorsum of metatarsal 5	Dorsiflexes and everts foot	Deep fibular nerve

TABLE 13.1 Continued

Name	Origin	Insertion	Action	Innervation
Muscles of the Posterior Leg				
Gastrocnemius	Condyles of femur	Calcaneus by the calcaneal tendon	Flexes leg, plantar flexes foot	Tibial nerve
Soleus	Posterior, proximal tibia; fibula	Calcaneus by the calcaneal tendon	Plantar flexes foot	Tibial nerve
Popliteus	Lateral condyle of femur	Proximal tibia	Flexes leg	Tibial nerve
Tibialis posterior	Interosseous membrane, tibia, fibula	Metatarsals 2–5, navicular, cuneiforms, cuboid	Plantar flexes and inverts foot	Tibial nerve
Flexor digitorum longus	Posterior tibia	Distal phalanges of digits 2–5	Flexes toes 2–5, plantar flexes and inverts foot	Tibial nerve
Flexor hallucis longus	Midshaft of fibula	Distal phalanx of hallux (big toe)	Flexes hallux, plantar flexes and inverts foot	Tibial nerve

the three adductor muscles proper. The **pectineus** is a small, triangular muscle that is the most proximal of the group. It is superficial to the adductor brevis muscle and proximal to the adductor longus muscle. Examine the material in the lab and compare the pectineus with figures 13.1 and 13.2.

The **adductor brevis** is the shortest adductor muscle; it is deep in the medial aspect of the thigh. It can be seen if the pectineus is reflected. The adductor brevis originates on the pubis and inserts on the proximal third of the thigh. The **adductor longus** also has a pubic origin, but its insertion is in the middle third of the thigh. The largest of the adductor muscles is the **adductor magnus,** and it originates along the length of the inferior hip bone (pubis and ischium) and inserts along the length of the femur. Examine these muscles in figure 13.2.

The muscles of the anterior aspect of the thigh belong to the **quadriceps femoris** group. All these muscles extend the leg, whereas only one member of the group flexes the thigh. The quadriceps muscles all have a common ligament of insertion called the patellar ligament. This ligament inserts on the tibial tuberosity. The most superficial muscle of the group is the rectus

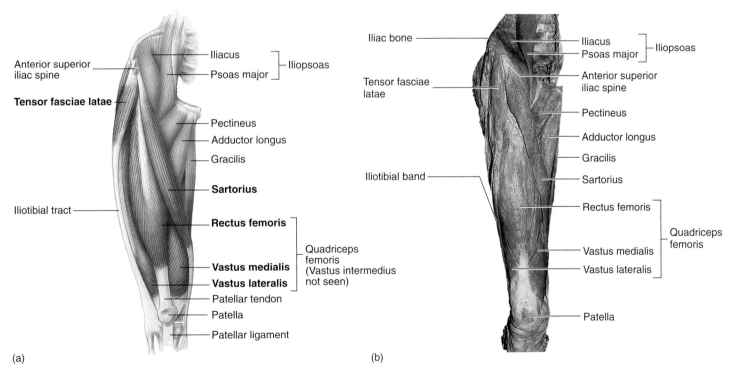

Figure 13.1 Muscles of the Right Thigh, Anterior View
(a) Diagram; (b) photograph.

femoris (figures 13.1 and 13.3). It originates directly inferior to the sartorius. The **rectus femoris** runs straight down the femur (*rectus* = straight). Because it crosses the hip joint, the rectus femoris is the only quadriceps muscle to flex the thigh.

The three other quadriceps muscles are the vastus muscles. The **vastus lateralis** is so named due to its lateral position. The **vastus intermedius** is deep to the rectus femoris, and the **vastus medialis** is the most medial of the vastus muscles. These muscles originate on the femur, so they do not have an action on the hip. They insert on the tibia and extend the leg. Locate these muscles in the lab and compare them with figures 13.1 and 13.3. Palpate the quadriceps femoris by extending your leg.

The **tensor fasciae latae** is a lateral muscle of the thigh that attaches to a thick band of connective tissue (the iliotibial band) that runs down the lateral aspect of the thigh like a stripe on a pair of tuxedo pants. This muscle abducts the thigh (figure 13.1).

The gluteus muscles in humans are different from those in other mammals, due to our walking upright. The largest of the gluteal muscles is the **gluteus maximus.** It extends the thigh when we stand up from a sitting position or when we climb stairs. The **gluteus medius** and the **gluteus minimus** both help keep our balance during walking in that they maintain the center of gravity. The gluteus medius is superficial to the gluteus minimus, but both of these muscles originate on the outer iliac blade. When we raise one leg to walk, gravity pulls the body in the direction of the lifted leg. The gluteus medius and minimus on the opposite side contract to maintain upright posture. The lateral rotators of the thigh are the **piriformis, superior gemellus, obturator internus, obturator exernus, inferior gemellus,** and the **quadratus femoris.** Compare these muscles with figure 13.4.

The **hamstring muscles** are so named because of an old practice of taking the ham muscles from a pig and tying them together by their tendons (the hamstrings) to hang in a smokehouse to cure. Three muscles make up the hamstrings: the biceps femoris, the semitendinosus, and the semimembranosus. All three of these muscles cross the hip joint and extend the thigh when walking and flex the leg.

The **biceps femoris** is the "two-headed muscle of the femur." One head originates with the other hamstring muscles on the ischial tuberosity, whereas the second head originates on the femur. The biceps femoris inserts laterally on the proximal leg. The **semitendinosus** also originates on the ischial tuberosity, and

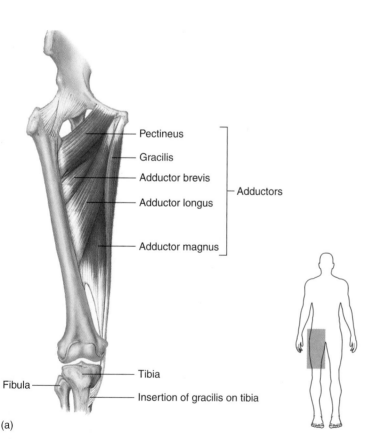

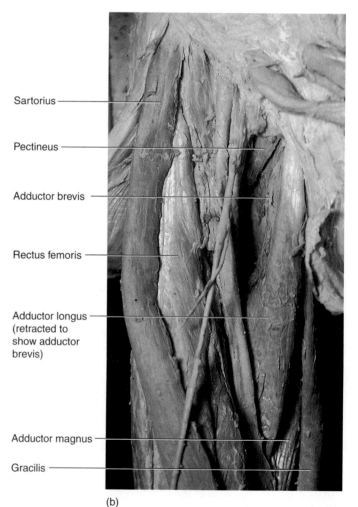

Figure 13.2 Adductor Muscles of the Right Thigh, Anterior View
(a) Diagram; (b) photograph.

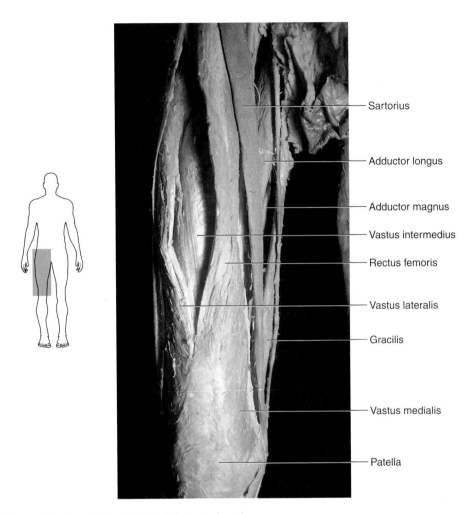

Figure 13.3 Quadriceps Muscles of the Right Thigh, Anterior View

it can be distinguished as a superficial muscle with a long, pencil-like distal tendon. As opposed to the biceps femoris, the semitendinosus inserts medially. The **semimembranosus** originates on the ischial tuberosity as well and has a broad, flat, membranous tendon on the proximal part of the muscle. The semimembranosus inserts medially on the leg, deep to the semitendinosus. Locate these muscles on the material in lab and compare them with figure 13.5.

Muscles of the Leg and Foot

Anterior Muscles

The **tibialis anterior** is located just lateral to the crest of the tibia on the anterior side of the leg. The tendon of the tibialis anterior crosses to the medial side of the foot and inserts on the first metatarsal and medial cuneiform. As the tibialis anterior contracts, it decreases the angle between the anterior tibial crest and the dorsum of the foot in an action known as dorsiflexion of the foot. Because the insertion of this muscle is medial, it also inverts the foot. Palpate the tibialis anterior, which is lateral to the tibia.

The **extensor digitorum longus** muscle is lateral to the tibialis anterior, and the tendons of this muscle splay out and insert on the middle and distal phalanges of all of the digits of the foot except the hallux. The **extensor hallucis longus** is the muscle that extends the hallux, as well as acting as a synergist to the extensor digitorum longus in dorsiflexion of the foot. Examine these muscles in the lab and compare them with figure 13.6.

The **fibularis (peroneus)** muscles are named for originating on and running the length of the fibula. The **fibularis longus** has a tendon that travels from the lateral side of the foot underneath to the medial side, crossing under the arch of the foot. The **fibularis brevis** parallels the fibularis longus, except that the tendon stops short and inserts on the lateral side of the foot at the fifth metatarsal. Both of these muscles have tendons that hook posterior to the lateral malleolus, and they both plantar flex and evert the foot. The third fibularis muscle, the **fibularis tertius,** does not arch behind the lateral malleolus, and it inserts on the dorsum of the fifth metatarsal. When it contracts, it dorsiflexes and everts the foot. Examine these muscles in figure 13.7.

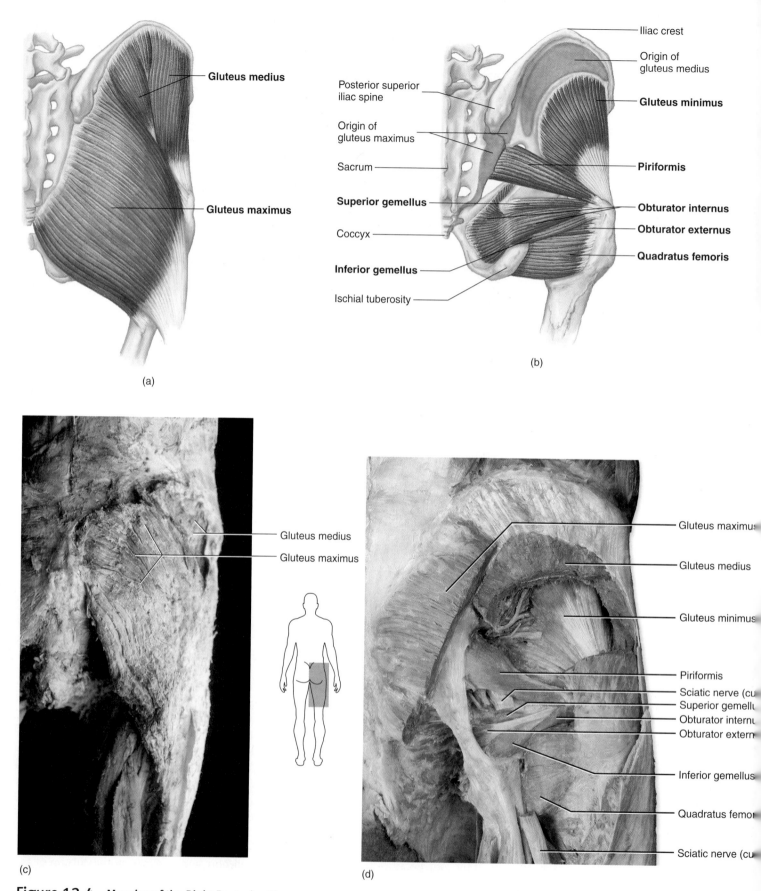

Figure 13.4 Muscles of the Right Posterior Hip
Diagram (a) superficial muscles; (b) deep muscles. Photograph (c) superficial muscles; (d) deep muscles.

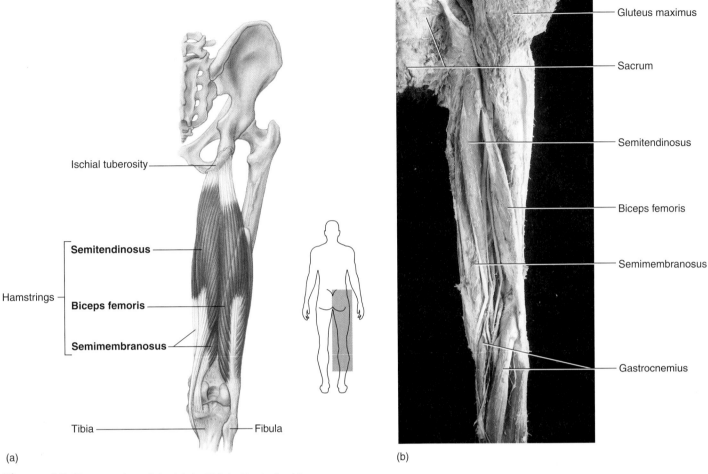

Figure 13.5 Muscles of the Right Thigh, Posterior View
(a) Diagram; (b) photograph.

Posterior Muscles

The **gastrocnemius** is a calf muscle that is the most superficial of the posterior leg group. The gastrocnemius crosses the knee joint, so it flexes the leg. It inserts on the calcaneus by way of the **calcaneal tendon,** plantar flexing the foot as well. The calcaneal tendon is also known as the Achilles tendon. Deep to the gastrocnemius is the **soleus** muscle. Unlike the gastrocnemius, the soleus does not originate on the femur; therefore, it does not cross the knee and has no action on the leg. It inserts on the calcaneus, sharing the calcaneal tendon with the gastrocnemius, and it plantar flexes the foot. Palpate the juction between the gastrocnemius and the soleus. The **popliteus** is a small muscle that crosses and unlocks the knee joint. Locate these muscles in figures 13.7 and 13.8. The **tibialis posterior** is a major muscle deep to the soleus that plantar flexes and inverts the foot. The tendon of the muscle runs along the medial aspect of the ankle. Near the tibialis posterior is the **flexor digitorum longus,** which is a muscle that inserts on the distal phalanges of all the digits of the foot except the hallux. The flexor digitorum longus has the action of flexing (curling) the toes in addition to plantar flexing and inverting the foot. The **flexor hallucis longus** flexes the hallux and aids the flexor digitorum longus in inverting the foot. Locate these muscles in the lab and compare them with figure 13.8.

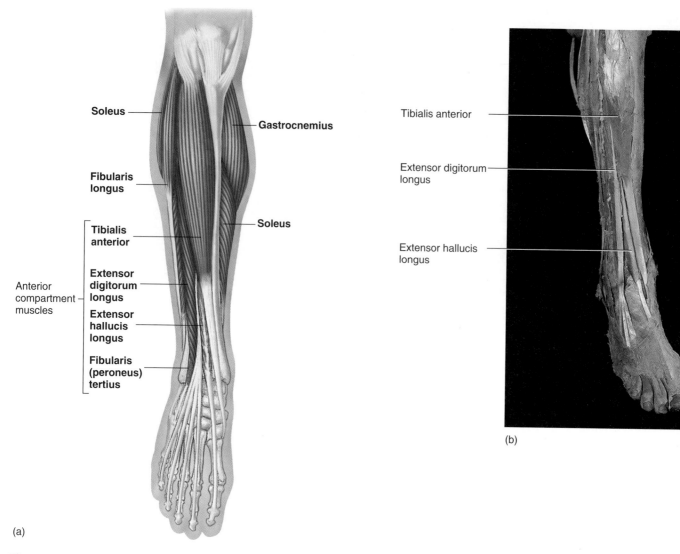

Figure 13.6 Muscles of the Right Leg and Foot, Anterior View
(a) Diagram; (b) photograph.

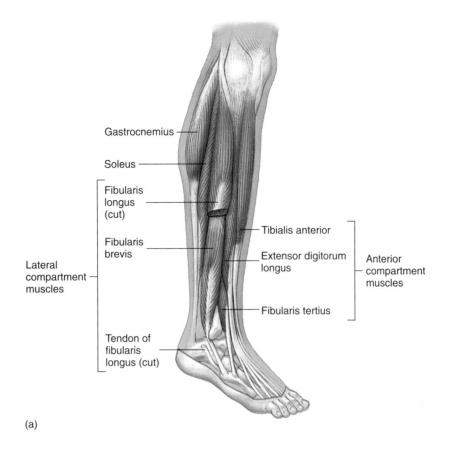

(a)

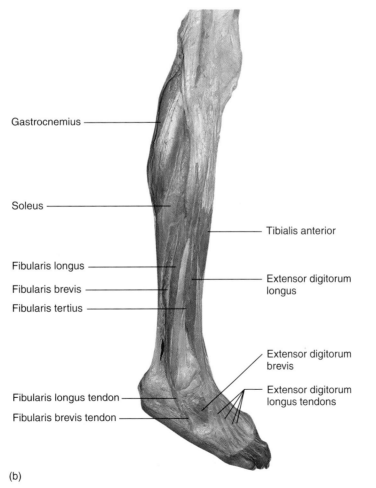

(b)

Figure 13.7 Muscles of the Right Leg and Foot, Lateral View
(a) Diagram; (b) photograph.

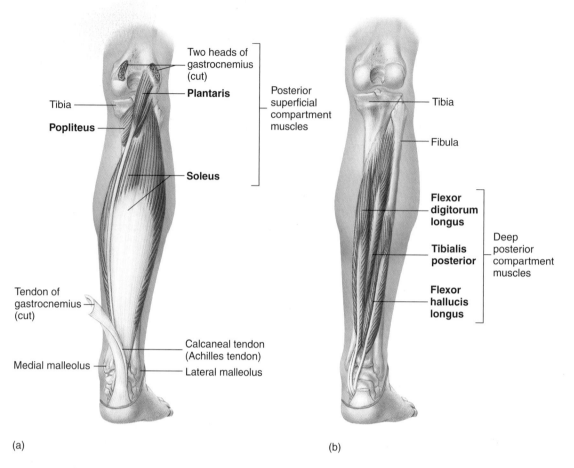

Figure 13.8 Muscles of the Right Leg and Foot, Posterior View

Diagram (a) superficial; (b) deep; (c) photograph.

Cat Dissection

If you have not done so already, remove the skin from the lower portion of the hind limb of the cat. Be careful cutting the skin from the distal portions of the leg so that you do not cut through the tendons of the foot. Be careful in the dissection of the cat muscles that you *leave the blood vessels and nerves intact* as you dissect your specimen. You will be looking at these structures in later exercises. Pay particular attention to the **great saphenous vein,** which runs under the skin and should be preserved. Remove any excess fat and fascia from the muscles as you dissect the material. If you need to cut a muscle, bisect it perpendicular to the fiber direction, so that half of it is attached to the origin side and half of it is attached to the insertion side.

Cat Musculature

The cat has many muscles of the hip and thigh that are good models for studying human muscles. The quadrupedal nature of the cat lends itself to having a different size or placement of some of the thigh muscles, which are illustrated in this section.

Exercise 13 Muscles of the Hip, Thigh, Leg, and Foot

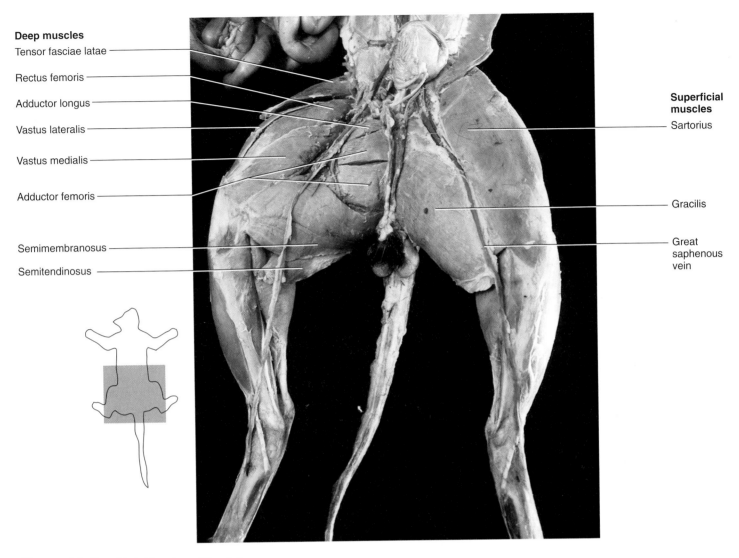

Figure 13.9 Thigh Muscles of the Cat
Superficial muscles (left side of cat); deep muscles (right side of cat).

Medial Muscles of the Thigh of the Cat

Two major muscles on the medial aspect of the thigh in the cat are the **sartorius** and the **gracilis** muscles. Locate these muscles in figure 13.9 and note that they are much broader in the cat than in the human.

Cut through the sartorius and gracilis and reflect the ends of these muscles. You should see the deeper muscles of the thigh, including the **tensor fasciae latae,** the **vastus medialis,** the **adductor femoris** (a large muscle specific to the cat), and the **semimembranosus.** The distal portion of the **semitendinosus** may also be seen from this view. Examine figure 13.9 for the deep muscles of the thigh and locate the **rectus femoris, vastus medialis,** and **vastus lateralis.** Bisect the rectus femoris in order to see the **vastus intermedius** muscle. The **adductor longus** is a thin muscle anterior to the **adductor femoris,** and it can be seen as a small muscle on the medial aspect of the thigh.

Lateral Muscles of the Thigh of the Cat

On the lateral aspect of the thigh are numerous muscles, including the **biceps femoris,** the **tensor fasciae latae,** the **gluteus** muscles, and a muscle specific to the cat, the **caudofemoralis.** The biceps femoris is the largest and most lateral muscle of the thigh. The **gluteus medius** is larger than the **gluteus maximus** in cats due to the lengthening of the pelvic girdle. The **semimembranosus** is a large muscle in cats (much larger than the semitendinosus), and it can be seen from the lateral side of the cat. The adductor magnus and adductor brevis are not found in the cat. Examine these muscles in figures 13.10 and 13.11.

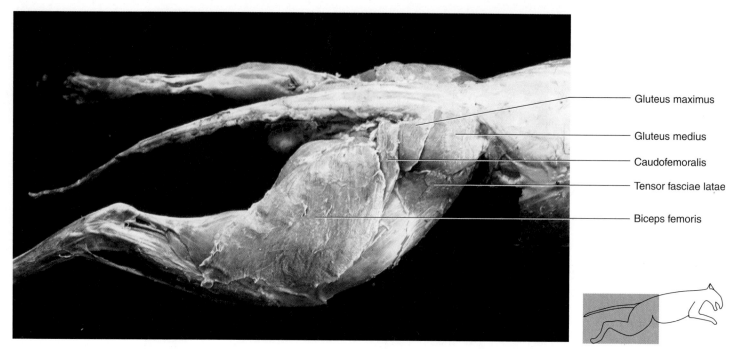

Figure 13.10 Lateral Thigh Muscles of the Cat
Superficial muscles of the right side.

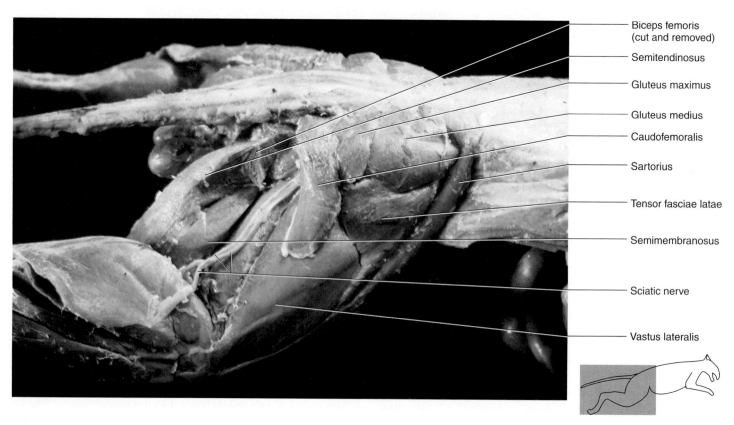

Figure 13.11 Lateral Thigh Muscles of the Cat
Deep muscles of the right side.

Exercise 13 Muscles of the Hip, Thigh, Leg, and Foot

Posterior Muscles

Examine the large **gastrocnemius** on the posterior aspect of the leg. The **soleus** is deeper to the gastrocnemius and inserts with the gastrocnemius on the calcaneus. In cats the soleus has only one point of origin, the fibula; in humans the soleus originates on the tibia and fibula. Compare figure 13.12 with your dissection. Cut through the calcaneal tendon and lift the gastrocnemius and soleus so that you can study the underlying muscles.

The **popliteus** is a small, triangular muscle that crosses the knee joint. Do not damage the nerves and blood vessels that pass over the popliteus because you will study them in the subsequent exercises.

The **flexor digitorum longus** runs along the medial side of the hind limb and flexes the digits of the cat. It joins with the **flexor hallucis longus,** which inserts on all the digits of the hind limb. The **tibialis posterior** is a narrow muscle that inserts on the tarsal bones of the foot. Locate these muscles in the cat and compare them with figure 13.13.

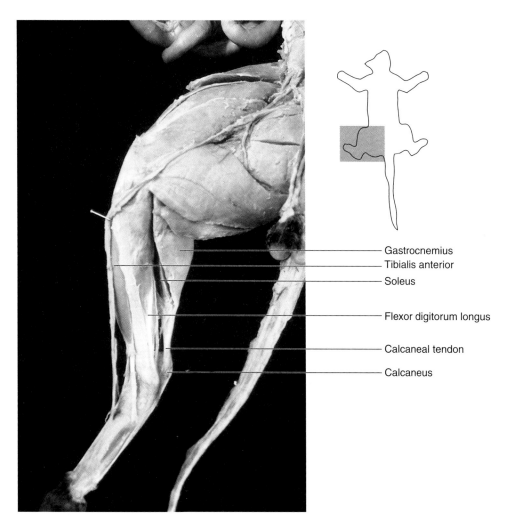

Figure 13.12 Superficial Muscles of the Right Leg of the Cat, Medial View

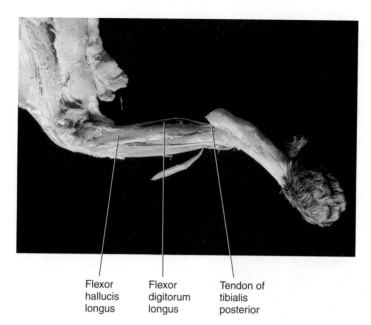

Figure 13.13 Deep Muscles of the Left Leg of the Cat, Lateral View

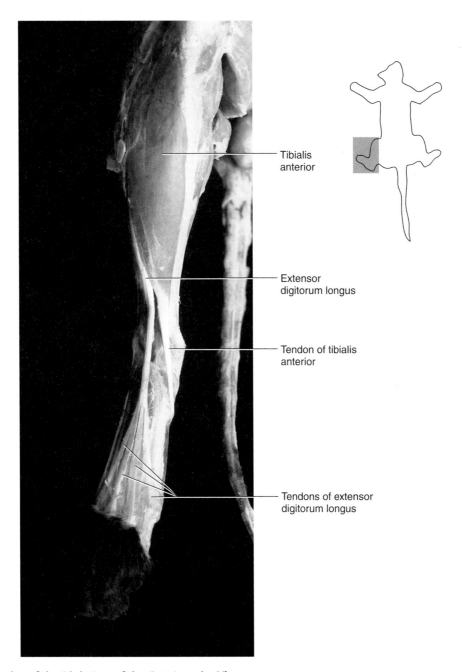

Figure 13.14 Muscles of the Right Leg of the Cat, Anterior View

Anterior Muscles

The **tibialis anterior** is a large muscle of the lower limb; it inserts on the dorsum of the foot. The **extensor digitorum longus** originates on the femur in cats and inserts on the distal phalanges in all of the digits in the cat. The extensor hallucis longus is not found in cats. These muscles can be seen in figure 13.14. Examine these muscles and isolate them in your dissection.

The **fibularis longus** extends along the length of the fibula with the **fibularis brevis.** The fibularis longus inserts at the base of the metatarsals, whereas the fibularis brevis inserts on the fifth metatarsal. The **fibularis tertius** inserts into the tendon of the extensor digitorum muscle. These muscles can be seen in figure 13.15.

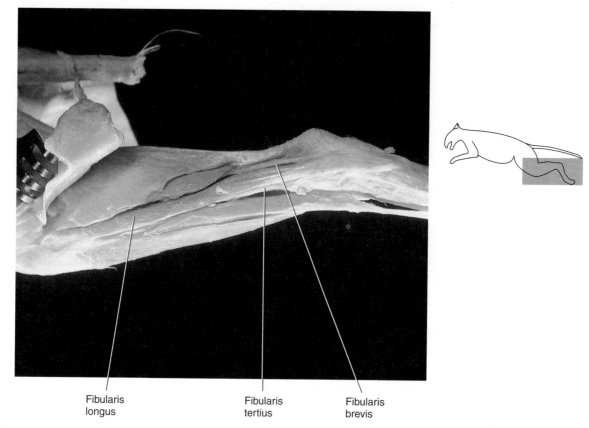

Figure 13.15 Muscles of the Left Leg of the Cat, Lateral View

Exercise 13 Review

Muscles of the Hip, Thigh, Leg, and Foot

Name: _____

Lab time/section: _____

Date: _____

1. If you were to ride a horse, what muscles would you use to keep your seat out of the saddle as you rode? _____
_____

2. How do the gluteus medius and gluteus minimus prevent you from toppling over as you walk? _____
_____

3. What is a muscle that is an antagonist to the biceps femoris muscle? _____

4. What are two muscles that are synergists with the biceps femoris muscle? _____
_____

5. What is the insertion of all the muscles of the quadriceps group? _____
_____

6. How does the action of the rectus femoris differ from that of the other quadriceps muscles? _____

7. How many adductor muscles are there? _____

8. List two muscles in this exercise that are responsible for thigh flexion. _____
_____

9. Where do the hamstring muscles originate as a group? _____

10. What is the action of the vastus lateralis? _____
_____

11. Which muscle group is located on the anterior part of the thigh? _____

12. Is abduction of the thigh movement away from or toward the midline? _____
_____

13. What muscle flexes the lumbar vertebrae as part of its action? _____
_____

14. Label the muscles in the following illustration using the terms provided.

    iliopsoas   sartorius   vastus lateralis
    vastus medialis   gracilis   tensor fasciae latae
    rectus femoris   pectineus

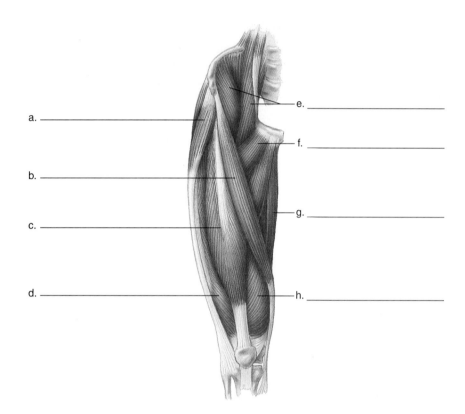

a. _____
b. _____
c. _____
d. _____
e. _____
f. _____
g. _____
h. _____

15. What is the origin of the gastrocnemius? _____

16. What is the insertion of the tibialis anterior in humans? _____

17. How does the action of the fibularis longus in humans differ from that of the fibularis tertius? _____
_____

18. What is the action of the extensor hallucis longus? _____

Exercise 13 Muscles of the Hip, Thigh, Leg, and Foot

19. Label the following illustration using the terms provided.

   vastus medialis   semimembranosus   sartorius
   gracilis   adductor femoris

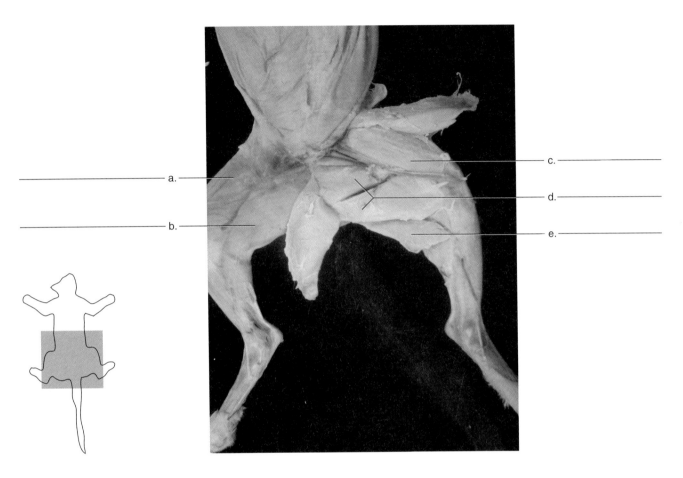

20. The calf is made up of what two major muscles? _____

21. Plantar flexion occurs by what muscles? _____

22. What muscle extends the toes? _____

23. Name a muscle in this exercise that dorsiflexes the foot. _____

24. Plantar flexion and eversion of the foot occur by the action of what muscle? _____

25. What is the insertion of the soleus? _____

26. What is the insertion of the fibularis tertius muscle? _____

27. What is the insertion of the flexor digitorum longus? _____

28. Label the muscles in the following illustration using the terms provided.
    extensor digitorum   soleus   tibialis anterior
    fibularis longus

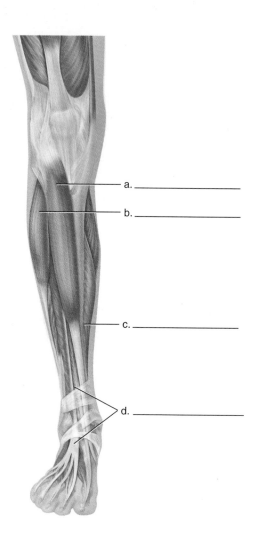

a. _____
b. _____
c. _____
d. _____

29. Label the muscles in the following illustration using the terms provided.
    extensor digitorum longus   gastrocnemius   fibularis longus
    soleus   tibialis anterior

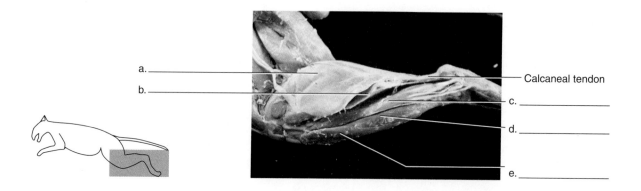

a. _____
b. _____
Calcaneal tendon
c. _____
d. _____
e. _____

Exercise 14
Muscles of the Head and Neck

INTRODUCTION

In this exercise, you continue your study of muscles with the muscles of the head and neck. These muscles can be grouped into a few functional classifications. Some muscles aid in chewing food (mastication) or manipulate the food in the mouth so that chewing or swallowing can occur. Other muscles function in facial expression or in the closing of the eyes or mouth. Still other muscles move the head or neck. Finally, some move the hyoid or the larynx during speech or swallowing. These muscles are discussed in the *Principles of Anatomy and Physiology* text in chapter 9, "Gross Anatomy and Functions of Skeletal Muscles."

OBJECTIVES

At the end of this exercise, you should be able to

1. locate the muscles of the head and neck on a torso model, chart, cadaver (if available), or cat;
2. list the origin, insertion, and action of each muscle presented;
3. list what nerve innervates each muscle;
4. name all the muscles that have a specific action at a joint, such as all the muscles that flex the head;
5. reproduce the actions of selected muscles of your own head or neck.

MATERIALS

Human torso model and head and neck model
Human muscle charts
Articulated skeleton
Cadaver (if available)
Cat
Cat wetting solution
Materials for cat dissection
    Dissection trays
    Scalpel and two or three extra blades
    Gloves (household latex gloves work well for repeated use)
    Blunt (Mall) probe
    String and tags
    Pins
    Forceps and sharp scissors
    First aid kit in lab or prep area
    Sharps container
    Animal waste disposal container

PROCEDURE
Human Muscles
Muscles of the Neck

Examine a model or chart of the human musculature and cadaver (if available) as you read the following descriptions. The trapezius and levator scapulae are muscles that have an action on the neck, but they are covered in Exercise 12 in this manual. The details of the muscles are listed in table 14.1. Examine a skull or an articulated skeleton and review the bony markings as you study the origins and insertions of the muscles in this exercise. Once you know the origins or insertions, the actions are easier to understand.

There are three scalene muscles on each side of the neck. These are the anterior, middle, and posterior scalenes. They are bounded by the sternocleidomastoid to the anterior and the levator scapulae to the posterior. They rotate the neck or elevate the ribs. Locate these muscles in lab and examine figure 14.1.

The **sternocleidomastoid** muscle (figures 14.1 and 14.2) rotates the head in a unique way. Place your hand on your right sternocleidomastoid and turn your head to the right. Notice how the muscle does not contract. Now turn your head to the left, and you can feel the muscle contract. The right sternocleidomastoid turns the head to the left and the left sternocleidomastoid turns the head to the right.

The **sternohyoid, sternothyroid,** and **omohyoid** are all named for their origins and insertions. In the sternohyoid and sternothyroid, the sternum anchors the stable part of the muscle (the origin), and the hyoid and thyroid cartilage of the larynx moves when the respective muscles contract. In the omohyoid (*omo* = shoulder), the scapula is the origin. The hyoid is depressed when the omohyoid contracts. Examine the material in lab and compare it with figure 14.1 for the muscles of the anterior neck.

The **digastric** is so named because it is a muscle with two bellies. Few muscles depress the mandible. The digastric is one that does. In addition to this, the digastric attaches to the hyoid, which is important in tongue movement during speech and swallowing. Deep to the digastric is the **mylohyoid,** a broad muscle of the floor of the mouth that pushes the tongue superiorly when swallowing. These muscles can be seen in figure 14.1.

The **platysma** is a broad, thin muscle that has a soft origin (on the fascia of the pectoral and deltoid muscles). It inserts on the mandible and skin of the lips and can be seen if you elevate your chin and subsequently pout. The thin wings that stick out on the side of your neck are the edges of the platysma muscle. Examine the platysma in figure 14.2 and in the models or on the cadaver in the lab.

TABLE 14.1 Muscles of the Head and Neck

Name	Origin	Insertion	Action	Innervation
Neck Muscles				
Scalenes (anterior, middle, and posterior)	Transverse process of cervical vertebrae	Ribs 1 and 2	Flexes and rotates neck, elevates ribs 1 and 2	Spinal nerves C4–8
Sternocleidomastoid	Manubrium and medial clavicle	Mastoid process, superior nuchal line	One: rotates and extends neck; both: flex neck	Accessory nerve (XI)
Sternohyoid	Manubrium of sternum	Hyoid	Depresses hyoid	Spinal nerves C1–3
Sternothyroid	Manubrium of sternum	Thyroid cartilage of larynx	Depresses larynx	Spinal nerves C1–3
Omohyoid	Superior border of scapula	Hyoid	Depresses and fixes hyoid	Spinal nerves C1–3
Digastric	Mastoid process of temporal (posterior belly)	Mandible near midline (anterior belly)	Elevates, protracts, and retracts hyoid; depresses mandible	Trigeminal (V) and facial (VII) nerves
Mylohyoid	Body of mandible	Hyoid	Elevates floor of mouth and tongue	Trigeminal nerve (V)
Muscles of the Face				
Platysma	Fascia covering pectoralis major and deltoid	Skin over inferior border of mandible	Depresses lower lip	Facial nerve (VII)
Frontalis	Galea aponeurotica	Skin superior to orbit	Raises eyebrows, draws scalp anteriorly	Facial nerve (VII)
Occipitalis	Occipital bone	Galea aponeurotica	Draws scalp posteriorly	Facial nerve (VII)
Orbicularis oculi	Frontal bone, maxilla	Skin of eyelid	Closes eye	Facial nerve (VII)
Orbicularis oris	Fascia of facial muscles, maxilla and mandible	Skin of lips	Closes lips	Facial nerve (VII)
Corrugator supercilii	Nasal bridge, orbicularis oculi	Skin of eyebrow	Depresses and pulls eyebrows together	Facial nerve (VII)
Zygomaticus (major and minor)	Zygomatic bone	Muscle and skin at angle of mouth	Elevates and abducts upper lip	Facial nerve (VII)
Risorius	Fascia of masseter, platysma	Skin at angle of orbicularis oris	Abducts corner of mouth	Facial nerve (VII)
Mentalis	Mandible	Skin of chin	Elevates lower lip	Facial nerve (VII)
Buccinator	Maxilla, mandible	Orbicularis oris at angle of mouth	Retracts angle of mouth, flattens cheek	Facial nerve (VII)
Depressor labii inferioris	Lower border of mandible	Skin of lower lip, orbicularis oris	Depresses lower lip	Facial nerve (VII)
Levator labii superioris	Maxilla	Skin and orbicularis oris of upper lip	Elevates upper lip	Facial nerve (VII)
Muscles of Mastication				
Temporalis	Temporal fossa	Coronoid process, mandibular ramus	Elevates and retracts mandible	Trigeminal nerve (V)
Masseter	Zygomatic arch	Lateral ramus of mandible	Elevates and protracts mandible	Trigeminal nerve (V)
Pterygoids (medial and lateral)	Pterygoid plate of sphenoid bone	Medial ramus and condylar process of mandible	Protracts, elevates, and depresses mandible (for chewing)	Trigeminal nerve (V)

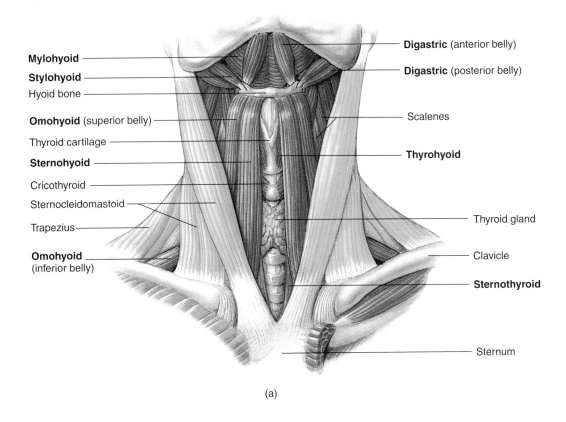

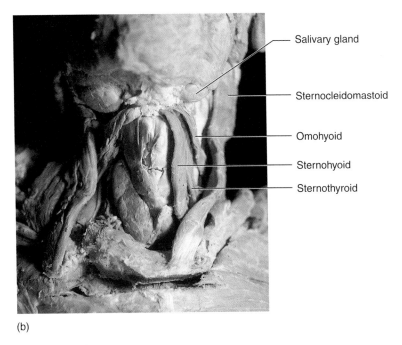

Figure 14.1 Muscles of the Anterior Neck
(a) Diagram; (b) photograph.

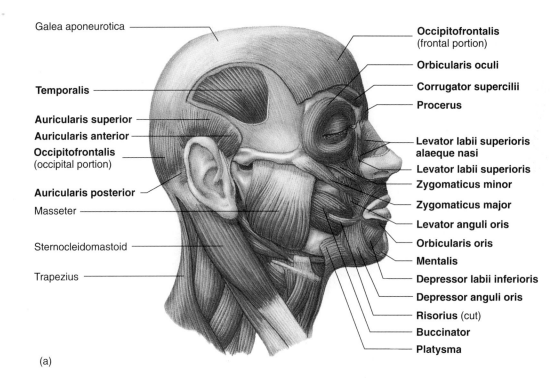

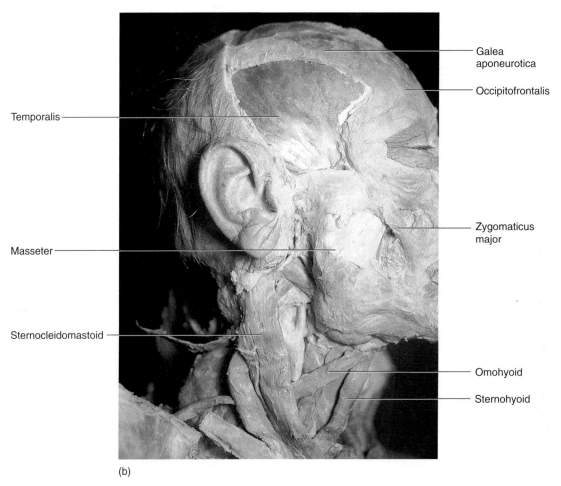

Figure 14.2 Muscles of the Head, Lateral View
(a) Diagram; (b) photograph; (c) diagram, anterior view.

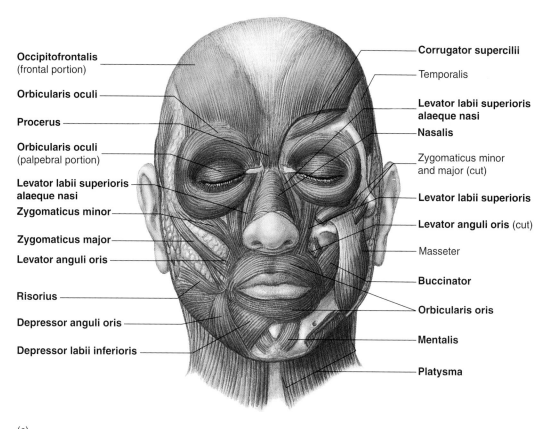

(c)

Figure 14.2—*Continued.*

Muscles of the Head

The **frontalis** muscle attaches to a broad, flat, tendinous sheet on the superior aspect of the skull known as the **galea aponeurotica.** Galea means helmet and aponeurotica refers to an aponeurosis, which is a wide, flat tendon. The insertion of the frontalis is on the eyebrow region. If the galea is held in place (fixed) by the occipitalis, then the frontalis raises the eyebrows. If the galea is not fixed then the scalp is brought forward, as in frowning. The **occipitalis** is a functional continuation of the frontalis muscle in that both muscles attach to the galea aponeurotica. The occipitalis pulls the scalp posteriorly. These muscles are sometimes discussed as one muscle, the **occipitofrontalis.** Examine these muscles in figure 14.2.

The **orbicularis oculi** and the **orbicularis oris** are sphincter muscles that close the eyes and the mouth, respectively. Sphincter muscles function like the strings of a drawstring purse. The orbicularis oculi has a medial, bony origin and an insertion on the eyelid. The muscles encircle the eye and close the eyelids. The orbicularis oris originates on fascia and facial muscles near the mouth and inserts on the skin of the lips, closing the lips. Examine the facial muscles in lab and in figure 14.2.

The **corrugator supercilii** muscle has a medial point of origin between the eyebrows and inserts laterally on the eyebrow. It furrows the eyebrows.

The **zygomaticus** (zīgō-mat´-i-kus) **major** and **zygomaticus minor** elevate the corners of the mouth by pulling them superiorly and laterally, as in smiling or laughing. They are named for their origin on the zygomatic bone. The **risorius** is known as the laughing muscle because it pulls the lips laterally. It does not have a bony point of origin but attaches to the fascia of the masseter (see following description) and platysma.

The **mentalis** originates on the chin (anterior mandible) and is another of the pouting muscles. The **buccinator** (buk´sĭ-nā´tōr) muscle of the cheek runs in a horizontal direction. It puckers the cheeks, as in trumpet playing, and pushes food toward the molars in chewing.

The **depressor labii** (lābē-ī) **inferioris** pulls the lower corners of the mouth inferiorly when pouting. It is named for its action (depressing the lower lip), whereas the **levator labii superioris** muscle elevates the skin of the lips and expands the nostrils, as in the expression of extreme disgust. Look at these muscles in lab and in figure 14.2.

The **temporalis** (tem´pŏ-rā-´līs) is a powerful muscle that elevates the mandible. It is a muscle of mastication. The temporal fossa is so named because it is a depression medial to the zygomatic arch (although when you examine the skull, the temporal fossa appears slightly domed). The temporalis muscle is deep to the zygomatic arch and inserts on the coronoid process and on the superior and medial ramus of the mandible.

The **masseter** (mă-sē-´ter) is a large muscle of the head whose synergist is the temporalis. It elevates the mandible. If you place your fingers on the ramus of the mandible and clench your teeth, you can feel the masseter tighten. These muscles are seen in figures 14.2 and 14.3.

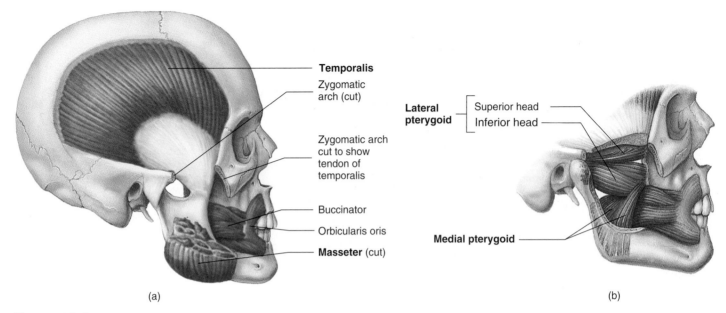

Figure 14.3 Muscles of Mastication
(a) Superficial view (platysma removed); (b) deep view.

The **pterygoid** (ter´-i-goyd) muscles are deep muscles that originate on the sphenoid bone and insert laterally on the mandible. They pull the jaw horizontally, which helps in rotatory chewing. The temporalis and masseter open the jaw, whereas the pterygoids provide the sideways movement characteristic of a person chewing gum. These muscles can be seen in figure 14.3.

Activities

1. When you turn your head to the left, which sternocleidomastoid muscle contracts?

   _____

2. Raise your head so that your chin is elevated and pout your lower lip. What muscle forms a thin membrane along the anterolateral neck? _____

3. Purse your lips and feel the buccinator muscle as it contracts in the cheek.

4. Clench your teeth and palpate the temporalis muscle and the masseter.

5. Examine figure 14.4 for surface views of facial muscles. Fill in which muscles are represented in the photographs.

Muscles of the Head and Neck of the Cat

In dissecting the muscles of the head and neck, take care not to cut through the digestive structures, such as the salivary glands (see figure 14.7), and the circulatory structures, such as the veins and arteries. You will study these structures in later exercises.

The **platysma** in the cat was removed during the skinning process, and it will not be seen unless you kept the skin with the cat. The **scalenes** can be dissected by reflecting a chest muscle called the pectoralis minor. Notice how the scalenes are composed of separate slips of muscle that run from the ribs to the neck. In humans, the muscle is more lateral than in cats. Compare your specimen with figure 14.5.

In the cat, the **sternocleidomastoid** consists of two muscles, the **sternomastoid** and the **cleidomastoid** (figure 14.6). Their origins and insertions are similar to the sternocleidomastoid in humans. The sternomastoid extends from the sternum to the mastoid process of the skull, and the cleidomastoid runs from the clavicle to the mastoid process. Underneath the sternomastoid is the most medial muscle of the neck group, the **sternohyoid.** The **sternothyroid** is deeper and more lateral than the sternohyoid. These muscles can be seen in figure 14.6.

The **digastric** muscle runs parallel to the lower edge of the mandible and underneath the submandibular gland. Dissect only one side of the head, leaving the structures on the other intact for study of the digestive system. Deep to the digastric is the **mylohyoid,** which is a broad muscle. Notice how the muscle fibers run perpendicular to the direction of the digastric. Compare your dissection with figure 14.7. The **occipitalis** is a posterior head muscle in the cat that attaches to the **galea aponeurotica** and to the anterior **frontalis** muscle. Lateral to these muscles are the temporalis and masseter muscles. The **temporalis** is located more dorsally than the masseter and can be dissected by removing the skin and fascia anterior to the ear. The **masseter** is a large, well-developed muscle in the cat that originates on the zygomatic arch and inserts on the lateral surface of the mandible. Examine these muscles in figure 14.8.

The **pterygoids** are usually not dissected because you have to cut through the ramus of the mandible to examine them. The muscles of facial expression are generally not studied in the cat. These muscles are small and are often removed with the skin. You should study these muscles on human models or a cadaver.

Exercise 14 Muscles of the Head and Neck 215

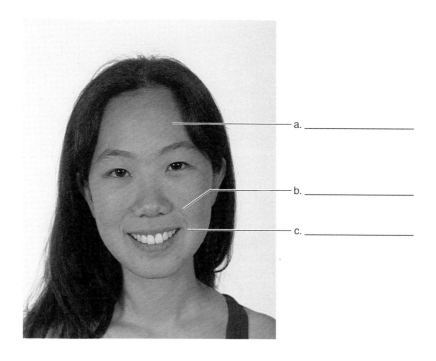

a. _____

b. _____

c. _____

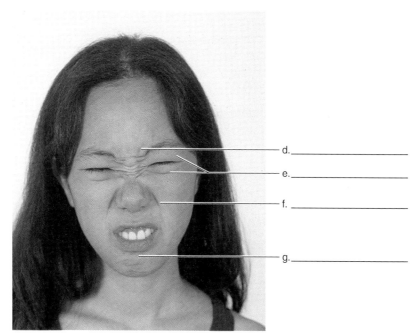

d. _____

e. _____

f. _____

g. _____

Figure 14.4 Muscles of Facial Expression

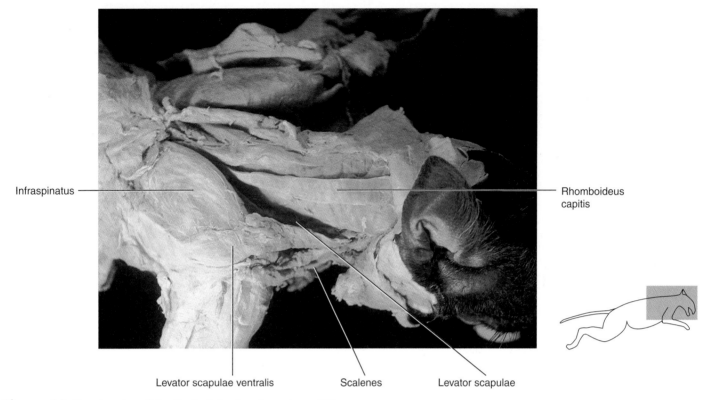

Figure 14.5 Muscles of the Neck of the Cat, Dorsolateral View

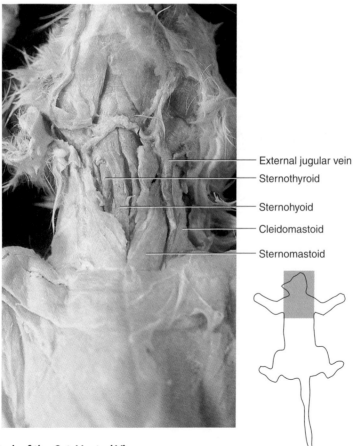

Figure 14.6 Muscles of the Neck of the Cat, Ventral View

Exercise 14 Muscles of the Head and Neck

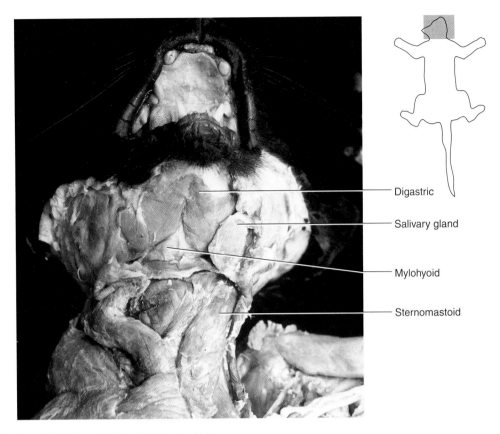

Figure 14.7 Muscles of the Head of the Cat, Ventral View

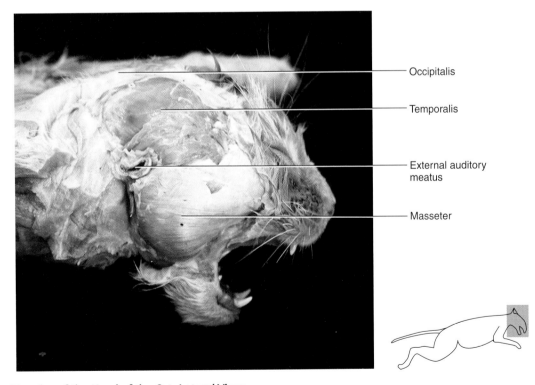

Figure 14.8 Muscles of the Head of the Cat, Lateral View

Exercise 14 Review

Muscles of the Head and Neck

Name: _____

Lab time/section: _____

Date: _____

Answer the following in terms of human muscles.

1. What is the origin of the masseter muscle? _____

2. What is the action of the risorius? _____

3. What kind of muscle is the orbicularis oculi or orbicularis oris muscle in terms of action? _____

4. Fill in the following illustration for the muscles of the neck using the terms provided.

    scalenes   mylohyoid   digastric   sternohyoid   sternocleidomastoid

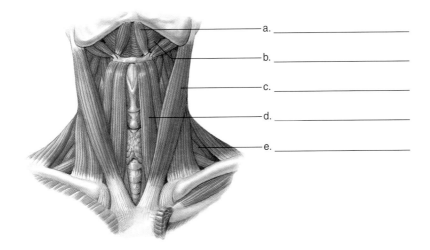

a. _____
b. _____
c. _____
d. _____
e. _____

5. What muscle originates on the temporal fossa? _____

6. Name two muscles that elevate the mandible. _____

7. Where does the sternocleidomastoid muscle insert? _____

8. What muscle closes the lips? _____

9. Where does the orbicularis oculi insert? _____

10. What is the insertion of the temporalis? _____

11. Name a muscle that closes the eye. _____

12. What is the action of the sternocleidomastoid if just one side contracts? _____

13. Label the muscles of the head in the following illustration using the terms provided.

masseter   buccinator   temporalis   occipitofrontalis   orbicularis oris   orbicularis oculi

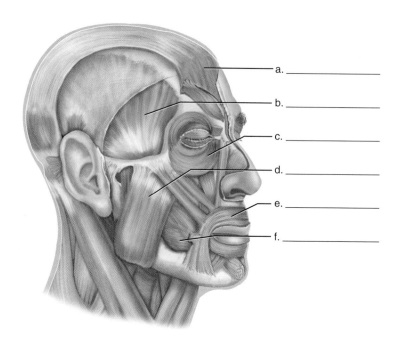

a. _____
b. _____
c. _____
d. _____
e. _____
f. _____

Answer the following in terms of cat muscles.

14. What two muscles in the cat compose the sternocleidomastoid in humans? _____

15. Which muscle is larger in cats, the temporalis or the masseter? _____

16. Which muscle is more lateral, the sternohyoid or the sternothyroid? _____

Exercise 15
Muscles of the Torso

INTRODUCTION

In this exercise, you continue your study of muscles with the muscles of the torso. These muscles can be grouped into a few functional areas. There are abdominal muscles, respiratory muscles, which assist in breathing, and postural muscles of the back. These muscles are discussed in the *Principles of Anatomy and Physiology* text in chapter 9, "Gross Anatomy and Functions of Skeletal Muscles."

OBJECTIVES

At the end of this exercise, you should be able to

1. locate the muscles of the torso on a model, chart, cadaver (if available), or cat;
2. list the origin, insertion, and action of each muscle;
3. describe what nerve controls each muscle;
4. name all the muscles that compress the abdomen.

MATERIALS

Human torso model
Human muscle charts
Articulated skeleton
Cadaver (if available)
Cat
Materials for cat dissection
    Dissection trays
    Scalpel and two or three extra blades
    Gloves (household latex gloves work well for repeated use)
    Blunt (Mall) probe
    String and tags
    Pins
    Forceps and sharp scissors
    First aid kit in lab or prep area
    Sharps container
    Animal waste disposal container

PROCEDURE

Review the muscle nomenclature and the actions as outlined in Exercises 10 and 12. Examine a torso model or chart in the lab and locate the muscles described next and in table 15.1. Correlate the shape or fiber direction of the muscle with the name, and visualize the muscles as you study the models or charts. You may want to look at an articulated skeleton as you study the origins and insertions of the muscles, so that you can better see the muscle attachment points. The descriptions in the text portion of this exercise can help you understand the nature of the muscle, whereas table 15.1 presents specific information about the muscle.

Once you have learned the origin and the insertion, you should be able to understand the action of the muscle. This is done, in part, by imagining how the bony attachments of the origin and insertion would come together if the muscle pulled them closer to one another.

Human Muscles
Abdominal Muscles

Most of the muscles of the abdomen compress the viscera, which aids in breathing, food regurgitation, and bowel movements. The **external abdominal oblique** is a broad, superficial muscle of the abdomen with fibers that run from a superior direction to an inferior, medial direction. The **internal abdominal oblique** is deep to the external abdominal oblique and has fiber directions that run perpendicular to the external abdominal oblique. Locate the abdominal muscles in the lab and compare them with figure 15.1.

The deepest of the abdominal muscles is the **transversus abdominis,** which has fibers running in a horizontal direction. The **rectus abdominis** (*rectus* = straight) runs vertically up the abdomen. The rectus abdominis has small connective tissue bands, called **tendinous intersections,** located horizontally across the muscle, dividing it into small segments. If the abdominal fat is minimal and the muscles are well developed, the "washboard stomach," or "six-pack," is apparent due to the muscle fibers increasing in girth while the tendinous intersections remain undeveloped. Another muscle in the area is the **quadratus lumborum,** which is a square muscle that runs from the iliac crest to the lower vertebrae and rib 12. These muscles can be seen in figures 15.1 and 15.2.

Postural Muscles

The **erector spinae** muscles make up most of the postural muscles of the back. The erector spinae are actually many muscles that are located between individual vertebrae or between the vertebrae and the ribs. These separate muscles are grouped into long strap muscles known collectively as the erector spinae. There are three major groups of erector spinae muscles: the **spinalis,** the **longissimus,** and the **iliocostalis.** Locate these muscles in lab and compare them with figure 15.2.

Thoracic Muscles

The **serratus anterior** is a broad, fan-shaped muscle that has slips of muscle originating on the upper ribs. These slips of muscle unite and insert on the medial border of the scapula. As the muscle contracts, it abducts the scapula by pulling it away from the spine. These muscles are shown in figure 15.1.

The intercostal muscles and the diaphragm are respiratory muscles. Normally, the diaphragm is responsible for about 60%

TABLE 15.1 Muscles of the Torso

Name	Origin	Insertion	Action	Innervation
Abdominal Muscles				
External abdominal oblique	Ribs 5–12	Inguinal ligament, iliac crest, rectus sheath	Compresses abdominal wall, laterally rotates trunk	Intercostal nerves from T7–12
Internal abdominal oblique	Inguinal ligament, iliac crest, lumbar fascia	Linea alba, ribs 10–12, rectus sheath	Compresses abdominal wall, laterally rotates trunk	Intercostal nerves from T7–12 and spinal nerve L1
Transversus abdominis	Inguinal ligament, iliac crest, ribs 7–12	Xiphoid process, linea alba, crest of pubis	Compresses abdominal wall, laterally rotates trunk	Intercostal nerves from T7–12 and spinal nerve L1
Rectus abdominis	Pubic crest, symphysis pubis	Inferior ribs, xiphoid process	Flexes vertebral column, compresses abdominal wall	Intercostal nerves from T6–12
Quadratus lumborum	Posterior iliac crest, lower lumbar vertebrae	T12, L1–4, rib 12	Laterally flexes vertebral column, depresses rib 12	T12, L1–4
Erector Spinae				
Iliocostalis	Vertebral column, ribs, ilium	Ribs, vertebral column	Extends laterally flexes, and rotates vertebral column	Spinal nerves, thoracic nerves
Longissimus	Vertebral column, ribs	Vertebral column, ribs, mastoid process	Extends head, neck, vertebral column	Cervical, thoracic, lumbar nerves
Spinalis	Vertebral column	Vertebral column	Extends vertebral column	Cervical, thoracic nerves
Thoracic Muscles				
Serratus anterior	Ribs 1–9	Medial border of scapula	Abducts scapula	Long thoracic nerve
Diaphragm	Xiphoid process, lower ribs, upper lumbar vertebrae	Central tendon	Inspiration	Phrenic nerve
External intercostalis	Inferior margin of a rib	Superior margin of rib below	Elevates ribs, inspiration	Intercostal nerves
Internal intercostalis	Superior margin of a rib	Inferior margin of rib above	Depresses ribs, expiration	Intercostal nerves
Splenius Muscles				
Splenius	C4–T6	Occipital and temporal bone	Extends and rotates head	Middle and lower cervical nerves
Transversospinal Muscles				
Semispinalis	C4–T6	Occipital bone	Extends and rotates head	Cervical nerves
Multifidus	Vertebral column, ilium	Vertebral column	Extends and rotates vertebral column	Numerous spinal nerves

C = cervical vertebrae, T = thoracic vertebrae, L = lumbar vertebrae, S = sacral vertebrae.

of the resting breath volume, whereas the external intercostals contribute to the remaining volume. The **diaphragm** is a domed muscle with a peripheral origin. The diaphragm inserts centrally at the base of the mediastinum. If you think of the diaphragm as a trampoline, the outer springs represent the origin, whereas the center (where you jump) represents the insertion. Examine the models in the lab and compare them with figure 15.3.

The intercostal muscles contribute to the breathing volume at rest, but they also contribute to a greater increase in the movement of the thorax during exercise. There is some debate as to the action of the intercostals. Some evidence suggests that both the intercostals are involved in inhalation. Other evidence suggests that the external intercostals are involved in inhalation, whereas the internal intercostals are involved in exhalation. This exercise treats the **external intercostals** as being responsible for inhalation (inspiration), whereas the **internal intercostals** are involved in exhalation (expiration). Locate the intercostal muscles and compare them with figure 15.3.

The deep muscles of the superior torso consist of the **splenius** and the **semispinalis** muscles. These extend and rotate the head and are illustrated in figures 15.2 and 15.4. The **multifidus** (figure 15.2) is a closely associated muscle with the erector spinae muscles (the spinalis, longissimus, iliocostalis, and multifidus or *s.l.i.m.* muscles as you move from medial to lateral and then inferior).

Exercise 15   Muscles of the Torso

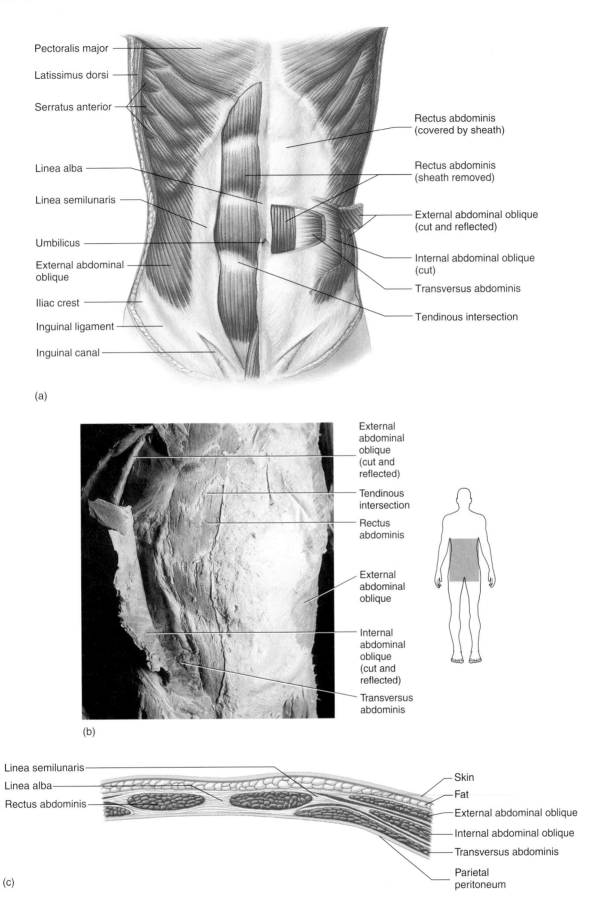

Figure 15.1 Muscles of the Abdomen, Anterior View
(a) Diagram; (b) photograph; (c) diagram of cross section.

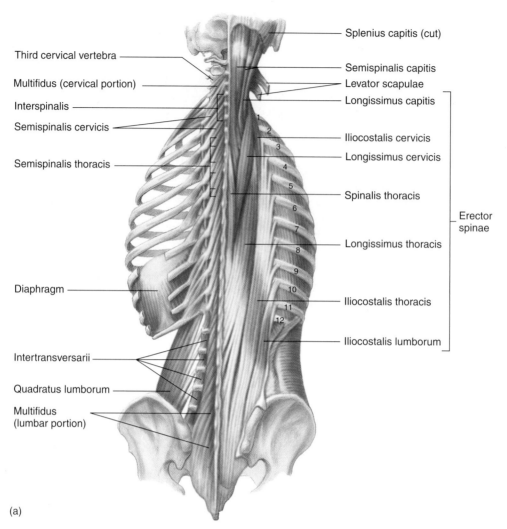

Figure 15.2 Deep Muscles of the Back
(a) Diagram; (b) photograph.

Cat Dissection

Abdominal Muscles

Place the cat on its back and examine the abdominal muscles. The abdominal muscles in the cat are similar to those in the human in that the **external abdominal oblique** is a broad, superficial muscle on the ventral abdomen. Carefully cut through the external abdominal oblique to reveal the **internal abdominal oblique,** as illustrated in figure 15.5. Deep to this is the **transversus abdominis,** and it can be seen by carefully dissecting the internal abdominal oblique. If you cut too deeply, you will enter the abdominal cavity, so be careful in this part of the dissection. The **rectus abdominis** is the muscle that runs from the pubic region to the sternum.

Move to the thoracic region and examine the muscle of the lateral thorax dorsal to the xiphihumeralis. This is the **serratus anterior** (actually, the serratus ventralis in the cat), and you should see the scalloped edges of the muscle. Carefully separate this muscle from the others and trace its insertion to the scapula. To see the intercostal muscles, you will have to bisect the superficial chest muscles. If you have not done so already, cut through the middle of the belly of the pectoral muscles, exposing the ribs of the cat. Carefully remove the outer layer of fascia from the muscle between the ribs and locate the **external intercostal** muscle. You should be able to cut part of this muscle away and expose the **internal intercostal** muscle. Note how the fibers run perpendicular to one another. Do not look for the diaphragm at this time. You can see it in Exercise 35, as you examine the lungs. Examine these thoracic muscles in figure 15.6.

Deep Muscles of the Back

Place the cat so that you can examine the dorsal surface. Locate the **splenius** and the **semispinalis** muscles. These are located in figure 15.7.

You will need to bisect the latissimus dorsi and the posterior portion of the external abdominal oblique to see the **erector spinae** muscles. The relative position of the cat erector spinae can be seen in figure 15.8. Compare this figure with your dissection. Locate the iliocostalis, longissimus, and spinalis of the erector spinae along with the multifidus in the cat.

Exercise 15 Muscles of the Torso

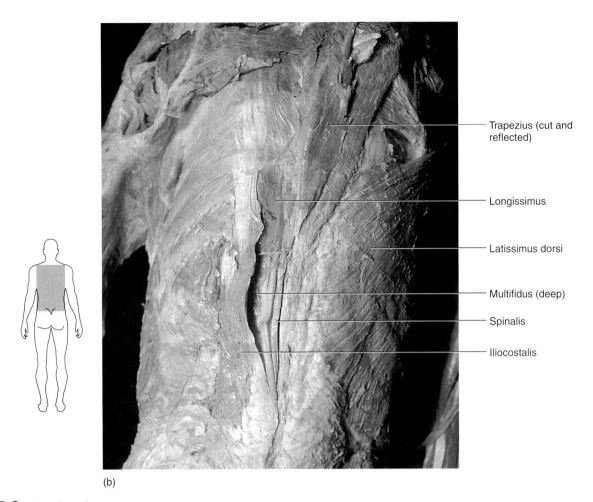

(b)

Figure 15.2—*Continued.*

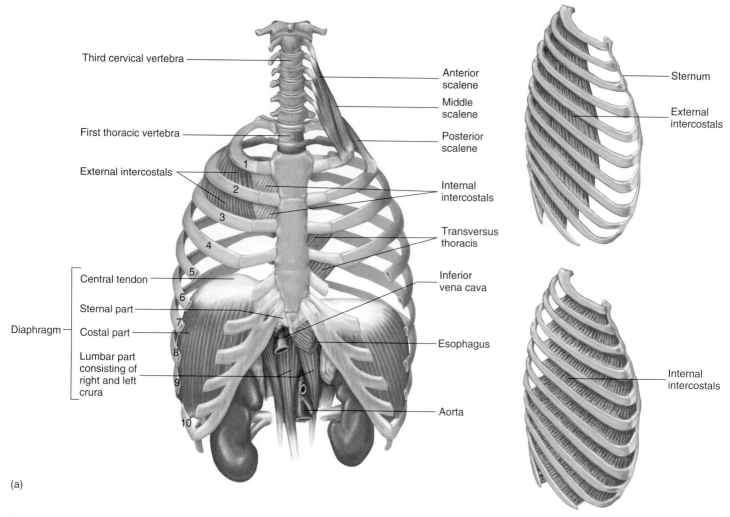

Figure 15.3 Muscles Involved in Respiration

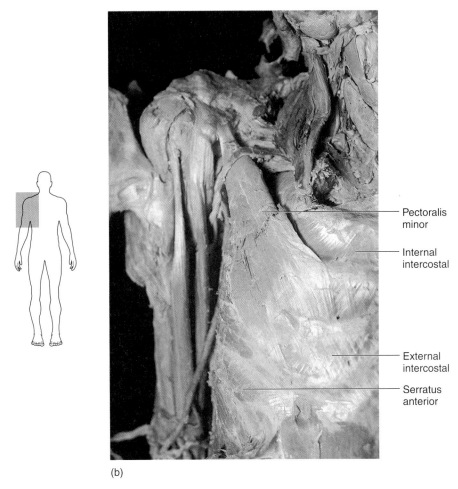

(b)

Figure 15.3—*Continued.*

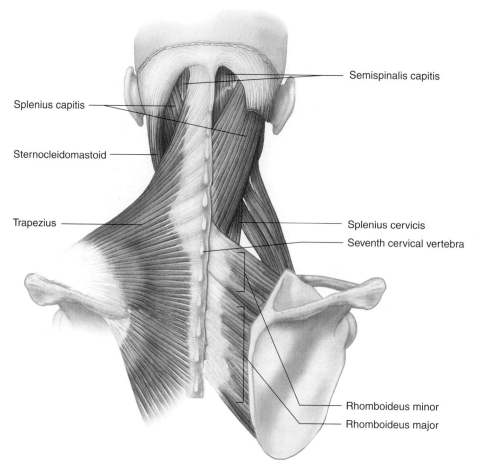

Figure 15.4 Muscles of the Neck, Posterior View

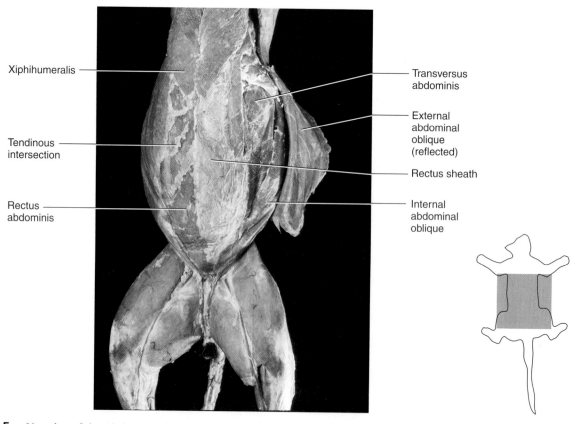

Figure 15.5 Muscles of the Abdomen of the Cat, Ventral View

Exercise 15 Muscles of the Torso

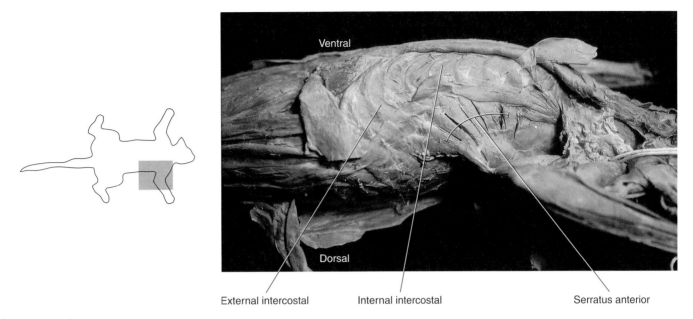

Figure 15.6 Muscles of the Thorax of the Cat, Lateral View

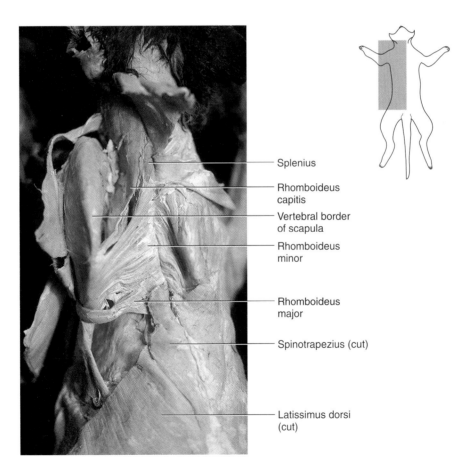

Figure 15.7 Anterior Muscles of the Back of the Cat, Dorsal View

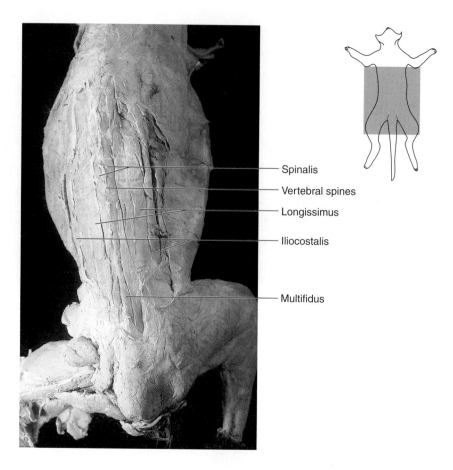

Figure 15.8 Posterior Muscles of the Back of the Cat, Dorsal View

Exercise 15 Review
Muscles of the Torso

Name: _____

Lab time/section: _____

Date: _____

1. What is the action of the serratus anterior muscle? _____

2. Name four muscles that extend the vertebral column. _____

3. How does the serratus anterior function as an antagonist to the rhomboideus muscles? _____

4. How does the action of the rectus abdominis differ from that of the other abdominal muscles? _____

5. What is the physical relationship of the intercostal muscles to each other? _____

6. What two muscles, originating on the neck, have an action to extend the head? _____

7. Extension and rotation of the vertebral column occur by what group of muscles? _____

8. What is the action of the intercostal muscles? _____

9. What muscle inserts on the central tendon? _____

10. What is the action of the quadratus lumborum? _____

11. The tendinous intersections are found in what muscle? _____

12. Flexion of the vertebral column occurs by what abdominal muscle? _____

13. Which is the deepest abdominal muscle? _____

14. Label the muscles in the following illustration using the terms provided.

rectus abdominis   transversus abdominis   external oblique
serratus anterior   internal oblique

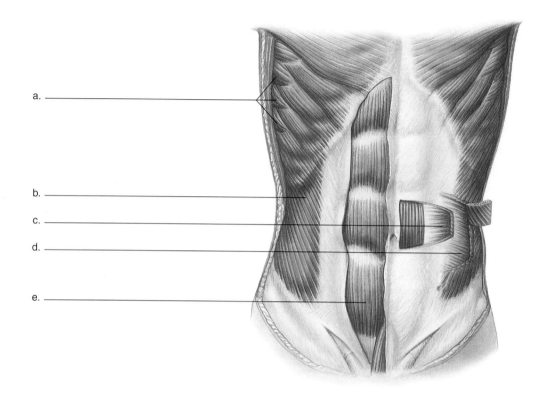

a. _____

b. _____

c. _____

d. _____

e. _____

15. How do the abdominal muscles in the cat compare with those in the human in terms of relative position? _____
_____

Exercise 16
Introduction to the Nervous System

INTRODUCTION

The functions of the nervous system, among other things, are **communication** between the various regions of the body, **coordination** of body functions (as in digestion or walking), **orientation** to the environment, and **assimilation** of information. These topics are further discussed in the *Principles of Anatomy and Physiology* text in chapter 10, "Functional Organization of Nervous Tissue."

In this exercise, you learn about the basic structure of the nervous system and the component cells of the nervous system. The functional unit of the nervous system is the neuron. It is the cell that carries out the activity of nervous tissue. Neurons are located in the nerves of the body, the spinal cord, ganglia, and the brain. Neuroglia, or glial cells, are also important cells of the nervous tissue. They help increase the speed of neuron transmission, provide nutrients to the neurons, affect synapse, and protect the neurons.

OBJECTIVES

At the end of this exercise, you should be able to

1. list the main divisions of the nervous system;
2. group the organs of the nervous system into the main divisions;
3. describe the three parts of the neuron;
4. draw a multipolar neuron and label the parts;
5. describe the functions of the various neuroglia.

MATERIALS

Charts or models of the nervous system
Charts or models of neurons
Microscopes
Prepared slides
    Spinal cord smear
    Longitudinal section of nerve
    Neuroglia

PROCEDURE
Divisions

The nervous system can be divided into two general divisions based on location. The **central nervous system (CNS)** consists of the brain and the spinal cord. The **peripheral nervous system (PNS)** is composed of the spinal nerves, the dorsal root ganglia, the somatic nerves (those that radiate into the extremities and other regions of the body), and the cranial nerves. Examine the models or charts in the lab for the central and peripheral divisions of the nervous system. Compare the material in the lab with figure 16.1.

The peripheral nervous system has a sensory, or afferent, division and a motor, or efferent, division. The **sensory division** conducts impulses *to* the central nervous system. The **motor division** conducts impulses *from* the central nervous system *to* other organs of the body. The motor division has two subunits—the **somatic motor nervous system,** which takes impulses from the CNS and innervates skeletal muscles, and the **autonomic nervous system (ANS),** which innervates glands, smooth muscle, cardiac muscle, and organs. The ANS has centers in the central nervous system (**midbrain, pons, medulla oblongata,** and **spinal cord**) and has peripheral branches. These function independently and provide automatic controls for activities normally under subconscious direction. When you walk up stairs, your heart rate increases automatically, along with your breathing rate, without your having to think about controlling these activities.

The activities of the ANS can be controlled consciously in some cases. The principle of biofeedback involves the conscious lowering of heart rate or blood pressure. Both of these activities are normally under the control of the ANS.

Neuron

Neurons consist of three main parts: the **axon,** the **dendrite,** and the **nerve cell body,** or **soma.** Examine figure 16.2 and models or charts in the lab to see the structure of neurons. The **neuron,** or **nerve cell,** is a remarkable cell, not only for its functional nature but also for the anatomical extremes it exhibits. The nerves in your thigh and leg are composed of neuron fibers (axons), and the neurons that pick up sensation in your toes continue as single cells up the leg and thigh to synapse (join) with other neurons in the lower back. When you look at prepared slides of neurons in the microscope in this exercise, remember that these neurons are very long in some cases. The receptive end of the neuron is the dendrite or the nerve cell body. Dendrites are so named because they have branching structures that resemble a tree (*dendros* = tree). Nerve cell bodies consist of the **neuroplasm** (cytoplasm of the neuron), the **Nissl substance** (rough endoplasmic reticulum of the neuron), and the **nucleus.** The triangular region of the nerve cell body that is devoid of Nissl substance is the **axon hillock,** and it leads to the axon that exits the nerve cell body. Axons in the PNS are elongated parts of the neuron wrapped in myelin sheaths. These sheaths are discussed later in the exercise.

Histology of the Neuron

Examine a prepared slide of a spinal cord smear under the microscope and locate the purple, star-shaped structures under low power. These are the nerve cell bodies of multipolar neurons. Switch to high power and locate the darker-staining Nissl substance, the nucleus, and the axon hillock. The axon is attached to the axon hillock. All the other processes that are attached to the nerve cell body are dendrites. The small nuclei that are scattered throughout the smear belong to

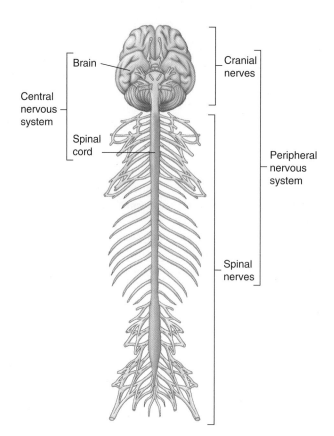

Figure 16.1 Divisions of the Nervous System

glial cells in the spinal cord. Compare your slide with figure 16.2b. Draw what you have seen in the microscope in the space provided.

Drawing of multipolar neuron:

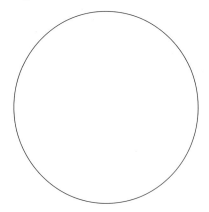

Functions of Neurons

There are three types of neurons based on function. **Sensory,** or **afferent, neurons** conduct impulses *to* the central nervous system. They convey information from the external or internal body environment to the spinal cord and/or the brain. **Motor,** or **efferent, neurons** conduct impulses *away from* the central nervous system. **Association neurons,** or **interneurons,** are located *between* sensory and motor neurons. They occur in the brain and are also in the PNS, where they transmit information to the brain for processing. These are illustrated in figure 16.3.

Neuron Shapes

Neurons can be classified according to shape. The majority of the neurons of the body (such as those in the brain and spinal cord) are multipolar neurons. Compare material in the lab with figure 16.4. A **multipolar neuron** consists of several dendritic processes, a single nerve cell body, and a single axon. In most neurons, dendrites take information to the cell body and axons take information away from the cell body. This is the case in multipolar neurons but it is not universal for all neurons. **Bipolar neurons** are so named because the nerve cell body has two poles. Dendrites receive information and conduct it to one pole of the nerve cell body. At the other pole, an axon leaves the nerve cell body and transmits the impulse away from the cell body. Bipolar neurons are located in nerves conducting the senses of smell and vision.

Unipolar, or **pseudounipolar, neurons** have a cell body with a single process attached to it (figure 16.4). Receptors receive impulses, which travel next to an axon. These axons transmit that information either to the cell body or via another axon directly to the spinal cord. Most of the common nerves of the body (somatic sensory nerves such as the ulnar nerve or femoral nerve) consist of unipolar neurons.

Synapses

Neurons transmit information electrochemically along the length of the axon to the **presynaptic terminal.** Neurons are not physically attached to one another but communicate by chemical signals that flow across a short space between the neurons. The space is called the **synapse,** and the chemicals, such as acetylcholine, that

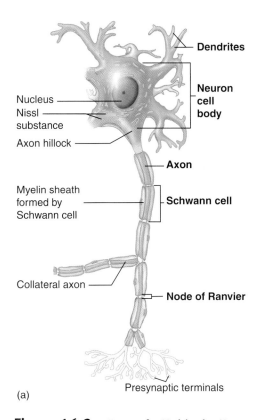
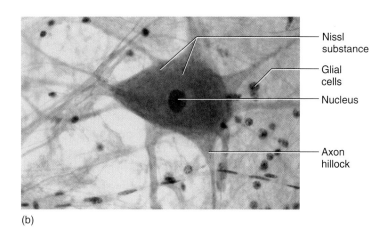

Figure 16.2 Parts of a Multipolar Neuron
(a) Diagram; (b) photograph.

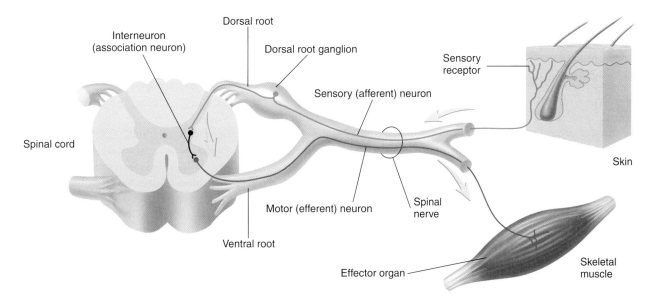

Figure 16.3 Sensory (Afferent) and Motor (Efferent) Neurons

move across the synapse are called **neurotransmitters.** Figure 16.5 illustrates a synapse.

Neuroglia

There are other types of cells that are important in the nervous system. Return to the spinal cord smear slide and locate the small purple cells outside of the multipolar neuron cells. These cells are called the **neuroglia,** or **glial cells.** About half of the weight of the brain is composed of glial cells and they occur as significant contributors to other parts of the nervous system. The common glial cell of the peripheral nervous system is the **Schwann cell,** or **neurolemmocyte** (figure 16.6). These glial cells wrap around the axon, leaving small gaps between successive cells called the **nodes of Ranvier,** as seen in figure 16.2. Neurolemmocytes are cells that wrap around the axon, much like a thin strip of paper wrapped around a pencil. The neurolemmocyte consists of a significant amount of a lipoprotein material called **myelin,** and the series of neurolemmocytes produces a **myelin sheath.**

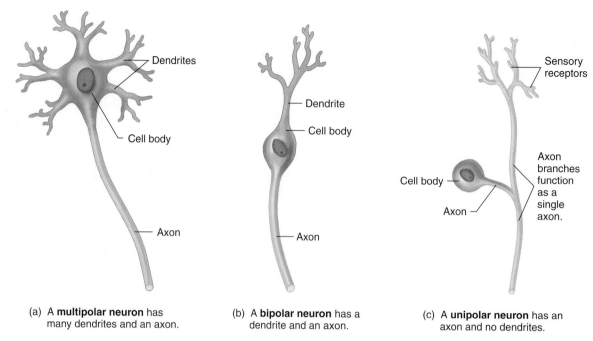

(a) A **multipolar neuron** has many dendrites and an axon.

(b) A **bipolar neuron** has a dendrite and an axon.

(c) A **unipolar neuron** has an axon and no dendrites.

Figure 16.4 Neuron Shapes
(a) Multipolar; (b) bipolar; (c) unipolar.

Myelinated nerve fibers appear white, so this type of nervous tissue is called **white matter.** The nodes of Ranvier allow for the nerve transmission to jump from node to node, increasing the transmission speed of the neuron. This type of jumping transmission is called **saltatory conduction. Unmyelinated fibers** and nerve cell bodies form the portion of the nervous tissue known as **gray matter.**

Histology of the Schwann Cell

Examine a prepared slide of a longitudinal section of nerve under high power. You should be able to see the axon fibers as long, dark threads in the microscope. The clear areas on each side of the axon fibers are the myelin sheaths. If you scan the slide closely, you should be able to see the junction of two Schwann cells (neurolemmocytes). The gap between them is the node of Ranvier. Compare your slide with figure 16.7. Draw the section of neuron in the space provided. Label the axon, neurolemmocytes, and node of Ranvier.

Drawing of nerve:

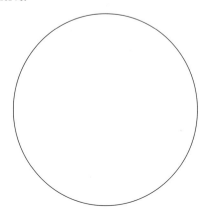

CNS Neuroglia

Schwann cells are glial cells in the PNS. Other types of neuroglia are located in the CNS and these are described in the following section:

- **Oligodendrocytes** are cells that produce myelin in the CNS. Unlike the neurolemmocytes, oligodendrocytes often wrap around several neurons (figure 16.8). The white matter of the spinal cord and brain is composed of lipoprotein sheaths of the oligodendrocytes.
- **Astrocytes** are branched glial cells that nourish neurons and provide a barrier between the nervous tissue and the blood. Astrocytes along with capillary endothelial cells are responsible for the **blood-brain barrier** that protects the nervous tissue from some bloodborne infections. This barrier also inhibits some medications from reaching the brain. Astrocytes also seem to play a role in the development of synapses between neurons. Astrocytes are illustrated in figure 16.9.
- **Microglia** are small glial cells that are phagocytic. The microglia digest foreign particles that invade the nervous tissue. Examine figure 16.10, which represents a microglial cell.
- **Ependymal cells** line the ventricles of the brain, line the central canal of the spinal cord, and serve as a barrier between the fluid in the area (the cerebrospinal fluid) and the nervous tissue. Ependymal cells are illustrated in figure 16.11.

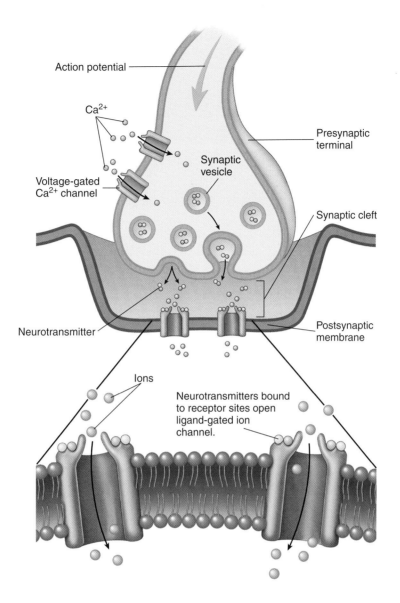

Figure 16.5 Synapse

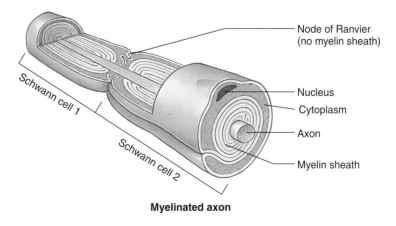

Figure 16.6 Schwann Cell

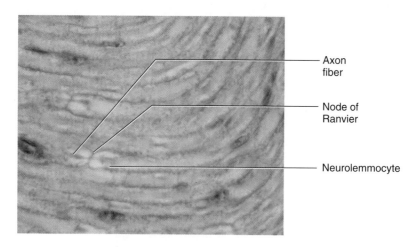

Figure 16.7 Nerve, Longitudinal Section (400×)

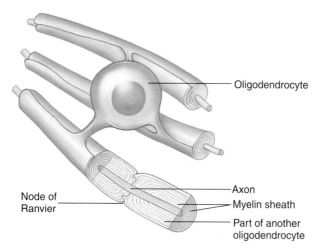

Figure 16.8 Oligodendrocyte

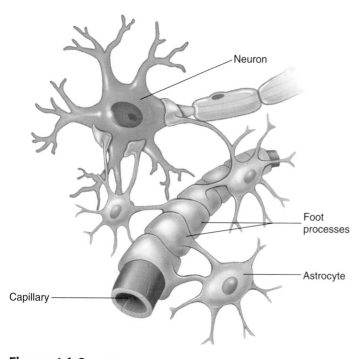

Figure 16.9 Astrocyte

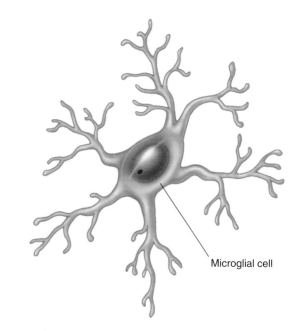

Figure 16.10 Microglia

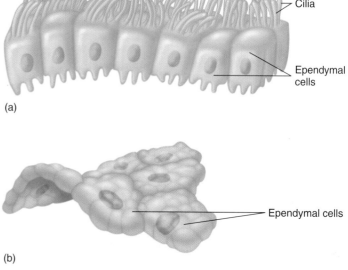

Figure 16.11 Ependymal Cells
(a) Ciliated; (b) squamous.

Exercise 16 Review

Introduction to the Nervous System

Name: _____
Lab time/section: _____
Date: _____

1. What kind of neuron (multipolar, bipolar, or unipolar) is represented by the drawing? Label the parts with the terms provided.

   axon　　　　nerve cell body
   dendrite　　Nissl substance

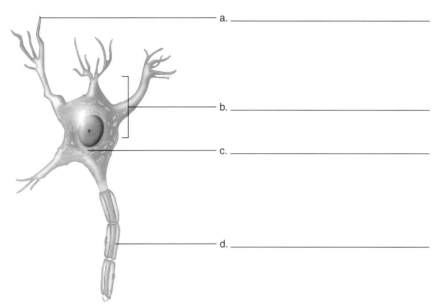

   a. _____
   b. _____
   c. _____
   d. _____

   Neuron type: _____

2. What kind of neuron (multipolar, bipolar, or unipolar) is represented by the drawing? Label the parts with the terms provided.

   axon　　　　nerve cell body
   dendrite　　nucleus

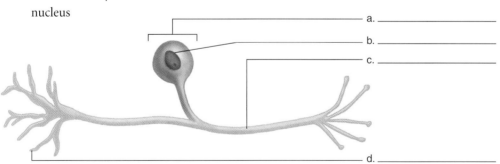

   a. _____
   b. _____
   c. _____
   d. _____

   Neuron type: _____

3. Describe the function of an astrocyte. _____

_____

_____

4. Describe the function of an ependymal cell. _____

_____

5. Describe the function of an oligodendrocyte. _____

_____

_____

6. The brain belongs to what division of the nervous system? _____

7. A spinal nerve belongs to what division of the nervous system? _____

8. To what major division of the nervous system does the spinal cord belong? _____

9. Which division of the nervous system is dedicated to subconscious function? _____

10. What kind of cell has an axon, a dendrite, and a nerve cell body? _____

11. In what part of the neuron is the nucleus found? _____

12. What is another name for an efferent neuron? _____

13. If a neuron has a soma with a dendrite on one side and an axon on the other, what kind of neuron is it? _____

_____

14. Two adjacent neurons communicate with one another across a space. What is this space called? _____

15. How are neuroglia different from neurons in terms of function? _____

16. In which of the nervous system divisions are neurolemmocytes located? _____

17. Myelin is made of what kind of material? _____

Exercise 17
Brain and Cranial Nerves

INTRODUCTION

Two specific traits distinguish humans from other animals. One is our upright posture, and the other is the extensive development of the brain. In this exercise, you examine the anatomy of the human brain and cranial nerves. The brain and spinal cord compose the central nervous system and are covered in the *Principles of Anatomy and Physiology* text in chapter 11, "Central and Peripheral Nervous System."

The brain is located in the cranial cavity of the skull and weighs approximately 1.4 kilograms (3 pounds). The brain is derived from three embryonic parts, each of which further develops into more specific areas. Cranial nerves are associated with the brain and should be studied after you know the anatomy of the brain.

You may also use a sheep brain as a model for understanding the human brain. There are many similarities between the organs in the two species. They have the same general parts, although sheep brains are smaller and some areas of the sheep brain are proportionally larger or smaller than the corresponding areas of human brains.

OBJECTIVES

At the end of this exercise, you should be able to

1. name the three meninges of the brain and describe their locations relative to one another;
2. name the three major regions of the brain;
3. locate the main structures in each of the three regions in intact and/or sectioned brains or brain models;
4. describe the function of the major structures in the brain;
5. trace the path of cerebrospinal fluid through the brain;
6. identify each of the 12 pairs of cranial nerves on an illustration of a brain or brain model;
7. state whether a cranial nerve is sensory, motor, or both sensory and motor;
8. identify the structure innervated by a particular cranial nerve.

MATERIALS

Models and charts of the human brain
Preserved human brains (if available)
Cast of the ventricles of the brain
Chart, section, or illustration of the brain in coronal and transverse sections
Sheep brains
Dissection trays
Scalpel and two or three extra blades
Gloves (household latex gloves work well for repeated use)
Blunt (mall) probe
First aid kit in lab or prep area
Sharps container
Animal waste disposal container

PROCEDURE
Meninges

Three layers, called meninges (mĕ-nin´-jez), surround the brain. The outermost of these is the **dura mater** (doó-ră mā´ter), a tough, dense connective tissue sheath that encircles the brain and has a series of shelves that extend into the brain. The dura mater is divided into an outer **periosteal dura** and an inner **meningeal dura.** The space deep to the dura is the **subdural space.** The next deeper layer is the **arachnoid** (ă-rak´noid) **mater,** which is a thin membrane that resembles a spider's web. Deep to the arachnoid is the **subarachnoid space,** which contains **cerebrospinal fluid (CSF).** The deepest layer is the **pia mater,** which is a membrane directly on the outer surface of the brain. Locate the meninges in preserved brains in the lab (if available) and compare them with figure 17.1.

Overview of the Brain

The brain can be subdivided into three major parts. The embryonic **prosencephalon** develops into the **forebrain,** which contains the cerebral hemispheres and diencephalon (see figure 17.2 and table 17.1). The **mesencephalon** develops into the **midbrain,** which is small in the adult brain. The embryonic **rhombencephalon** becomes the **hindbrain** and it consists of the **pons, medulla oblongata,** and **cerebellum.** The hindbrain is the most inferior portion of the brain and connects to the spinal cord at the foramen magnum. Table 17.1 lists the major structures of the brain covered in this exercise and the part of the brain in which they are located. Examine a model of the brain and find the major parts.

Ventricles of the Brain

The hollow neural tube that develops in the first trimester of pregnancy becomes the ventricles of the brain in the adult. The two ventricles that occupy the center of each cerebral hemisphere are known as the **lateral ventricles.** Tufts of capillaries called **choroid plexuses** secrete cerebrospinal fluid into the ventricles. In preserved brains, the choroid plexuses are small, brown areas at the superior portions of the ventricles. Fluid from the lateral ventricles flows through the **interventricular foramina (foramina of Monro)** into the **third ventricle.** The third ventricle is located in the thalamus and receives CSF from choroid plexuses in that area.

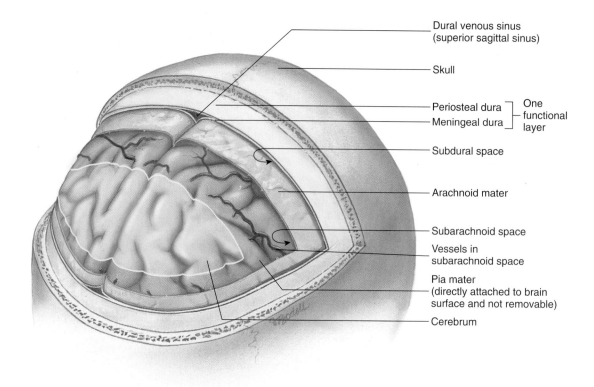

Figure 17.1 Meninges of the Brain

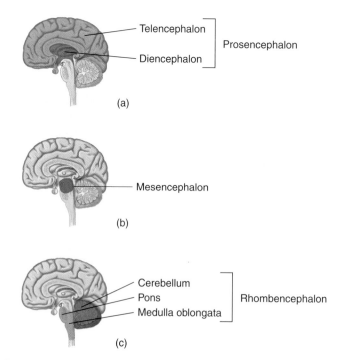

Figure 17.2 Embryonic Derivatives
(a) Forebrain, or prosencephalon; (b) midbrain, or mesencephalon; (c) hindbrain, or rhombencephalon.

TABLE 17.1 Parts of the Brain

Forebrain

Telencephalon
    Cerebrum (cerebral hemispheres)
    Cerebral cortex (gray matter)
    Basal nuclei (gray matter)
    Corpus callosum

Diencephalon
    Pineal body
    Thalamus
    Hypothalamus
    Pituitary gland
    Mammillary bodies

Midbrain

Peduncles

Tectum

Corpora quadrigemina
    Superior colliculus
    Inferior colliculus

Hindbrain

Metencephalon
    Pons
    Cerebellum

Myelencephalon
    Medulla oblongata

The third ventricle drains into the **fourth ventricle** by way of the **cerebral (mesencephalic) aqueduct,** which is also known as the **aqueduct of Sylvius.** If this duct becomes occluded, then CSF accumulates in the lateral and third ventricles. This increases the size of the ventricles, causing a condition known as **hydrocephaly.** Normally, the cerebral aqueduct is open and CSF passes through the midbrain. Inferior to the cerebral aqueduct is the fourth ventricle, which occupies a space anterior to the cerebellum. The fourth ventricle also has a choroid plexus that secretes CSF. Cerebrospinal fluid flows from the fourth ventricle to the central canal of the spinal cord, as well as into the subarachnoid space of the spinal cord and brain. The CSF cushions and provides buoyancy to the brain. CSF flows through the arachnoid granulations under the dura mater of the skull to the venous sinuses, and the fluid returns to the rest of the cardiovascular system by the **internal jugular veins.** There are approximately 150 mL of CSF in the central nervous system, and it takes about 6 hours to circulate through the system. Locate the ventricles of the brain in figure 17.3.

Surface View of the Brain

Examine a model or chart of the brain and locate its major surface features. Of the major parts of the brain, you will be able to easily see the **forebrain** and the **hindbrain.** In the forebrain, you should examine the large **cerebrum,** with its folds and ridges known as **convolutions.** The ridges of the convolutions are **gyri**

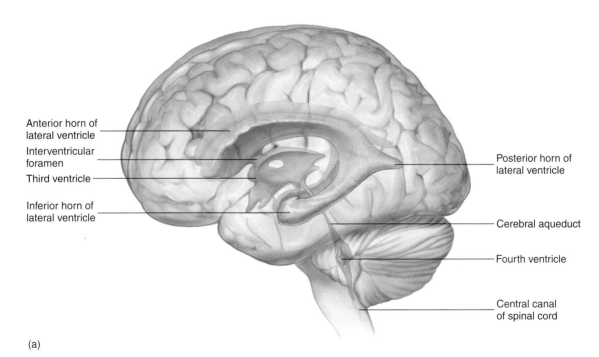

(a)

Figure 17.3 Ventricles of the Brain
(a) Lateral view; (b) schematic.

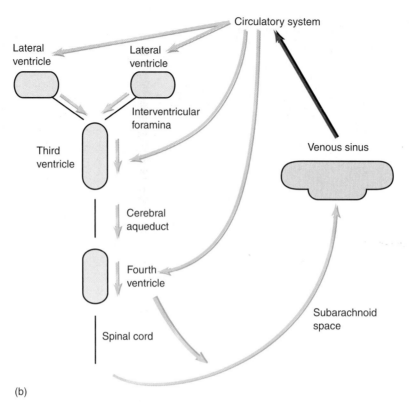

(b)

Figure 17.3—*Continued.*

(sing. **gyrus**), and the depressions are either **sulci** (sing. **sulcus**) or **fissures.** Usually, fissures are deeper than sulci. The lateral view of the brain allows you to see the major lobes of each cerebral hemisphere. These lobes are named for the bones of the skull under which they lie. They are the **frontal, parietal, occipital,** and **temporal lobes.** The **lateral fissure** separates the temporal lobe from the frontal and parietal lobes of the brain. These features can be seen in figure 17.4. In the surface regions of the cerebrum, **association areas** receive information from the **primary sensory areas** and then make sense of the information. For example, when you see an object, the neural signal first goes to the primary sensory area. How you interpret that object is part of the association area.

Frontal Lobe

The **frontal lobe** is anterior to the **central sulcus.** It is responsible for many of the higher functions associated with being human. The frontal lobe is involved in intellect, abstract reasoning, creativity, social awareness, and language. An important area responsible for controlling the formation of speech is called **Broca's area,** or the **motor speech area.** This area controls the muscles involved in speech. It is usually located in the left frontal lobe. Locate the frontal lobe in figure 17.4.

To find the central sulcus, look for two convolutions that run from the superior portion of the cerebrum to the lateral fissure, more or less continuously. The gyrus anterior to the central sulcus is part of the frontal lobe and is known as the **precentral gyrus,** or the **primary motor cortex.** This cortex is important for directing a part of the body to move and has been mapped, as in figure 17.5a. By constructing an image of the body on the brain, this figure produces the image of a person known as a homunculus. A **motor homunculus** is seen in figure 17.5a.

How much of the precentral gyrus is dedicated to the face?

_____

How much of the gyrus is dedicated to the hands?

_____

How much is dedicated to the trunk? _____

Parietal Lobe

The gyrus posterior to the central sulcus is known as the **postcentral gyrus,** or the **primary sensory cortex.** This is part of the **parietal lobe** and is involved in receiving sensory information from the body. This area has also been mapped, and you can see a **sensory homunculus,** represented in figure 17.5b. The primary sensory cortex receives information, yet the material is integrated just posterior to the sensory cortex in **association areas.** The primary sensory cortex pinpoints the part of the body affected, and the association area interprets the sensation (pain, heat, cold, etc.). Locate the association areas of the parietal lobe in figure 17.4.

Wernicke's area, located in both the parietal and temporal lobes, is involved in the formation of language, such as the recognition of written and spoken language, and in the formation of coherent sentences.

Occipital Lobe

Posterior to the parietal lobe is the **occipital lobe** of the brain. The occipital lobe is considered the **visual area** of the brain, and

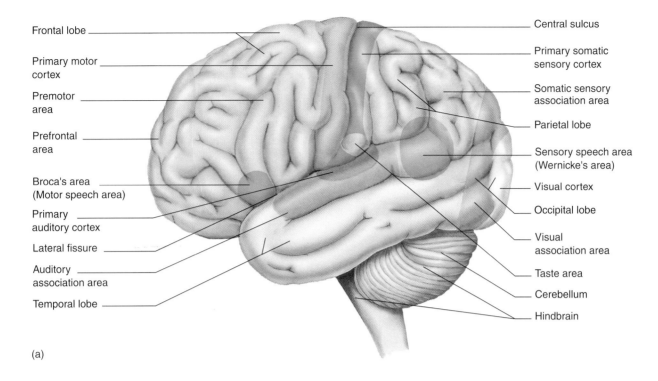

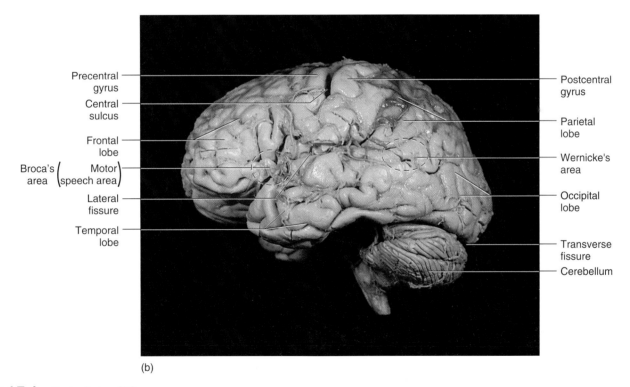

Figure 17.4 Brain, Lateral View
(a) Diagram; (b) photograph.

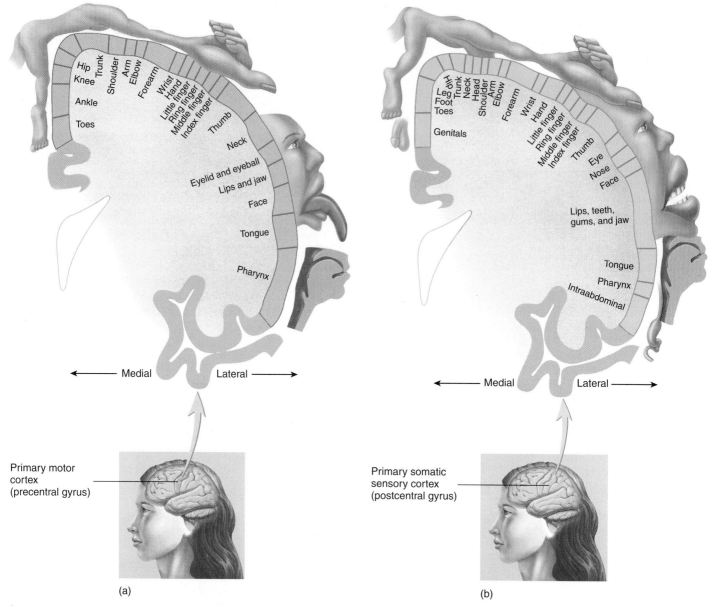

Figure 17.5 Primary Motor and Sensory Cortex
(a) Motor cortex (precentral gyrus); (b) sensory cortex (postcentral gyrus).

damage to this lobe can cause blindness. Shape, color, and distance of objects are perceived here. Recollection of past visual images occurs here as well. As you are reading these words, your occipital lobe is receiving the information and transferring it to other regions, which convert the words to thought. Between the occipital lobe and the cerebellum is a **transverse fissure** that separates these two regions of the brain. Locate the occipital lobe and the cerebellum in figure 17.4.

Temporal Lobe

The **temporal lobe** is separated from the frontal and parietal lobes by the **lateral fissure,** as seen in figure 17.4. The temporal lobe contains an area, known as the **primary auditory cortex,** that interprets hearing impulses sent from the inner ear. This auditory cortex distinguishes the nature of the sound (music, noise, speech), as well as the location, distance, pitch, and rhythm. The primary auditory cortex translates words into thought. The temporal lobe also has centers for the sense of smell (**olfactory centers**) and taste (**gustatory centers**). Deep to the temporal lobe is a small mass of cortical material called the **insula.**

Cerebral Hemispheres

If you rotate the brain so that you are looking at it from a superior view, you should be able to see the **longitudinal fissure** that separates the cerebrum into the left and right cerebral hemispheres. The **left cerebral hemisphere** in most people is involved in language and reasoning.

The **right cerebral hemisphere** of the brain is involved in space and pattern perceptions, artistic awareness, imagination, and music comprehension. The specialization of cerebral hemispheres for different tasks is known as cerebral asymmetry, or hemispheric dominance. Examine the surface features of the brain in figure 17.6.

Inferior Aspect of the Brain

Forebrain

Examine the frontal lobes of the cerebrum and the temporal lobes from an inferior view. You may be able to see the **pituitary gland** if it has not been removed. The **optic chiasma** (*chiasma* = cross) is anterior to the pituitary. It transmits visual impulses from the optic nerves to the brain. Two small processes posterior to the pituitary are the **mammillary bodies,** which function in olfactory reflexes.

Hindbrain

Locate the **pons, medulla oblongata,** and **cerebellum** from the inferior aspect. The cerebellum has much finer folds of neural tissue called **folia.** The medulla oblongata is located inferior to the cerebellum and connects to the spinal cord at the level of the foramen magnum. The enlarged portion of the brain anterior to the medulla is the pons, which serves as a relay center for information. Examine these structures in figure 17.7.

Midsagittal Section of the Brain

Forebrain

Examine a midsagittal section of a brain, as illustrated in figure 17.8. This section is seen by cutting the brain through the longitudinal fissure. Locate the C-shaped **corpus callosum,** which connects the two cerebral hemispheres. The posterior portion of the corpus callosum is known as the **splenium,** and the anterior portion is the **genu.** Just inferior to the corpus callosum is the **septum pellucidum,** which separates the lateral ventricles from each other. If the septum pellucidum is missing, you will be able to look into the lateral ventricle without obstruction.

Examine the material in the lab and locate the **diencephalon,** which consists of, in part, the thalamus and the hypothalamus. The **thalamus** forms the lateral wall around the third ventricle and is a relay center that receives almost all the sensory information from the body and sends it to the cerebral cortex.

Below the thalamus is the **hypothalamus,** which directs some parts of the ANS and is involved with the **pituitary gland** (in the hypothalamopituitary axis) in many endocrine functions. Centers for thirst, water balance, pleasure, rage, sexual desire, hunger, sleep patterns, temperature regulation, and aggression are located in the hypothalamus.

Locate the **mammillary bodies** on the inferior portion of the diencephalon and the **optic chiasma** just anterior to it. The **pineal gland** is located posterior to the thalamus and is an endocrine gland that secretes melatonin, a hormone that regulates daily rhythms. Both the pineal gland and the pituitary gland are covered in more depth in Exercise 24.

Midbrain

The midbrain is a small area posterior to the diencephalon. This small area consists of the **cerebral peduncles,** which occupy an area superior to the pons. The **cerebral aqueduct** passes through the midbrain. The peduncles are anterior to the **tectum,** which is a roof posterior to the cerebral aqueduct. Locate these features in the material in the lab

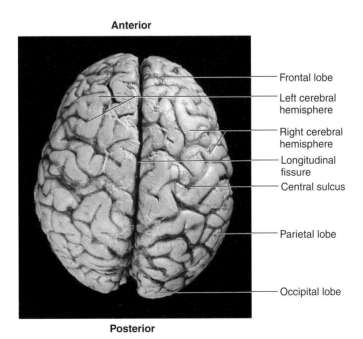

Figure 17.6 Brain, Superior View

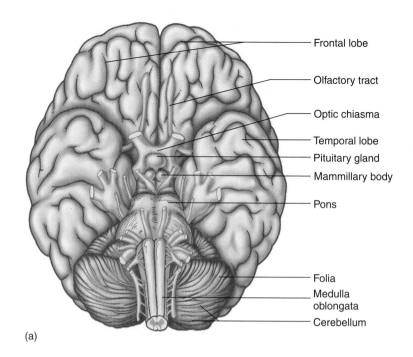

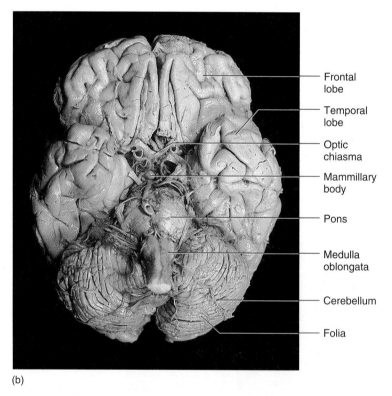

Figure 17.7 Brain, Inferior View
(a) Diagram; (b) photograph.

and in figure 17.8. Posterior to the tectum are four hemispheric processes known as the **corpora quadrigemina,** which consist of the **superior colliculi** (areas of visual reflexes) and the **inferior colliculi** (areas of auditory reflexes). The midbrain also houses a center known as the **substantia nigra** (not seen in midsagittal sections), which, when not functioning properly, causes Parkinson disease.

Hindbrain

The hindbrain consists of an anterior bulge known as the **pons,** a terminal **medulla oblongata,** and the highly folded **cerebellum** (ser-e-bel´ŭm) (figure 17.8). The pons is a relay center shunting information from the inferior regions of the body through the thalamus and to other areas of the brain. The pons

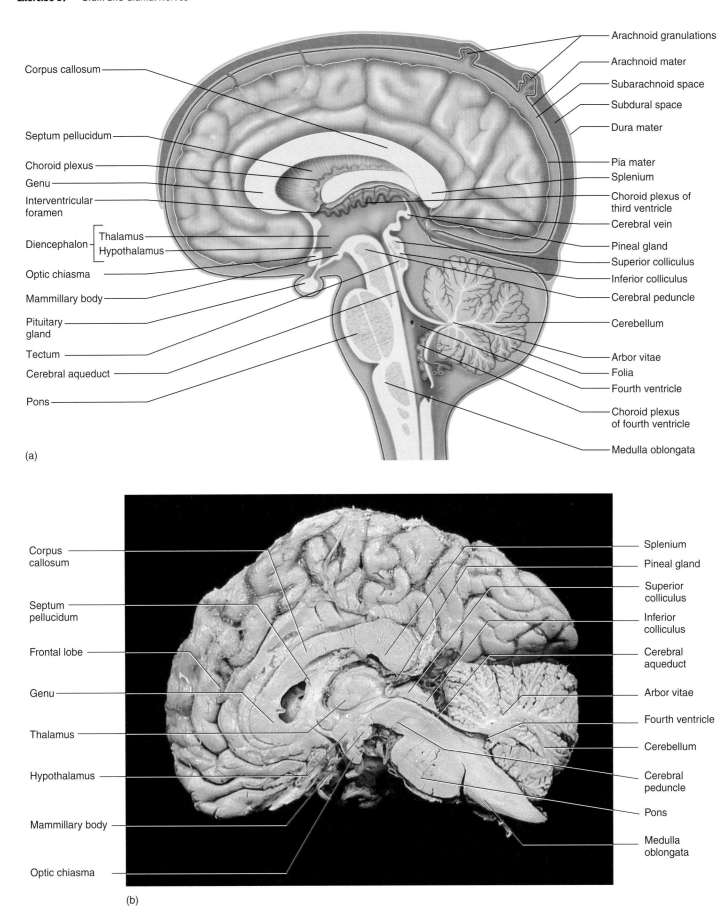

Figure 17.8 Brain, Midsagittal Section
(a) Diagram; (b) photograph.

has important **respiratory centers** that are involved in controlling the breathing rate.

The medulla oblongata has centers for respiratory rate control, centers for blood pressure control, and other vital centers. Some information from the right brain crosses over to the left side of the body in the medulla oblongata. Information from the left brain crosses over to the right side of the body. The area where motor tracts cross over in the medulla is known as the **decussation of the pyramids.** The medulla oblongata terminates at the foramen magnum and becomes the cervical region of the spinal cord. The cerebellum is a location primarily noted for muscle coordination and maintenance of posture. The cerebellum consists of an outer **cerebellar cortex** and an inner, extensively branched pattern of white matter known as the **arbor vitae.** The **folia** are folds in the cortex of the cerebellum and are seen in this section. The triangular space anterior to the cerebellum is the **fourth ventricle** and can be seen in figure 17.8. Examine the features of the hindbrain as described here and seen in material in the lab.

Coronal Section of the Brain

The brain consists of **unmyelinated** gray matter and **myelinated** white matter. The **gray matter** of the brain is extensive and forms the **cerebral cortex.** Most of the active, integrative processes of the brain occur in the cerebral cortex (figure 17.9). The cerebral cortex is approximately 4 mm thick and occupies the superficial regions of the brain. The cortex is where humans do most of their "thinking." It is the main metabolic area of the brain. Deep to the cortex is the **white matter** of the brain, which consists of tracts that take information from one region of the brain to the cerebral cortex for processing. Sensory information coming from the spinal cord moves through the inferior regions of the brain and through the white matter for integration in the cerebral cortex. White matter can also take information from one region of the cerebral cortex to another for integration or from the cerebral cortex back to the spinal cord and to other parts of the body for action. Gray matter is not restricted to the cerebral cortex, however. Deep islands of gray matter in the brain compose the **basal nuclei,** also seen in this section. Basal nuclei serve a number of functions in the brain, many of which involve subconscious processes, such as the swinging of arms while walking or the regulation of muscle tone. Basal nuclei not only are found in the cerebrum but surround the thalamus and midbrain as well.

Limbic System

Part of the limbic system is seen in a coronal section. The limbic system is a complex region of the brain involved in the formation of memory, mood, and emotion and has centers for feeding, sexual desire, fear, and satisfaction. The inferior portion of the limbic system has neural fibers that come from the olfactory regions of the brain. These are best seen in a model of the limbic system or in a transected brain. Examine figure 17.10 for the major features of the limbic system.

Brainstem

The brainstem consists of the midbrain, the pons, and the medulla oblongata. Look at a model or section of brain that has

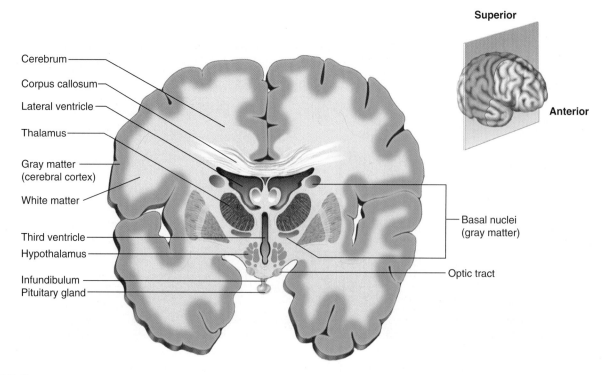

Figure 17.9 Coronal Section of the Brain

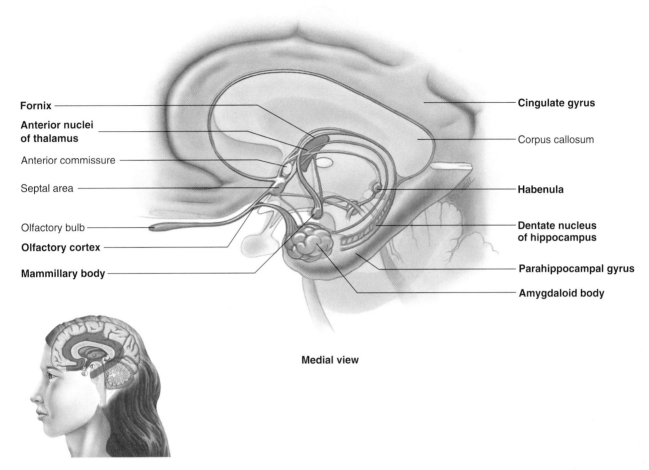

Figure 17.10 Limbic System

had the cerebrum removed. Locate the corpora quadrigemina (figure 17.11), along with the medulla oblongata and the pons.

Cranial Nerves

There are 12 pairs of cranial nerves. The cranial nerves are part of the PNS, but they are often studied along with the brain. The cranial nerves are listed by Roman numeral, and you should know the nerve by name *and* by number. Examine a model of the brain along with figure 17.12 and note that all the cranial nerves except nerve XII are in sequence from anterior to posterior. Nerves may be sensory, motor, or mixed (both sensory and motor). Sensory nerves take information to the CNS, whereas motor nerves conduct information away from the CNS, resulting in some kind of action (usually skeletal muscle contraction). When you study nerves, you should examine the anatomy of the brain to see where the nerve emerges from the brain. If you know, for example, that the abducent is found between the pons and the medulla oblongata, then you can use that nerve as a way to locate other nerves.

The **olfactory nerves** (*olfaction* = smell) pass through the ethmoidal bone and turn into the olfactory tract, which runs along the anterior base of the brain at the inferior aspect of the frontal lobe and transmits the sense of smell to the brain. The **optic nerve** (*optic* = sight) is a sensory nerve from the eye that leads to the base of the brain and forms the **optic chiasma.** Some tracts from the optic nerve lead to one side of the brain, whereas others cross to the other side. The **oculomotor nerve** (*oculo* = sight; *motor* = the function of the nerve) controls eye muscles and is located anterior to the pons, more or less in the midline of the brain, whereas the **trochlear nerve** (*trochlea* is in reference to the trochlea that the superior oblique muscle passes through) is found at about a 45 degree angle from midline on the lateral aspect of the pons. The large **trigeminal nerve** (*trigeminal* = triplets as the nerve divides into three parts) is at a 90 degree angle to the pons and is located on the lateral aspect of the pons. It is a mixed nerve that innervates much of the face. The **abducent nerve** (*abduce* = to pull away; the muscle controlled by this nerve pulls the eye laterally) is located at the midline junction of the pons and medulla oblongata, whereas the **facial nerve** (named for its innervation of the face) is more lateral. Posterior to the facial nerve is the **vestibulocochlear nerve** (the vestibule and cochlea are parts of the ear), and the nerve directly behind that is the **glossopharyngeal nerve** (*glosso* = tongue; *pharyngeal* = pharynx). The **vagus nerve** (*vagus* = wandering) is seen

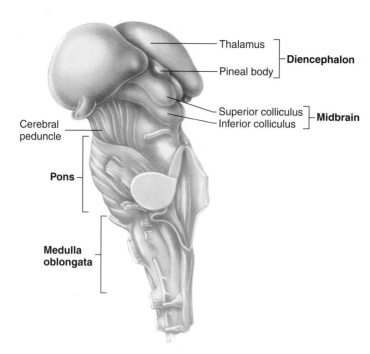

Figure 17.11 Brainstem, Posterolateral View

as a large nerve or large cluster of fibers on the lateral aspect of the medulla oblongata. The vagus travels to the thorax and abdomen. Posterior to the vagus nerve is the **accessory nerve,** which controls muscles of the neck. More toward the midline of the medulla oblongata is the **hypoglossal nerve** (*hypo* = below; *glossus* = tongue), which is the last of the cranial nerves. Locate these nerves in figure 17.12 and note their details in tables 17.2 and 17.3. There is a mnemonic to help you remember the sequence of cranial nerves. If you use the first letter of the name of the cranial nerve, you can learn the sequence of these nerves: "Old Oliver Ogg Traveled To Africa For Very Good Vacations And Holidays."

The cranial nerves are listed in table 17.4 as sensory nerves, motor nerves, or both sensory and motor nerves. A mnemonic device is also listed in the table to help you remember the name and function of each nerve.

Dissection of the Sheep Brain

Take a sheep brain to your table, along with a dissecting tray and appropriate dissection tools. If the brain still has the **dura mater,** examine this outermost of the three meninges. Look for the pituitary gland before you remove the dura. Cut through this layer to examine the other meninges underneath it. Deep to the dura mater is a filmy layer of tissue. This is the **arachnoid mater.** If you tease some of the membrane away from the brain, you will see that it has material extending into the **subarachnoid space,** which contains blood vessels. The subarachnoid space may contain some fluid, which is the **CSF.** Underneath this layer and adhering directly to the brain convolutions is the **pia mater,** which is the surface lining of the **convolutions** of the brain. Work in pairs during the dissection of the sheep brain.

Find the major lobes of the cerebrum of the sheep brain, the cerebellum, pons, and medulla oblongata (figure 17.13). Since sheep are quadrupeds, the flexure of the brain does not occur in them as it does in humans. Sheep have a horizontal spinal cord, whereas humans have a vertical one. Sheep also have a reduced cerebrum. Examine the inferior surface of the sheep brain. You should see the olfactory bulbs and tracts and the optic nerve and optic chiasma easily from this view. The pituitary gland will probably not be attached, but you should locate the infundibulum, caudad to the optic chiasma. Locate these structures in figure 17.14. If the sheep brain is intact, you will need to decide which brain will be sectioned in the midsagittal plane (figure 17.15) and which will be sectioned in the coronal plane (figure 17.16). For the midsagittal section, divide the brain along the length of the longitudinal fissure. Your cut should reflect a section illustrated in figure 17.15. Note that the sheep brain has an enlarged corpora quadrigemina, compared with that of humans. Locate the corpus callosum, lateral ventricles, third ventricle, hypothalamus, pineal gland, superior and inferior colliculi, cerebellum, arbor vitae, pons, medulla oblongata, cerebral aqueduct, and fourth ventricle. Make a coronal section about at the level of the optic chiasma of the cerebrum. Locate the cerebral cortex, cerebral medulla, lateral ventricles, corpus callosum, third ventricle, thalamus, and hypothalamus.

Exercise 17 Brain and Cranial Nerves

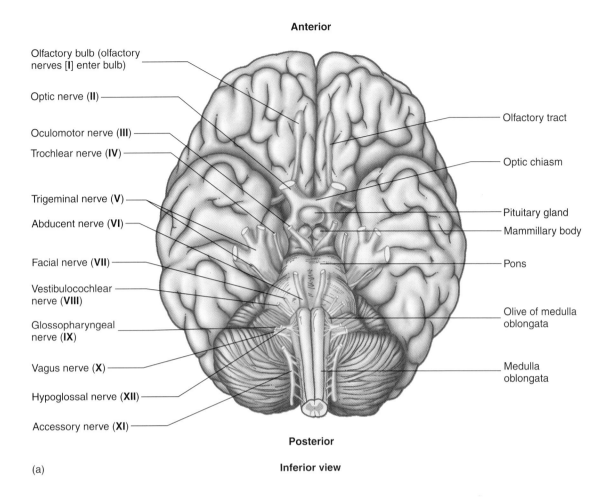

(a)

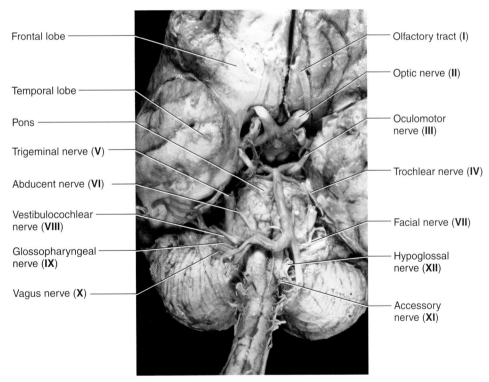

(b)

Figure 17.12 Cranial Nerves, Inferior View
(a) Diagram; (b) photograph.

TABLE 17.2 Cranial Nerves—Function

Number	Name	Function
I	Olfactory	Receives sensory information from the nose
II	Optic	Receives sensory information from the eye, transmitting the sense of vision to the brain
III	Oculomotor	Transmits motor information to move the eye muscles, particularly to the medial, superior, and inferior rectus muscles and to the inferior oblique muscle
IV	Trochlear	Transmits motor information to move the eye muscles, particularly the superior oblique muscle
V	Trigeminal	A three-branched nerve; transmits both sensory information from and motor information to the head
VI	Abducent	A motor nerve; moves the eye muscles, particularly the lateral rectus muscle
VII	Facial	Receives sensory information from the anterior tongue and takes motor information to the head muscles
VIII	Vestibulocochlear	Receives sensory information from the ear; the vestibular part transmits equilibrium information and the cochlear part transmits acoustic information
IX	Glossopharyngeal	A mixed nerve of the tongue and throat; receives information on taste
X	Vagus	Receives sensory information from abdomen, thorax, neck, and root of tongue; transmits motor information to pharynx and larynx; controls autonomic functions of heart, digestive organs, spleen, and kidneys
XI	Accessory	A motor nerve to the muscles of the neck; moves the head
XII	Hypoglossal	A motor nerve to the tongue

TABLE 17.3 Cranial Nerves—Location

Number	Name	Function
I	Olfactory	Begins in the upper nasal cavity and passes through the cribriform plate of the ethmoid bone. It synapses in the olfactory bulbs on each side of the longitudinal fissure of the brain. The fibers take information on the sense of smell and pass it through the olfactory tracts to be interpreted in the temporal lobe of the brain.
II	Optic	Takes sensory information from the retina at the back of the eye and transmits the impulses through the optic canal in the sphenoid bone. Some fibers cross at the optic chiasma and pass via the optic tracts to the occipital lobe, where vision is interpreted.
III	Oculomotor	Emerges from the surface of the brain near the midline and just anterior to the pons. It passes through the superior orbital fissure, innervates the inferior oblique muscle and the medial, superior, and inferior rectus muscles; it carries parasympathetic fibers to the lens and iris.
IV	Trochlear	Is at the sides of the pons at about a 45° angle from the midline of the brain. It passes through the superior orbital fissure to the superior oblique muscle.
V	Trigeminal	Is at a 90° angle from the midline at the lateral sides of the pons. The trigeminal has three branches: (1) The ophthalmic branch passes through the superior orbital fissure; (2) the maxillary branch passes through the foramen rotundum of the sphenoid bone; (3) the mandibular branch passes through the foramen ovale of the sphenoid bone and enters the mandible by the mandibular foramen and exits by the mental foramen.
VI	Abducent	Begins at the midline junction between the pons and the medulla oblongata and passes through the superior orbital fissure to carry motor information to the lateral rectus muscle of the eye.
VII	Facial	Begins as the first of a cluster of nerves on the anterolateral part of the medulla oblongata. It passes through the internal auditory meatus and through the inner ear to the stylomastoid foramen of the temporal bone to innervate facial muscles and glands. It receives sensory information from the anterior tongue. Sensory information of the tongue is interpreted in the temporal lobe of the brain.
VIII	Vestibulocochlear	Comes from the inner ear and passes through the internal auditory meatus. The conduction passes to the pons, and hearing and balance are interpreted in the temporal lobe.
IX	Glossopharyngeal	Passes through the jugular foramen to innervate muscles of the throat (pharyngeal branches) and the tongue (glossal branches). Motor portions of the nerve control some muscles of swallowing and salivary glands, whereas sensory nerves receive information from the posterior tongue and from baroreceptors of the carotid artery.
X	Vagus	Passes through the jugular foramen and along the neck to the larynx, heart, and abdominal region. The sensory impulses travel in this nerve from the viscera in the abdomen, the thorax, the neck, and the root of the tongue to the brain.
XI	Accessory	Multiple fibers arise from the lateral sides of the medulla oblongata and pass through the jugular foramen to numerous muscles of the neck.
XII	Hypoglossal	Begins at the anterior surface of the medulla and passes through the hypoglossal canal to innervate the muscles of the tongue.

Exercise 17   Brain and Cranial Nerves

TABLE 17.4 Cranial Nerves—Types

Number	Name	Name Mnemonic	Type	Mnemonic
I	Olfactory	Old	Sensory	Sally
II	Optic	Oliver	Sensory	Sells
III	Oculomotor	Ogg	Motor*	Many
IV	Trochlear	Traveled	Motor	Monkeys
V	Trigeminal	To	Both	But
VI	Abducent	Africa	Motor	My
VII	Facial	For	Both	Brother
VIII	Vestibulocochlear	Very	Sensory	Sells
IX	Glossopharyngeal	Good	Both	Bigger
X	Vagus	Vacations	Both	Better
XI	Accessory	And	Motor	Mega
XII	Hypoglossal	Holidays	Motor	Monkeys

*Many of the motor nerves have sensory fibers that come from proprioreceptors in the muscles they innervate. Information about the tension of the muscle is sent back to the brain to make adjustments in contractile rate. Since the main function of these nerves is motor, they are listed as motor nerves, even though they have some sensory capabilities.

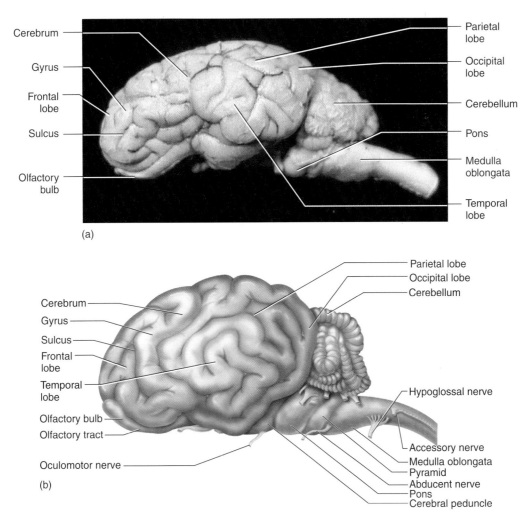

Figure 17.13   Brain of the Sheep, Lateral View
(a) Photograph; (b) diagram.

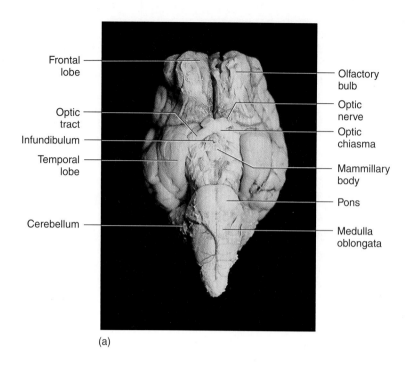

(a)

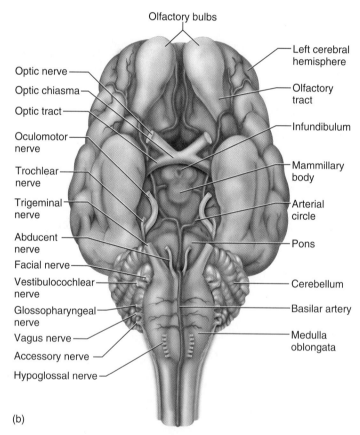

(b)

Figure 17.14 Brain of the Sheep, Inferior View
(a) Photograph; (b) diagram.

Exercise 17 Brain and Cranial Nerves

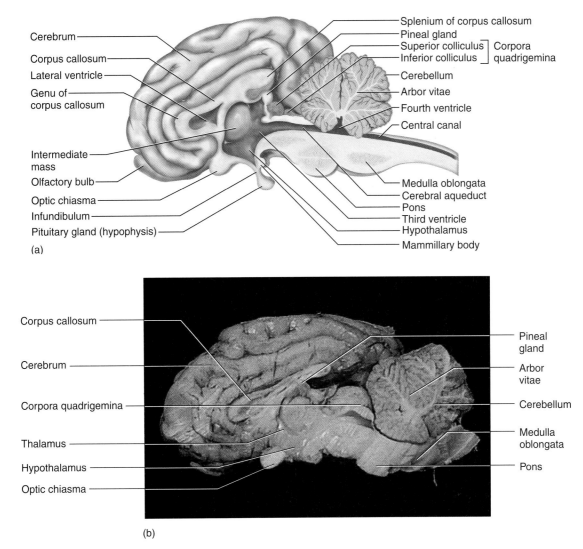

Figure 17.15 Brain of the Sheep, Longitudinal Section
(a) Diagram; (b) photograph.

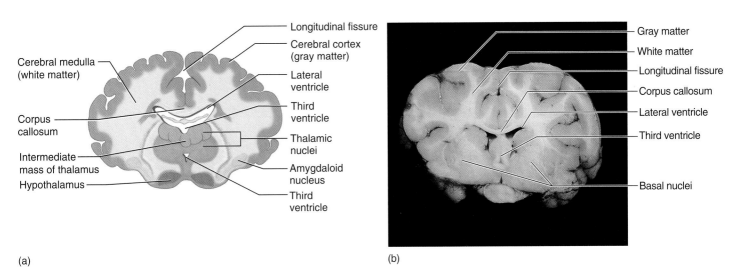

Figure 17.16 Brain of the Sheep, Coronal Section
(a) Diagram; (b) photograph.

Exercise 17 Review

Brain and Cranial Nerves

Name: _____

Lab time/section: _____

Date: _____

1. Which one of the meninges is just superficial to the brain? _____

2. What fluid is found in the ventricles of the brain? _____

3. Into what space does fluid flow from the cerebral aqueduct? _____

4. What is the difference between a gyrus and a sulcus? _____

5. Name all the lobes of the cerebrum. _____

6. What is the function of the precentral gyrus? _____

7. What sense does the temporal lobe alone interpret? _____

8. What physical depression separates the temporal lobe from the parietal lobe? _____

9. What structure connects the cerebral hemispheres? _____

10. Name the major regions of the midbrain. _____

11. What is the function of the cerebellum? _____

12. John "pulled a no-brainer" by hitting his forehead against a wall. What damage might he have done to the function of his brain, particularly the functions associated with the frontal lobe? _____

13. If a stroke had affected all the sensations interpreted by the brain concerning only the face and the hands, how much of the postcentral gyrus would be affected? _____

_____

14. Describe what effect the loss of an entire cerebral hemisphere would have on specific functions, such as spatial awareness or the ability to speak. _____

_____

_____

_____

15. Aphasia is loss of speech. Different types of aphasia can occur. If Broca's area were affected by a stroke, would the content of the spoken word be affected, or would the ability to pronounce the words be affected? _____

_____

16. Label the following illustration using the terms provided.

    pons   pineal gland   medulla oblongata   corpora quadrigemina
    corpus callosum thalamus   pituitary gland   hypothalamus
    cerebral aqueduct arbor vitae   fourth ventricle

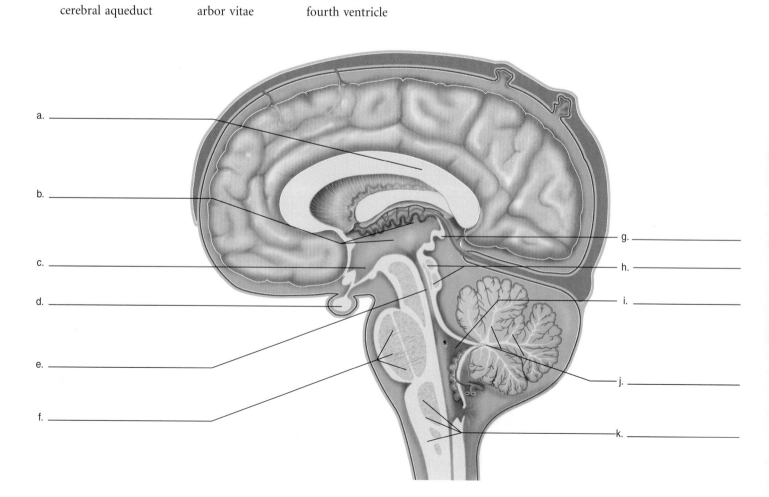

Exercise 17 Brain and Cranial Nerves

17. Label the brain using the terms provided.

    pons   frontal lobes
    pituitary gland   optic nerve
    medulla oblongata   temporal lobe

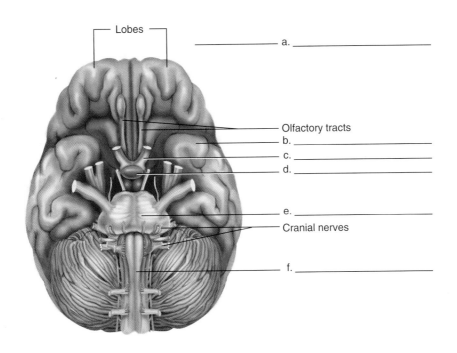

18. Label the following illustration using the terms provided.

    olfactory nerve   vagus nerve
    trochlear nerve   optic nerve
    trigeminal nerve   facial nerve
    abducent nerve   oculomotor nerve

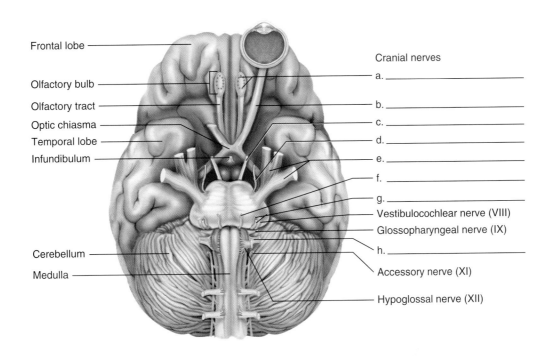

19. Describe the following nerves in terms of function (sensory, motor, or both).

   Optic nerve _____

   Trochlear nerve _____

   Glossopharyngeal nerve _____

   Hypoglossal nerve _____

   Vagus nerve _____

20. Name the cranial nerve that innervates each of the following areas.

   Anterior tongue _____

   Ear _____

   Mandible _____

   Retina of the eye _____

   Stomach _____

   Lateral rectus muscle of the eye _____

Exercise 18
Spinal Cord and Somatic Nerves

INTRODUCTION

The spinal cord begins at the foramen magnum of the skull and ends approximately at the second lumbar vertebra. This is because the vertebrae continue to grow after the spinal cord has reached its maximum length. The spinal cord receives sensory information from and transmits motor information to the spinal nerves, which radiate into the body as somatic nerves. In this exercise, you learn the major features of the spinal cord in longitudinal aspect and in cross section, as well as the nerves that are associated with the spinal cord.

The peripheral nervous system associated with the spinal cord consists of the somatic nerves and their ganglia. Somatic nerves are those that take sensory information from the body to the spinal cord and motor information from the spinal cord to the body. Nerves such as the radial nerve and the femoral nerve are somatic nerves. These nerves run throughout the body, receiving sensory information and sending it to the CNS or taking motor information from the CNS to skeletal muscles. The material in this lab is covered in the *Principles of Anatomy and Physiology* text in chapter 11, "Central and Peripheral Nervous System."

OBJECTIVES

At the end of this exercise, you should be able to

1. describe the anatomy seen in the longitudinal aspect of the spinal cord;
2. identify the major regions in a cross section of spinal cord;
3. list the major nerves that arise from each plexus;
4. name all the major nerves of the upper and lower extremities;
5. describe what region is innervated by a particular nerve.

MATERIALS

Models or charts of the central and peripheral nervous systems
Cadaver (if available)
Prepared slide of a spinal cord in cross section
Model or chart of a spinal cord in cross section and longitudinal section

PROCEDURE
Spinal Cord

Longitudinal Aspect of the Spinal Cord

Examine a model or chart in the lab of a longitudinal view of the spinal cord and locate the major features illustrated in figure 18.1. The spinal cord terminates inferiorly as the **conus medullaris** at approximately vertebra L2 and is attached to the coccyx by a thin thread of connective tissue known as the **filum terminale** (fī´lŭm tér-mi-nal´-ĕ), which is an extension of the pia mater. In the vertebral canal of the lumbar and sacral regions, the axons of the spinal cord continue as a parallel cluster of nerve fibers resembling a horse's tail called the **cauda equina.**

The spinal cord has a couple of expansions in two locations. The **cervical enlargement** is located at C3 through T2 because of nerves that supply the upper limb. The **lumbar enlargement** is located at about T7 through T11, and this expanse is due to the nerves that supply the lower limb. Compare the material in the lab with figure 18.1.

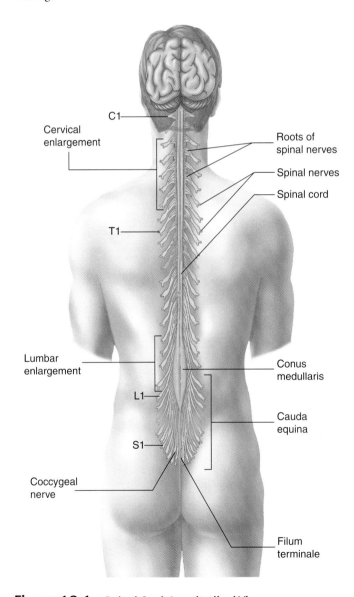

Figure 18.1 Spinal Cord, Longitudinal View

Cross Section of Spinal Cord

Examine a model or chart in the lab of a cross section of the spinal cord. Locate the **gray matter,** which appears in the shape of an H or butterfly in the middle of the spinal cord. The **white matter** is located on the periphery of the cord. Note how the distribution of gray and white matter in the spinal cord differs from that in the brain.

In a cross section of the spinal cord, you should see the gray matter divided into two narrow horns and two rounded horns. The narrow horns are known as the **posterior horns,** and these areas receive sensory information from the somatic nerves. The rounded horns are the **anterior horns,** and these send motor signals to the spinal nerves. In some parts of the spinal cord, there are additional sections of gray matter known as the **lateral horns.**

Each side of the gray matter is connected to the other by a crossbar known as the **gray commissure.** In the middle of the gray commissure is the **central canal,** which runs the length of the spinal cord. The white matter of the cord is divided into **tracts,** or **fasciculi** (fă-sik´-ū-lī), which take sensory information to the brain or motor information from the brain. The tracts that take sensory information to the brain are called **ascending tracts,** and those that receive motor information are called **descending tracts.** Tracts can be bundled into larger units called **columns.** You should also see a depression in the posterior surface of the spinal cord. This is the **posterior median sulcus.** The deeper depression on the anterior side is known as the **anterior median fissure.** Examine a model or chart in the lab and compare it with figure 18.2a.

Histology of the Spinal Cord

Examine a prepared slide of the spinal cord under low power and locate the **anterior** and **posterior horns,** the **gray commissure,**

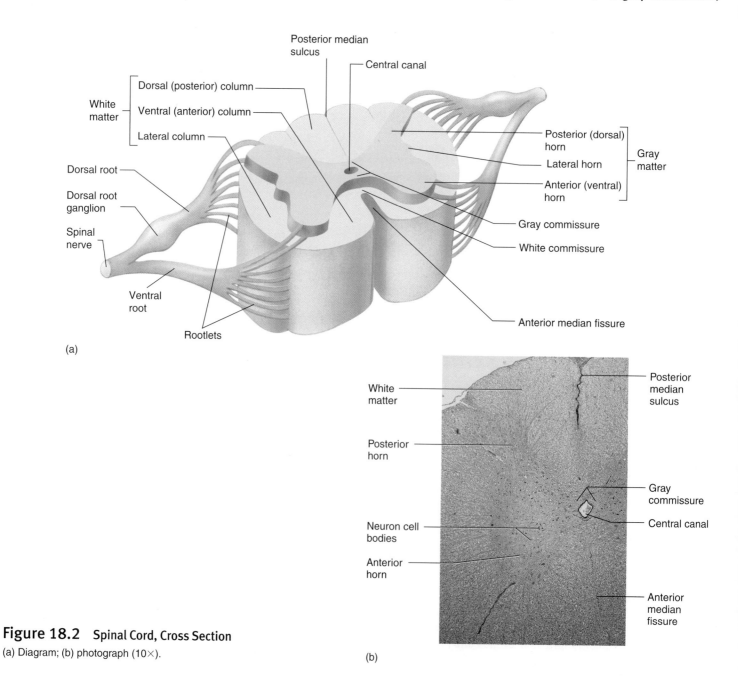

Figure 18.2 Spinal Cord, Cross Section
(a) Diagram; (b) photograph (10×).

the **central canal,** the **posterior median sulcus,** and the **anterior median fissure.** Examine the anterior horn of the spinal cord and look for the nerve cell bodies of the **multipolar neurons** there. Compare your slide with figure 18.2b.

Meninges

The spinal cord is covered by **meninges,** similar to the brain. The outer covering of the cord consists of the **dura mater.** Between the dura mater and the bone of the spinal canal is the **epidural space,** a site used for the injection of anesthetics. The next layer deeper to the dura mater is the **arachnoid mater.** Deep to the arachnoid is the **pia mater,** which is the innermost of the meninges and a thin cover on the spinal cord proper. Between the pia mater and the arachnoid is the **subarachnoid space,** which contains the **cerebrospinal fluid.** Examine figure 18.3 for an illustration of the meninges of the spinal cord.

Nerves Associated with the Spinal Cord

Nerve Structure

Parallel neuron fibers (axons) that carry information from one area to another are called **nerves** in the peripheral nervous system or **tracts** if they occur in the central nervous system. Nerves have a number of connective tissue wrappings that envelop the individual nerve fibers, clusters of fibers, and the entire nerve. The sheath that wraps around single nerve fibers is the **endoneurium,** and the sheath that wraps around groups of nerve fibers is the **perineurium.** The wrapping that covers the entire nerve is called the **epineurium.** Examine figure 18.4 for these layers.

Spinal Nerves and Plexuses

From the periphery of the vertebral column, the **dorsal ramus,** a branch of sensory and motor nerve fibers, unites with the **ventral ramus** to form a **spinal nerve.** The spinal nerve is located in the intervertebral foramen and divides into a **dorsal root ganglion** and a **ventral root.** The dorsal root ganglion contains the nerve cell bodies of the **sensory nerves,** and the **dorsal root** carries sensory information to the dorsal horn of the spinal cord. The nerve cell bodies of the motor nerves are located in the ventral horn of the spinal cord and exit via the ventral root. Examine the material in the lab and compare it with figure 18.5.

There are 31 pairs of **spinal nerves** that exit from the spinal cord. The spinal nerves are **mixed nerves** carrying both sensory and motor information. The spinal nerves are named according to their region of origin. There are **8 cervical nerves, 12 thoracic nerves, 5 lumbar nerves, 5 sacral nerves,** and **1 coccygeal nerve.** Some of these nerves exit the spinal cord and branch out to parts of the body individually, whereas others exit the spinal cord and form a branching network with other spinal nerves. These networks are called **plexuses,** and there are four generally recognized plexuses. The composition of the plexuses is outlined in table 18.1 and illustrated in figure 18.6.

Cervical and Brachial Plexus Nerves

The cervical plexus exits from the upper spinal nerves of the neck. An important nerve that comes from the cervical plexus is the **phrenic nerve.** This nerve runs to the diaphragm and is responsible for its contraction in breathing. Examine the nerves of the **cervical plexus** in figure 18.7. Many nerves from the cervical plexus innervate the muscles and skin of the neck and the skin of the ear. Since breathing is controlled by the phrenic nerve, the observation that a person is breathing is an obvious test that at least one nerve of the plexus is working! You can also test the function of the nerves by lightly pinching the sides of the neck or the ear for sensory perception.

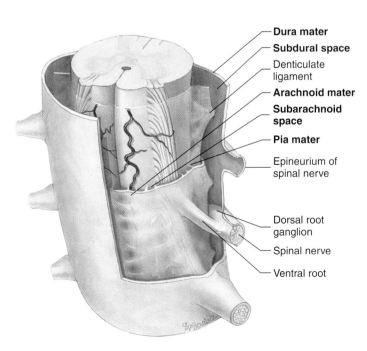

Figure 18.3 Spinal Meninges

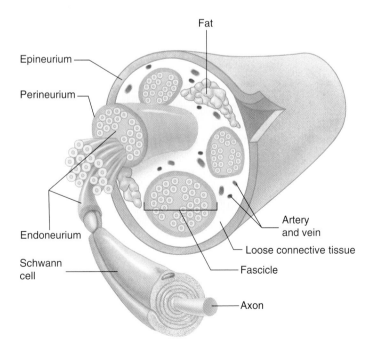

Figure 18.4 Coverings of the Nerve

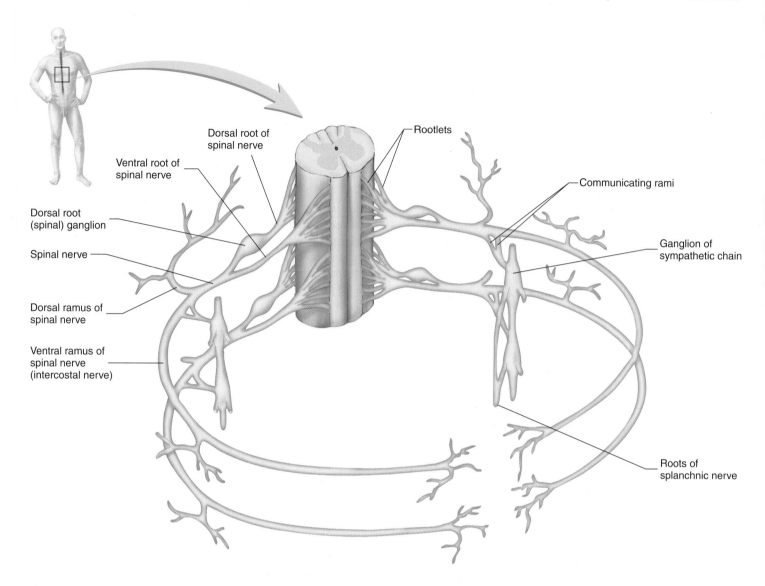

Figure 18.5 Nerves Associated with the Spinal Cord

TABLE 18.1	Somatic Nerves	
Name	Contributing to Plexus	Major Nerves of Plexus
Cervical	C1–4	Phrenic
Brachial	C5–T1	Radial, median, ulnar, musculocutaneous, axillary
Lumbar	L1–4	Femoral, obturator
Sacral	L4–S4	Sciatic (tibial and common peroneal)

The **brachial plexus** has branches from C5 to T1 and leads to nerves that primarily innervate the upper extremities. The plexus is illustrated in figure 18.8. The innervations of muscles were covered in the lab exercises on muscles. A major nerve from the brachial plexus is the **axillary nerve,** which innervates the upper shoulder and can be seen in figure 18.9. Another nerve, the **radial nerve,** mostly innervates the extensors of the hand (figure 18.10). The **musculocutaneous nerve** innervates the muscles that flex the arm and forearm, as seen in figure 18.11. The **ulnar nerve** crosses behind the medial epicondyle of the humerus and is commonly known as the "funny bone." The ulnar nerve involves some flexors and many hand muscles (figure 18.12), whereas the **median nerve** runs the length of each upper extremity and serves important hand and forearm flexors, as illustrated in figure 18.13. The nerves of this plexus can be tested by pinching the fingers, the medial and lateral aspects of the forearm, and the anterior and posterior aspects of the arm. Sensations from these areas are conducted to the brain via the brachial plexus.

Thoracic Nerves

Many nerves are not associated with a plexus. The thoracic nerves are a good example of this. Many of the **thoracic nerves** exit through the intervertebral foramina of the vertebral column and

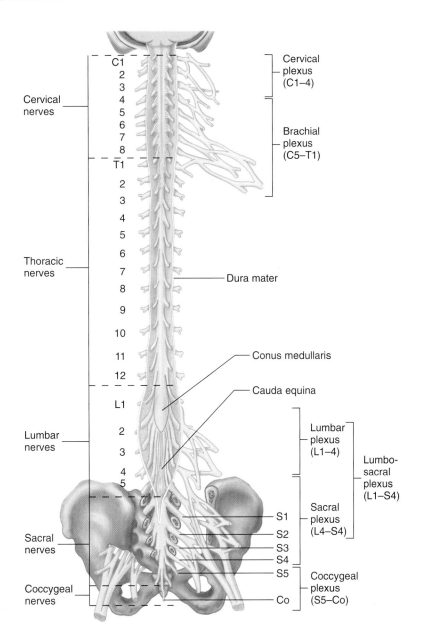

Figure 18.6 Spinal Cord and Nerve Plexuses

innervate the ribs, muscles, and other structures of the thoracic wall. These nerves can be seen in figure 18.5.

Lumbar and Sacral Plexus Nerves

The lumbar and sacral plexus take sensory information from and motor information to the lower limbs. Some authors combine the two plexuses into the **lumbosacral plexus,** as seen in figure 18.14. One of the nerves of the lumbar plexus is the **obturator nerve.** It innervates the adductor muscles of the thigh, as seen in figure 18.15. The **femoral nerve** is another nerve arising from the **lumbar plexus.** This large nerve passes anteriorly across the inguinal ligament and mostly innervates the muscles of the anterior thigh (figure 18.16). To test the nerves of this plexus, you can lightly pinch the anterior thigh for the femoral nerve and the medial thigh for the obturator nerve.

There are many nerves that come from the **sacral plexus.** Many innervate the pelvis and the muscles that move the hip, thigh, and leg. Two of the nerves from this plexus, the **tibial** and **common fibular (peroneal) nerve,** unite to form the **sciatic nerve** (see figure 18.14). These nerves innervate the leg and foot, as indicated in figures 18.17 and 18.18.

Test for the sciatic nerve by lightly pinching the posterior aspect of the thigh. Examine the nerves on models, in charts, or in a cadaver in the lab and compare them with the illustrations.

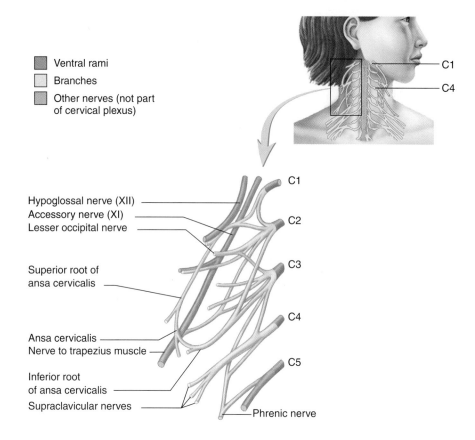

Figure 18.7 Nerves of the Cervical Plexus

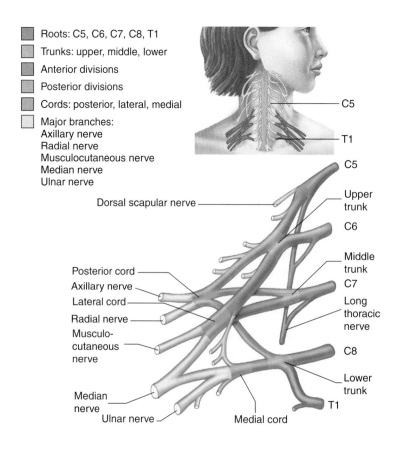

Figure 18.8 Nerves of the Brachial Plexus

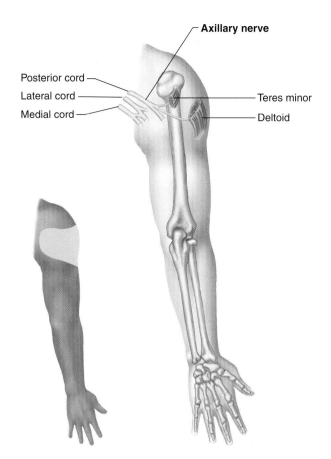

Figure 18.9 Axillary Nerve

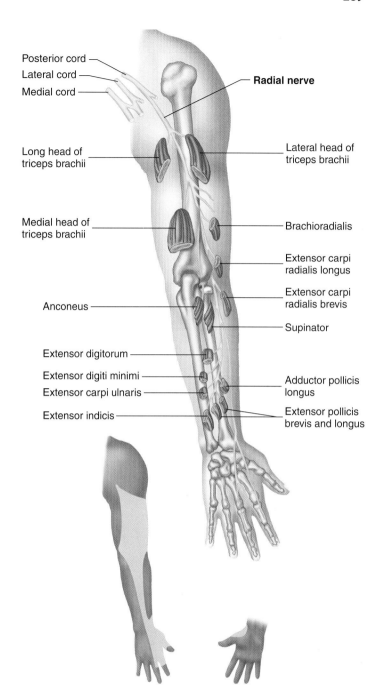

Figure 18.10 Radial Nerve

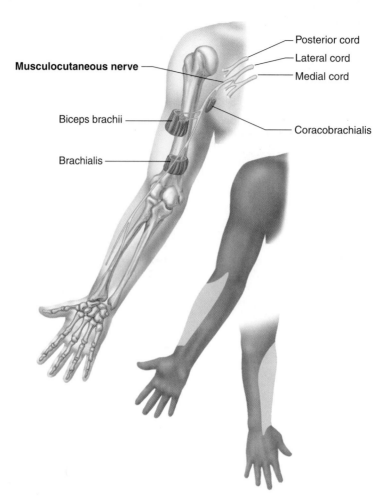

Figure 18.11 Musculocutaneous Nerve

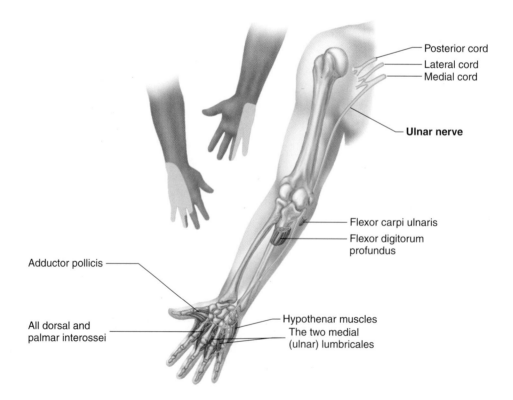

Figure 18.12 Ulnar Nerve

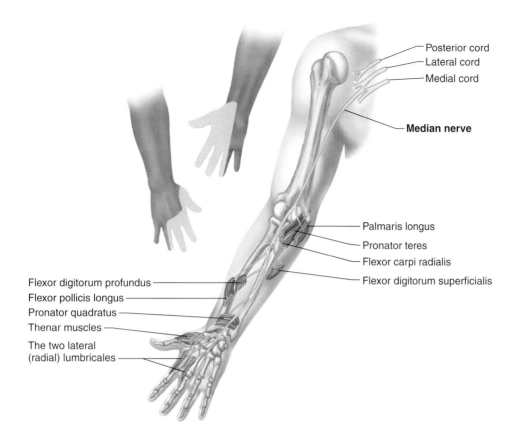

Figure 18.13 Median Nerve

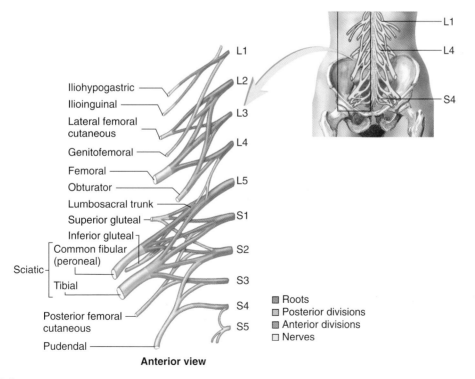

Figure 18.14 Lumbosacral Plexus

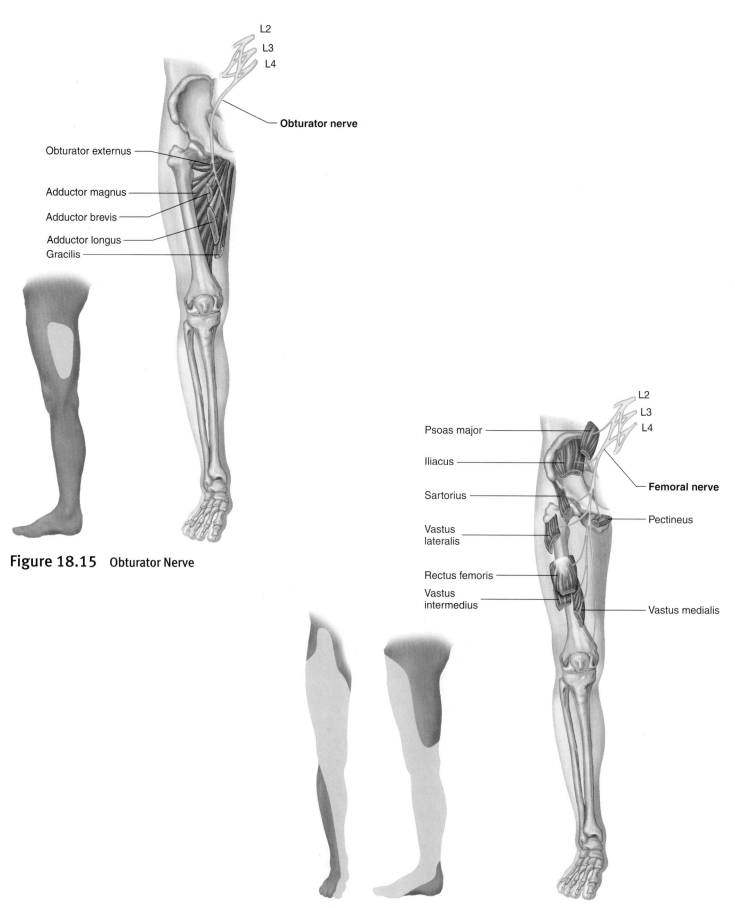

Figure 18.15 Obturator Nerve

Figure 18.16 Femoral Nerve

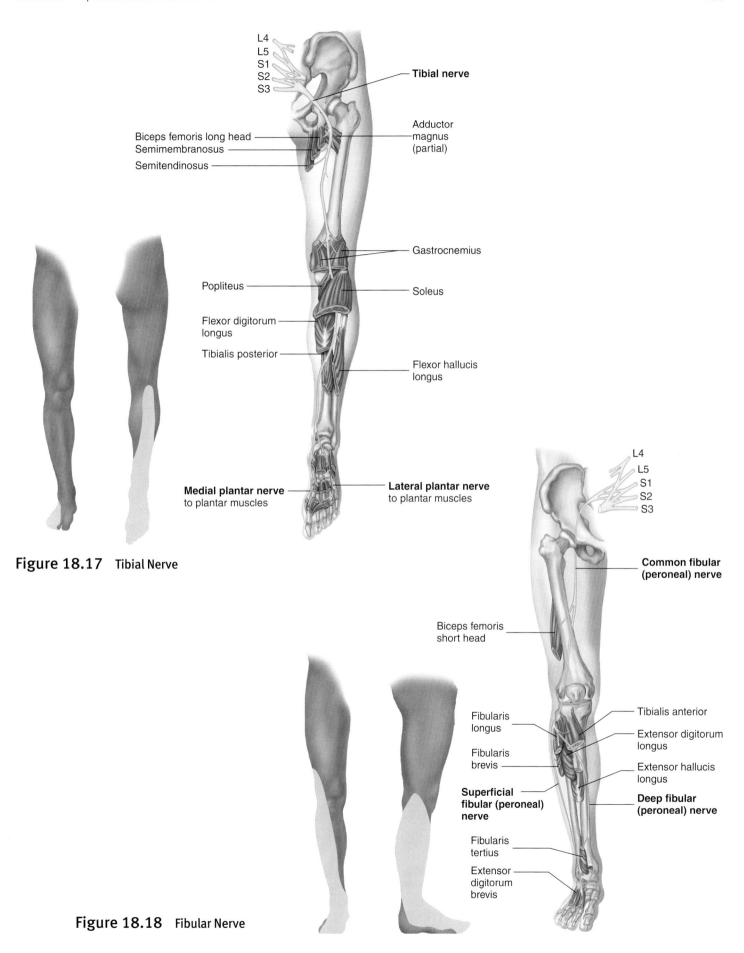

Figure 18.17 Tibial Nerve

Figure 18.18 Fibular Nerve

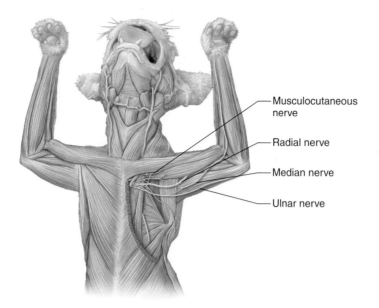

Figure 18.19 Brachial Plexus of the Cat

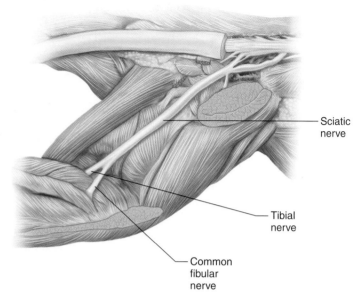

Figure 18.20 Sacral Plexus, Posterolateral View of Right Side

Cat Dissection

Brachial Plexus

The dissection of the brachial plexus involves careful dissection in the axillary region. Your cat should be placed ventral side up as you begin the dissection. Remove any skin, if you have not done so already. Bisect the pectoralis major and pectoralis minor muscles and carefully fold them back to see the brachial plexus that is deep to these muscles. Remove adipose tissue and fascia to expose the blood vessels and the brachial plexus. Do not damage the blood vessels, as you will study these in future labs. Examine figure 18.19 as you study the cat.

The large and anterior nerve of the brachial plexus is the **musculocutaneous nerve.** It innervates the skin of the brachium and is the motor nerve of the biceps brachii muscle.

The **radial nerve** is the largest nerve of the plexus and is posterior to the musculocutaneous nerve. It is located on the posterior of the distal arm, where it innervates the muscles on the dorsal side of the forelimb, including many of the extensor muscles.

The **median nerve** is located alongside the brachial artery and is in the midline of the brachium and antebrachium. It innervates many of the flexor muscles.

The **ulnar nerve** is the most posterior of the brachial plexus nerves. It travels posterior to the medial epicondyle of the humerus to innervate the muscles on the ulnar side of the antebrachium.

Sacral Plexus

The sacral plexus is best seen from the dorsal view. Examine figure 18.20 and locate the large **sciatic nerve** by separating the biceps femoris on the dorsal thigh. The sciatic nerve is composed of two nerves, the medial **tibial nerve** and the lateral **common fibular nerve.** These can be seen at the distal part of the thigh.

Exercise 18 Review

Spinal Cord and Somatic Nerves

Name: _____
Lab time/section: _____
Date: _____

1. In the spinal cord, what type of impulse (sensory/motor) travels through the

   a. anterior gray horn? _____

   b. posterior gray horn? _____

   c. ascending spinal tracts? _____

   d. descending spinal tracts? _____

2. What causes the cervical enlargement of the spinal cord? _____

3. Where is the filum terminale located? _____

4. What is the conus medullaris? _____

5. What is the cauda equina? _____

6. In the spinal cord, which is deep to the other, the white matter or the gray matter? _____

7. What is the area of gray matter that is found between the lateral halves of the spinal cord? _____

8. The subarachnoid space is filled with what fluid? _____

9. What major nerves arise from the following plexuses? _____

   a. Cervical _____

   b. Brachial _____

   c. Lumbar _____

   d. Sacral _____

10. In terms of function, how does the dorsal spinal root vary from the ventral spinal root? _____

11. What is the endoneurium? _____

12. How do tracts differ from nerves? _____

13. What is a mixed nerve? _____

14. The diaphragm's contractions are regulated by what nerve? _____

15. The muscles of the arm, such as the biceps brachii, have what innervation? _____

16. The extensor muscles of the hand are controlled by what nerve? _____

17. The sciatic nerve is composed of two nerves. What are they? _____

18. Fill in the illustration using the terms provided.

    anterior median fissure   posterior median sulcus

    gray commissure   central canal

    anterior horn   posterior horn

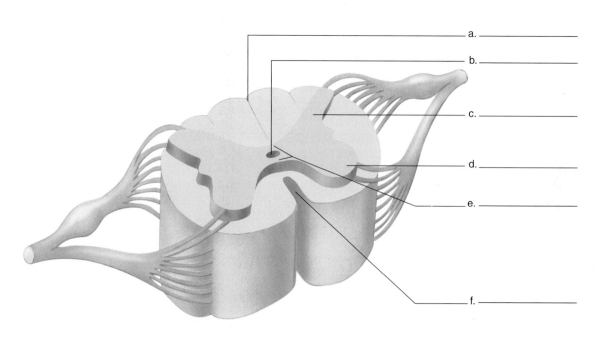

a. _____
b. _____
c. _____
d. _____
e. _____
f. _____

Exercise 19
Nervous System Physiology: Stimuli and Reflexes

INTRODUCTION

Nervous tissue shows two fundamental properties: irritability and conductivity. **Irritability** is the potential of a nerve to respond to some kind of stimulus (chemical, mechanical, electrical), and **conductivity** is the movement of a nerve impulse along the length of the neuron. Nerves receive information from a particular area (sense organ, regions in the CNS, etc.) and carry signals to either the brain for interpretation or an effector for some kind of action. The nature of nerve impulses and reflexes is covered in the *Principles of Anatomy and Physiology* text in chapter 10, "Functional Organization of Nervous Tissue."

In this exercise, you study the basic properties of neuronal conduction and sensitivity to various stimuli. In addition to studying the neurons, you experiment with nerves as they form reflex arcs in several parts of the body.

OBJECTIVES

At the end of this exercise, you should be able to

1. describe the threshold nature of nerve responses;
2. list three things that cause a nerve to be stimulated;
3. name one substance that stimulates nerves and one that inhibits them;
4. describe reflex arcs;
5. list all the parts of a monosynaptic and a polysynaptic reflex arc;
6. define *hyporeflexic* and *hyperreflexic*.

MATERIALS

Nerve Physiology Section

Ph.I.L.S. CD and compatible computer
Frog
Latex or plastic gloves
Glass rod with hook at one end
Hot pad or mitt
Bunsen burner
Matches or flint lighter
Frog Ringer's solution in dropper bottles
Ice bath
Microscope slide or small glass plate
Filter paper or paper towel
Dissection equipment for live animals
Scalpel or scissors
Cotton sewing thread
Stimulator apparatus with probe
Myograph transducer
Physiograph or physiology computer
5% sodium chloride solution
0.1% hydrochloric acid solution (1 mL of concentrated HCl in 1 liter of water)
Procaine hydrochloride solution
Gauze squares
Small beakers (50 mL)
Sterile cotton applicator sticks

Reflex Section

Patellar reflex hammer
Rubber squeeze bulb
Models or charts of spinal cord and nerves

PROCEDURE

Virtual Experiments in Neurophysiology

Load the Physiology Interactive Lab Simulations (Ph.I.L.S.) disc. You will need a minimum of either a Windows XP, Vista or Pentium III computer or a Macintosh 10.2, G3 processor and Adobe Flash 8.x Player to run the program.

For all of the virtual muscle experiments you should follow the standard procedure that comes up when you open the CD-ROM.

1. Click on the number of the exercise that you want to complete. These are under the **Resting Potentials** or **Action Potentials** headings in the CD.

After you have selected your choice you should:

2. Read the objectives and introduction to the lab simulation. You can click on highlighted terms to see pictures or animations of the material in question.
3. Take the pre-lab quiz. If you get a question wrong, the program will let you know and provide you with the correct answer.
4. Click on the **Wet Lab** tab and read the material. The highlighted terms open up videos of an actual procedure.
5. Click on **Continue,** which opens up the **Laboratory Exercise.**
6. Perform the exercises as outlined in the paragraphs that follow. Print out any information that your instructor directs after you have performed the experiment, or download the information to a portable data storage device.
7. Take the post-lab quiz to determine your understanding of the lab.

The information for the specific lab exercises follows.

Resting Potentials 8. Resting Potential and External [K$^+$]

Open the **Resting Potential and External** [K$^+$] program and start the procedure by initiating a recording on the control panel. Insert the microelectrode into the muscle and then click on the Journal Entry to get an initial reading.

On the left side of the Data Acquisition Unit there is a black circle with a gray sphere in the middle. This is the micromanipulator. If you move the cursor over this area, two arrows will show up. These are the up and down arrows, and they control the position of the microelectrode in the muscle. As you penetrate the muscle for the first time you will see a change in the voltage on the right screen.

If you move the mouse to the left side of the data acquisition unit you will bring up an "Overhead View" of the microelectrode in the muscle. If you continue to move the cursor to the left you will see two gray cylinders. These control where the microelectrode is moved in the muscle tissue. Every time you want to record data you must move the microelectrode to a new location. You must enter all the data in the journal for all of the concentrations of saline before you plot the graph.

What impact does increasing the potassium ion concentration have on membrane depolarization? Record your answer here.

Effect of increasing potassium ion concentration: _____
_____

Resting Potentials 9. Resting Potential and External [Na$^+$]

Follow the same procedure as in number 8 on the CD-ROM for the **Resting Procedure and External Sodium Concentration** program. The set up is the same except that you will be increasing the sodium ion concentration rather than the potassium ion concentration.

Comparing the two solutions, which has a greater impact on membrane dynamics, sodium or potassium? Record your answer here.

Greater impact on membrane dynamics: _____

Action Potentials 10. The Compound Action Potential

Place all of the color-coded cables on the appropriate attachment points of the chamber holding the nerve. The drain tap is the small circular wheel located at the top of the chamber on the left-hand side. You must click and hold the knob in order for it to drain.

If you examine the graph you will see the stimulus on the left side of the graph followed by a compound action potential on the right. As with the muscle physiology experiment, if you move the cursor to the graph you will see crosshairs as you move the cursor. Place the crosshairs on the top of the curve and click. Click again on the baseline of the curve. When you click on the **Journal** icon it will enter the data into the journal.

You must click the X on the journal in order to proceed with obtaining more data. The threshold voltage is the voltage below which you get no response from the nerve. What was your threshold voltage? Record your answer here.

Threshold voltage: _____

Action Potentials 11. Conduction Velocity and Temperature

In this exercise you examine the speed of nerve conduction by altering the temperature of the nerve. The experiment runs at room temperature (22° C) and cold temperature (10° C). Follow the instructions as presented. Make sure that you click on both power buttons to turn on the equipment. Place all of the color-coded cables on the attachment points on the chamber holding the nerve. The tap is the small circular wheel located at the top of the chamber on the left-hand side. You must click and hold the knob in order for it to drain. After you run the first experiment at room temperature make sure that you enter your data into the journal by clicking on the **Journal** icon. Run the second experiment and compare the results of nerve conduction at different temperatures.

In which sequence (room temperature or cold) did you see a slowing down of the nerve conduction? Record your answer here.

Slower nerve response: _____

How does this response correlate to decreasing enzyme function based on temperature? Record your answer here.

Correlation: _____

Action Potential 12. The Refractory Period

In this exercise, begin by increasing the voltage until there is no greater peak produced with increasing voltage. This value is the maximum recruitment for the nerve, and it means that all of the axons of the nerve are firing. With less voltage only some of the axons are firing and there is less of a response.

What is the maximum recruitment voltage that you obtained? Record your answer here.

Maximum recruitment voltage: _____

In the next section you can alter the frequency between the shocks and determine the absolute refractory period and the relative refractory period. Enter these data here.

Absolute refractory period: _____
Relative refractory period: _____

Review the structure of the neuron in Exercise 16 for descriptions of the dendrites, nerve cell body, and axon and the anatomy of the nerve in Exercise 18. Make sure you read through all of this exercise before performing the experiments. Wear latex gloves as a general precaution when working with fresh specimens, such as frogs.

Frog Nerve Conduction

In the first part of this lab exercise you determine if nerves respond to only a specific stimulus or if they are more general and respond to many stimuli. You will also determine if certain materials or

Exercise 19 Nervous System Physiology: Stimuli and Reflexes

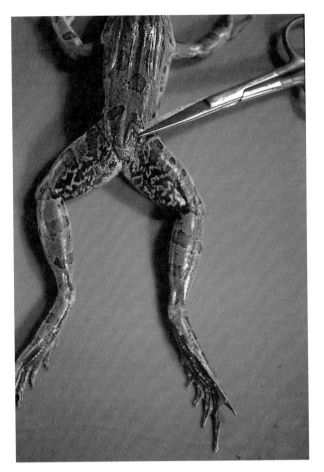

Figure 19.1 Removal of the Skin and Nerve Preparation

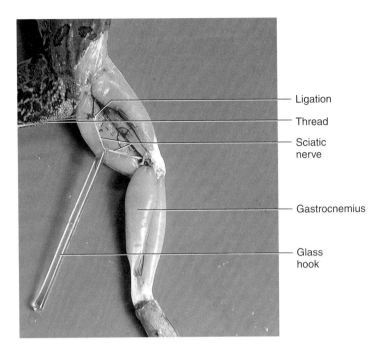

Figure 19.2 Ligation of the Sciatic Nerve of the Frog

environmental conditions inhibit nerve response. In this exercise, you observe the process of nerve impulse conduction by experimenting on frogs or watching a demonstration, depending on the wishes of your instructor. If you are to experiment on frogs, obtain a doubly pithed frog, dissection equipment, frog Ringer's solution, and various test solutions and take them to your table. Keep the frog nerve preparation moist with Ringer's solution and cotton gauze during the entire experiment. Prepare the frog by cutting the skin away from the hip (figure 19.1). Do not cut, pinch, or otherwise damage the sciatic nerve on the posterior side of the thigh.

Gently remove the nerve from between the muscles with a glass rod (figure 19.2). Do not stretch the nerve; leave it intact alongside the muscles. You can attach the tendon of the gastrocnemius to a myograph transducer, as you did in Exercise 11, or just examine the muscle to see if there is a response.

Nerve Response to Physical Stimuli

Flush the nerve with Ringer's solution and make sure it remains moist. In this part of the exercise, you examine the effects of physical stimuli on nerve impulse conduction. You measure the effects of the nerve stimulation by the contraction of the gastrocnemius muscle. If the nerve is stimulated, then the gastrocnemius muscle should contract. This is due to the neuromuscular junction where the nerve impulse crosses the junction and causes the muscle to contract.

Cut a small (10 cm) section of cotton thread and gently slip it under the sciatic nerve with a pair of fine forceps. Gently move the thread up toward the hip. When you reach the place where the nerve descends into the muscle, loop the thread and ligate (tie off) the nerve close to the sacrum (figure 19.2) while watching the gastrocnemius muscle.

As the thread begins to tighten on the nerve, record the response in the space provided.

Response of the nerve to physical stimulation: _____

_____

_____

After you have ligated the nerve, cut it from the anterior side, leaving the nerve attached to the gastrocnemius muscle. Moisten the nerve with Ringer's solution and prepare for the next experiment.

Nerve Response to Electrical Activity

Place a stimulator probe connected to a stimulator underneath the sciatic nerve, lifting the nerve away from the gastrocnemius muscle. Keep the nerve moist as you determine the minimum voltage (**threshold voltage**) required for nerve conduction. Set the stimulator to a frequency of two pulses per second and a duration of 10 milliseconds. The voltage should be set at zero (with the knob on 0.1 volt). Slowly increase the voltage until you see the gastrocnemius twitch. As soon as you see the gastrocnemius twitch at the lowest voltage, record this as threshold voltage in the space provided. Turn the voltage to zero, flush the nerve with Ringer's solution, and let it rest for a moment.

Threshold voltage: _____

Continue to increase the voltage until the muscle contracts maximally. Record this as the **maximum recruitment voltage.** This voltage is obtained when all the neurons of a particular nerve are stimulated.

Maximum recruitment voltage: _____

Nerve Response to Chemical Stimuli

In the following two experiments, you will test the response of the nerve to different chemical agents. Make sure you observe the nerve as soon as you apply the solution and rinse it as soon as the observation is made.

Acid Solution

Apply a 0.1% hydrochloric acid solution to a cotton applicator stick and touch the applicator gently to the nerve. Record the nerve response.

Response to hydrochloric acid: _____
_____
_____

Flush the nerve with Ringer's solution and let it rest for a moment.

Salt Solution

Gently apply a 5% sodium chloride solution to the nerve with a new cotton applicator stick. Record the response. Flush the nerve with Ringer's solution and let it rest for a moment.

Response to sodium chloride solution: _____
_____
_____

Nerve Response to Anesthetics

Apply a solution of **procaine hydrochloride (Novocain)** to the nerve by soaking a small square of gauze or cotton with procaine solution and placing it on the nerve for a moment. As the gauze remains on the nerve, set up the stimulator apparatus and place the nerve over the stimulator probes. Remove the gauze and stimulate the nerve with a single pulse stimulus at the voltage that produced a maximum recruitment voltage in the previous experiment. If the nerve responds to the stimulus, leave the procaine hydrochloride on longer. When the nerve does not respond, remove the gauze and stimulate the nerve once every 30 seconds until it recovers from the local anesthetic. Keep the nerve moist at all times with frog Ringer's solution. Record the recovery time.

Recovery time: _____

Nerve Response to Changes in Temperature

Gently touch the nerve with a glass rod at room temperature. Record the response.

Response to gentle touch: _____
_____
_____

Place the glass rod in an ice bath. As it equilibrates in the ice water, place a small chip of ice on the nerve and let it stay there for a moment. Gently touch the nerve with the cold rod and record the response. Flush the nerve with room temperature Ringer's solution and let it rest for a moment.

Response to gentle touch with cold stimulation: _____
_____
_____

Take another glass rod in a hot pad or mitts and heat one end of it in a Bunsen burner. Touch the nerve with the hot end of the glass rod. What is the response? Record your result.

Response to gentle touch with hot stimulation: _____
_____
_____

Clean Up
Make sure you clean your station before continuing. Place the specimen in the appropriate container, and use care when cleaning sharp instruments, such as scalpels or razor blades.

Reflexes

A reflex is defined as a motor response to a stimulus without conscious thought. Reflexes are involuntary, predictable responses to stimuli. Reflexes occur through **reflex arcs,** and these arcs have the following structure:

1. **Receptor** (the structure that receives the stimulus)
2. **Afferent (sensory) neuron** (the neuron taking the stimulus to the CNS)
3. **Integrating center** (the brain or spinal cord)
4. **Efferent (motor) neuron** (the neuron taking the response from the CNS)
5. **Effector** (the structure causing an effect)

If the effector is skeletal muscle, it is a **somatic reflex.** If the effector is a gland, smooth muscle, or cardiac muscle, it is called a **visceral,** or **autonomic, reflex.**

Most reflexes involve many neurons with many synapses and are called **polysynaptic reflex arcs.** A few reflexes involve only two neurons, a sensory neuron and a motor neuron with one synapse between them, and these are called **monosynaptic reflex arcs.** A polysynaptic reflex arc is illustrated in figure 19.3.

These reflexes depend on a **stimulus,** or an environmental cue, a receptor that is sensitive to the stimulus, a sensory neuron, a motor neuron, and an effector. The polysynaptic reflex arc has these structures, as well as an **association neuron,** or **interneuron,** located between the sensory and motor neurons. Interneurons can take information to the brain via ascending tracts in the spinal cord.

Testing for reflexes is very important for the clinical evaluation of the condition of the nervous system. Decreased response or exaggerated response to a stimulus may indicate disease or damage to the nervous system.

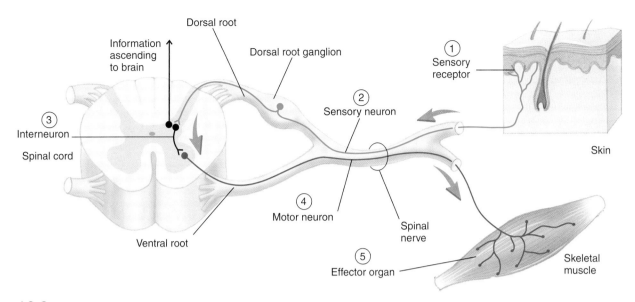

Figure 19.3 Reflex Arc
A stimulus is picked up by the sensory receptor (1), which transmits the stimulus to the sensory neuron (2). The stimulus travels either through the interneuron (3) or to the motor neuron (4), where it stimulates the effector (5) for a response.

In this experiment, you test several reflexes and determine if the response is **normal** (movement of an inch or 2), **hyporeflexic** (showing less than average response), or **hyperreflexic** (showing an exaggerated response). In clinical settings, stretch reflexes are often used to determine one's neurologic condition or to identify conditions such as hypothyroidism. In hypothyroidism, the reflexes are depressed. In some CNS-damaged patients, the reflexes are exaggerated.

Stretch Reflexes

For stretch reflexes, receptors are in the muscle spindle. Stretching the muscle causes an increase in action potentials in the sensory neuron. The impulse travels to the motor neuron, causing contraction of the muscle that is stretched. A typical example of a stretch reflex is the patellar reflex, in which striking the patellar ligament stretches the quadriceps muscles. Sensory neurons transmit this information to the spinal column where motor neurons stimulate the quadriceps muscle to contract, thus extending the leg. Stretch reflexes work best when the subject is relaxed.

Patellar Reflex

The **patellar reflex** tests the conduction of the femoral nerve. It is the most commonly performed reflex test in clinical settings. Sit down on the lab table and, with your leg hanging over the edge of the table, have your lab partner tap you on the patellar ligament with the blunt side of a patellar reflex hammer. The percussion should be placed 3 to 4 cm below the inferior edge of the patella, and it should be firm but not hard enough to hurt. Look for extension of the leg as a response to the patellar reflex (figure 19.4).

The tap stimulates the stretch receptors in the tendon and is representative of a **monosynaptic reflex arc** (one with a sensory neuron directly synapsing with a motor neuron). Place a check mark next to the term that describes the degree (hyperreflexic, normal, hyporeflexic) of the response.

Patellar Reflex
_____ Normal (foot moves a few inches)
_____ Hyperreflexic (foot moves extensively)
_____ Hyporeflexic (foot moves little or not at all)

Figure 19.4 Patellar Reflex

Figure 19.5 Triceps Reflex

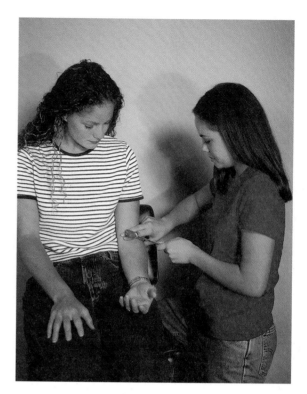

Figure 19.6 Biceps Brachii Reflex

Triceps Brachii Reflex

The **triceps brachii reflex** tests the radial nerve. Sit on a chair or lie down on your back on a clean lab table or cot and place your forearm on your abdomen. Have your lab partner tap the distal tendon of the triceps brachii muscle about 2 inches proximal to the olecranon process. Look for the triceps muscle to twitch (figure 19.5). If you do not obtain good results with the triceps reflex you should hold the subject's arm at a 90-degree angle from anatomic position so that the arm is abducted. From this position the forearm should be at a 90-degree angle so the forearm dangles from the arm. As the experimenter, support the arm so that the subject is relaxed. Tap the tendon as described previously and look for movement. Record your result by placing a check mark next to the term that describes the degree of the reflex response.

Triceps Brachii Reflex
_____ Normal
_____ Hyperreflexic
_____ Hyporeflexic

Biceps Brachii Reflex

The **biceps brachii reflex** tests the musculocutaneous nerve. Sit comfortably and have your lab partner place his or her fingers on the biceps tendon just proximal to the antecubital fossa (figure 19.6).

Your lab partner should tap his or her fingers with the reflex hammer while they remain on the tendon and look for the biceps brachii muscle contraction. Record your results by placing a check mark next to the term that describes the degree of the reflex response.

Biceps Brachii Reflex
_____ Normal
_____ Hyperreflexic
_____ Hyporeflexic

Calcaneal (Achilles) Tendon Reflex

To test the **calcaneal tendon reflex,** kneel on a chair with your foot dangling over the edge (figure 19.7). Have your lab partner tap the calcaneal tendon in order to test the tibial nerve.

As your lab partner taps your calcaneal tendon, look for plantar flexion of the foot. You may see an initial movement of the foot due to the depression of the tendon by the reflex hammer, but there should be a slight pause and then another quick movement of the foot. Record your results.

Calcaneal Tendon Reflex
_____ Normal
_____ Hyperreflexic
_____ Hyporeflexic

Eye Reflexes

The automatic blinking of the eye is important to keep material, such as dust, away from the outer layer of the eye, known as the cornea. In the first part of this experiment, have your lab partner try to make you blink by carefully flicking his or her fingers near your

Figure 19.7 Calcaneal Reflex

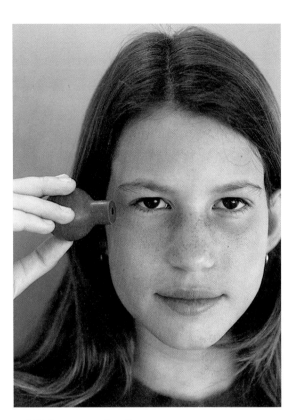

Figure 19.8 Corneal Reflex

eyes. They should not come close enough to touch your eyes! Can you prevent the blinking reflex? Record your answer.

Control of Blink Reflex

_____ Yes

_____ No

Now have your lab partner take a *clean* rubber squeeze bulb (a large pipette bulb works well) and squirt a blast of air across the surface of the eye (figure 19.8). Use either new bulbs or ones that are free of debris to avoid damage to the eye. Can you inhibit the corneal reflex? Record your response.

Control of Corneal Reflex

_____ Yes

_____ No

Plantar Response

The extension of the hallux (big toe) in reponse to stroking the plantar surface of the foot is the **plantar response** or **Babinski reflex.** This response is perfectly normal in infants under the age of 1. The plantar response in infants is due to incomplete myelination of some nerve fibers. In adults this stimulation usually results in flexion of the toes. A plantar response in adults may occur due to damage to the pyramidal tracts and is important in determining spinal damage.

Using the metal end of the patellar hammer, stroke the foot from the heel along the lateral, inferior surface and then toward the ball of the foot (figure 19.9). The pressure should be firm but not uncomfortable.

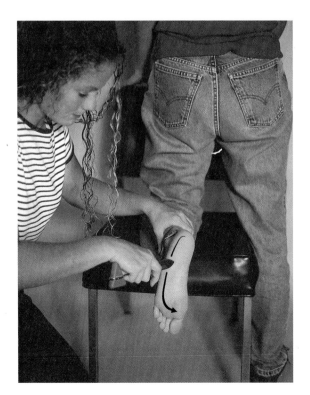

Figure 19.9 Plantar Response

Exercise 19 Review

Nervous System Physiology: Stimuli and Reflexes

Name: _____

Lab time/section: _____

Date: _____

1. Define *threshold voltage* in nerve conduction. _____
   _____

2. Define *maximum recruitment voltage* in nerve conduction. _____
   _____

3. What structure receives a stimulus from the external environment and relays it to the sensory neuron? _____

4. What is another name for an efferent neuron? _____

5. Define *reflex*. _____
   _____
   _____

6. In what kind of reflex do you have just two neurons? _____

7. Polysynaptic reflexes have a neuron specific to them. What is the name of that neuron? _____

8. In terms of numbers of synapses, what kind of reflex is a patellar reflex? _____

9. After surgery, patients leave the operating room and are transferred to an area called the "recovery room." Correlate the meaning of the word *recovery* in this context with what you have learned in this exercise about the recovery of nerves. _____
   _____
   _____
   _____

10. Draw a reflex arc in the space provided. Label your illustration with the terms provided.

    association neuron   simulus   motor neuron   effector   sensory neuron   synapse

11. What action occurs with a hyperreflexic response? What action occurs with a hyporeflexic response? _____

12. List the positive responses obtained in the frog experiment, and correlate this with the specificity of neuronal sensitivity. _____

13. What was the threshold voltage observed in the nerve response that you obtained by experimentation? _____

14. What was the maximum recruitment voltage in the nerve response that you obtained by experimentation? _____

Exercise 20
Introduction to Sensory Receptors

INTRODUCTION

The gateway to understanding our world comes from our ability to sense the environment around us. There are many types of sense receptors in the body, and each responds to specific stimuli (for example, light, sound, touch). Sense receptors are not uniformly distributed throughout the body but are absent, or few in number, in some areas, while densely clustered in others. This pattern of uneven distribution is called **punctate distribution.**

For the sensory system to operate, several factors need to be present. There can be no perception without environmental input or stimulus which are listed by type, or **modality.** Examples of modalities are light, heat, sound, pressure, and chemicals. **Receptors** are the receiving units of the body that respond to stimuli. They transform the stimulus to neural signals that are transmitted by sensory nerves and neural tracts to the somatosensory cortex or other areas of the brain, which interpret the message. If any link in this sensory chain is broken, the perception of stimuli cannot occur.

Receptors are sensitive to specific modalities and can be classified according to them. **Photoreceptors** detect light (for example, the retina of the eye); **thermoreceptors,** located in the skin, detect changes in temperature; **proprioceptors** detect changes in tension, such as those in joints; **pain receptors,** or **nociceptors,** present as naked nerve endings in the skin or stomach; **mechanoreceptors** perceive mechanical stimuli (for example, touch receptors or receptors that determine hearing or equilibrium in the ear); **baroreceptors** respond to changes in pressure, such as blood pressure; and **chemoreceptors** respond to changes in the chemical environment (for example, taste and smell). These receptors are discussed in the *Principles of Anatomy and Physiology* text in chapter 13, "The Special Senses."

The skin has several types of receptors and therefore makes a good starting point for understanding sense organs. There are receptors for pressure, pain, temperature, and light touch at your fingertips.

OBJECTIVES

At the end of this exercise, you should be able to

1. define the terms *modality* and *receptor;*
2. list the major receptor types in the body;
3. define the punctate distribution of sensory receptors;
4. distinguish between tonic and phasic receptors;
5. define *adaptation* in reference to a stimulus;
6. distinguish between relative and absolute determination of stimuli;
7. define *referred pain.*

MATERIALS

Microscopes
Prepared slides of thick skin (with Pacini and Meissner corpuscles)
Blunt metal probes
Three lab thermometers
Dishpan, or large finger bowls, of ice water (2 L)
Dishpan of room temperature water
Dishpan of warm water (45°C)
Towels
Small centimeter ruler
Tweezers
Hand lotion
Black, fine-tipped, washable felt markers
Red, fine-tipped, washable felt markers
Blue, fine-tipped, washable felt markers
Two-point discriminators (or a mechanical compass)
von Frey hairs (horse hair glued onto a wooden stick)

PROCEDURE
Touch Receptors

Examine a slide of thick skin for touch corpuscles. Two of these light touch corpuscles are the **Meissner corpuscles,** in the upper portion of the dermis, and **Merkel discs,** located in the upper dermis and lower epidermis. These two receptors allow for the perception of very light touch stimuli (such as a fly walking over your cheek). In addition, deep touch or pressure receptors are found in the dermis, farther away from the epidermis. **Pacini (Pacinian or lamellated) corpuscles** sense pressure, as when you lean against a wall or feel a vibration (figure 20.1). Other receptors in the skin are warm receptors and cool receptors. When you are at a comfortable temperature, both of these receptors are firing. If you increase or decrease the skin temperature beyond the perception of these receptors, pain receptors are stimulated. Pain receptors are naked nerve endings in the dermis that respond to many environmental stimuli. Locate the Meissner corpuscles and Pacini corpuscles in a prepared slide of skin.

Draw the corpuscles in the following space.

Meissner corpuscle | Pacini corpuscle

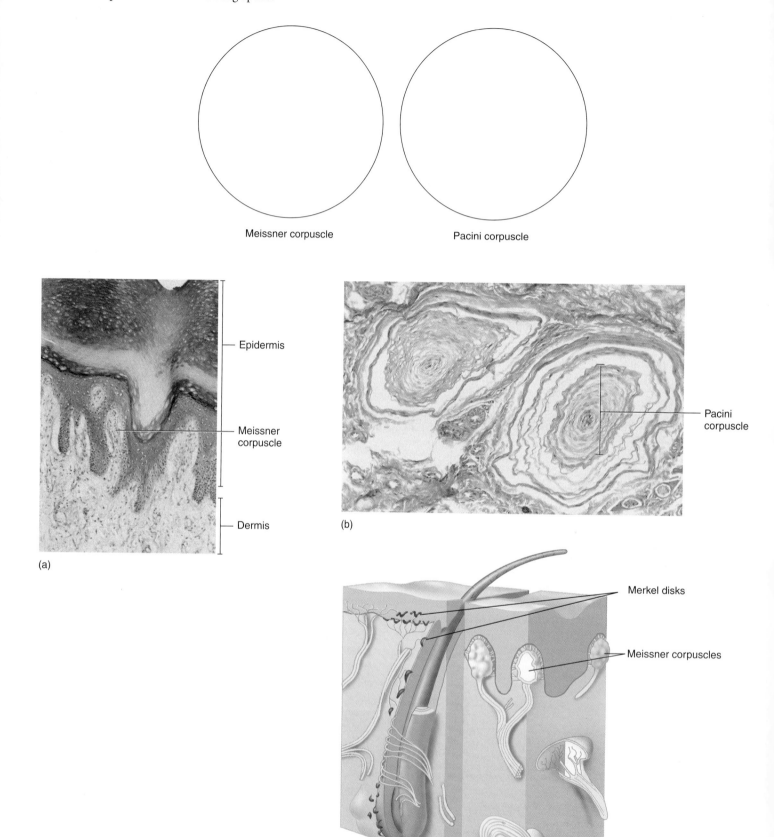

Figure 20.1 Skin Receptors (100×)
(a) Meissner corpuscle; (b) Pacini corpuscle;
(c) diagram of receptors.

Mapping Light-Touch Receptors

You can map light-touch receptors by testing the ability of your lab partner to distinguish fine touch. Using washable markers, draw a square, 2 cm on a side, on the *anterior* surface of the *forearm*. Using a stiff bristle hair attached to a matchstick, or a von Frey hair, map the number of areas in the square that your lab partner can perceive. Press only until the hair bends a little to stimulate the touch corpuscles. If you push too hard, you will stimulate other receptors. Use a washable black felt pen to record the location of each positive result. How many positive responses did you get in the square on the forearm? Record your result.

Number of anterior forearm responses: _____

Repeat the experiment on the *lateral* surface of the *arm* using another square, 2 cm on a side. How does the number of receptors here compare with those on the forearm? Record your results.

Number of responses on lateral side of the arm: _____

Two-Point Discrimination Test

The ability to distinguish touch depends on the type and number of nerve endings in the skin. You can easily map the relative density of the receptors in the skin by performing a two-point discrimination test. Have your lab partner sit with eyes closed and his or her hand, palm up, resting on the lab counter. Using the two-point discriminator or a mechanical compass, touch your lab partner's fingertip with both points of the instrument and see if he or she can sense one or two points. This is illustrated in figure 20.2. Determine the *minimum* distance apart your lab partner senses as two points. Begin the procedure by determining the minimum distance on the finger. To establish an accurate reading, make sure you gently touch both points of the discriminator at the same time. If you sequentially touch the two-point discriminator on the finger, your lab partner may perceive two points in time and not in space. One good way to establish accuracy is to occasionally touch just one of the points to your lab partner's finger. Another way is to vary the spread of the discriminator. You might begin with 2 cm and then adjust it to 0.5 cm followed by a 3 cm spread. Record the minimum distance perceived as two points.

Minimum distance on fingertip: _____

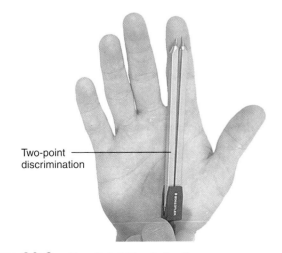

Figure 20.2 **Two-Point Discrimination**

Now move to the posterior forearm. Establish the minimum distance that is perceived as two points by your lab partner. Record your results.

Distance on posterior forearm: _____

Is there a difference in distance between the fingertip and the posterior forearm? _____

If there is, what might be the explanation for the difference in terms of the number of nerve endings per unit area?

_____

Now try the palm and then the back of the shoulder or neck. Record the results.

Distance on palm: _____

Distance on posterior shoulder or neck: _____

Warm and Cool Receptors

The skin has receptors that are sensitive to cool or warm temperatures. To determine the difference between the two, take two blunt metal probes and place the end of one in an ice water bath and one in a warm water bath (45° C). Let the probes reach the temperature of each respective bath, which should take a minute or two. Have your lab partner close his or her eyes and rest an arm on the lab counter. Remove one of the probes and quickly wipe it on a clean towel. Test the ability of your lab partner to distinguish between cool and warm by using the handle of the probe on your lab partner's forearm. Repeat the experiment for a total of five cool trials and five warm trials. Frequently return the metal probe handles to their respective water baths to maintain the appropriate temperature. Record your results.

Number of accurate cool recordings: _____

Number of accurate warm recordings: _____

Mapping Temperature Receptors

Mark off a square that is 2 cm on a side on the anterior forearm of your lab partner using a felt marker. If you apply a little hand lotion to the skin before doing this test, the ink comes off more easily after the experiment. Now take the *pointed end* of the blunt probe and place it in cold water. Let the probe sit in the cold water for a few minutes. Quickly wipe the probe and systematically test areas in the square. When your lab partner perceives cold (not just touch) in a location, mark it with the blue marker. Retest the area with the warm probe and place a red mark where warm is perceived.

Number of cool receptors in square: _____

Number of warm receptors in square: _____

What is the ratio of blue to red (cold/warm receptors) in the square? _____

Adaptation to Touch

Receptors may be classified according to the length of time that they can perceive stimuli. **Tonic receptors** constantly perceive stimuli, whereas **phasic receptors** adapt to stimuli. In this experiment, you try to determine if the sense of light touch is tonic or phasic. Cut out a small piece of paper, about 2 cm on a side, and crumple it into a small ball the size of a pea. Have your lab partner place his or her hand, anterior side up, comfortably on the lab desk. With a pair of tweezers, place the ball of paper on your lab partner's palm. Is the paper ball perceived after a few seconds? Record your results in the following space and determine whether the sense of light touch is tonic or phasic.

Perception of paper ball: _____

Locating Stimulus with Proprioception

In this exercise, you use washable markers of two colors. Have your lab partner close his or her eyes and rest a forearm on the lab counter. Touch your lab partner's forearm with a felt marker and have him or her try to locate the same spot with a felt marker of another color. Test at least five locations on various parts of the forearm, and repeat each location at least twice. Now try the fingertip and palm of the hand and record the result.

Maximum distance error on the forearm: _____

On the palm: _____

On the fingertip: _____

Another method to test proprioception is to close your eyes and *gently* try to touch the lateral corner of your own eye with your fingertip. Have your lab partner watch you and determine the accuracy of your attempt. While your eyes are still closed, bring your hand far behind your head and then try to touch the bottom part of your earlobe or the exact tip to your chin. Record the error distance, if any, for each location.

Corner of eye: _____

Earlobe: _____

Tip of chin: _____

Temperature Judgment

In this exercise, you examine the adaptation of thermoreceptors to temperature and the ability to determine temperature by absolute value or by relative value.

On the lab counter, place three dishpans or large finger bowls full of water. One bowl, located on the right, should be marked "Cold" (it should be about 10°C); another bowl, located on the left, should be marked "Warm" (it should be about 45°C); and the middle bowl should be marked "Room Temperature." Place one hand in the cold dish and the other in the warm dish and let them adjust to the temperature for a few minutes. If your hand begins to ache in the cold water, you may remove it for a short time, but try to keep it in the cold water for as long as possible during the adjustment time. After your hands have equilibrated, place them both in the room temperature water and describe to your lab partner the temperature (cold, warm, hot) of the water as sensed by each hand. How does the hand that was in cold water feel in the room temperature water, and how does the hand that was in warm water feel in the room temperature water? Record your results.

Cold hand perception: _____

Warm hand perception: _____

If the determination of temperature is absolute, then the room temperature water should feel the same whether you are testing it

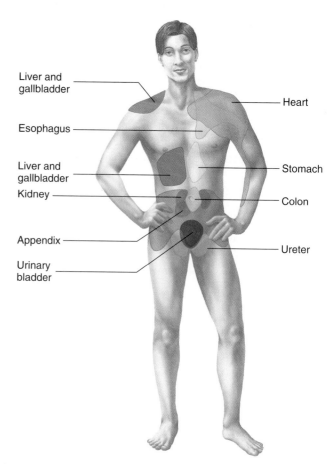

Figure 20.3 Regions of Referred Pain

with your warm hand or your cold hand. If it is relative, the room temperature water should feel warmer with your cold hand and colder with your warm hand.

Is the determination of the temperature of water absolute or relative? _____

How did your experiment prove this? _____

Referred Pain

Referred pain is the perception of pain in one area of the body when the pain is actually somewhere else (see chapter 12 in the text). An example of referred pain is the pain felt in the left shoulder and arm when a person is suffering from a heart attack or chest pain (angina pectoris; figure 20.3). Referred pain may be caused by impulses from different receptors traveling along the same neural tract to the brain. When one area sends a message of pain, the brain may interpret the sensation as coming from more than one location. Place your elbow into a dish of ice water and leave it there for 2 painful minutes. Describe the sensation you feel and the location of the sensation. Record your results.

Description of sensation: _____

Initial location of sensation: _____

Sensation after 2-minute period: _____

Exercise 20 Review

Introduction to Sensory Receptors

Name: _____
Lab time/section: _____
Date: _____

1. An area with a great number of nerve endings is the upper lip. What can you predict about the ability of the upper lip to distinguish two points? _____

2. Distinguish among the functions of Pacini corpuscles, Meissner corpuscles, and pain receptors in the skin. _____

3. When you extend the temperature beyond the level of cool and warm, the pain receptors in the skin are activated. Cool receptors are activated between 12°C and 35°C. Warm receptors are activated between 25°C and 45°C. You or your lab partner may have had an experience with very cold conditions, such as when cleaning out a freezer or holding dry ice. What perception is sensed? _____

4. Are there more cool receptors or warm receptors in the skin? What adaptive advantage might there be for an unequal number of receptors of one kind? _____

5. Adaptation occurs in some sensory stimulation. Why do you think this is important? _____

6. Why do pain receptors function as tonic receptors? _____

7. In terms of receptor density, describe why it is difficult to find the same location on the forearm when your eyes are closed.

   _____

   _____

   _____

8. In reference to the sense organs, what is a modality? _____

9. What kinds of receptors are sensitive to the following modalities?

   a. light _____

   b. touch _____

   c. temperature _____

   d. sound _____

   e. smell _____

10. What kind of receptor is responsive to extremely hot sensations? _____

    _____

11. Meissner corpuscles respond to what kind of sensation? _____

    _____

12. What kind of receptor determines the weight of an object when you pick it up? _____

13. Which kind of receptor (phasic/tonic) adapts to low light in a darkened movie theater? _____

14. When you drink a liquid that is burning hot, the "chest pain" felt in the region of the sternum does not really occur there. What is this kind of pain called? _____

Exercise 21
Taste and Smell

INTRODUCTION

We take for granted our senses of taste and smell. They are not fully appreciated unless they are lost. People who have lost the sense of smell find food difficult to eat, since they derive little or no pleasure from the act of eating. In this exercise, you examine the structure and function of these two important senses.

Both taste and smell are examples of chemoreception in which specific chemical compounds are detected by the sense organs and interpreted by various regions of the brain. The sense of taste, or **gustation,** is received predominantly by taste buds in the tongue, although there are also receptors in the soft palate and pharynx. The sense of taste is transmitted by the facial and glossopharyngeal nerves and interpreted in the postcentral gyrus of the parietal lobe of the brain. The sense of smell, or **olfaction,** originates when particles stimulate hair cells in the olfactory epithelium (a specialized neuroepithelium in the upper nasal cavities) and is transmitted by the olfactory nerves. This transmission occurs through the cribriform plate of the ethmoid bone to the olfactory bulb at the base of the frontal lobe of the brain. From there, the sense of smell is transmitted to two regions, one in the limbic system and another in the temporal lobe of the brain.

The sense of taste and the sense of olfaction are covered in the *Principles of Anatomy and Physiology* text in chapter 13, "The Special Senses."

OBJECTIVES

At the end of this exercise, you should be able to

1. list the two major chemoreceptors in the head;
2. draw a diagram of a taste bud;
3. trace the sense of smell from the nose to the integrative areas of the brain;
4. list the five tastes perceived by humans;
5. describe what happens in an olfactory reflex;
6. compare and contrast the senses of taste and smell.

MATERIALS

Microscopes
Prepared slides of taste buds
Roll of household paper towels
Small bowl of salt crystals (household salt)
Small bowl of sugar crystals (household granulated sugar)
Flat toothpicks
Sterile cotton-tipped applicators
Five dropper bottles containing one of the following:
    Solution of salt water (3%) labeled "salty"
    Tonic water labeled "bitter"
    Vinegar solution (household vinegar or 5% acetic acid solution) labeled "acidic"
    Sugar solution (3% sucrose) labeled "sweet"
    Umami solution (15 g MSG in 500 mL water) labeled "Umami"
Biohazard bag
Small bottle (100 mL) of household ammonia
Four small vials colored red and labeled "wild cherry" filled with benzaldehyde solution
Four small vials of "almond" essence
Several small vials (10–20 mL), screw-cap bottles with cotton saturated with essential oil labeled "peppermint," "almond," "wintergreen," and "camphor" (keep vials in separate wide-mouthed jars to prevent cross-contamination of scents)
One vial of dilute perfume (one part perfume, five parts ethyl alcohol)
Selection of four or five fruit nectars (such as Kern's nectars), two cans each: apricot, coconut/pineapple, strawberry, mango, peach, apple
Small, 3 oz paper cups (89 mL), five per student (or student pair)
Marking pens
Noseclips and alcohol swabs
Napkins

PROCEDURE
Examination of Taste Buds

Examine the prepared slide of taste buds and compare them with figure 21.1. Note how they are located on the sides of the papilla on the tongue. The taste buds appear lighter than the surrounding tissue (like microscopic onions cut in long section). Taste buds are composed of neural tissue and epithelial tissue. Draw the taste buds you see in the following space.

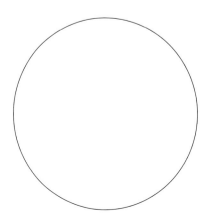

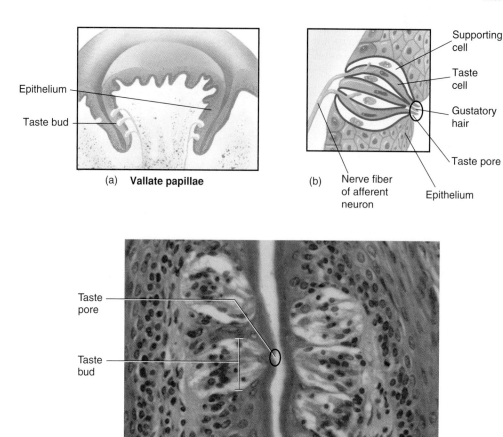

Figure 21.1 Taste Buds
(a) Taste buds on sides of a tongue papilla; (b) details of taste buds; (c) photomicrograph of taste buds (100×).

Taste Determination of Solid Materials

For something to be tasted, it must be in solution. This allows the fluid to run down the sides of the tongue papilla, where the taste buds are located. Blot your tongue thoroughly with a paper towel. Make sure the surface is relatively dry. Have your lab partner select either the sugar crystals or the salt crystals and place a small scoop (with the end of a flat toothpick) on your tongue. Keep your mouth open and do not swirl saliva around. Can you determine what the sample is? Close your mouth and determine the nature of the sample.

Mapping the Tongue for Taste Receptors

There are five primary tastes: sweet, sour, salty, bitter, and umami. Umami gives meat and cheese their particular tastes and it can be referred to as the taste of "savory." We have varying degrees of sensitivity to these five tastes. Some of us find bitter tastes to be especially objectionable, whereas others do not seem to mind them as much. As you perform the taste experiments, determine if members of your class are equally sensitive to the same tastes as you are.

Caution
If you have food allergies, you may wish to omit this part of the lab. People with allergies, migraines, or heart problems should avoid the tasting of umami, which is derived from monosodium glutamate (MSG).

Using a sterile cotton-tipped applicator stick and a dropper bottle, apply one of the tasting solutions (salty, bitter, acidic, umami, or sweet) to the cotton. Saturate the cotton swab. Remember to keep track of the kind of substance that is on your applicator stick. Do *not* reuse the applicator sticks for more than one experiment. Dab the entire surface of your lab partner's tongue and determine by nod or hand signals when perception of the taste occurs. Do not have your lab partner tell you about the taste at first, since the movement of the tongue will wash the solution to other areas of the tongue and you may get a false reading. Use figure 21.2 and note where you perceive the specific sensations of taste on the illustration of the tongue. Make sure you swab under

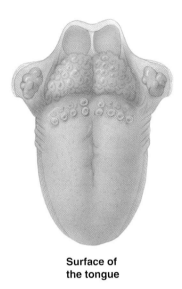

Figure 21.2 Map of Taste Receptors on the Tongue
Indicate the regions of your tongue that are sensitive to the specific tastes of sweet, sour, salty, bitter, and umami.

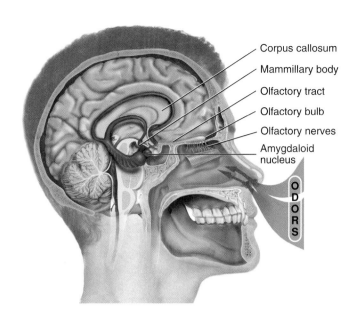

Figure 21.3 Olfactory Transmission

the sides of the tongue. Are there regions of the tongue that are more sensitive to some tastes? If so, note these areas on figure 21.2. **Throw the applicator stick in the biohazard bag.** Repeat the test with a new applicator stick for each of the five solutions. Do not contaminate the solutions by reusing the applicator stick with the same or different solutions after it has been in your lab partner's mouth! Compare the taste map of members of the class with yours. Are they the same? If not, do the regions or relative areas vary significantly?

Transmission of Sense of Olfaction to Brain

The sense of smell begins with material inhaled through the nose, where the particles come into contact with the olfactory epithelium. From here the sense is transmitted as nerve impulses via the **olfactory nerve fibers,** through the **cribriform plate** of the ethmoid bone, to the **olfactory bulbs, olfactory tracts,** and then interpreted in the **limbic system** (specifically the **amygdaloid nucleus**) and the **temporal lobe** of the brain. Examine a model, chart, or diagram of a midsagittal section of the head or a model of the brain and locate the olfactory nerve fibers, cribriform plate, olfactory bulb, and olfactory tract. Compare the lab charts or models with figure 21.3 and trace the pathway of the sense of smell.

Olfactory Reflex

Place a small bottle of household ammonia quickly under the nose of your lab partner. Have your lab partner take a brief sniff from the bottle. If there is a visible movement of the head in a posterior direction, then your lab partner demonstrates an olfactory reflex to smell. Record the results of your experiment in the following space.

Olfactory Reflex

_____ yes

_____ no

What might be the adaptive benefit for people having an olfactory reflex? _____

_____

_____

Visual Cues in Smell Interpretation

In Shakespeare's *Romeo and Juliet*, Juliet says, "that which we call a rose by any other name would smell as sweet." That may be true, but what about color? Is our perception of smell, our interpretation, the same if the rose is black or green? In this section, you examine the influence of visual cues on the interpretation of smell. You seek to determine whether the color of a substance has any effect on what you perceive the smell to be.

Have your lab partner show you a small vial labeled "almond" and then smell it. Then examine and smell the small red vial labeled "wild cherry." Do you perceive these as two separate smells? _____ (yes/no)

Close your eyes and have your lab partner select a vial for you. Can you tell which one it is? _____ (yes/no)

Olfactory Discrimination

The ability to tell one scent from another is **olfactory discrimination**. In this part of the exercise you try to distinguish one smell from another. Obtain four vials of different scents—peppermint, almond, wintergreen, and camphor. While keeping your eyes closed, try to determine the name of each essential oil as your lab partner presents it to you. Record how many of the smells you get correct out of the four. If you are hypersensitive to smells, have your lab partner do this section of the experiment.

Number correct: _____

Adaptation to Smell

Have you ever walked into a room and smelled something delicious and then 5 or 10 minutes later not noticed the smell? Getting accustomed to a scent is **adaptation** to smell. Adaptation to smell by the olfactory receptors occurs very rapidly, but the adaptation by the receptors is incomplete. Complete adaptation to smell probably occurs by additional CNS inhibition of the olfactory signals. In this section of the experiment, you are trying to determine approximately how long olfactory adaptation takes. Have your lab partner record the time when you begin the experiment and how long it takes for the perception of smell to decrease significantly. Close your eyes and plug one nostril. Inhale the scent from a vial of dilute perfume, or one of the vials from the olfactory discrimination test, until the smell decreases significantly. How long does this take?

Length of time to significant reduction of the smell: _____

What might be the evolutionary advantage of adaptation to smell?

_____

_____

Predict whether adaptation to one smell causes adaptation to another smell. Record your prediction that the smell of one material does/does not (select one) cause adaptation to another smell.

Now smell the wintergreen or peppermint vial. Does the adaptation of one smell cause the olfactory receptors to adapt to other smells? _____

Taste and Olfaction Tests

This experiment demonstrates the dependence of the sense of smell as a component of what we call taste. Obtain four or five small cups (3 oz) and, using a marking pen, label each with the name of the fruit juice or nectar it will contain. Have your lab partner select several types of fruit nectars and pour each into the proper cup.

Caution
If you have a food allergy, notify your instructor. You may wish to omit part or all of this test.

Sit with your eyes and nose closed (use clean noseclips or pinch off your nostrils with your finger and thumb) and try to identify the sample presented to you by your lab partner. Your lab partner should place the sample cup (sample unknown to you) in your hand and you should guess what fruit nectar is in the cup. After you have "tasted" the sample, try to name it. Test all of the samples with your nose closed. After you make your determination, release your nostrils but still keep your eyes closed and taste the samples again. Record the results. How accurate is your comparison?

Trial Number	Sample	Accuracy (Yes/No) Nose Closed	Nose Open
1	Apricot		
2	Mango		
3	Coconut/pineapple		
4	Strawberry		
5	Peach		
6	Apple, other		

Exercise 21 Review

Taste and Smell

Name: _____
Lab time/section: _____
Date: _____

1. Why does material have to be in solution for it to be sensed as taste? _____
   _____
   _____

2. What are the primary tastes? _____
   _____

3. Did everyone in your lab have the same reaction to tastes, such as sweet, sour, or bitter? _____

4. What nerves transmit the sense of smell to the brain? _____

5. What nerves transmit the sense of taste to the brain? _____

6. Where are the taste buds located? _____
   _____

7. What is the exact region of the nasal cavity that is sensitive to smell stimuli? _____

8. What is the adaptation for having taste buds that determine unpleasant bitter compounds in many plant species?
   _____
   _____
   _____

9. Some people with severe sinus infections can lose their sense of smell. How can an infection that spreads from the frontal or maxillary sinus impair the sense of smell? What structure or structures might be affected? _____
   _____
   _____

10. Material must be in solution for it to be tasted. What process would be used (olfaction or gustation) to perceive a lipid-based food? _____
    _____

11. How does a cold (rhinovirus) influence our perception of taste? _____

_____

12. Does adaptation to one smell influence adaptation to another smell? _____

13. Some smells that we perceive as two separate smells are actually identical. What other cues do we use to distinguish between these two smells? _____

14. Toxic chemicals, acids, and other environmental dangers may sometimes be noticed by smell. How would the olfactory reflex be an evolutionary adaptation to these substances? _____

_____

Exercise 22
Eye and Vision

INTRODUCTION

Eyesight accounts for much of our accumulated knowledge. In this exercise, you learn the features of the eye and some of the functions of the eye. The anatomy and physiology of the eye are discussed in the *Principles of Anatomy and Physiology* text in chapter 13, "The Special Senses."

The importance of the eye can be inferred in that 4 of the 12 pairs of cranial nerves are dedicated, at least in part, to either receiving visual stimuli or coordinating the eyes.

Anatomically, the eye consists of an anterior portion, visible as we look at a person's face, and a posterior portion, located in the orbit of the skull. Light travels through a number of transparent structures before it strikes the retina, which is the receptive layer of the eye that converts light energy to neural impulses. Nerve impulses travel from the eyes through the optic nerves to the brain, where they are interpreted as sight in the occipital lobes of the brain.

OBJECTIVES

At the end of this exercise, you should be able to

1. identify the major structures of the mammalian eye;
2. describe the six extrinsic muscles of the eye and their effect on the movement of the eye;
3. describe the position of the choroid in reference to the retina and the sclera;
4. distinguish between the pupil and the iris of the eye;
5. describe the function of the rods and the cones of the eye;
6. define the near point of the eye;
7. determine the visual field for both eyes;
8. demonstrate the Snellen vision tests and those for accommodation and astigmatism.

MATERIALS

Models and charts of the eye
Microscopes
Prepared microscope slides of the eye in sagittal section
Paper with small print (8- or 9-point font)
Ruler (approximately 35 cm and meter stick)
Paper card with a simple, colored image (red circle, blue triangle) printed on it
Vision Disk (Hubbard) or large protractor
Snellen charts
3- by 5-inch cards
Astigmatism charts
Ophthalmoscope and batteries
Penlight
Ishihara color book or colored yarn
Preserved sheep or cow eyes
Dissection trays
Dissection gloves (latex or plastic)
Scalpel or razor blades
Animal waste disposal container
Desk lamp with 60-watt bulb
First aid kit

PROCEDURE
External Features of the Eye

The external anatomy of the eye and the accessory structures are illustrated in figure 22.1. Examine the eye of your lab partner and compare it with the figure. The **pupil** is located in the center of the eye and is surrounded by the colored **iris.** The **sclera** is the white of the eye and it is covered by a membrane, known as the **conjunctiva,** that continues underneath the eyelids. The eyelids join together at the **lateral commissure** and the **medial commissure.** There is a small piece of tissue near the medial commissure known as the **lacrimal caruncle.** Examine the **upper eyelid** and **eyelashes** and the **lower eyelid** and eyelashes, which prevent material from entering the eyes and (in the case of the eyelids) reduce visual stimulation when we sleep. Look also at the **eyebrows** located superficial to the supraorbital ridge. The sclera is a protective portion of the eye that is an attachment point for the muscles of the eye and helps maintain the **intraocular pressure** (the pressure inside the eye). The pressure maintains the shape of the eye and keeps the retina attached to the posterior wall of the eye. Numerous blood vessels traverse the sclera, and if they become dilated anteriorly they give the eye the appearance of being "bloodshot." The sclera is continuous with the transparent **cornea** of the anterior eye.

Attached to the sclera are the **extrinsic muscles** of the eye. There are six extrinsic muscles that, in coordination, move the eye in quick and precise ways. Locate these muscles on a model and compare them with figure 22.2. These muscles and their actions on the eye are listed in table 22.1.

A structure important in the maintenance of the exterior of the eye is the **lacrimal apparatus.** This consists of the **lacrimal**

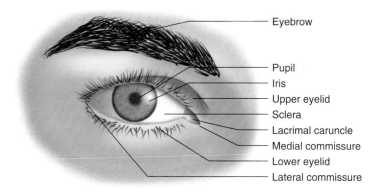

Figure 22.1 External Anatomy of the Right Eye

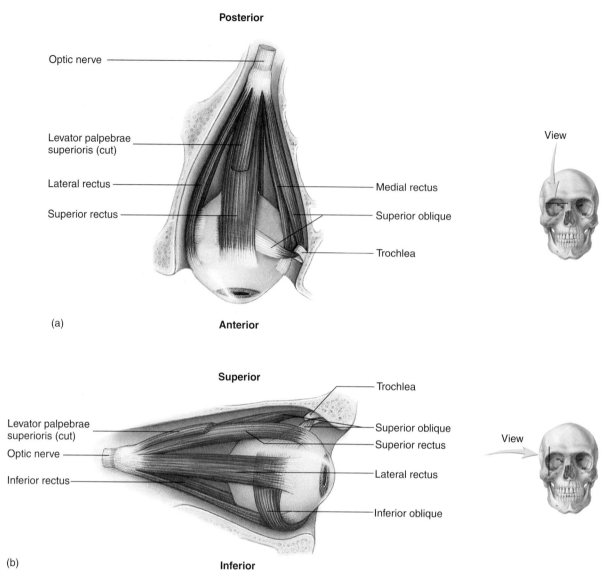

Figure 22.2 Right Eye, Extrinsic Muscles
(a) Superior view; (b) lateral view.

TABLE 22.1	Extrinsic Muscles of the Eye	
Muscle Name	Innervation	Direction Eye Turns
Lateral rectus	VI (abducent)	Laterally
Medial rectus	III (oculomotor)	Medially
Superior rectus	III (oculomotor)	Superiorly
Inferior rectus	III (oculomotor)	Inferiorly
Inferior oblique	III (oculomotor)	Superiorly and laterally
Superior oblique	IV (trochlear)	Inferiorly and laterally

gland located superior and lateral to the eye (figure 22.3). Lacrimal secretions (tears) bathe and protect the eye and clean dust from its surface. The fluid drains through the **nasolacrimal duct** into the nasal cavity.

Interior of the Eye

From the anterior of the eye, the first layer covering the sclera is the **conjunctiva.** The conjunctiva is composed of epithelial tissue and is an important indicator of a number of clinical conditions (for example, conjunctivitis). In the center of the eye is the transparent **cornea** (figure 22.4). The cornea is the structure of the eye most responsible for the bending of light rays that strike the eye. It is composed of dense connective tissue and is avascular. Why would the presence of blood vessels in the cornea be a visual liability?

The interior of the eye consists of two main compartments, the **anterior compartment,** which is anterior to the lens, and the **posterior compartment,** which is posterior to the lens. What can be confusing is that the anterior compartment is composed of both the **anterior chamber,** between the cornea and the iris, and the **posterior chamber,** between the iris and the lens. So do not confuse the posterior chamber of the anterior compartment with the posterior compartment. The anterior compartment is filled with

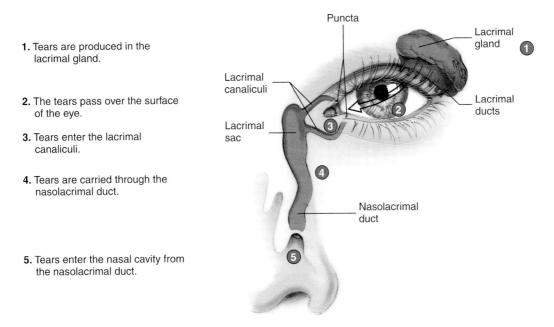

1. Tears are produced in the lacrimal gland.
2. The tears pass over the surface of the eye.
3. Tears enter the lacrimal canaliculi.
4. Tears are carried through the nasolacrimal duct.
5. Tears enter the nasal cavity from the nasolacrimal duct.

Figure 22.3 Lacrimal Apparatus

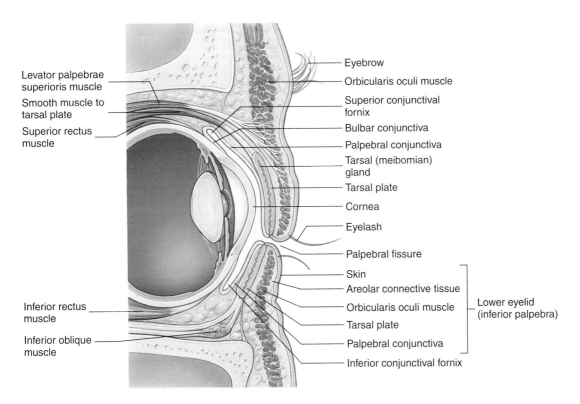

Figure 22.4 Sagittal Section of the Anterior Portion of the Eye

aqueous humor, which is produced by the **ciliary body.** Only a few milliliters of aqueous humor are produced each day, and this amount is absorbed by the **scleral venous sinus (canal of Schlemm),** as illustrated in figure 22.5.

The **iris** is what gives us a particular eye color. People with blue or gray eyes are more sensitive to ultraviolet light than those with brown eyes, due to more of the protective pigment **melanin** found in brown eyes. The **circular muscles** form the **sphincter pupillae** of the iris and constrict in bright light, reducing the diameter of the **pupil** (the space enclosed by the iris), and the **radial muscles** form the **dilator pupillae** and constrict in dim light, increasing the diameter of the pupil. Behind the pupil is the **lens,** which is made of proteins called **crystallines.** The **ciliary muscle** in the **ciliary body** contracts and the **suspensory ligaments** that

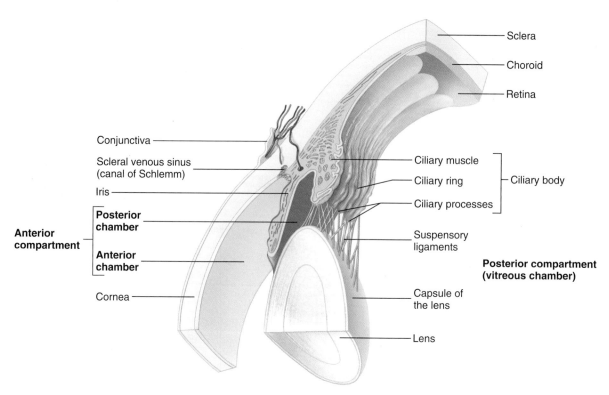

Figure 22.5 Eye Compartments and the Ciliary Body

attach to the lens loosen, decreasing the pull on the lens. The lens becomes more round for focusing on closer images. The lens is more pliable in youth and becomes less elastic as a person ages, which accounts for the need for reading glasses later in life.

Behind the lens is the **posterior compartment,** or **vitreous chamber** (figure 22.5). This compartment occupies most of the posterior portion, or **fundus,** of the eye. The posterior compartment is filled with **vitreous humor,** a clear, jellylike fluid that maintains the shape of the eyeball. Most of the posterior compartment is bounded by three layers of the eye. The outermost one, the **fibrous layer** which includes the **sclera,** is composed of dense irregular connective tissue. Inside this is the **vascular layer** containing the pigmented **choroid.** The blood vessels found in this layer nourish the eye and the pigmentation prevents light from scattering and blurring vision. The layer closest to the vitreous humor is the **nervous layer,** which contains the **retina.** These structures are seen in figure 22.6.

The retina converts light energy into nerve impulses. Light strikes the photoreceptive cells at the posterior portion of the retina, causing these cells to transmit signals to the **bipolar neurons.** The bipolar neurons synapse with the **ganglion cells;** thus, the visual stimulation that occurs in the posterior portion of the eye is transmitted anteriorly (toward the vitreous humor) to the ganglion cells. The axons of the ganglion cells exit the eye as the **optic nerve.** From there, the neural impulse is transmitted to the occipital region of the brain and integrated further in the temporal lobe (figure 22.7).

Posterior Wall of the Eye

Obtain a microscope and a prepared slide of the retina and compare what you see in the slide with figure 22.8. The retina is composed of three layers: the **ganglionic, bipolar,** and **photoreceptive** layers. The photoreceptive layer is composed of **rods** and **cones.** Rods are important for determining the shape of objects and for seeing in dim light. Cones are involved in color vision and in visual acuity (determining fine detail).

Examine a model or chart of the eye and locate the **macula** (mak′ū-lă) at the posterior region of the eye. In the center of the macula is the **fovea centralis,** a region where the concentration of cones is greatest. In the fovea, the cone cells are not covered by the neural layers, as they are in the other parts of the retina. When you focus on an object intently, you are directing the image to the fovea. Locate the fovea in models or charts available in the lab and compare them with figure 22.9.

Visual Tracking

Eye movements are very precisely controlled by the six extrinsic muscles of the eye. Impaired visual tracking may be used to help diagnose drug use, schizophrenia, or Parkinson's disease. Have your lab partner follow your finger as you move it in front of the eyes. Is the movement of the eye smooth (which indicates normal function), or do the eyes move in a jerky fashion (an abnormal condition)? Record your results in the following space.

Eye movement (select one): smooth/jerky

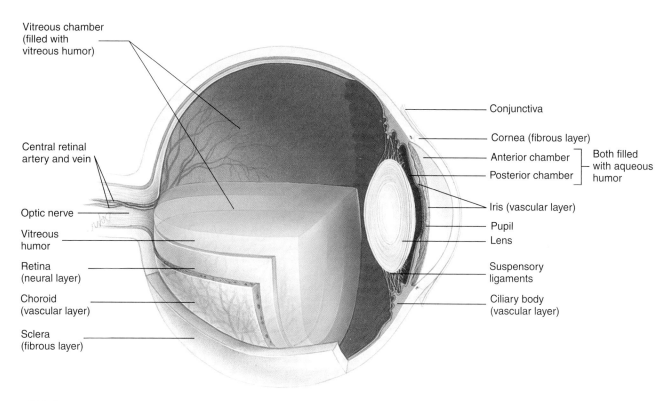

Figure 22.6 Sagittal Section of the Eye with Layers

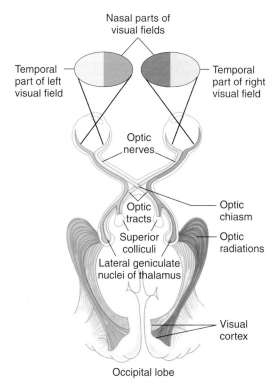

Figure 22.7 Visual Pathway to the Brain

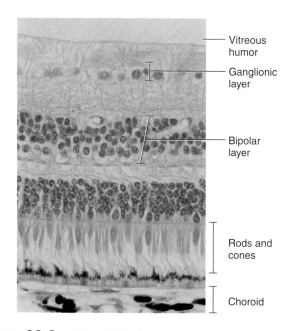

Figure 22.8 Retina (400×)

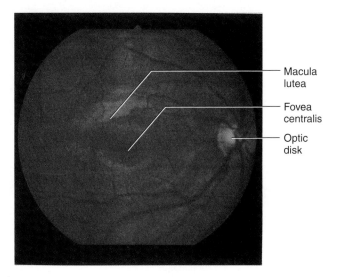

Figure 22.9 Eye, Posterior View

- Macula lutea
- Fovea centralis
- Optic disk

Determination of the Near Point

The *minimum* distance an object can comfortably be held in focus is called the **near point.** The eye's ability to focus, called **accommodation,** is due to the elasticity of the lens. This decreases with age. A 10-year-old can focus 8 to 10 cm away from the eye, yet a 65-year-old may not be able to focus closer than 80 to 100 cm. The stiffening of the lens becomes noticeable around 40 to 50 years of age, when many people find reading glasses necessary because their aging lenses do not accommodate as well as the lenses of a younger person. You can measure your near point by holding a piece of paper with fine print (8- or 9-point font) at arm's length in front of you. Close one eye and slowly move the paper closer until the print becomes blurry. Have your lab partner measure the distance, in centimeters, from your eye to the paper. This is the near point distance. Measure the near point for both eyes, in centimeters, and record the data.

Near point of right eye: _____

Near point of left eye: _____

Measurement of the Distribution of Rods and Cones

To determine the distribution of rods and cones, have your lab partner look straight ahead. Slowly move a small, colored object (a pen, piece of chalk, comb, etc.), without letting your lab partner see it ahead of time, from the back of your lab partner's head, around the side, toward the front. With your lab partner still looking straight ahead, note the angle when your lab partner is able to see the object (the use of rods). Continue to move the object forward slowly until your partner recognizes the color of the object. Note the angle from the tip of the nose. You can record the angle by using an apparatus called a Vision Disk or by placing a protractor above your lab partner's head while you make your measurements.

Angle where object is perceived: _____

Angle where color is determined: _____

What is the difference in the distribution of rods and cones in the eye?

Rod distribution: _____

Cone distribution: _____

When you are intently focusing on a subject, what cell type (rods or cones) are you using? _____

Measurement of Binocular Visual Field

Not all animals have the same visual field. Some prey species (such as deer and sheep) have wide visual fields with little binocular vision. On the other hand, many predators, birds, and arboreal animals have a more limited visual field, yet they have greater **binocular,** or **stereoscopic, vision.** Binocular vision allows arboreal animals to perceive depth, which is vital when judging how far away the next branch is. You can determine your visual field with the use of a Vision Disk or protractor. If you are using a Vision Disk, follow the instructions enclosed. If not, sit down and have your lab partner stand behind you. Close your right eye and look straight ahead with the left one. While your right eye is closed, have your lab partner move an object (pen, paper, disk, etc.) from behind your head from the left until you can just see the object while looking straight ahead. Measure the angle from the tip of the nose to where you first saw the object. This is illustrated in figure 22.10 as angle A.

With the same eye directed ahead, have your lab partner continue moving the object until it is out of view (to the right of your nose somewhere). This is illustrated as angle B. Determine the angle from your nose to where the object disappears.

Angle A (where object appears): _____

Angle B (where object disappears): _____

Now close your left eye and repeat the exercise on the right side. This is illustrated as angles C and D in figure 22.10*b*.

Angle C: _____

Angle D: _____

Determine the total visual field for each eye (the sum of angles A and B for the left eye and the sum of angles C and D for the right eye) and then for both eyes (the sum of angles A and C) and the degree of overlap between the eyes (the sum of angles B and D; figure 22.11). Record these data.

Visual field for left eye (A and B): _____

Visual field for right eye (C and D): _____

Complete visual field for both eyes (A and C): _____

Degree of overlap (B and D): _____

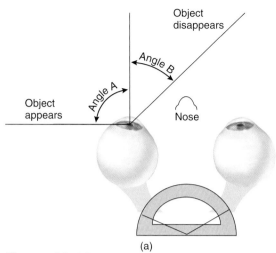
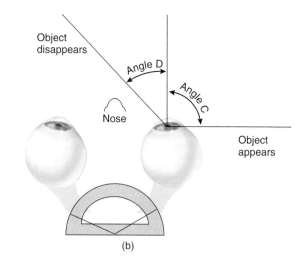

Figure 22.10 Measuring the Visual Field
(a) Left eye; (b) right eye.

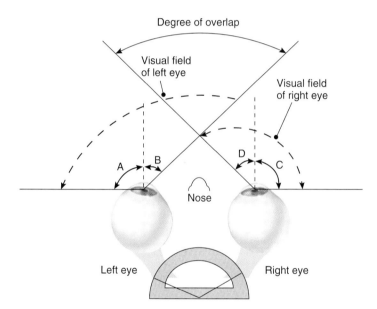

Figure 22.11 Visual Field

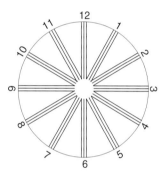

Figure 22.12 Astigmatism Chart

that has the smallest print *in which you made no errors.* Your lab partner should record the numbers at the side of the line. This number refers to your **visual acuity.** Now switch the card to the other eye and repeat the test. Record your results.

Left eye: _____

Right eye: _____

Measurement of Visual Acuity (Snellen Test)

A vision of 20/20 is considered normal. In 20/20 vision, you can see the same details at 20 feet that most other people can see at the same distance. If your vision is 20/15, then you can see at 20 feet what most people can see at 15 feet. If your vision is 20/100, then you see at 20 feet what most people see at 100 feet.

Face the Snellen eye chart from 20 feet away and have your lab partner stand next to the chart. Cover one eye with a 3- by 5-inch card and have your lab partner point to the largest letter on the chart. Do *not* read the chart with both eyes open. Your lab partner should then progressively move down the chart and note the line

Astigmatism Tests

If the cornea, or lens, of the human eye were perfectly smooth, the incoming image would strike the retina evenly and there would be no blurry areas. In most people, the cornea, or lens, is not perfectly smooth, and this is known as astigmatism. Various degrees of **astigmatism** occur. In a normal optical exam, the distance correction is made first (to determine visual acuity) and then, with the corrective lenses in place, an astigmatism test is performed.

If you do not normally wear corrective lenses, then cover one eye and examine the astigmatism chart (similar to figure 22.12) from 20 feet away. The chart consists of a series of parallel lines

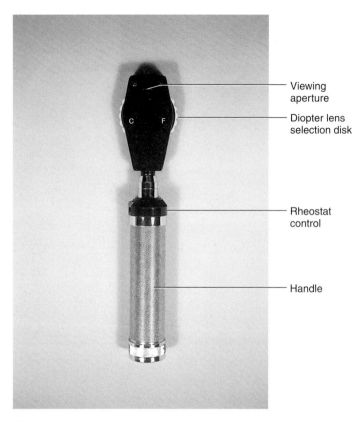

Figure 22.13 Ophthalmoscope

Figure 22.14 Use of Ophthalmoscope

radiating away from the center. Stare at the center of the chart and determine which of the sets of parallel lines, if any, appears light or blurry. Have your lab partner note the corresponding number on the chart and record the number.

Astigmatism numbers: _____

If you wear corrective lenses, then not only is the condition of nearsightedness or farsightedness corrected but astigmatism is corrected also. To test for astigmatism, move about 12 feet from the chart, hold your glasses slightly away from your face, and then rotate them 90°. If your glasses correct for astigmatism, some of the lines on the chart will be blurry as you rotate your glasses.

Ophthalmoscope

Clinical use of the ophthalmoscope is important not only for the diagnosis of variances in the eyeball but also as a potential indicator of diseases, such as **diabetes mellitus,** that affect the eye. Familiarize yourself with the parts of the ophthalmoscope (figure 22.13). Locate the **ring** (rheostat control) at the top of the handle. Depress the button and turn the ring until the light comes on. The **head piece** of the ophthalmoscope consists of a **rotating disk** of lenses. You can change the **diopter** (strength) of the lenses by rotating the disk clockwise or counterclockwise. As the numbers get progressively larger, you are looking through more convex lenses. If you rotate the dial in the other direction, the lenses become more concave. At the zero reading, there are no lenses in place.

Observation with the Ophthalmoscope

Caution
Examine the posterior region of the eye only for a short period of time. Extensive use of the ophthalmoscope can damage the eye.

Sit facing your lab partner. Select the left eye to examine. Have the ophthalmoscope setting at zero and use your left hand to hold the ophthalmoscope. Use your left eye to look through the scope and move in close to examine your lab partner's eye (figure 22.14). Look into the pupil and examine the back of the eye. If the back appears fuzzy, rotate the dial clockwise (positive diopters) and see if it comes into focus. If a positive number provides a clear view of the eye, then your lab partner has **hyperopia (hypermetropic vision),** or **farsightedness.** In farsightedness, the eyeball is short and the image is focused posterior to the retina. Positive diopter lenses focus the image on the retina. If the image is indistinct, then adjust the dial counterclockwise to obtain a negative diopter reading. If the image is clear with a negative number, then your lab partner has **myopia (myopic vision), or nearsightedness.** In nearsightedness, the eyeball is long and lenses with negative diopters focus the images farther back on the retina. This procedure is based on the assumption that your vision, as the examiner, is normal.

Pupillary Reactions

Have your lab partner sit in a dark room for a minute or two. Examine his or her eyes in dim light. Are the pupils dilated or constricted?

Pupil diameter in dim light: _____

Now shine a penlight briefly in the right eye. Record what happens to the pupil diameter.

Right pupil diameter: _____

As you shine the light into the right eye, what occurs in the left pupil? This is called a **consensual reflex.** Record the effect.

Left pupil diameter: _____

Color Blindness

Color vision is dependent on three separate cone cell sensitivities. Cones may be **red, green,** or **blue sensitive.** Changes in the genes on the X chromosome are the most common cause of color blindness. Some people may be unable to see a particular color at all, whereas others may have a reduced ability to see a particular color. **Color blindness** is most common in males and relatively rare in females. The male chromosomal makeup is XY. The Y chromosome does not carry the gene for color vision. If the X chromosome carries a gene for color blindness, then the male exhibits color blindness. On the other hand, if a female carries the gene for color blindness on the X chromosome, the chances are that she will have a normal gene on the other X chromosome. The normal gene is expressed, cone pigments are produced, and the female has normal color vision.

You can test for color blindness by matching various samples of colored yarn with a presented test sample or you can use Ishihara color charts. If you use the color charts, flip through the book with a lab partner and record which charts were accurately viewed and which plates, if any, were missed. You can calculate the type and degree of color blindness by following your instructor's directions or those that come with the color chart.

Your degree of color vision: _____

Dissection of a Sheep or Cow Eye

Rinse a sheep or cow eye in running water and place it on a dissection tray. Obtain a scalpel, forceps, and a blunt probe. Be careful with the sharp instruments and cut *away from* the hand holding the eye. Wear latex or plastic gloves while you perform the dissection. Remove the fat and muscles carefully from the eyeball (figure 22.15a). Examine the eye and find the cornea, extrinsic eye muscles, sclera, and optic nerve. Using a scissors or a scalpel, make a coronal section of the eye behind the cornea (figure 22.15b). Do not squeeze the eye

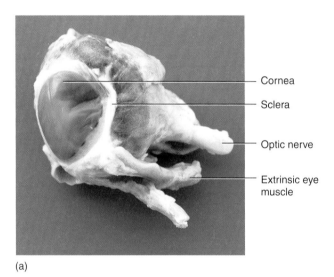

(a)

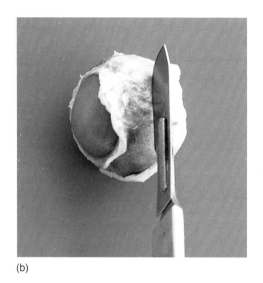

(b)

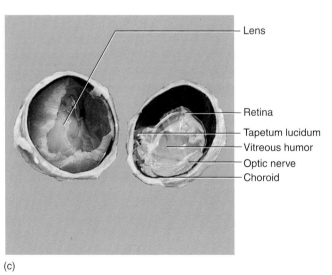

(c)

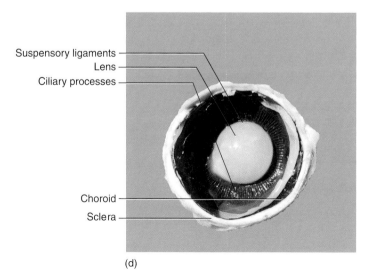

(d)

Figure 22.15 Dissection of a Sheep Eye
(a) External features; (b) coronal section; (c) eye with vitreous humor; (d) anterior eye without vitreous humor.

with force or thrust the blade sharply because you may squirt yourself with vitreous humor. Cut through the eye entirely and note the jellylike material in the posterior compartment (figure 22.15c). This is the **vitreous humor.** Look at the posterior portion of the eye. Note the beige **retina,** which may have pulled away from the darkened **choroid.** The choroid in humans is very dark, but you may see an iridescent color in your specimen. This is the **tapetum lucidum,** and it improves night vision in some animals. The tapetum lucidum produces the "eye shine" of nocturnal animals. Also examine the tough, white **sclera,** which envelops the choroid.

Now examine the anterior portion of the eye (figure 22.15d). Is the **lens** in place? The lens in your specimen probably will not be clear. Normally, the lens is transparent and allows for light penetration.

Note that the lens is held to the **ciliary body** by the **suspensory ligaments** (see figure 22.5). These ligaments pull on the lens and alter its shape for distant vision. Locate the ciliary body at the edge of the suspensory ligaments and the ciliary processes.

Now turn the eye over. Is there any **aqueous humor** left in the **anterior compartment?** Make an incision through the **conjunctiva** and **cornea** into the anterior compartment. Find the **anterior chamber** and the **posterior chamber,** which make up the anterior compartment. Locate the **iris** and the **pupil.** When you are finished, dispose of the specimen in a designated waste container and rinse your dissection tools.

Afterimages

The photosensitive pigment of the eye is called **rhodopsin,** which is composed of light-sensitive **retinal** (a metabolite of vitamin A) and the protein **opsin.** When light strikes the retina, the purple-colored rhodopsin splits into its two component parts and becomes pale (a process known as "bleaching"). You can test the time for separation and reassembly of the photopigments by staring at a colored image on a card under a moderately bright (60-watt) lightbulb for a few moments. Stare long enough to get an image (about 10 to 20 seconds) and then shut your eyes. Have your lab partner record the time. You should see a colored image against a dark background. This is known as a **positive afterimage,** which is due to the photoreceptors continuously firing. After a few moments, you should see the reverse of the original image (dark against a light background). This is known as a **negative afterimage.** A negative afterimage reflects the bleaching effects of rhodopsin.

Determination of the Blind Spot

The **optic disk** is a region where the retinal nerve fibers exit from the back of the eye and form the **optic nerve.** The mass exit of the nerve fibers leaves a small circle at the back of the eye devoid of photoreceptors. This region, the optic disk, is also known as the **blind spot.** You can locate the blind spot by holding this lab manual at arm's length. Use your right eye (close your left eye) and stare at the following X. It should be in line with the middle of your nose. Slowly move the manual closer to you and stare only at the X. Notice that at a particular distance the dot disappears.

X

You can test for the blind spot in your left eye as well.

Exercise 22 Review

Eye and Vision

Name: _____
Lab time/section: _____
Date: _____

1. "Eye shine" in nocturnal mammals is different from the "red eye" seen in some flash photographs. Eye shine is the reflection of light off the tapetum lucidum. What visual mechanism might explain red eye? _____
_____

2. Label the following illustration using the terms provided.

   lens   retina   choroid
   sclera   optic nerve   anterior cavity (anterior chamber)

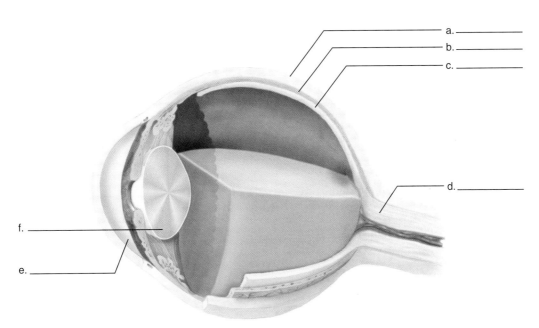

a. _____
b. _____
c. _____
d. _____
e. _____
f. _____

Sagittal section of the eye

3. Since the lens is made of protein, what effect might the preserving fluid used in lab have on the structure of the lens? How might this affect the clarity? _____
_____

4. What is the consensual reflex of the pupil? _____
_____

5. How does vitreous humor differ from aqueous humor in terms of location and viscosity? _____
_____

6. What layer of the eye converts visible light into nerve impulses? _____

7. What nerve takes the impulse of sight to the brain? _____

8. What is another name for the sclera? _____

9. Something catches your eye and you look to the right. What muscles would each eye be using? _____
_____

10. What gland produces tears? _____

11. What is the name of the transparent layer of the eye in front of the anterior chamber? _____

12. The iris of the eye has what function? _____
_____

13. Where is vitreous humor located? _____

14. What is the middle layer of the eye called? _____

15. Is the lens anterior or posterior to the iris? _____

16. Which retinal cells are responsible for vision in dim light? _____

17. How would you define the near point of the eye? _____
_____

18. What do the numbers 20/100 mean for visual acuity? _____
_____

19. What is astigmatism? _____
_____

20. In what area of the eye is the blind spot located? _____

21. Which way would you look if the inferior oblique muscle of the right eye contracted? _____

22. Name all of the structures and spaces that light passes through from the outside to the retina. _____
_____

Exercise 23
Ear, Hearing, and Balance

INTRODUCTION

The ear is a complex sense organ that performs two major functions, hearing and balance. It consists of three regions: an external ear, a middle ear, and an inner ear. The structure and function of the ear are covered in the *Principles of Anatomy and Physiology* text in chapter 13, "The Special Senses."

Hearing is considered **mechanoreception** because the ear receives mechanical vibrations (sound waves) and translates them into nerve impulses. This process begins with the vibrations reaching the external ear and ends up being interpreted as sound in the temporal lobe of the brain. Short sound waves are sound waves of high frequency, and they produce sound of high pitch. The height of the wave determines the loudness, which is measured in units called decibels. Longer sound waves have lower frequency and produce low pitch. Balance, on the other hand, involves receptors in the inner ear, visual cues, and **proprioception** (the perception of gravity or of forces applied to a structure). Two types of balance are sensed by the inner ear: **static balance** and **dynamic balance** (or **kinetic balance**). In static balance, a person can determine his or her nonmoving position (such as standing upright or lying down). In dynamic balance, motion is detected. Sudden acceleration, abrupt turning, and spinning are examples of dynamic balance.

OBJECTIVES

At the end of this exercise, you should be able to

1. explain how mechanical sound vibrations are translated into nerve impulses;
2. list the structures of the external, middle, and inner ear;
3. describe the structure of the cochlea;
4. perform conduction deafness tests, such as the Rinne and Weber tests;
5. compare dynamic and static balance and the structures involved in their perception.

MATERIALS

Models and charts of the ear
Microscope
Prepared slides of the cochlea
Tuning fork (256 Hz)
Rubber reflex hammer
Audiometer
Model of ear ossicles
Meter stick
Ticking stopwatch
Bright desk lamp

PROCEDURE
Anatomy of the Ear

The ear can be divided into three regions: the external, middle, and inner ear. The **external ear** consists of auditory structures superficial to the **tympanic membrane** (eardrum). The **middle ear** contains the tympanic cavity, ear ossicles, and auditory tube, and the **inner ear** consists of the cochlea, vestibule, and semicircular canals (figure 23.1).

Structure of the External Ear

The external ear consists of the **auricle,** or **pinna,** which can further be subdivided into the **helix** and the **lobule,** or earlobe. The helix is composed of stratified squamous epithelium overlying elastic cartilage. This cartilage allows the ears to bend significantly. Deep to the pinna, the external ear forms the **external auditory meatus,** which penetrates into the temporal bone. The **tympanic membrane** is the deep border of the external ear. It is composed of connective tissue covered by epithelial tissue. The membrane is sensitive to sound and vibrates as sound is funneled down the external auditory meatus.

Structure of the Middle Ear

The middle ear consists of a main cavity known as the **tympanic cavity;** three small bones, or **ossicles;** and the **auditory, eustachian,** or **pharyngotympanic** (fă-ring´-go´-tim-pan´-ik) **tube** (figure 23.1).

The ossicle that is attached to the tympanic membrane is the **malleus** (*malleus* = hammer). As the membrane vibrates, the malleus rocks back and forth, carrying the sound waves. The malleus is attached to the **incus** (*incus* = anvil), which is attached to the **stapes** (*stapes* = stirrup; figure 23.2).

The ear ossicles magnify sound about 20 times. Sound consists of pressure waves. As these waves strike the tympanic membrane, it vibrates. This vibration is conducted by the ossicles to the oval window. The process of moving from a large-diameter structure (tympanic membrane) to a smaller-diameter structure (oval window) magnifies the sound.

Examine the ossicles in figure 23.2 and on models in the lab. In addition to the ossicles, the auditory tube is in the middle ear. This tube connects the middle ear to the nasopharynx and provides for the equalization of pressure between the middle ear and the external environment when pressure changes occur (such as during changes of elevation). The auditory tube can be a conduit for microorganisms that travel from the nasopharynx to the middle ear and lead to middle-ear infections, particularly in young children.

Structure of the Inner Ear

The inner ear is encased in two labyrinthine structures and filled with two separate fluids. The outermost structure is the **bony**

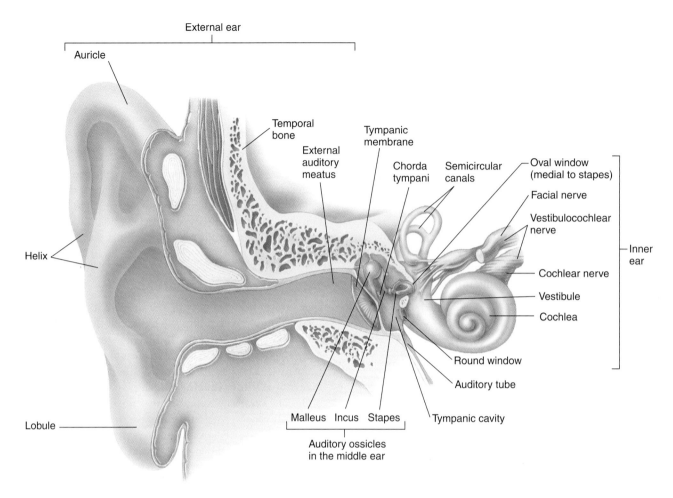

Figure 23.1 Anatomy of the Ear

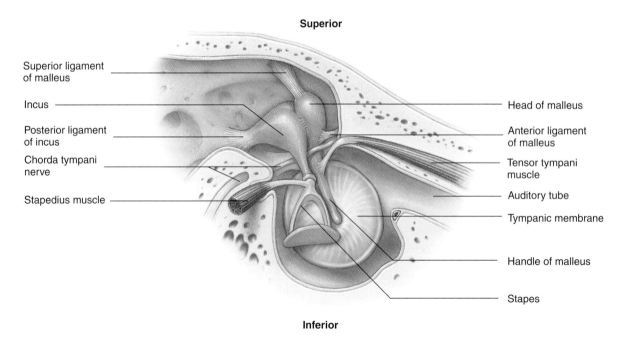

Figure 23.2 Middle Ear, Medial View

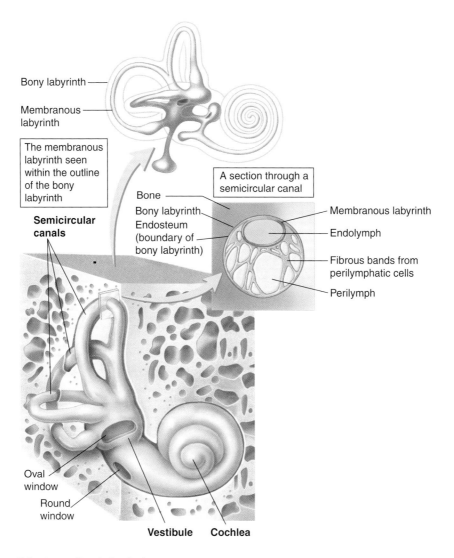

Figure 23.3 Anatomy of the Inner Ear, Labyrinths

labyrinth. Inside the bony labyrinth is **perilymph,** a clear fluid that is external to the **membranous labyrinth.** The fluid enclosed by the membranous labyrinth is the **endolymph,** which is important in both hearing and balance.

The inner ear is a complex structure composed of three separate regions: the cochlea, the vestibule, and the semicircular canals (figures 23.1 and 23.3).

Cochlea

The **cochlea** (figures 23.1 and 23.4) is involved in hearing. As the sound waves travel down the external auditory meatus, they cause the tympanic membrane to vibrate. This vibration rocks the ear ossicles, which are connected to the inner ear. As the stapes vibrates, it moves back and forth in the **oval window,** and this causes fluid to move back and forth in the cochlea. The cochlea also has a **round window,** which allows the vibration from the ossicles to move fluid back and forth. Without the round window, the fluid that transmits the sound would not move as easily and hearing would be greatly reduced. Sound waves are measured by their wavelengths. These wavelengths are measured in cycles per second, also known as hertz (Hz).

High-pitched sounds with vibrations of short wavelength, up to 20,000 Hz, stimulate the region of the cochlea closest to the middle ear. Low-pitched sounds with longer wavelengths, down to 20 Hz, stimulate the region of the cochlea farther from the middle ear. In this way, the cochlea can perceive sounds of varying wavelengths at the same time.

Microscopic Section of Cochlea

If you examine a microscopic section of the cochlea in cross section, you will see a number of chambers. The chambers are clustered in threes. Find the **scala vestibuli, scala media (cochlear duct),** and **scala tympani** on the microscope slide. Compare them with figures 23.4 and 23.5. Note the **spiral organ,** or the **organ of Corti,** which is in the area between the **vestibular membrane** and the **basilar membrane.** The spiral organ is seen here in cross section, but remember that it runs the length of the cochlea. The spiral organ is sensitive to sound waves. As a particular region of the spiral organ is stimulated, the basilar membrane and **tectorial membrane** vibrate

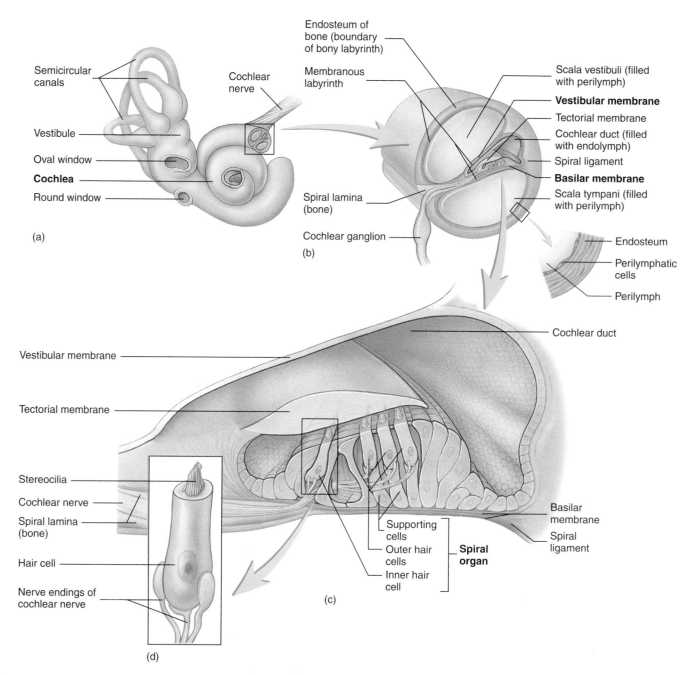

Figure 23.4 Anatomy of the Inner Ear
(a) Overview; (b) details of cochlear chambers; (c) spiral organ; (d) enlarged hair cell.

independently of one another. This produces a tugging on the cilia that attach the hair cells to the tectorial membrane. The hair cells are mechanosensitive channels that open potassium gates and send impulses to the cochlear branch of the **vestibulocochlear nerve.** These impulses travel to the **auditory cortex** of the **temporal lobe,** where they are interpreted as sound (figure 23.6).

Vestibule

Another part of the inner ear is the vestibule. The **vestibule** consists of the **utricle** and the **saccule** (figure 23.7). These two chambers are involved in the interpretation of static balance and acceleration. The utricle and saccule have regions known as **maculae,** which consist of **cilia** grouped together with an overlying gelatinous mass and calcium carbonate stones, called **otoliths.** These can be seen in figure 23.7. As the head is accelerated or tipped by gravity, the otoliths cause the cilia to bend, indicating that the position of the head has changed. Static balance is perceived not only from the vestibule but from visual cues as well. When the visual cues and the vestibular cues are not synchronized, then a sense of imbalance or nausea can occur.

Semicircular Canals

The third part of the inner ear consists of the **semicircular canals,** which are involved in determining dynamic balance. There are

Exercise 23 Ear, Hearing, and Balance

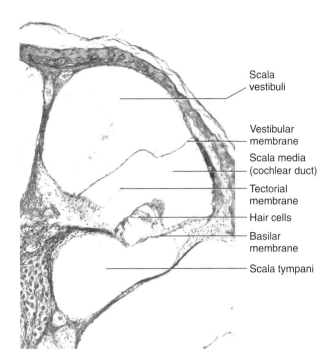

Figure 23.5 Photomicrograph of Cross Section of the Cochlea (100×)

1. Sensory axons from the cochlear ganglion terminate in the cochlear nucleus in the brainstem.
2. Axons from the neurons in the cochlear nucleus project to the superior olivary nucleus or to the inferior colliculus.
3. Axons from the inferior colliculus project to the medial geniculate nucleus of the thalamus.
4. Thalamic neurons project to the auditory cortex.
5. Neurons in the superior olivary nucleus send impulses to axons to the inferior colliculus, back to the inner ear, or to motor nuclei in the brainstem that send efferent fibers to the middle ear muscles.

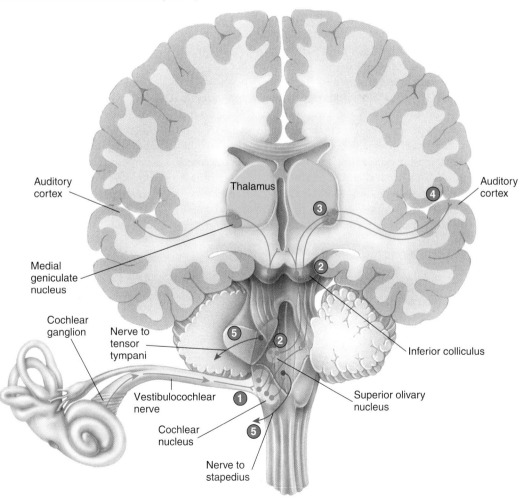

Figure 23.6 Interpretive Pathway of Hearing

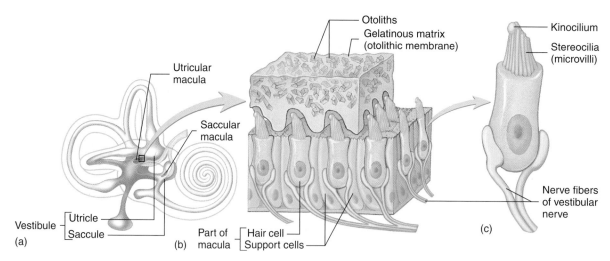

Figure 23.7 Vestibule
(a) Inner ear; (b) macula when head is upright; (c) hair cell.

three semicircular canals, each at 90 degree to one another (in the horizontal, sagittal, and coronal planes; figure 23.8). Each semicircular canal is filled with endolymph and is expanded at the base into an **ampulla.** Inside each ampulla is a **crista ampullaris,** a cluster of **hair cells** (cilia) with an overlying gelatinous mass called the **cupula.** The endolymph that is in the membranous labyrinth has inertia; that is, it tends to remain in the same place. As the head is turned, the cupula bends against the endolymph, the hair cells bend, and **angular,** or **rotational, acceleration** is perceived, as shown in figure 23.9. The three semicircular canals are placed at right angles to each other. If motion in the forward plane occurs (such as by doing back flips), then the **anterior semicircular canals** are stimulated. If you were to turn cartwheels, then the **posterior semicircular canals** would be stimulated. If you were to spin around on your heels, then the **lateral semicircular canals** would pick up the information. Motion that occurs in between these areas is picked up by two or more of the canals and interpreted as movement due to a combination of impulses from the semicircular canals.

Hearing Tests

Hearing tests can be administered by either a ticking stopwatch and a meter stick or an audiometer. Select one of the two following experiments based on your equipment in lab.

Obtain a meter stick and a ticking stopwatch. Have your lab partner sit in a quiet room and slowly move the ticking watch away from one of his or her ears until the sound can no longer be heard. Record the distance in centimeters (when the sound can no longer be heard) in the space provided. Check the other ear and record the data.

Maximum distance for right ear: _____

Maximum distance for left ear: _____

Audiometer Test

Ask your instructor to demonstrate how to use an audiometer to test hearing. In a typical audiometer, the red side is for the right ear and the blue side is for the left ear. Hearing normally decreases somewhat with age, although hearing loss is greatly accelerated by loud music, moderate to extensive stereo headphone use, and exposure to machines that operate at high decibel (db) levels. Another common source of hearing loss is the use of firearms without hearing protection.

One way to test hearing is with the use of an audiometer to determine the threshold of hearing for standard frequencies. The threshold for hearing is the lowest level at which your lab partner can hear the tone in at least 50% of the trials. To measure this value, set the audiometer to 125 Hz and then lower the level in 10 db increments from 50 db to the point where your lab partner can no longer hear the tone. Record the threshold for hearing. You can test other frequencies as well. Hearing loss is significant when the hearing threshold is at 20 db or more at any two frequencies in an ear or 30 db at any particular frequency. Record the minimum sound levels.

	Threshold (in Decibels) for Each Ear	
	Left Ear	Right Ear
125:	_____	_____
250:	_____	_____
500:	_____	_____
1000:	_____	_____
2000:	_____	_____
4000:	_____	_____
8000:	_____	_____

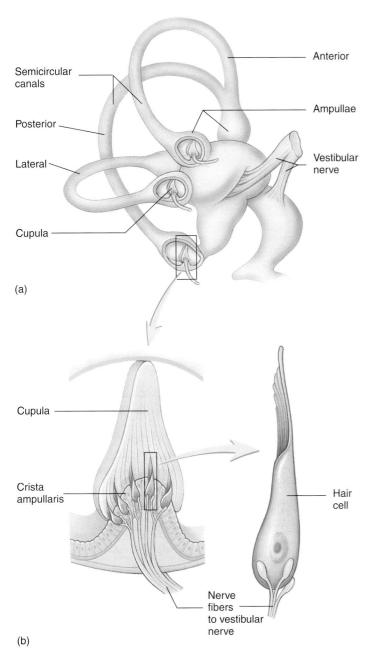

Figure 23.8 Semicircular Canals
(a) Overview; (b) details of canals, crista ampullaris, and cupula.

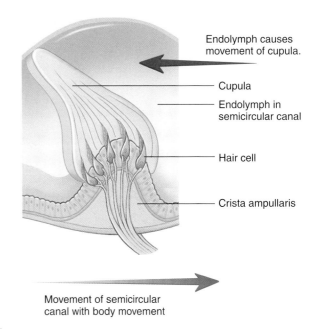

Figure 23.9 Movement of Semicircular Canals

There are two kinds of hearing loss, **conduction deafness** and **sensorineural deafness.** Conduction deafness involves the outer or middle ear where the mechanical vibrations in the external environment do not reach the cochlea. This may be due to a number of causes, including a perforated tympanic membrane, a blocked external auditory meatus, or damage from middle-ear infections. Sensorineural hearing loss takes place in the cochlea, in the vestibulocochlear nerve, or in the auditory centers of the temporal lobe of the brain. There are a couple of tuning fork tests that can indicate either conduction or sensorineural hearing loss. These are the Weber test and Rinne test.

Weber Test

Strike a tuning fork (at 256 Hz, or middle C) with a patellar hammer or strike the fork on your knee or elbow. Hold the handle of the tuning fork to the forehead of your lab partner (figure 23.10) while the tines (the forked portion) vibrate. If there is asymmetrical hearing loss, the sound will be louder in one ear. If your lab partner has conductive hearing loss, the sound will be louder in the ear with the hearing loss. You can demonstrate this by having your lab partner plug one ear and do the test. The sound will be louder in the ear that is plugged. Record the results of your test in the following space.

Sound perception (right/left or equal in both ears): _____

_____

You can interpret your test in the following way.

1. Normal—sound is the same in both ears
2. Conductive hearing loss—sound is louder in the ear with conductive loss
3. Sensorineural hearing loss—sound is louder in the ear without hearing loss

Rinne Test

Strike the tuning fork as you did in the Weber test and place the handle of the fork on your lab partner's mastoid process for a few seconds (figure 23.11). Once the sound diminishes significantly, hold the fork a few centimeters from the external auditory canal. Can your lab partner hear the sound? If hearing is normal, your lab partner should hear the sound through the external auditory canal (air conduction) louder than when the tuning fork was held to the mastoid process (bone conduction). If bone conduction is better than air conduction, then there is conductive hearing loss.

Figure 23.10 Weber Test

Figure 23.11 Rinne Test

This is a negative Rinne test. Do the test for both ears and record your results.

Right ear: _____

Left ear: _____

Sound Location

Have your lab partner sit with eyes closed. Strike the tuning fork with a rubber reflex hammer above his or her head. Have your lab partner describe to you where the sound is located. Strike the tuning fork behind the head, to each side, in front, and below the chin of your lab partner. Strike the tuning fork at the location to be tested. Do not strike the tuning fork and move it to the spot for location of sound. Record the results.

Where Sound Was Struck	Where Sound Was Perceived
Above head	
Behind head	
Right side	
Left side	
In front of head	
Below chin	

Balance

Postural Reflex Test

Postural reflexes are important for maintaining the upright position of the body. These reflexes are negative-feedback mechanisms. If, for example, you lean slightly to the left, your left foot abducts to regain the center of balance.

1. Select an area that is free from any obstacles.
2. While you read this lab manual, stand on the tips of your toes. Your lab partner should give you a little nudge to the left or right (not enough to knock you off your feet) to push you off balance. The postural reflex is reflected in the movement of the foot on the opposite side of your lab partner. If you are nudged to the left, then your left or right foot should move to the side to correct against the direction of the contact. Did your postural reflex work?
3. Record your result in the following space.

    Postural reflex: _____

Barany's Test

Barany's test examines visual responses to changes in dynamic balance. As the head turns, one of the reflexes that occurs is movement of the eyes in the opposite direction of the rotation. Nerve impulses from the semicircular canals innervate the eye muscles and cause the eye movement. When the head rotates in one direction, the eyes move in the opposite direction, so that there is enough time for a visual image to be fixed. The volunteer for this test should not be subject to dizziness or nausea.

1. Place the subject in a swivel chair with four or five students close by and equally spaced around the chair in case the

subject loses balance and begins to fall. The student chosen for this exercise should grab onto the chair firmly so as not to fall off. The subject should tilt his or her head forward about 30 degree, which will place the lateral semicircular duct horizontally for maximum stimulation.

2. One member of the group should spin the chair around about 10 revolutions while the subject keeps his or her eyes open.
3. Stop the chair and have the subject look forward. The twitching of the eyes is called **nystagmus** and is due to the stimulation of the endolymph flowing in the semicircular ducts. When the chair is stopped, the fluid in the endolymph will have overcome the inertia and will continue to flow in the ducts. Did the eyes move in the direction of the rotation or in the opposite direction? Record your results.

    Direction of chair rotation relative to subject: _____

    Direction of eye rotation: _____

Romberg Test

The Romberg test involves testing the static balance function of the body.

1. Place the subject in front of a chalkboard or any surface on which you can see the shadow of the subject. Do not lean against the chalkboard or against the wall.
2. Have the subject stand in that direction for 1 minute and determine if there are any exaggerated movements to the left or right. The feet should be close together.
3. Record your results in the following space.

    Amount of lateral sway: _____

You should then have your lab partner close his or her eyes and repeat the test. Is the swaying motion greater or less than before? Record your results in the following space.

    Amount of lateral sway with eyes closed: _____

A positive Romberg's test (indicated by significant swaying) may indicate inner ear problems or nerve, brain, or proprioceptor dysfunctions in muscles or joints.

Exercise 23 Review

Ear, Hearing, and Balance

Name: _____

Lab time/section: _____

Date: _____

1. What are the three general regions of the ear? _____

2. The pinna of the ear consists of what two main parts? _____
   _____

3. The ear is what kind of receptor? _____

4. The ear performs two major sensory functions. What are they? _____
   _____

5. What structure separates the external ear from the middle ear? _____

6. Name the three ear ossicles. _____
   _____

7. From the following choices, select the function of the cochlea.
   a. static balance   b. taste
   c. hearing   d. dynamic balance

8. What area is found between the scala vestibuli and the scala tympani? _____

9. What is the name of the nerve that takes information about balance and hearing to the brain? _____

10. What units are used to measure sound energy? _____

11. What part of the inner ear is involved in perceiving static balance? _____

12. Name the parts of the ear that might be impaired if a person demonstrates conduction deafness. _____
    _____

13. What two diagnostic tests are used to determine conduction deafness? _____
    _____

14. What is the name of the tube that runs from the auricle to the tympanic membrane? _____

15. The auditory tube connects what two cavities? _____

16. What tube is responsible for the equalization of pressure when you change elevation? _____

17. What is the name of the space that encloses the ear ossicles? _____

18. Place the ear ossicles in sequence from the tympanic membrane to the oval window. _____

19. What is the function of the semicircular canals? _____

20. Background noise affects hearing tests. In the ticking watch test or audiometer test, what kind of result, in terms of auditory sensitivity, would you have recorded if moderate background noise were present? _____

21. In the Weber test, the ear that perceives the sound as being louder is the deaf ear. Why is this the case? _____

22. Label the following illustration of the ear using the terms provided.

    cochlea   ear ossicles   external auditory meatus
    auditory tube   tympanic membrane   semicircular canals

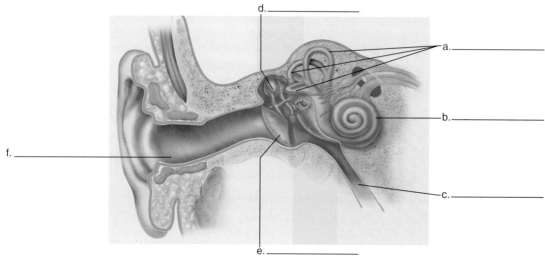

Exercise 23 Ear, Hearing, and Balance

23. Label the following illustration of the cross section of the cochlea using the terms provided.

   hair cells   scala tympani   tectorial membrane
   scala vestibuli   vestibular membrane   scala media (cochlear duct)

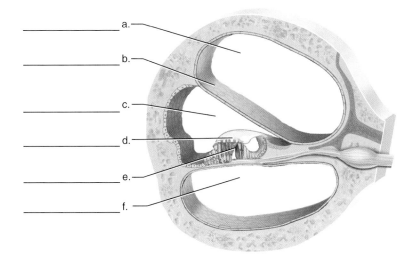

Exercise 24
Endocrine System

INTRODUCTION

The **endocrine system** consists of glands that produce chemical messengers called **hormones,** which are picked up by blood capillaries. This type of release is called a "ductless secretion," and it is characteristic of the endocrine system. Glands that secrete material in tubules or ducts (such as sweat and salivary glands) are known as **exocrine glands.** The secretions of the endocrine glands enter the interstitial fluid and then travel by blood vessels, which act as highways to carry the hormones throughout the body. Areas that are receptive to hormones are called **target areas;** these may be organs or tissues. The target areas are sensitive to hormones because they have cells with receptor sites for specific hormones. If a hormone has an effect on a gland (such as the thyroid gland), it is due to the receptor on the cells for that specific hormone. Cells in the liver may come into contact with the hormone, but if their cells do not have receptors for that hormone, then there is no effect on those cells. Hormones can have many effects. Some of these actions are growth, changes and development, maturation, metabolism, sexual development, regulation of the sexual cycle, and homeostasis. In terms of homeostasis some hormones have opposing effects. For example, if blood calcium levels are low, the parathyroid glands secrete a hormone that increases blood calcium levels. If blood calcium levels are high, the thyroid gland secretes a hormone to lower blood calcium levels. In this way the body maintains the proper amount of calcium in the blood. This is just one example of many hormones that oppose one another, thus keeping the body within normal limits. Many organs of the body, such as the stomach, heart, and kidney, produce hormones and thus have endocrine functions. This exercise focuses on the glands that have a major endocrine component. Hormones and the endocrine glands that produce them are covered in *Principles of Anatomy and Physiology* in chapter 15, "Endocrine System."

OBJECTIVES

At the end of this exercise, you should be able to

1. discuss how the secretions of the endocrine glands differ from those of the exocrine glands;
2. list the major endocrine organs of the human body;
3. identify endocrine organs in histological slides;
4. name the hormones produced by the endocrine organs;
5. list the hormones from the anterior and posterior pituitary gland;
6. describe the detection of hormones by monoclonal antibodies.

MATERIALS

Models and charts of endocrine glands
Model or chart of midsagittal section of head
Microscopes
Microscope slides
    Thyroid
    Pituitary
    Adrenal gland
    Pancreas
    Testis
    Ovary
Urine samples from women in the middle of their ovarian cycle.
Ovulation test kit
Disposable gloves (latex or plastic) and biohazard container

PROCEDURE
Anatomy of the Major Endocrine Organs

Locate the major endocrine glands in charts or models in lab and compare them with figure 24.1, including the hypothalamus, pineal gland, pituitary gland, thyroid, parathyroids, thymus, pancreas, adrenals, and gonads (testes or ovaries). Once you have noted their location, you can proceed with a more detailed study.

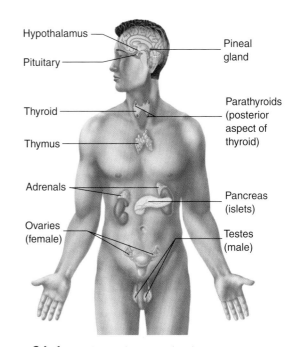

Figure 24.1 Major Endocrine Glands

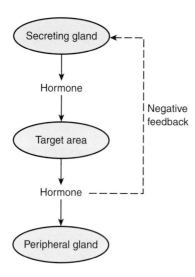

Figure 24.2 Negative Feedback Regulation of Hormones

Negative Feedback Regulation

In the case of the hypothalamus and pituitary, regulation of the secretion of hormones takes place by **negative feedback inhibition.** Hormones secreted by the hypothalamus cause the release of hormones from other glands that inhibit the additional release of hypothalamus hormones. In this way the level of the hormones are controlled. This is illustrated in figure 24.2.

Hypothalamus

The **hypothalamus** has a major role in controlling endocrine functions. The hypothalamus secretes both stimulating and inhibitory hormones that either cause the release of hormones from their target areas or prevent the release of hormones from these areas. Locate the hypothalamus in figure 24.3.

Pineal Gland

The **pineal gland (pineal body)** develops from the diencephalon of the brain. The gland is so named because it resembles a pine nut. Locate the pineal gland in a model or chart of a midsagittal section of the head and compare it with figure 24.3. One of the hormones the pineal gland secretes is **melatonin,** which is an inhibitory hormone and probably regulates circadian rhythms. Melatonin inhibits the release of hormones in the hypothalamus that stimulate the production of sex hormones. Melatonin has a psychologically depressing effect in some people. It has been named the "hormone of darkness" because it is produced during times of low light levels and its manufacture is inhibited in bright light.

Pituitary Gland

The **pituitary gland,** or **hypophysis,** is seen in figure 24.3. It is divided into the **anterior pituitary,** or **adenohypophysis,** and the **posterior pituitary,** or **neurohypophysis.** The pituitary is suspended from the base of the brain by a stalk called the **infundibulum.**

Anterior Pituitary

The anterior pituitary originates from the roof of the oral cavity during embryonic development. It can be divided into three parts: the pars tuberalis, the pars intermedia, and pars distalis (figure 24.4). The cells of the anterior pituitary have the same embryonic origin, yet they have differentiated into several types of specialized cells. The result of this development can be seen if you examine a prepared slide of the pituitary gland. In the histological section, note that the adenohypophysis is composed of **epithelial cells** (figure 24.4). These cells produce a number of hormones that have broad effects throughout the body. Draw a section of the anterior pituitary in the following space.

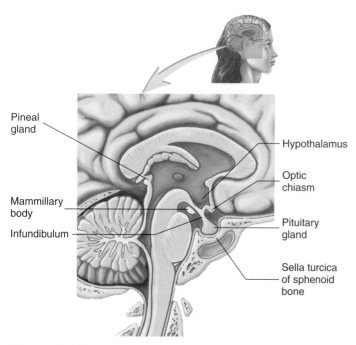

Figure 24.3 Endocrine Glands of the Head (Midsagittal Section)

Some of the hormones produced in the anterior pituitary and their functions are as follows:

- **Thyroid-stimulating hormone (TSH),** or **thyrotropin,** stimulates the thyroid gland to produce thyroid hormones.
- **Growth hormone (GH),** or **somatotropin,** promotes the growth of most of the cells and tissues of the body.

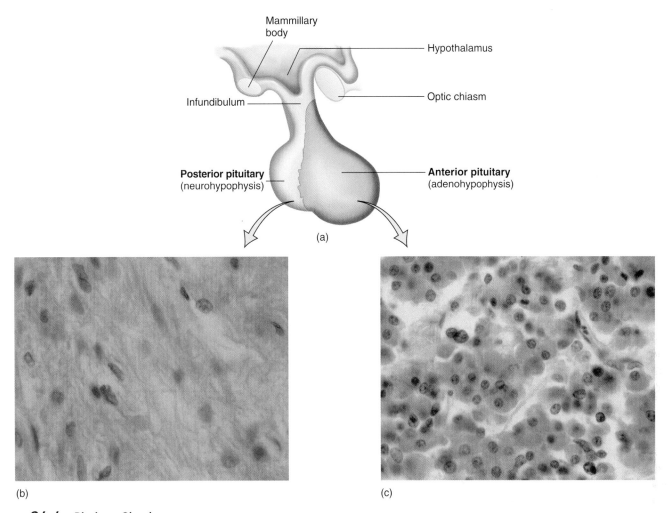

Figure 24.4 Pituitary Gland
(a) Regions of the entire gland; (b) photomicrograph of the posterior pituitary (400×); (c) photomicrograph of the anterior pituitary (400×).

- **Prolactin,** or **luteotropin (LTH),** stimulates mammary glands to begin the production of milk.
- **Gonadotropins** stimulate ovaries and testes.
  —**Follicle-stimulating hormone (FSH)** stimulates the production of sex cells, causes ovarian follicles to mature (develop and produce egg cells), and stimulates testes to produce spermatozoa.
  —**Luteinizing hormone (LH)** stimulates ovaries to produce other hormones (for example, progesterone), causes the maturation of Graafian follicles and ovulation, and stimulates testosterone production in testes.
- **Adrenocorticotropin (ACTH)** regulates hormone production in the adrenal cortex.

Posterior Pituitary

The tissue of the posterior pituitary is very different from that of the adenohypophysis. The neurohypophysis is composed of **nervous tissue** that originated from the base of the brain. Compare the tissue of the neurohypophysis in a prepared slide with figure 24.4. Draw a representative image of the tissue in the following space.

Hormones released from the posterior pituitary are secreted by the **hypothalamus** and flow through axons to be stored in the posterior pituitary. These hormones and their actions include the following:

- **Antidiuretic hormone (ADH)** stimulates the reabsorption and retention of water by the kidneys. It is also known as **vasopressin** because it causes arterioles to constrict, which elevates blood pressure.

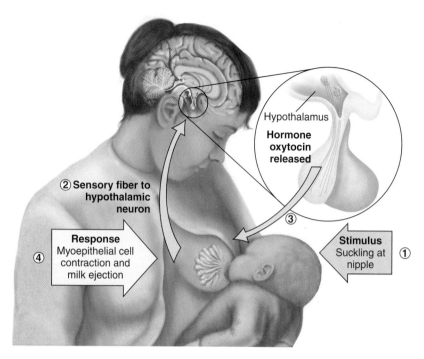

Figure 24.5 External Influences on Hormonal Action

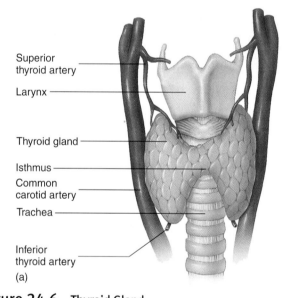

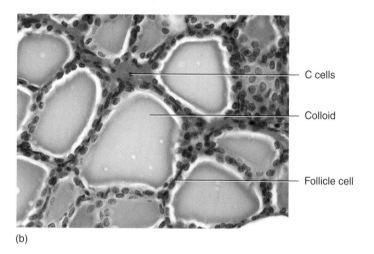

Figure 24.6 Thyroid Gland
(a) Diagram; (b) photomicrograph (100×).

- **Oxytocin** stimulates the contraction of the cells of the mammary glands, resulting in the release of milk, and causes uterine contractions. External influences on hormone actions can be seen in the release of oxytocin. Look at figure 24.5 and note that the sucking action of an infant on the mother's breast sends impulses to the hypothalamus. The hypothalamus stimulates the posterior pituitary to release oxytocin, which causes milk ejection.

Thyroid

Figure 24.6 illustrates the **thyroid gland,** which is named for its location inferior to the thyroid cartilage of the larynx. The thyroid gland has two lateral **lobes** and a medial **isthmus** that connects them.

The thyroid is easy to recognize histologically. Examine a slide of thyroid gland and compare it with figure 24.6. Locate the **follicle cells** that surround the **colloid,** a storage material for thyroid hormones. Colloid is mostly composed of thyroglobulin,

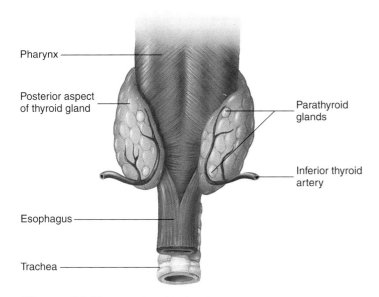

Figure 24.7 Parathyroid Glands

which is a large molecular weight compound. Thyroid hormones are manufactured in the thyroid and bound to the larger thyroglobulin molecule. They are cleaved from it when needed by the follicular cells. Also locate the **parafollicular cells (C cells),** which are located in the spaces between the follicles.

The thyroid gland secretes three specific hormones. Two of these are T_3, or **triiodothyronine** (a molecule that includes three iodine atoms), and T_4, or **thyroxine** (containing four iodine atoms per molecule). These hormones increase **basal metabolic rates** and are stored in the colloid. Another hormone of the thyroid gland is **calcitonin.** Calcitonin causes the deposition of calcium in bone by decreasing osteoclast activity. Calcitonin is produced by the parafollicular cells. Calcitonin is antagonistic to parathyroid hormone (discussed next).

Parathyroid Glands

In the posterior portion of the thyroid are typically two pairs of organs called the **parathyroid glands. Chief cells** in the parathyroid gland secrete **parathyroid hormone (PTH),** or **parathormone,** which is responsible for increasing calcium levels in the blood. It accomplishes this by increasing calcium uptake in the intestines, increasing the kidneys' reabsorption of calcium, and releasing calcium from bone. Examine the location of the parathyroid glands embedded in the posterior surface of the thyroid gland (figure 24.7).

Thymus

The **thymus** is active in young people and plays an important part in immunocompetency. The thymus is located anterior and superior to the heart (see figure 24.1). You will not find the thymus on models of adults. It secretes a hormone, **thymosin,** that causes the maturation of T cells. These cells originate as stem cells in the bone marrow and migrate to the thymus. Under the influence of thymosin, the cells mature to provide **cell-mediated immunity** against **antigens** (foreign material, such as viruses, and tumors, that cause immune reactions). The T cells migrate from the thymus predominantly to the lymph nodes and the spleen to carry out their functions.

Pancreas

The **pancreas** is a mixed gland in that it has an exocrine function and an endocrine function. Locate the pancreas in figures 24.1 and 24.8. The exocrine function is digestive because pancreatic juice contains both buffers and digestive enzymes. These are secreted by the pancreatic acinar cells. The endocrine function of the pancreas consists of the secretion of the hormones including **insulin** and **glucagon,** which regulate blood glucose levels. When blood glucose levels drop, glucagon converts glycogen (a starch-like storage product) to glucose. Glucagon is produced in specialized cells **(alpha cells)** on the periphery of clusters called **pancreatic islets (islets of Langerhans;** figure 24.8b). Pancreatic islets also produce insulin, which lowers the blood glucose level. Insulin, which is produced in **beta cells,** stimulates the conversion of glucose to glycogen. The pancreatic islets also contain **delta cells** (not illustrated), which secrete **somatostatin,** a hormone that inhibits insulin and glucagon secretion. Locate the pancreatic islets on a microscope slide and compare them with those in figure 24.8. Draw an islet in the space provided.

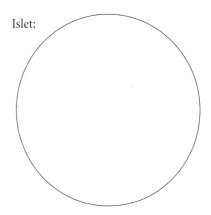

Adrenal Glands

The **adrenal glands** (*ad* = next to, *renal* = kidney) are located superior to the kidneys. Each adrenal gland is composed of an outer **cortex** and an inner **medulla.** Examine a microscope slide of an adrenal gland and locate the cortex and the medulla. Compare this with figures 24.1 and 24.9.

The hormones secreted from the adrenal cortex are called **corticosteroid hormones.** They are important in water and electrolyte balance (Na^+, K^+) in the body. They are also important for carbohydrate, protein, and fat metabolism, as well as stress management. The cortex can be divided into three regions. The outermost is the **zona glomerulosa** (figure 24.9), which consists of clusters of cells that secrete **mineralocorticoids** (especially **aldosterone**). Inside this layer (closer to the medulla) is the **zona fasciculata,** which consists of parallel bundles of cells that secrete **glucocorticoids.** The deepest cortical layer is the **zona reticularis,** which consists of a branched pattern of cells that produce both glucocorticoids and **sex hormones**

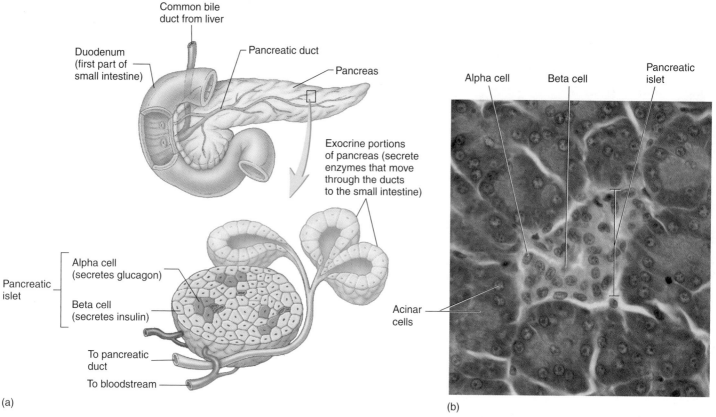

Figure 24.8 Pancreas
(a) Diagram; (b) photomicrograph (400×).

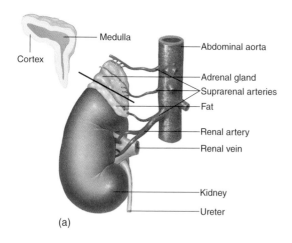

Figure 24.9 Adrenal Gland
(a) Diagram; (b) photomicrograph of cortex (100×).

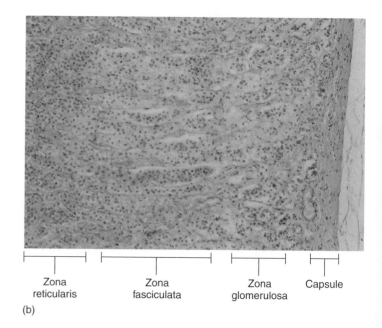

(**androgens** and **estrogens**). The hormone **dehydroandrostenedione** (DHEA) is a weak androgen produced in both males and females. DHEA is secreted from the adrenal glands but is converted to testosterone in other tissues. The major source of androgens in males is the testes. The hormones **epinephrine** and **norepinephrine** are produced in the adrenal medulla. Stimulation of the adrenal glands by the sympathetic nervous division causes the release of epinephrine and norepinephrine from the gland.

Gonads

The testes and ovaries not only produce the sex cells but also hormones so they are endocrine glands. Both testes and ovaries are stimulated by follicle-stimulating hormone (FSH) from the anterior pituitary, which causes the production or maturation of the sex cells (**spermatozoa** or **oocytes**). They are also under the influence of luteinizing hormone (LH), which increases the level of hormone production.

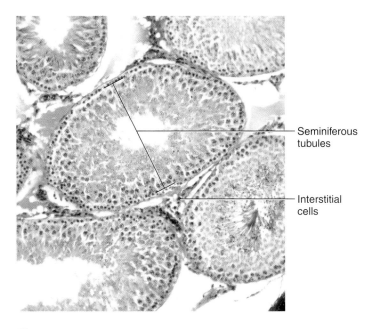

Figure 24.10 Histology of the Testis (100×)

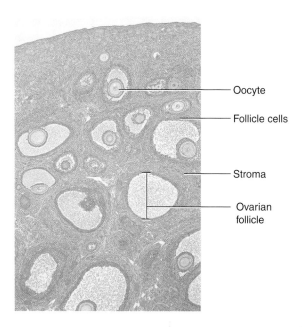

Figure 24.11 Histology of the Ovary (40×)

Testes

In males, the **testes** produce **testosterone,** which is a hormone responsible for the development of the male genitalia during embryonic and fetal development and for secondary sex characteristics, such as the development of facial and body hair, the expansion of the larynx (which produces a deeper voice), and the increased muscle and bone mass seen in males. Testosterone also aids FSH in the production of **spermatozoa.** The testes are illustrated in figure 24.1. The exocrine function of the testis is explored in Exercise 42, on the male reproductive system.

Examine a microscope slide of a testis and find the **seminiferous tubules** and **interstitial cells** (figure 24.10). The interstitial cells are found between the tubules, and they produce testosterone. Draw the seminiferous tubule and interstitial cells in the space provided.

Ovaries

In females, the **ovaries** produce oocytes (eggs) and have an endocrine function in that they produce **estrogen** and **progesterone** (figure 24.1). The ovaries are illustrated in figure 24.11. Estrogen is produced by the **follicle cells** and progesterone is produced in the corpus luteum. Female hormones are also responsible for secondary sex characteristics in women, such as the development of breasts, an additional subcutaneous adipose layer, and a high voice. Estrogen and progesterone also influence the development of the endometrium, cause the maturation of the oocytes, and regulate the menstrual cycle. *Estrogen* is a generic term for several hormones produced by the female, including **estradiol.** During pregnancy, the placenta functions as an endocrine gland in the secretion of estrogen and progesterone. Draw an ovarian follicle in the following space.

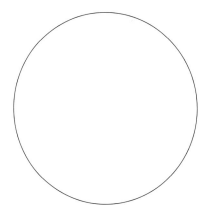

Endocrine Physiology—Detection of Hormones

Hormones are secreted in small but detectable amounts in the bloodstream. Some of these hormones may be filtered in the urine and detected from a urine sample. **Luteinizing hormone (LH)** is one of these hormones. LH stimulates the final maturation of the oocyte and causes ovulation. Normally, small amounts of LH occur in a woman's body but an increase, or spike, in the levels of LH occurs about 24–36 hours prior to ovulation (figure 24.12). This increase lasts typically for 10 to 30 hours. In this part of the exercise you will use a test kit to determine the presence of LH. Your instructor may provide an unknown sample of urine, or female students in the middle of their ovarian cycle may be needed as volunteers to test for the presence of LH.

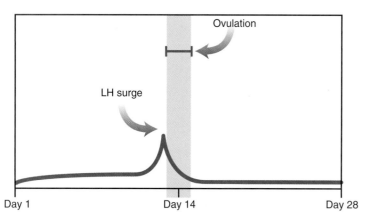

Figure 24.12 Surge of Luteinizing Hormone (LH) Prior to Ovulation
Day 1 of the menstrual cycle begins with menstrual flow.

Caution
Remember to treat bodily fluids as if they carry pathogens. Wear protective gloves and eyewear when handling fluids! Dispose of all contaminated disposable materials in the biohazard container. The reusable material should be placed in a 10% bleach solution.

The test kit uses **monoclonal antibody** techniques to determine the presence of LH. The development of a monoclonal antibody starts when a mouse (or other mammal) is injected with human LH. Since this hormone is foreign to the mouse, specific white blood cells called beta lymphocytes make antibodies against human LH. These can be removed from the mouse in a blood sample and fused with tumor cells. Tumor cells are used because they grow rapidly. The hybrid tumor-lymphocyte cells replicate, producing more cells that have the antibody against LH. The antibodies are isolated and absorbed by a solid substrate (the test strip). An enzyme that bonds to the LH-antibody is also attached to the substrate. This enzyme is attached to a dye marker and it changes color when attached to the antibody. The test strip contains antibodies to LH, an enzyme that will bind to the LH-antibody complex, and a dye that changes color when the enzyme binds to the LH-antibody complex. If a woman has significant amounts of LH then the test strip produces a color (usually blue). The generic name for tests that use antibodies and enzymes to determine the presence of biological material is called enzyme-linked immunosorbent assay (ELISA). In some cases, pregnancy, endometriosis, hyperthyroidism, and ingestion of some prescription drugs may produce false test results.

Select the unknown urine sample or have three women who are in the middle of their ovarian cycle perform the test. The test should be positive for women who are just about ready to ovulate. This is 2 weeks prior to the next anticipated menstrual cycle.

Obtain an ovulation test kit and, following the instructions provided in the kit or from your instructor, determine if the urine sample provided as an unknown, or the sample from female students in the class, has significant levels of LH. Typically the test involves collecting a urine sample and dipping the absorbent sampler in the urine for a specific amount of time. The sample is then compared to a reference line to indicate ovulation.

Sample _____ (Kit/student)

Presence of LH _____

Once you have studied the endocrine glands and material, complete table 24.1. You should fill in the blank spaces where needed.

TABLE 24.1

Organ	Hormones Produced	Effect of Hormones
_____	Antidiuretic hormone	_____
Thyroid	Thyroxine	_____
_____	Corticosteroid hormones	Regulate electrolyte balance
Ovary	_____	Regulates ovarian cycle
_____	Melatonin	Regulates sleep cycles, inhibits release of reproductive hormones
_____	Luteinizing hormone	_____
Neurohypophysis	Oxytocin	_____
Parathyroid	_____	Increases calcium in blood
_____	Glucagon	Increases blood glucose levels
Pancreas	_____	Decreases blood glucose levels
_____	Calcitonin	_____

Exercise 24 Review

Name: _____

Lab time/section: _____

Date: _____

Endocrine System

1. What is the general name for organs that produce hormones? _____

2. What name is given to regions that are receptive to hormones? _____
   _____

3. Melatonin is secreted by what gland? _____

4. In what specific part of what gland is ADH stored? _____
   _____

5. What is the effect of TSH, and where is it produced? _____
   _____

6. What does glucagon do as a hormone, and where is it produced? _____
   _____

7. Which hormones in the adrenal gland control water and electrolyte balance? _____
   _____

8. What is the primary gland that secretes epinephrine? _____

9. Where is growth hormone produced? _____

10. What is another name for T_3? _____

11. What connects the two lobes of the thyroid gland? _____

12. Does parathormone increase or decrease calcium levels in the blood? _____

13. Label the endocrine glands in the following illustration using the terms provided.

ovaries   adrenal glands   pancreas   thymus
testes   thyroid gland   parathyroid glands

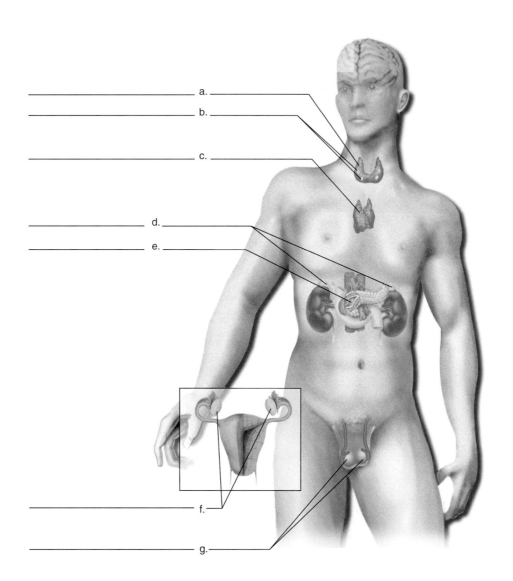

a. _____
b. _____
c. _____
d. _____
e. _____
f. _____
g. _____

14. Identify the parts of the pituitary, as seen in the following illustration, and label two hormones located in each.

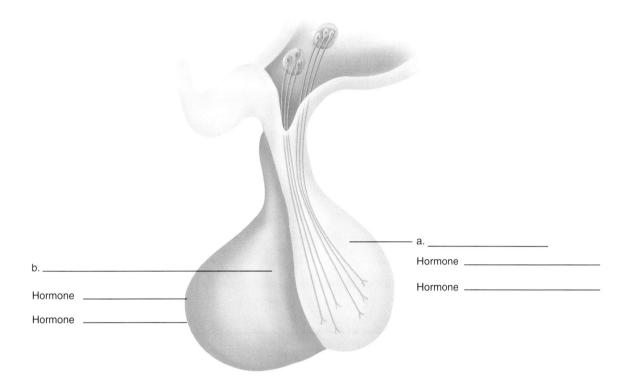

a. _____
Hormone _____
Hormone _____

b. _____
Hormone _____
Hormone _____

15. Identify the three layers of the adrenal cortex, as illustrated, and list the hormones produced by each layer.

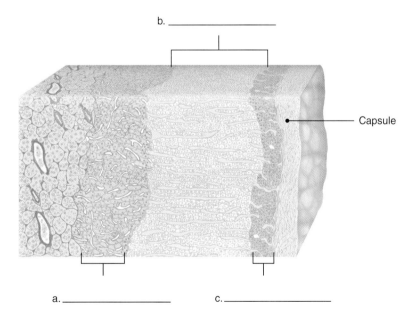

a. _____
b. _____
c. _____
Capsule

16. Interstitial cells of the testis produce which hormone? _____

17. Target cells must have the correct receptor site in order for hormones to influence those cells. Cut out the following 12 outlines, which represent either hormones or an effect caused by a hormone. In the structures, hormones are circled above the endocrine glands that produce them. The effects caused by each hormone are not circled, and the organ where the effect occurs is listed below the effects. Assemble the hormones and effects according to the sequence in which they fit. There may be hormones or cells that do fit the sequence. Once the sequences are assembled, color each one with a different color. Two of the sequences represent an axis, which is a relationship among the hypothalamus, pituitary, and a more distant gland. There are three axes in the body: the hypothalamo-pituitary-gonadal axis (HPG), the hypothalamo-pituitary-thyroid axis (HPT), and the hypothalamo-pituitary-adrenal axis (HPA). Write HPG, HPT, or HPA on the corresponding sequence. Notice that, in some sequences, a hormone can have more than one effect (though not all the effects of the hormones are listed on this page). Select a hormone and target gland from your lab text and make a sequence of your own.

Exercise 25
Blood

INTRODUCTION

Blood is made of two major components, formed elements and plasma. Formed elements make up about 45% of the blood volume. They can be subdivided into red blood cells (erythrocytes), white blood cells (leukocytes), and platelets (thrombocytes). Plasma, which constitutes about 55% of the blood volume, contains water, lipids, dissolved substances, colloidal proteins, and clotting factors.

The study of blood is significant both because it is the fluid medium of the cardiovascular system and because it has significant clinical importance. Changes in the numbers and types of blood cells can be used as indicators of disease. In this exercise, you examine the nature of blood cells, which are discussed in *Principles of Anatomy and Physiology* in chapter 16, "Blood."

Caution

The risk of bloodborne diseases is significantly minimized by the use of sterilized human blood and nonhuman mammal blood in this exercise. However, use the same precautions as if you were handling fresh and potentially contaminated human blood. Your instructor will determine whether to use animal blood (from nonhuman mammals), sterilized blood, or synthetic blood. In all these cases, follow strict procedures for handling potentially pathogenic material.

1. Wear protective gloves during the procedures.
2. Do not eat or drink in lab.
3. If you have an open wound, do not participate in this exercise, or make sure the wound is *securely* covered.
4. Your instructor may elect to have you use your own blood. Due to the potential for disease transmission, such as HIV or hepatitis B and C, in fresh blood samples, **make sure you keep away from other students' blood and keep them away from your blood.**
5. Do not participate in this part of the exercise if you have recently skipped a meal or if you are not well.
6. Place all disposable material in the biohazard bag.
7. Place all used lancets in the sharps container and all glassware that is to be reused in a 10% bleach solution.
8. After you have finished the exercise, clean and disinfect the countertops with a 10% bleach solution.

OBJECTIVES

At the end of this exercise, you should be able to

1. discuss the composition of blood plasma;
2. distinguish among the various formed elements of blood;
3. describe hematopoiesis;
4. determine the percentage of each type of leukocyte in a differential white cell count;
5. describe what an elevated level of a particular white blood cell may indicate about a disease state or an allergic reaction.

MATERIALS

Prepared slides of human blood with Wright's or Giemsa stain
Compound microscopes
Lab charts or illustrations showing the various blood cell types
Latex gloves
Roll of paper towels
Vial of mammal blood or
    Sterile cotton balls,
    Alcohol swabs,
    Sterile, disposable lancets, and
    Adhesive bandages
Dropper bottle of Wright's stain
Squeeze bottle of distilled water or phosphate buffer solution
Large finger bowl or staining tray
Toothpicks
Clean microscope slides
Coverslips
Pasteur pipette and bulbs
Hand counter
Biohazard bag or container
10% bleach container
Sharps container

PROCEDURE
Plasma

Plasma is the fluid portion of blood and is about 91% water. The remainder consists mostly of proteins, such as albumins, globulins, and fibrinogen. **Albumins** are produced by the liver and make up the majority of the plasma proteins. Some **globulins** are made by the plasma cells, and these make up the next largest share of proteins. **Fibrinogen** is a clotting protein; this and other clotting factors are produced by the liver. Plasma also contains electrolytes (Na^+, K^+, and Cl^-), nutrients, hormones, and wastes.

Examination of Blood Cells

In this part of the exercise, you need to distinguish among red blood cells, white blood cells, and platelets. If you are using a prepared slide of blood, you can move to the section entitled "Microscopic Examination of Blood." If you are to make a smear, then read the following directions.

Preparing a Fresh Blood Smear

Your lab instructor will direct you to use either fresh, nonhuman mammal blood or your own blood. If you are using provided mammal blood, wear latex gloves and withdraw a small amount of blood from the vial with a clean Pasteur pipette. Place a drop of the blood on a clean microscope slide. If you are using your own blood, make sure you follow the safety precautions as discussed at the beginning of the exercise.

Withdrawal of Your Own Blood

Obtain the following materials: a clean, sterile lancet; an alcohol swab; an adhesive bandage; rubber gloves; sterile cotton balls; two clean microscope slides; and a paper towel.

1. Arrange the materials on the paper towel in front of you on the countertop.
2. Clean the end of the donor finger with the alcohol swab and let the hand from which you will withdraw blood hang by your side for a few moments to collect blood in the fingertips. You will be puncturing the pad of your fingertip (where the fingerprints are located).
3. Peel back the covering of the lancet and hold on to the blunt end as you withdraw it from the package. Do not touch the sharp end or lay the lancet on the table before puncturing your finger.
4. Jab your finger quickly and wipe away the first drop of blood that forms with a sterile cotton ball. Never reuse the lancet or set it down on the table. Place the lancet in the sharps container.
5. Place the second drop of blood that forms on the microscope slide about 2 cm away from one of the ends of the slide (figure 25.1) and place another cotton ball on your finger.
6. You can now put an adhesive bandage on your finger.
7. Whether you are using prepared blood or your own blood, use another clean slide to spread the blood by touching the drop with the edge of the slide and push the blood across the slide (figure 25.1). This should produce a smooth, thin smear of blood. Let the blood smear dry completely.
8. After the slide is dry, place it in a large finger bowl elevated on toothpicks or on a staining tray.
9. Cover the blood smear with several drops of Wright's stain from a dropper bottle. Let the stain remain on the slide for 1 to 2 minutes.
10. After this time, add water or a prepared phosphate buffer solution to the slide. You can rock the slide gently with gloved hands or gently blow on it to stir the stain and water. A metallic green material should come to the surface of the slide. Let the slide remain covered with stain and water for 3 to 8 minutes.
11. Wash the slide gently with distilled water until the material is light pink, and then stand it on edge to dry. You can also stain blood using an alternate stain (such as Giemsa stain), following your lab instructor's procedure.

Place all blood-contaminated disposable material in the biohazard container. Place the slide used to spread the blood in the 10% bleach solution and make sure the lancets are in the sharps container. Once the slide is completely dry, it can be examined under the high-power or oil immersion lens of your microscope.

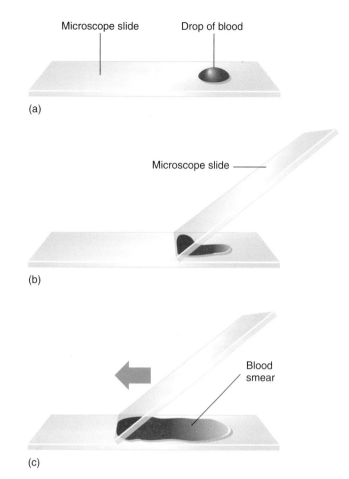

Figure 25.1 Making a Blood Smear
(a) Placing blood on a slide; (b) touching glass slide to front of blood drop; (c) spreading blood across slide.

Microscopic Examination of Blood

Overview

As you examine the blood slide, refer to figure 25.2 and to the photomicrographs of the blood cells. You will have to look at many blood cells in order to see all of the cells that are included in this exercise.

Red Blood Cells

Examine a slide of blood stained with either Wright's or Giemsa stain. Red blood cells are the most common cells you will find on the slide. There are about 5 million red blood cells per cubic millimeter. They do not have a nucleus but appear as pink, biconcave disks (as if you had placed two dinner plates back to back). This shape increases surface area and provides for greater oxygen-carrying capabilities. Blood formation is called **hematopoiesis,** or **hemopoiesis.** The specific production of red blood cells is called **erythropoiesis.**

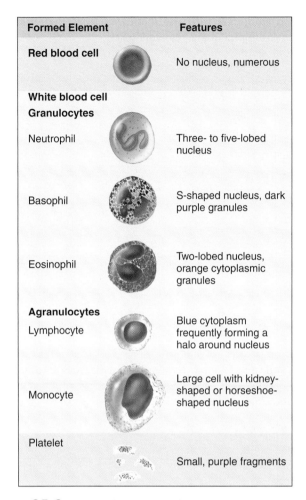

Figure 25.2 Formed Elements of the Blood

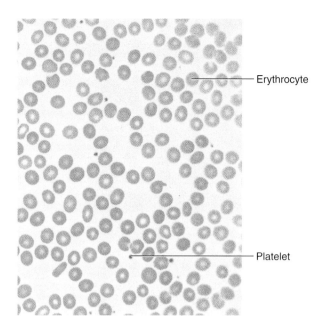

Figure 25.3 Blood Smear (400×)

Red blood cells are about 7.5 μm in diameter, on average, and have a life span of about 120 days, after which time they are broken down by the spleen or liver. Compare what you see under the microscope with figure 25.3.

Platelets

There are about 250,000 to 400,000 platelets per cubic millimeter of blood. Platelets, or thrombocytes, are involved in clotting and consist of small fragments of megakaryocytes. Examine the slide for small, purple fragments that may be single or clustered and compare them with the platelets in figure 25.3.

White Blood Cells

There are far fewer white blood cells, or leukocytes, in the blood than red blood cells. The number of white blood cells in a healthy adult is about 7000 cells per cubic millimeter, with a range of 5000 to 9000 cells per cubic millimeter of blood. White blood cells are formed in bone marrow tissue. The life span of a white blood cell varies from a few hours to several months. Many are capable of ameboid movement as they squeeze between cells (a process termed *diapedesis*), engulfing foreign particles and cellular debris.

White blood cells can be divided into two groups based on the presence or absence of granules in their cytoplasm. These two groups are the **granular** and **agranular leukocytes.** Examine the slide under high power or oil immersion and identify the different white blood cells as presented in the following text.

Granular Leukocytes

Granular leukocytes (granulocytes) are so named because they have granules in their cytoplasm. They are also known as **polymorphonuclear (PMN) leukocytes** due to the variable shapes of their nuclei (which are lobed, not round).

Neutrophils are the most common of all white blood cells. Neutrophils typically live for about 10 to 12 hours and have ameboid capabilities. They move by diapedesis between blood capillary cells and in the interstitial areas between the cells of the body. Neutrophils move toward infection sites and destroy foreign material. They are the major phagocytic white blood cells. The granules of neutrophils absorb very little stain, but neutrophils can be distinguished from other white blood cells by their three- to five-lobed nuclei. They are about one and a half times the size of red blood cells. Examine your slide for neutrophils, compare them with figure 25.4a and draw one in the following space.

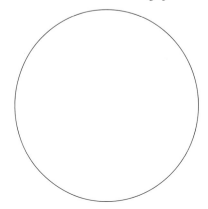

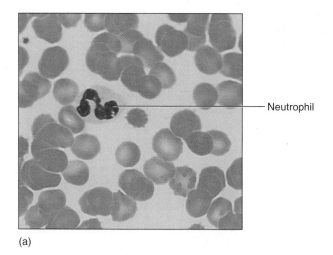

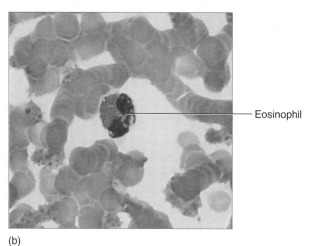

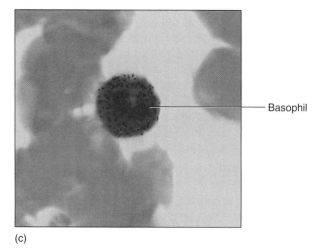

Figure 25.4 Granular Leukocytes
(a) Neutrophil (1,000×); (b) eosinophil (1,000×); (c) basophil (1,500×).

Eosinophils typically have a two-lobed nucleus with pink-orange granules in the cytoplasm. The term *eosinophil* actually means eosin-loving (eosin is an orange-pink stain and is picked up by the granules). They are about twice the size of red blood cells.

Compare figure 25.4b with your slide as you locate the eosinophils. Draw an eosinophil in the following space.

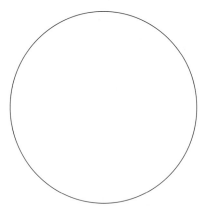

Basophils are rare. The granules stain very dark (blue-purple), and sometimes the nucleus is obscured because of the dark-staining granules. The nucleus is S-shaped, and the cell is about twice the size of red blood cells. Their granules contain histamines (vasodilators) and heparin, an anticoagulant that allows the movement of other white blood cells out of the capillaries to infection sites. You may have to look at 200 to 300 white blood cells before finding a basophil. Compare the basophil in your slide with figure 25.4c and draw one in the following space.

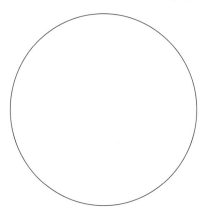

Agranular Leukocytes

Agranular leukocytes (agranulocytes) are so named because they lack cytoplasmic granules. The nuclei are not lobed but may be dented or kidney bean-shaped. There are two types of agranular white blood cells, lymphocytes and monocytes.

Lymphocytes have a large, unlobed nucleus that usually has a flattened or dented area. The cytoplasm is clear and may appear as a blue halo around the purple nucleus. In terms of function, lymphocytes do not need prior exposure to recognize antigens, and they are remarkable in that separate cells have specific antibodies for specific antigens. There are two groups of lymphocytes, the B cells and the T cells. These cells cannot normally be distinguished from one another in standard histological preparations (for example, Wright's stain), and they are simply considered lymphocytes in this exercise. Both B cells and T cells arise from fetal bone marrow. **B cells** probably mature in the fetal liver and spleen, and **T cells** mature in the thymus gland. B cell lymphocytes mature into

plasma cells, which make **antibodies.** Plasma cells provide **antibody-mediated immunity** (plasma cells secrete antibodies that travel in the fluid portion of the blood).

T cells, on the other hand, provide **cell-mediated immunity.** In cell-mediated immunity, the cells themselves (not antibodies in the blood plasma) move close to and destroy some types of bacteria or virus-infected cells. T cells also attack tumors and transplanted tissues. Most of the T cells are found in the lymph nodes, thymus, and spleen. They enter the bloodstream via the lymphatics. Compare your slide with figure 25.5a for lymphocytes and draw one in the following space.

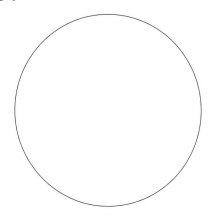

Monocytes are very large and have a kidney bean or horseshoe-shaped nucleus. They are about three times the size of red blood cells and are activated by T cells. Locate the large cells on your slide and compare them with figure 25.5b and draw one in the following space.

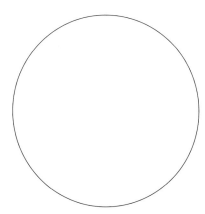

Differential White Blood Cell Count

Review the formed elements in table 25.1. Once you have identified the various white blood cells on the blood slide, conduct a differential white blood cell count. This is a very important count. Changes in the relative percentages of white blood cells may indicate the presence of a disease.

1. Use a hand counter and count 100 leukocytes. One lab partner should look into the microscope and methodically call out the names of the different types of leukocytes seen.
2. The other lab partner should record how many of each type are found and keep track of the overall number with the use of a hand counter until 100 cells are counted.

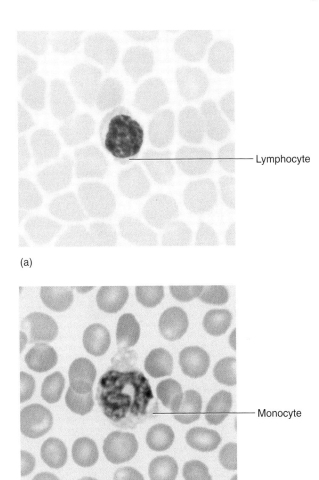

Figure 25.5 Agranular Leukocytes (1000×)
(a) Lymphocyte; (b) monocyte.

Figure 25.6 Counting Leukocytes

3. Scan the slide in a systematic way so that you do not count any cell twice. One method is illustrated in figure 25.6. Tally your results.

   Neutrophils: _____

   Eosinophils: _____

   Basophils: _____

   Lymphocytes: _____

   Monocytes: _____

Neutrophils represent about 60–70% of all of the white blood cells. Their numbers increase in cases of appendicitis or acute bacterial infections.

TABLE 25.1 Summary of Formed Elements in Blood

Formed Element	Size	Number	Characteristics
Erythrocytes	7.7 μm	5 million/mm^3	Live 120 days on average No nucleus
Platelets	2–4 μm	130,000–360,000/mm^3	Cell fragments
Leukocytes		7000/mm^3	Nucleus present
Granular leukocytes			
Neutrophils	9–12 μm	60–70% of leukocytes	Three- to five-lobed nucleus Granules indistinct
Eosinophils	10–14 μm	2–4% of leukocytes	Two-lobed nucleus Orange granules
Basophil	8–10 μm	0.5–1% of leukocytes	S-shaped nucleus Large, dark granules
Agranular leukocytes			
Lymphocytes	5–17 μm	20–25% of leukocytes	Nucleus appearing dented Thin rim of cytoplasm in some
Monocytes	12–15 μm	3–8% of leukocytes	Large, kidney-shaped nucleus

Eosinophils represent about 2–4% of all white blood cells. Eosinophils increase in number during allergic reactions and parasitic infections (for example, trichinosis).

Basophils represent about 0.5–1% of all white blood cells. They increase in number during allergic reactions and when a person is exposed to radiation.

Lymphocytes make up 20–25% of the white blood cells. Their numbers increase in cases of viral infection, antibody–antigen reactions, and infectious mononucleosis.

Monocytes make up about 3–8% of the white blood cells. Their numbers increase during chronic infections, such as tuberculosis.

Fill in the following table with what you know about the formed elements. You should read all of the information given before filling in the table.

Clean Up

Make sure the lab is clean after you finish. Place any slide with fresh blood on it or any material contaminated with bodily fluid in the bleach solution. Place all gloves and contaminated paper towels in the biohazard container. Place all sharps material (broken slides or coverslips, lancets, etc.) in the sharps container. Clean the counters with a paper towel and a 10% bleach solution. If you used immersion oil on the microscope, make sure the objective lenses are wiped clean (use clean lens paper only).

Characteristics of Formed Elements

Formed Element	Granules (If Present)	Shape of Nucleus (If Present)	Cause for Increase
Erythrocyte	_____	_____	_____
_____	Not obvious	_____	Mononucleosis
_____	Orange-staining	_____	Parasitic infections
_____	_____	Three- to five-lobed	_____
_____	Not obvious	Kidney bean	_____
Basophil	_____	_____	_____

Exercise 25 Review
Blood

Name: _____

Lab time/section: _____

Date: _____

1. Formed elements consist of three main components. What are they? _____
   _____

2. What is the most common plasma protein? _____

3. What is another name for a thrombocyte? _____

4. Which is the most common blood cell? _____

5. What is another name for a white blood cell? _____

6. What white blood cell is most numerous in a normal blood smear? _____

7. How many red blood cells are normally found per cubic millimeter of blood? _____

8. What is an average number of white blood cells found per cubic millimeter of blood? _____

9. B cells and T cells belong to what class of agranular leukocytes? _____

10. What value is there to a change in the percentage of white blood cells to diagnostic medicine? _____

11. In counting 100 white blood cells, you are accurately able to distinguish 15 basophils. Is this a normal number for the white blood cell count, and what possible health implications can you draw from this? _____
    _____
    _____
    _____

12. What is the function of the platelet? _____
    _____
    _____
    _____

13. Formed elements constitute what percentage of the total blood volume? _____

14. Label the formed elements in the following illustration.

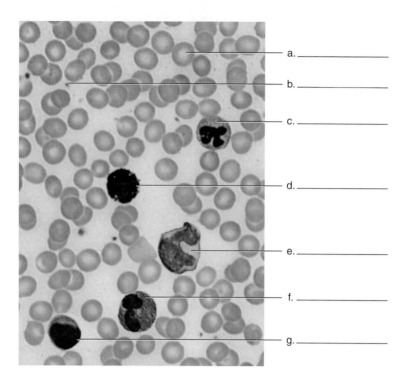

a. _____
b. _____
c. _____
d. _____
e. _____
f. _____
g. _____

15. In terms of volume, does the blood normally contain more plasma or more formed elements? _____

Exercise 26
Blood Tests and Typing

INTRODUCTION

Blood is a complex medium that has multiple functions in the body. In Exercise 25, the visual characteristics of blood were explored. In this exercise, you perform selected tests that allow you to evaluate different samples of blood. This is discussed in *Principles of Anatomy and Physiology* in chapter 16, "Blood."

Many clinical tests are performed on blood to determine the number of red or white blood cells in a known volume of blood, a person's blood type, hemoglobin concentration, or other information. As in the previous exercise, handle the samples with care. The caution statement from the previous exercise is repeated here because of the importance of handling bodily fluids safely.

Caution
The risk of bloodborne diseases has been significantly minimized by the use of sterilized human blood and nonhuman mammal blood in this exercise. However, use the same precautions as if you were handling fresh and potentially contaminated human blood. Your instructor will determine whether to use animal blood (from nonhuman mammals), sterilized blood, or synthetic blood. In any of these cases, follow strict procedures for handling potentially pathogenic material.

1. Wear protective gloves during the procedures.
2. Do not eat or drink in lab.
3. If you have an open wound, do not participate in this exercise, or make sure the wound is *securely* covered.
4. Your instructor may elect to have you use your own blood. Due to the potential for disease transmission, such as HIV or hepatitis B and C, in fresh blood samples, **make sure you keep away from other students' blood and keep them away from your blood.**
5. Do not participate in this part of the exercise if you have recently skipped a meal or if you are not well.
6. Place all disposable material in the biohazard bag.
7. Place all used lancets in the sharps container and all glassware that is to be reused in a 10% bleach solution.
8. After you have finished the exercise, clean and disinfect the countertops with a 10% bleach solution.

OBJECTIVES

At the end of this exercise, you should be able to

1. determine the antigens (agglutinogens) present in a particular ABO blood type;
2. list the antibodies (agglutinins) present in a particular ABO blood type;
3. relate Rh-positive or Rh-negative blood to antigens present;
4. perform specific diagnostic tests, such as a hematocrit test;
5. demonstrate the procedure for typing blood.

MATERIALS

5 mL of fresh, nonhuman mammal blood (dog, sheep, or cow)
Sterilized human blood (Carolina #k3-70-0120 or other supply company), labeled 1 to 4
Blood typing antisera (anti-A, anti-B, anti-D), test cards, and toothpicks
Rh warming tray
Heparinized blood microcapillary tubes
Capillary tube centrifuge
Hematocrit reader (Criticorp Micro-hematocrit tube reader, Damon Micro-capillary reader, or mm ruler)
Seal-ease®
Latex gloves
Goggles
Contamination bucket with autoclave bag
Container with 10% household bleach solution
Sharps container
ELISA test kits such as:
    Wards Lyme Disease #36 V 8907
    Carolina ELISA Simulation Kit #211248
    Fisher Scientific HIV/AIDS Test Simulation #S66656

PROCEDURE
Blood Typing

An understanding of blood type is critical in clinical work. Proper matching of blood between donor and recipient is a vital process, and you learn the essentials of this process in this exercise. There are many different typing systems that match blood, such as the Kell, Lewis, MNS, and Duffy blood groups, but the ABO and Rh systems are the two most common in clinical settings. These are antigen/antibody groups, and they are often used in forensic studies. Blood types are genetically determined.

Blood cells have surface membrane molecules (**glycoproteins**) that are **antigens (agglutinogens).** If this blood is injected into a person with **antibodies (agglutinins)** against that blood, then the injected blood clumps, or **agglutinates.** In the ABO system, no prior exposure to the agglutinogen is needed for granulation, or clumping, to occur. For example, if a person has **type A blood,** that person has agglutinins against **type B blood.** If the person with type A blood receives a transfusion of type B blood, then the **anti-B agglutinins** attack the agglutinogens in the blood

345

TABLE 26.1	ABO Blood System	
Blood Type	Agglutinogens (Antigens)	Agglutinins (Antibodies)
A	A	Anti-B
B	B	Anti-A
AB	A and B	None
O	None	Anti-A and anti-B

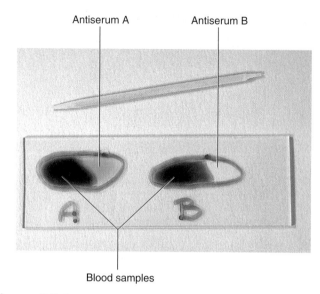

Figure 26.1 Blood Testing Procedure

introduced into the system. This causes a **transfusion reaction,** which is the agglutination and **hemolysis** of the transfused red blood cells. If the reaction is severe enough, death may occur. Table 26.1 gives details of the ABO blood system.

In a normal blood transfusion, donated blood is matched to the exact blood type of the recipient (for example, type AB matched with type AB). In emergencies, **type O blood,** with no antigens, can be used for individuals with A, B, or AB blood. Therefore, a person with O blood is considered a **universal donor.** A **universal recipient** has which blood type?

Universal recipient: _____

Procedure for Blood Typing

1. Read the entire blood typing procedure before you begin this part of the exercise.
2. Obtain a sample of sterilized blood as directed by your instructor. The vial should be labeled 1, 2, 3, or 4.
3. Record the sample number you are using in the space provided.
4. Place two separate drops of the sample blood on a blood test card or on a very clean glass microscope slide (figure 26.1).
5. Place **antiserum A** on one drop and **antiserum B** on the other drop.
6. Use separate toothpicks to stir each sample of blood and antiserum.
7. Keep the antisera separate from one another. Examine the sample for clumping, or granulation, within 2 minutes. Make sure that you dispose of your toothpicks in the biohazard bag. If both blood samples coagulate, you have a sample of type AB blood. If neither sample agglutinates, you have a sample of type O blood. If the blood with antiserum A agglutinates, you have a sample of type A blood, and if blood with antiserum B agglutinates, you have a sample of type B blood.
8. Compare your sample with figure 26.2.

Blood sample number: _____

Blood type: _____

Common ABO blood types for various groups in the United States are listed by percentages in table 26.2.

Determination of Rh Factor

The other blood system of clinical significance is the Rh system. *Rh* stands for *rhesus monkey* (the animal in which the Rh system was

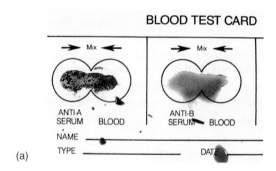

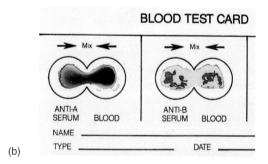

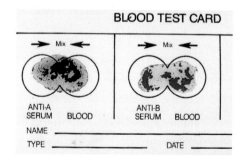

Figure 26.2 Blood Types
(a) Type A; (b) type B; (c) type AB.

TABLE 26.2	Percent U.S. Population by Blood Type			
Blood Type	Caucasians	African Americans	Asian Americans	Native Americans
A	41	27	28	16
B	9	20	23	4
AB	3	7	13	1
O	47	46	36	79
Total	100	100	100	100

discovered). About 85% of the U.S. population is Rh-positive; 15% is Rh-negative. An Rh-positive individual carries the **Rh antigen** (surface membrane marker). The Rh-negative individual does not carry the Rh antigen but develops **antibodies** to the antigen after exposure to the Rh antigen.

Determination of the Rh type is a much more subtle test than ABO typing. The **Rh antiserum (anti-D)** is more fragile in shipping and storage and should not be considered clinically relevant in this exercise. Use the same precautions as outlined at the beginning of the exercise regarding bodily fluids.

1. Place one drop of blood on a slide and place a drop of antiserum-D (Rh antiserum) on the slide.
2. Use a new toothpick and stir the two drops together.
3. Place this mixture on a warming tray and gently rock the slide.
4. Examine the slide for clumping after a minute or two. If clumping occurs, then the sample is Rh-positive (the Rh antiserum reacted with the Rh antigen in the blood). If no clumping occurs, then the sample is Rh-negative. This test is more difficult to determine, so look for slight granulation. Rh antiserum should be used fresh.
5. Record your results.

   Rh factor determination: _____

Rh determination is very important if a pregnant woman is Rh-negative. If her unborn child is Rh-positive, then she may develop antibodies against the Rh antigen in the fetal blood during pregnancy or delivery (a time when fetal and maternal blood often mix). If her second child is Rh-positive, then the antibodies that she produced during her first pregnancy may cross the placenta and cause severe reactions in the fetus. This is called **hemolytic disease of the newborn (HDN),** or **erythroblastosis fetalis.** Rho-GAM, an Rh immune globulin, prevents antibody formation in the mother against the Rh antigen and is often given to the mother during pregnancy and again after delivery. A woman who is Rh-positive does not have the antibodies for the Rh factor and therefore will not produce antibodies against the developing fetus, regardless of the Rh factor of the fetus.

Blood Typing Problems

Extensive blood transfusions can lead to problems for the recipient of the transfusions. Blood that is donated from one person carries antibodies, and these may react with the recipient's blood. There are other blood types in addition to the ABO and Rh systems, and sensitivities may develop due to antibodies present in the recipient's blood.

Hematocrit

Hematocrit, or **packed cell volume (PCV),** is the percentage of red blood cells (or blood cells in general) in the total blood volume. The percentage of red blood cells, or total cell volume (red blood cells and white blood cells), can be calculated after centrifuging a sample of blood. The cells end up as a large sediment in the bottom of the microcentrifuge tube, leaving the less dense plasma on top. Normal hematocrit values for females are 38–48%, and values of 44–54% are normal for males. An increase in the hematocrit above normal is known as **polycythemia** and can exceed 65%. When blood is lost faster than it is replaced or when the production of red blood cells is low, **anemia** occurs. In anemia, the hematocrit may drop to 15% or less. Anemia may also be due to low levels of hemoglobin in the blood. **Hemoglobin** is a complex molecule composed of an iron-containing **heme** group and the protein **globin.** You will next compare the hematocrit of sterilized human blood with that of another mammal.

Procedure for Determining Hematocrit

Refer to the caution statement at the beginning of the exercise concerning working with blood.

1. While wearing latex gloves, fill two capillary tubes with blood. One should have commercially prepared, sterilized human blood and the other should have nonhuman mammal blood.
2. Fill each tube by touching the red end of the capillary tube to the blood sample (figure 26.3). Let the blood flow up the tube by capillary action.
3. Once the tube is filled, place your finger on the other end of the tube (the nonred end) to prevent blood from flowing back out. Fill each tube to about three-quarters of the tube length. Both tubes should have about the same volume of blood in them.

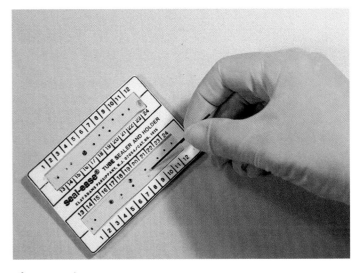

Figure 26.3 Preparation of a Hematocrit Tube

4. Seal the red end of the tube with Seal-ease® or modeling clay. Be careful—the capillary tubes are both fragile and sharp. They can break and puncture the skin.
5. Place the capillary tube in the centrifuge with the clay plug against the outer rubber ring.
6. Place your tubes opposite each other and record the number of each tube. Make sure you keep track of the sample from each tube and do not mix them up.

   Mammal blood tube #: _____

   Sterilized blood tube #: _____

After many tubes have been placed into the centrifuge, make sure that each has the clay seal facing the rubber gasket. If the seals are facing the center of the centrifuge, the blood will spray out of the tube when the centrifuge begins turning. The tubes should be opposite one another, so that the centrifuge is balanced. **Make sure the metal top is on the centrifuge and screwed in place.** Close the cover and spin the capillary tubes for 3 to 5 minutes.

Once the centrifuge has come to a complete stop, remove each tube and calculate the percentage of red blood cells in the tube. This can be done with hematocrit readers or a ruler. Your instructor can help you use the hematocrit readers. If you use a ruler, you need to measure the total height of the blood column in millimeters and subsequently the total height of the red blood cells in the column. If you use a ruler to measure the hematocrit, record your results here.

Sterilized Blood

   Red blood cell height (mm): _____

   Total column height (mm): _____

Mammal Blood

   Red blood cell height (mm): _____

   Total column height (mm): _____

The hematocrit can be figured by the following equation:

$$\frac{\text{Millimeters of red blood cell}}{\text{Millimeters of total blood}} \times 100 = \text{Hematocrit}$$

Record the hematocrit.

   Sterilized blood: _____

   Mammal blood: _____

   How do these values compare? _____

   _____

   _____

   What commercial procedure might lead to a difference in the hematocrit between these two samples? _____

   _____

Plasma with red coloration indicates hemolysis of the red blood cells. Normally, the plasma should be straw-colored or light yellow. Does either of your samples show hemolysis?

   _____

Note the buff-colored layer between the red blood cell layer and the plasma layer. This is the white blood cell layer.

Hemoglobin Determination

The number of red blood cells may be normal in a sample of blood, yet a person may be anemic due to the lack of hemoglobin in the blood. Hemoglobin can be measured a number of ways, such as by the Tallquist method, by colorimetry, or with the use of a hemoglobinometer. Hemoglobin (Hb) levels are expressed as grams Hb/100 mL of blood. The average value for humans is 12 to 16 g/100 mL of blood. In males, the normal range is 14 to 18 g/100 mL; in females, it is 12 to 16 g/100 mL.

To measure the percent hemoglobin, follow the directions provided with the hemoglobinometer. Be sure to clean the glass slides with alcohol before and after use. Disinfect the apparatus with bleach afterwards.

   Hemoglobin from sterile blood sample: _____

   Hemoglobin from mammal blood sample: _____

Blood Cell Counts

Another way to determine the number of blood cells is to count the number of cells in a known volume and subsequently determine the total number of cells per cubic millimeter (mm^3). Modern evaluation of red blood cell and white blood cell counts is performed by injecting blood samples into an optical computer system, which automatically calculates the cell counts. Most modern hospitals today use computer-driven optical systems to count cells (figure 26.4). Since these systems are fully automated we will not perform these tests in lab.

Figure 26.4 **Optical Cell Counter**

ELISA Tests

As described in Exercise 24 under the "Endocrine Physiology—Detection of Hormones" section, ELISA (Enzyme-Linked Immunosorbent Assay) tests involve the use of antibodies and enzymes to determine the presence of specific biological material. ELISA tests have been developed for hundreds of tests, from detecting cancer, infectious diseases, and allergic reactions to determining fertility, diabetes, or hormone levels in the blood. The basic test commonly involves an antibody attached to a solid surface. The antibody used is specific to the material of interest (such as a particular hormone or bacterium), and it forms an antibody complex. A dye-linked enzyme is also present that will bind to the antibody complex. If it does bind, then the enzyme releases a dye, indicating a positive result.

There are several ELISA test kits or simulations kits available such as those available from Ward's Natural Science, Carolina Biological, or Fisher Scientific. Select the kit available in your lab. Run the tests and record which test subjects were positive in the space below.

Positive results:

___

Clean Up
Make sure the lab is clean before you leave. Wipe any blood from the microscope objective lenses with lens cleaner and lens paper. Make sure you do not leave glass or blood on the lab tables. Use a 10% bleach solution and a paper towel to clean off your lab table before you leave.

Exercise 26 Review
Blood Tests and Typing

Name: _____

Lab time/section: _____

Date: _____

1. What is the name of a surface membrane molecule on a blood cell that causes an immune reaction? _____

2. What ABO blood type is found in a person who is a universal donor? _____

3. What is the average range of hematocrit for a normal female? _____

4. What is the average range of hematocrit for a normal male? _____

5. What percentage of the blood volume consists of formed elements? _____

6. A person with blood type B has what kind of agglutinins (antibodies)? _____

7. A person has antibody A and antibody B in his or her blood with no Rh antibody. What blood type does this person have? _____

8. A person with blood type B negative is injected with type A positive blood. From an immunologic (antigen/antibody) standpoint, what will happen after the injection? _____

9. Using the following illustration, calculate the hematocrit of the individual. Determine if it falls within normal limits.

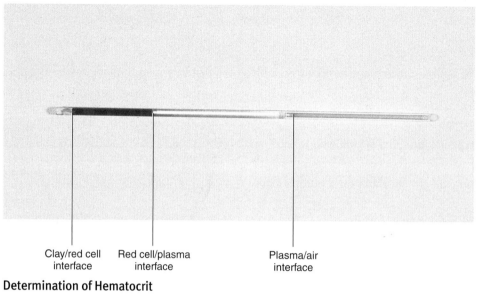

Determination of Hematocrit

10. Define *anemia*. _____

_____

_____

_____

_____

11. Explain the possible erroneous results that you might get if you used just one toothpick to stir the various blood types in the ABO blood test. _____

Exercise 27
Structure of the Heart

INTRODUCTION

In this exercise, you study the structure of the heart, which is covered in *Principles of Anatomy and Physiology* in chapter 17, "Heart." In the general circulatory pattern, vessels that carry blood from the heart are called **arteries,** and those that return blood to the heart are called **veins.** The **coronary arteries** carry blood from the heart (thus, they are arteries), but they carry the blood to the heart muscle. The **cardiac veins** take blood from the heart muscle back to the heart for recirculation.

In this exercise, you examine models of the heart, preserved sheep hearts, and human hearts, if available. As you look at the preserved material, determine how the structure of the heart relates to its function.

OBJECTIVES

At the end of this exercise, you should be able to

1. list the three layers of the heart wall;
2. describe the position of the heart in the thoracic cavity;
3. describe the significant surface features of the heart;
4. describe the internal anatomy of the heart;
5. find and name the anatomical features on models of the heart and in the sheep heart;
6. describe the blood flow through the heart and the function of the internal parts of the heart;
7. discuss the functioning of the atrioventricular valves and the semilunar valves and their role in circulating blood through the heart.

MATERIALS

Models and charts of the heart
Preserved sheep hearts
Preserved human hearts (if available)
Blunt probes (mall probes)
Dissection pans
Scalpels
Sharps container
Disposable gloves
Waste container
Microscopes

PROCEDURE
Heart Wall

The heart is located deep in the thorax between the lungs in a region known as the mediastinum. The **mediastinum** contains the heart, the coverings of the heart (the pericardia), and other structures, such as the esophagus and descending aorta. The mediastinum is located between the sternum, the lungs, and the thoracic vertebrae and is illustrated in figure 27.1.

If you open the chest cavity, the first structure you see is the **fibrous pericardium.** This tough, outer connective tissue sheath encloses the heart. The inner lining of the fibrous pericardium is the **parietal pericardium.** Deep to the parietal pericardium is the **pericardial cavity,** which contains a small amount of **serous pericardial fluid.** This fluid reduces the friction between the outer surface of the heart and the parietal pericardium. The layer closest to the heart is the **serous pericardium.** Locate these pericardial layers in figures 27.1 and 27.2.

The heart wall is composed of three major layers. The outermost layer is the **epicardium,** or **serous pericardium,** which is composed of epithelial and connective tissue. The middle layer, the **myocardium,** is the thickest of the three layers. It is mostly made of cardiac muscle. You may wish to review the slides of involuntary cardiac muscle and note the intercalated disks, branching fibers, and fine striations of cardiac tissue (as described in Exercise 4). The cardiac muscle is arranged spirally around the heart, providing a more efficient wringing motion to the heart. The inner layer of the heart wall is the **endocardium,** which is a serous membrane and consists of **endothelium** (simple squamous epithelium) and connective tissue. These layers are seen in figure 27.2.

It is best to examine heart models before dissecting a sheep heart, unless your instructor directs you to do otherwise. Heart models are color-coordinated and labeled to make the structures easier to locate.

Examination of the Heart Model

Overview

The heart is a four-chambered pump with two atria and two ventricles. Blood enters the heart in the right atrium (figure 27.3) and flows into the right ventricle. From the right ventricle, the contraction of the ventricular wall sends blood to the lungs. The blood is oxygenated in the lungs and returns to the heart by entering the left atrium. Blood moves from the left atrium to the left ventricle and is then pumped from there to the rest of the body.

Exterior of the Heart

Examine the heart model and notice that the heart has a pointed end, or **apex,** and a blunt end, or **base.** The apex of the heart is inferior, and the great vessels leaving the heart are located at the base (therefore, in the case of the heart, the base is superior to the apex). Compare the model with figure 27.4 and see how the **aorta** curves to the left in an anterior view of the heart and is posterior to the **pulmonary trunk.**

Locate the anterior features of the heart. The heart is composed of two large inferior **ventricles** and two smaller and

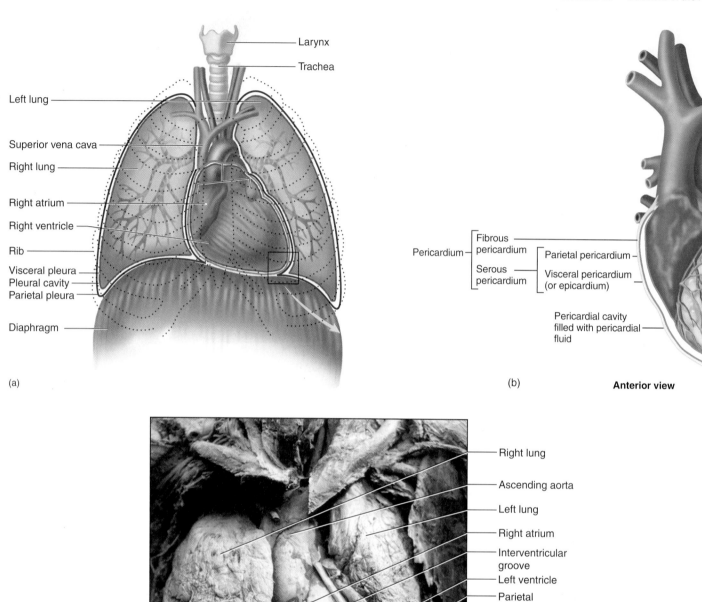

Figure 27.1 Heart in Thoracic Cavity

(a) Diagram of heart in thoracic cavity; (b) diagram of heart in pericardium; (c) photograph.

Exercise 27   Structure of the Heart

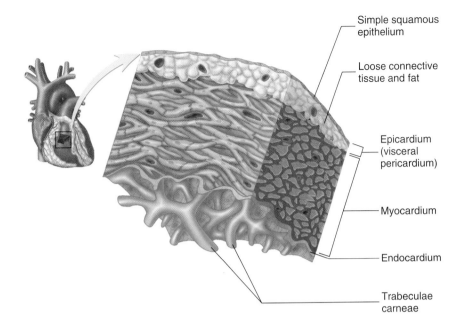

Figure 27.2   Layers of the Heart Wall

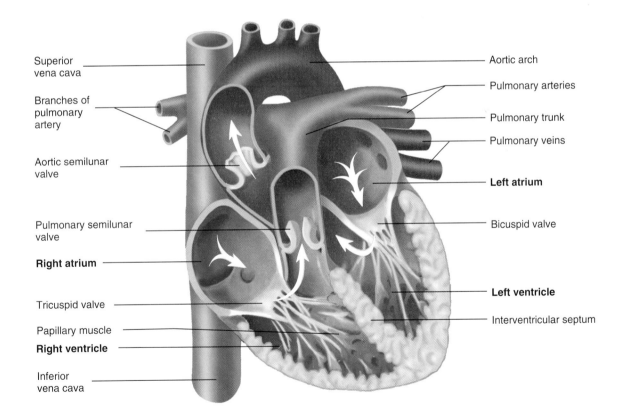

Figure 27.3   Chambers of the Heart

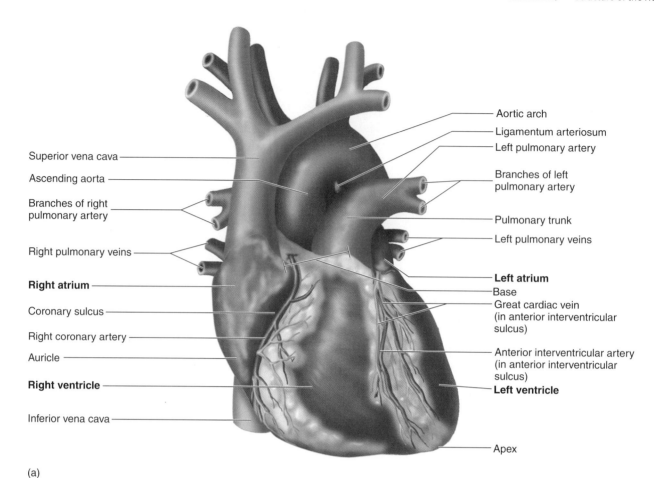

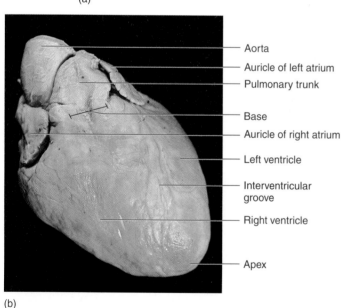

Figure 27.4 Surface Anatomy of the Heart, Anterior View
(a) Diagram; (b) photograph.

superior **atria.** The **left ventricle** extends to the apex of the heart and is delineated from the **right ventricle** by the **interventricular sulcus,** or **groove.** In this interventricular sulcus are some of the **coronary arteries** and **cardiac veins,** which are discussed later. The left ventricle is larger than the right ventricle. Note the two earlike flaps on the anterior, superior region of the heart. These structures are the **auricles,** which are part of the atria.

If you examine the heart from the posterior side, you will see the atria more clearly. At the junction of the **right atrium** and the right ventricle is the **atrioventricular sulcus,** or **groove.** The **coronary sinus,** a large venous chamber that carries blood from the cardiac veins to the right atrium, is in this sulcus. Locate the **superior vena cava** and the **inferior vena cava,** two vessels that also return blood to the right atrium. Locate the **pulmonary veins,** which carry blood from the lungs to the left atrium. Compare the heart model with figure 27.5.

Exercise 27 Structure of the Heart

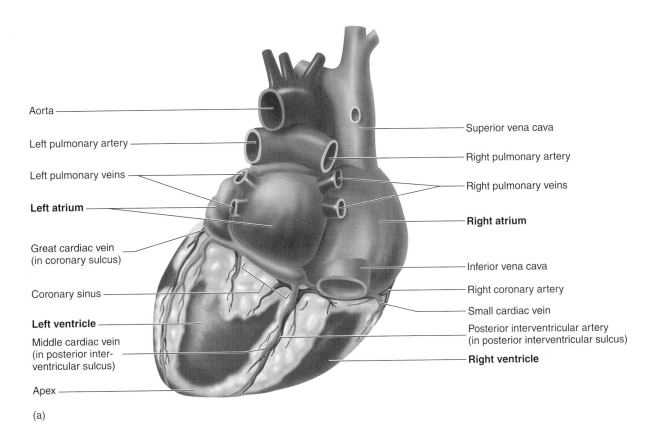

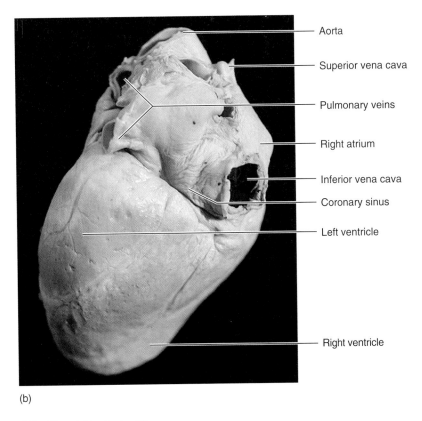

Figure 27.5 Surface Anatomy of the Heart, Posterior View
(a) Diagram; (b) photograph of cadaver heart.

The major vessels of the heart are illustrated in figures 27.4 and 27.5. Locate the **pulmonary trunk, pulmonary arteries, ligamentum arteriosum** (between the pulmonary trunk and aortic arch), **ascending aorta, pulmonary veins, superior vena cava, inferior vena cava, coronary arteries,** and **cardiac veins.** The heart tissue is nourished by coronary arteries. The **left coronary artery** arises from the **ascending aorta** (figure 27.6) and then branches into the **anterior interventricular artery** and the **circumflex artery.** The **right coronary artery** also arises from the ascending aorta and branches to form the **posterior interventricular artery** and the **right marginal artery.** These major arteries of the heart supply blood to the myocardium. On the return flow from the heart muscle, the **great cardiac vein** follows the depression of the interventricular groove and the atrioventricular groove to the **coronary sinus.** On the right side of the heart, the **small cardiac vein** leads to the coronary sinus, which empties into the right atrium. Locate these vessels on the external surface of the model of the heart and compare them with figure 27.6.

Interior of the Heart

Examine a model of the interior of the heart and locate the right and left ventricles. Note that the **right ventricle** is much thinner-walled than the left ventricle. The ventricles are separated by the **interventricular septum,** which forms a wall between the two ventricular chambers. Compare the model with figure 27.7.

Examine the **right atrium** and note how thin the wall is, compared with the ventricles. Examine the medial wall of the atrium, known as the **interatrial septum,** and locate a thin, oval depression in the atrial wall. This depression is the **fossa ovalis** (figure 27.8). In fetal hearts, this is the site of the **foramen ovale,** but the foramen usually closes just after birth. Note the extensive **pectinate muscles** on the wall of the atrium. These provide additional strength to the atrial wall. Blood in the superior vena cava, the inferior vena cava, and the coronary sinus returns to the right atrium. Examine the features of the right atrium in figure 27.8.

The blood from the right atrium flows into the right ventricle. Now examine the valve between the right atrium and right ventricle. This is the **right atrioventricular valve,** or **tricuspid valve,** and it prevents the return flow of blood from the right ventricle into the right atrium during ventricular contraction. Examine the valve for three flat sheets of tissue. These are the three **cusps** of the tricuspid valve (figures 27.7 and 27.8). The tricuspid valve has thin, threadlike attachments called **chordae tendineae.** These tough cords are attached to larger **papillary muscles,** which are extensions from the wall of the ventricle. The right ventricle wall has small extensions called **trabeculae carneae,** which, like the pectinate muscles of the atria, strengthen the ventricle wall. The blood from the right ventricle flows into the pulmonary trunk toward the lungs.

Locate the **pulmonary semilunar valve.** It appears as three small cusps between the right ventricle and the pulmonary trunk and keeps blood from flowing backwards from the pulmonary trunk into the right ventricle during ventricular relaxation. Examine the details of the right ventricle in models in the lab and in figures 27.7 and 27.8.

Blood from the pulmonary trunk flows through the pulmonary arteries before entering the lungs. Blood in the lungs releases carbon dioxide and picks up oxygen. The **pulmonary veins** carry

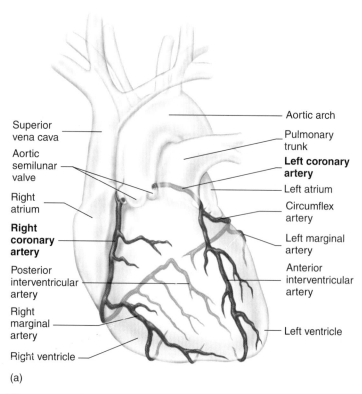

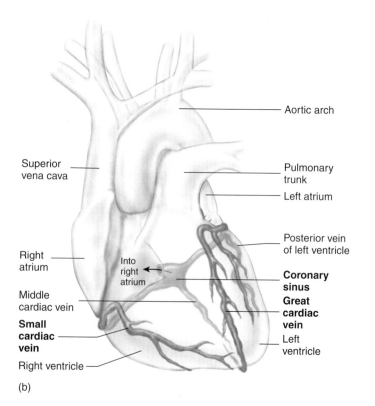

Figure 27.6 Vessels of the Heart
(a) Arterial circulation; (b) venous circulation.

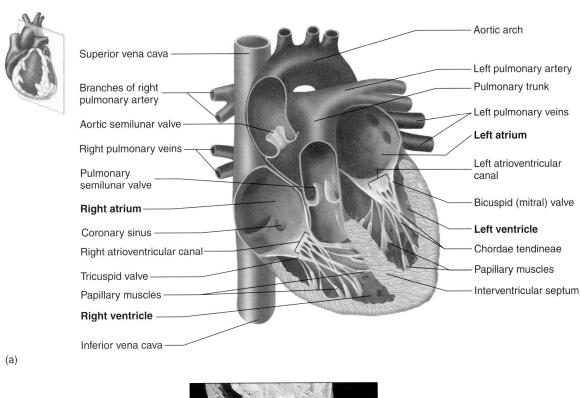

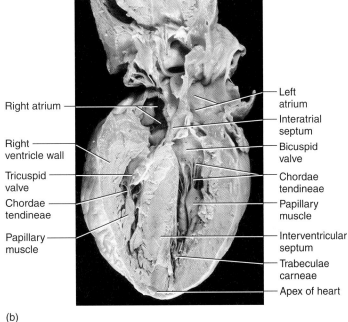

Figure 27.7 Coronal Section of the Heart
(a) Diagram; (b) photograph of cadaver heart.

oxygenated blood from the lungs into the **left atrium.** These vessels are located in the superior, posterior portion of the left atrium. Blood from the left atrium flows into the left ventricle. Locate the two large cusps of the **bicuspid valve** between the left atrium and left ventricle. The bicuspid valve is also known as the **mitral valve, or left atrioventricular valve.** It also has attached chordae tendineae and papillary muscles, which you should locate in the models in the lab. Note the thickness of the left ventricle wall, compared with the wall of the right ventricle. Compare the left side of the heart with figures 27.7 and 27.9.

The **aortic semilunar valve** is located at the junction of the left ventricle and the ascending aorta. It has the same basic structure and general function as the pulmonary semilunar valve in that it prevents the flow of blood from the aorta into the left ventricle. Blood from the left ventricle moves into the aorta and subsequently to the rest of the body.

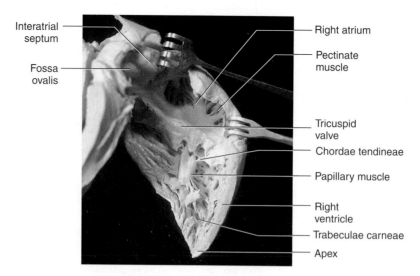

Figure 27.8 Details of the Right Atrium and Ventricle

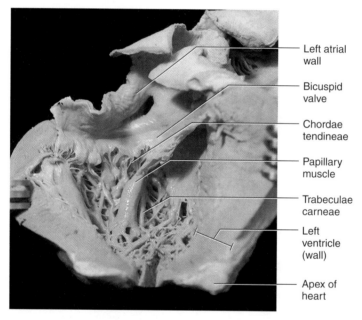

Figure 27.9 Left Atrium and Ventricle

Caution

Be careful when handling preserved materials. Put on protective gloves and ask your instructor for the proper procedure for working with preserving fluid and for handling and disposal of the specimen. Do not dispose of animal material in the sinks. Place it in an appropriate waste container.

Dissection of the Sheep Heart

The sheep heart is similar to the human heart and usually is readily available as a dissection specimen. Dissection of anatomical material is valuable in that you can examine structures that are represented more accurately in preserved material than in models. Also, the preserved material has greater flexibility and is more easily manipulated. There are differences between sheep hearts and human hearts, especially in the position of the **superior** and **inferior venae cavae.** In sheep, these are called the anterior and posterior venae cavae, but we refer to them using the human terminology.

If your sheep heart has not been dissected, then you will need to open the heart. If your sheep or other mammalian heart has been previously dissected, then you can skip the next three paragraphs.

Place the heart under running water for a few moments to rinse off the preserving fluid. Examine the external features of the heart. Note the fat layer on the heart. The amount of fat on the human or sheep heart varies.

Locate the **left ventricle,** the **right ventricle,** the **interventricular sulcus,** the **right atrium,** and the **left atrium.** Note the **auricles** that extend on the anterior surface of the atria.

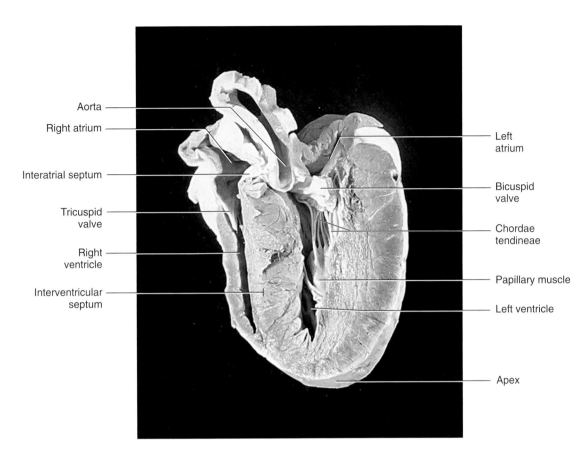

Figure 27.10 Sheep Heart, Coronal Section

Using a sharp scalpel, make an incision along the right side of the heart (lateral side) from the apex of the heart to the lateral side of the right atrium. If you are unsure about how to proceed during any part of the dissection, ask your instructor for directions. Make another long cut from the lateral side of the left atrium through the lateral side of the left ventricle. You will have made a coronal section of the heart if you have cut through the **interventricular septum.** Once you have opened the heart, compare the structures of the sheep heart with figure 27.10.

You can locate the vessels of the heart by inserting a blunt metal probe into the vessels and determining which chamber the vessel goes to or comes from. Place the heart in anatomic position and insert the probe into the large, anterior vessel that exits toward the specimen's left side. The blunt end of the probe should enter the **right ventricle.** This vessel is the **pulmonary trunk.** The pulmonary trunk may still have the **pulmonary arteries** attached. Locate the large vessel directly behind the pulmonary trunk (figure 27.11). This is the **ascending aorta.** If the vessels are cut farther away from the heart, you can see the **aortic arch.** Insert the probe into this vessel and into the left ventricle. The coronary arteries are small holes that exit the ascending aorta. Insert a blunt probe into these holes to locate the coronary arteries.

Turn the heart to the posterior surface and locate the **superior vena cava** and **inferior vena cava.** Insert the probe into the superior and inferior venae cavae, pushing the probe into the **right atrium.** If you find only one large opening in the atrium, you may have cut through either the superior or inferior vena cava during your initial dissection. The probe can be felt through the wall more easily here than in a ventricle because the atrial walls are thinner than those of the ventricles. On the left side, the **pulmonary veins** may appear either as four separate veins or as a large hole on each side of the left atrium if the vessels were cut close to the atrial wall. Locate the same structures in the sheep heart that you found on the model and compare them with figure 27.12.

Cut into the right atrium and use your blunt probe to locate the opening of the **coronary sinus** in the posterior, inferior portion of the atrium. It is small and somewhat difficult to find.

Examine the opening between the right atrium and the right ventricle to locate the **tricuspid valve.** You may have dissected through one of the cusps as you opened the heart. Locate the major features of the right ventricle. Find the **chordae tendineae** and the **papillary muscles.** You can find the **pulmonary trunk** by inserting a blunt probe into the superior portion of the right ventricle. Make an incision in the pulmonary trunk near the right ventricle to expose the three thin cusps of the **pulmonary semilunar valve.** Note how the cusps press against the wall of the pulmonary trunk when the probe is pushed against them in a superior direction. These cusps close when blood begins to flow back into the right ventricle as the ventricle relaxes.

Locate the **left atrium, bicuspid valve,** and **left ventricle.** In the left ventricle, you should find the papillary muscles, chordae tendineae, and **trabeculae carneae.** You can find the **aorta** and **aortic semilunar valve** by inserting a blunt probe toward the superior end of the left ventricle toward the middle of the heart.

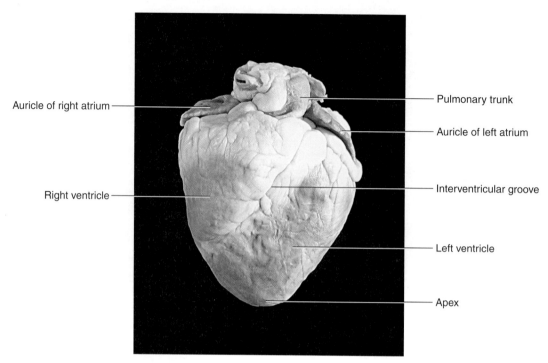

Figure 27.11 Sheep Heart, Anterior View

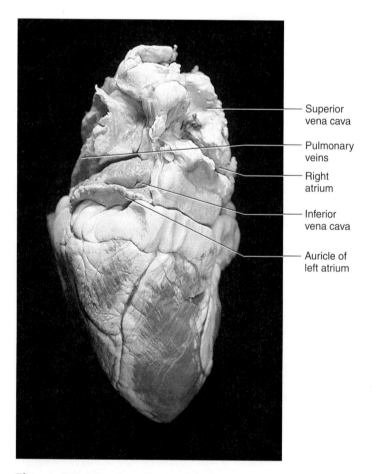

Figure 27.12 Sheep Heart, Posterior View

Clean Up
When you have finished your study of the sheep heart, make sure you clean your dissection equipment with soap and water. Be careful with sharp blades. Place the sheep heart either back in the preserving fluid or in the appropriate waste container, as directed by your instructor.

Exercise 27 Review

Name: _____

Lab time/section: _____

Date: _____

Structure of the Heart

1. The heart is located between the lungs in an area known as the _____.

2. What is the name of the layer that is superficial to the pericardial cavity? _____

3. What is the innermost layer of the heart wall called? _____

4. Is the apex of the heart superior or inferior to the rest of the heart? _____

5. What is the name of the depression between the two ventricles on the anterior surface of the heart? _____

6. Are auricles extensions of the atria or the ventricles? _____

7. What three vessels take blood to the right atrium? _____
   _____

8. Where does the great cardiac vein and the small cardiac vein take blood? _____
   _____

9. What blood vessels nourish the heart tissue? _____
   _____

10. What structure separates the left atrium from the right atrium? _____
    _____

11. What is the name of the thin spot between the atria? _____
    _____

12. The bicuspid valve is located between what two chambers of the heart? _____
    _____

13. Name the structure between the atrioventricular valve and the papillary muscle. _____

14. What is the function of the aortic semilunar valve? _____
    _____

15. What is another name for the tricuspid valve? _____

16. What cell type makes up most of the myocardium? _____

17. What adaptation do you see in the walls of the left ventricle being thicker than those of the right ventricle? _____

18. How does cardiac muscle resemble skeletal muscle? _____

19. In terms of function, how is cardiac muscle different from skeletal muscle? _____

20. Label the following illustration using the terms provided.

    right atrium   left atrium   interatrial septum   apex
    right ventricle (wall) chordae tendineae   left ventricle (wall)   interventricular septum

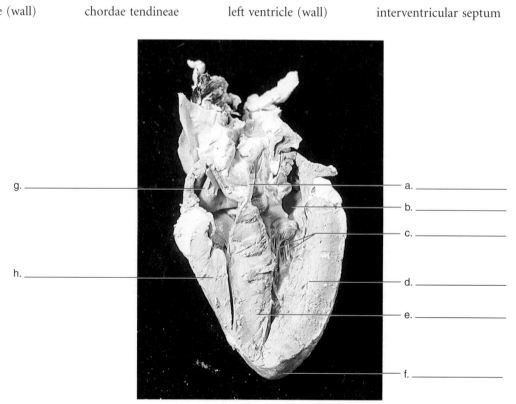

g. _____
h. _____
a. _____
b. _____
c. _____
d. _____
e. _____
f. _____

Exercise 28
Electrical Conductivity of the Heart

INTRODUCTION

The living heart is a phenomenal organ. It contracts an average of 72 times per minute for the length of an average person's lifetime. In this exercise, you observe the electrical activity of the heart and correlate it to the mechanical functioning of the heart. You learn the conductive structures of the heart and make a recording of its electrical activity. These topics are covered in *Principles of Anatomy and Physiology* in chapter 17, "Heart."

Two concepts are important in understanding the electrical activity of the heart. One is that the heart produces **electrochemical impulses** in a way similar to the production of impulses in the nervous system; the other is that these impulses can travel through the medium of the body and be recorded from areas distant from the heart. The cells of the body are bathed in a saline solution of a little less than 1% salt. This solution is an excellent conducting medium for electric impulses. As these impulses are generated in the heart, they travel through the body, including the distal regions of the limbs.

OBJECTIVES

At the end of this exercise, you should be able to

1. distinguish between systole and diastole;
2. list the structures in the conductive system of the heart;
3. describe the sequence of electrical conductivity of the heart;
4. associate the P wave, QRS complex, and T wave of an ECG with electrical events that occur in the heart;
5. relate how the electrical activity of the heart is transferred to the ECG recording;
6. calculate heart rate, PQ interval, QRS interval, and QT interval from an ECG recording;
7. determine if these ECG values are in the normal range.

MATERIALS

Electrocardiograph machine, physiograph, or physiologic computer unit
Cot or covered lab table
Alcohol swabs
Electrode jelly, paste, or saline pads
Watch with accuracy in seconds

PROCEDURE

The function of the heart is dependent on the structure of the heart. Review the anatomy of the heart in Exercise 27 before beginning this section.

Muscular Function of the Heart

The heart has been described as a muscular pump or, more accurately, two pumps acting in unison. As a pump, the heart has a contraction mode and a relaxation mode. **Systole** is contraction of the heart muscle; it can be described more specifically as **atrial systole** and **ventricular systole** (contraction of the atria and ventricles, respectively). **Diastole** is the relaxation of a heart chamber. Atrial and ventricular diastole are the respective relaxation of the atria and ventricles. Muscular contraction of the heart is preceded by electrical activity that stimulates the muscle to contract.

Electrical Functioning of the Heart

The heart has its own **cardiac conducting system** that produces and propagates electrical impulses that stimulate cardiac contraction. The system is composed of specialized muscle cells.

The initiation of the electric impulse in the heart begins at the **sinoatrial (SA) node,** which is commonly known as the **pacemaker.** Cells of the sinoatrial node use the sodium-potassium pump to generate a difference in electrical voltage across the cells' membranes, causing the cells to be **polarized.** The sinoatrial node spontaneously **depolarizes,** which causes a change in the total amount of charge across the cell membrane. This change in membrane potential occurs approximately 72 times per minute (the average heart rate). The impulse from the sinoatrial node is conducted across the atria, causing the muscles of the atria to contract. The impulse that spreads out across the atria reaches the **atrioventricular (AV) node.** The impulse has a slight delay (0.1 second) in the node before being conducted farther. This delay allows the atrial cardiac muscle to contract prior to ventricular firing (figure 28.1).

The electric impulse then travels from the AV node to the **atrioventricular bundle (bundle of His),** to the **right** and **left bundle branches,** and finally to the **Purkinje (conduction) fibers.** The Purkinje fibers stimulate the cardiac muscle of the ventricles to contract. The ventricles are thus stimulated from the apex toward the base, and the contraction proceeds from the inferior end of the ventricles toward the atria. Locate the structures of the heart's conduction system in figure 28.1.

Nature of the ECG

An **electrocardiograph** is a machine that can measure the electrical activity of the heart. The electrocardiograph produces a chart paper recording called an **electrocardiogram** (**ECG** or **EKG**). Before you record an ECG, examine the ECG in figure 28.2 and read the accompanying description.

Time is measured along the horizontal axis (each millimeter is equal to 0.04 second), and voltage difference is measured along the vertical axis (in millivolts). The ECG typically has three major events. The first event, the **P wave,** is a small bump called a **deflection wave.**

1. Action potentials originate in the sinoatrial (SA) node and travel across the wall of the atrium (*arrows*) from the SA node to the atrioventricular (AV) node.

2. Action potentials pass through the AV node and along the atrioventricular (AV) bundle, which extends from the AV node, through the fibrous skeleton, into the interventricular septum.

3. The AV bundle divides into right and left bundle branches, and action potentials descend to the apex of each ventricle along the bundle branches.

4. Action potentials are carried by the Purkinje fibers from the bundle branches to the ventricular walls.

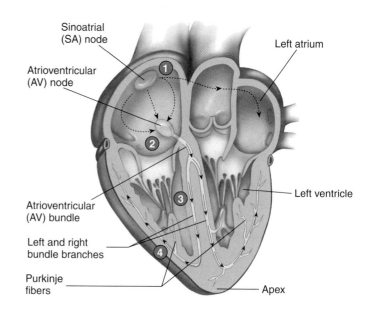

Figure 28.1 **Conduction System of the Heart**
Conduction starts at 1 and finishes at 4.

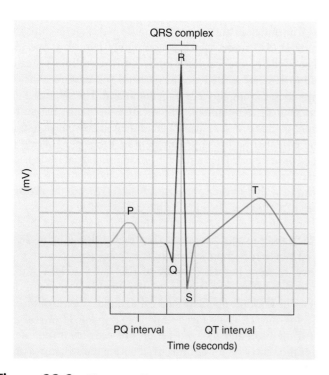

Figure 28.2 **Electrocardiogram**

The P wave represents **atrial depolarization.** The **QRS complex** represents **ventricular depolarization,** and the **T wave** represents **ventricular repolarization.** A small electrical event called **atrial repolarization** occurs during ventricular depolarization and is masked by the large QRS complex.

The electric impulse generated by the atria and ventricles depolarizing and repolarizing can be recorded using an electrocardiograph because of the conductivity of the intercellular fluid. Some other bodily functions, such as nerve impulse conduction and skeletal muscle contraction, also occur by transmitting low-voltage signals. The "average" potential difference of **−90 millivolts** is a little less than one-tenth of a volt; therefore, the body is operating electrically at slightly less than 0.1 volt. As this electricity is produced, it passes through the saline bodily fluid and can be picked up by **sensors (electrode plates)** placed in particular locations on the body. In many college physiology labs, the ECG electrode plates are attached to four areas. These are the medial side of the **left** and **right ankles** and the anterior surface of the **left** and **right wrists.** The attachment of the electrode plate to the right ankle serves as an **electrical ground** and is not used for measurement. The ECG is measured as the potential voltage difference between selected electrode plates.

Lead I is the potential voltage difference across the top of the heart. It measures the conductivity from the electrode plate on the right arm to one on the left arm (figure 28.3).

Lead II measures the potential voltage difference across the body as received from the electrode plate on the right arm and one on the left leg.

Lead III indicates the potential voltage differences between the electrode plates on the left arm and left leg.

Procedure for ECG

Make sure the ECG machine, physiograph, or physiologic computer is plugged in and turned on. Make sure that the ECG machine is on "Standby" when you attach the electrodes. Accidental grounding of the electrodes may damage the equipment. Older ECG machines should warm up for a few minutes prior to running the experiment. Remove metal watches or any jewelry that might interfere with the electric signal between the heart and the extremities. Have your lab partner scrub the inside of the ankle, about 2 cm above the medial malleolus, with an alcohol swab to remove dust and skin oil. Clean the anterior sides of the wrist in the same way. Saline pads or electrode jelly may be applied to these areas. Attach the electrodes firmly, but not so tight that you cut off blood flow, using the straps provided as follows:

LA attaches to left arm.
RA attaches to right arm.
RL attaches to right leg.
LL attaches to left leg.

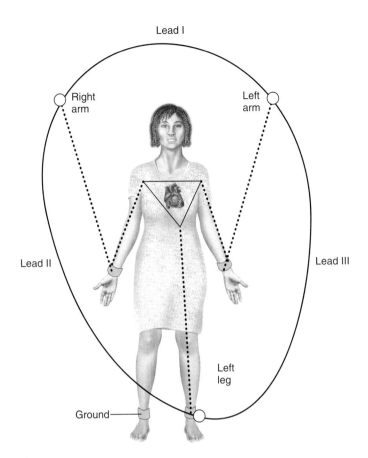

Figure 28.3 Leads of a Standard ECG

plays a part in electrical activity. If a patient is nervous or excitable, this will show in the ECG. Background electrical noise in the ECG varies from person to person, but you should be able to get a good ECG in the lab.

Analysis of the ECG

ECG recordings are important in assessing the health of the heart. **Conduction problems, myocardial infarcts (heart attacks),** and **heart blocks** are a few of the problems that may show up as variances on an ECG. The interpretation of ECGs in this exercise is for educational purposes only. Clinical evaluations of ECGs should be done only by trained health-care workers. Locate the P wave, the QRS complex, and the T wave.

The standard rate of paper travel in an ECG machine is 25 mm/second. Therefore, each millimeter is equal to 0.04 second. The small squares are millimeters (mm), and the darker lines indicate 5 mm, or 0.5 cm. Calculate the resting heart rate by counting how many millimeters ("Y" mm) occur between the peaks of the R waves. Record the number of millimeters in the following space.

Millimeters: _____

Calculate the number of beats per minute by using the following formula. Record your results in table 28.1.

$$\frac{1 \text{ beat}}{\text{"Y" mm}} \times \frac{25 \text{ mm}}{1 \text{ sec}} \times \frac{60 \text{ sec}}{1 \text{ min}} = \frac{\text{beats}}{\text{min}}$$

TABLE 28.1 Adult Heart Values

	Beats/Minute	Your Values
Normal range	60–100 beats/minute	
Tachycardia	Above 100 beats/minute	
Bradycardia	Below 60 beats/minute	
ECG Values		
PQ interval	0.16 second	
Potential heart block	Longer than 0.2 second	
QRS complex	0.8–0.10 second	
Potential bundle branch block	Longer than 0.12 second	
QT interval	0.3 second	

Your instructor will explain how to operate your specific electrocardiograph. Have your lab partner rest comfortably on a cot or table for a few moments before monitoring the ECG. Some ECG machines are automatic and you do not need to manually switch leads. If you have an older machine, turn the knob on the ECG machine from "Standby" to "Run" and record lead I for about 10 cycles. Turn the knob to "Standby" after recording each lead. Then record lead II and lead III for about 10 cycles each. Tear the ECG from the machine and make sure you write your lab partner's name on the paper and write the appropriate lead on the paper. If you are using a computer, make sure that you save the recorded data in an appropriately labeled file. It is important that you provide a relaxed atmosphere for your lab partner to produce a good ECG recording. People who are fidgeting or nervous generate extraneous interference and do not produce a level baseline recording. You can demonstrate this electrical "noise" by the following activity.

Interference in an ECG

While your lab partner is still attached to the ECG machine, run three or four cardiac cycles and then have your lab partner clench his or her fists. Note the electrical interference that occurs as the skeletal muscles depolarize and repolarize. Emotional state also

Exercise Effects on the ECG

Clinically, an exercise ECG, or "exercise stress test," is used to measure ischemic heart disease, which is due to a decrease in the diameter of the coronary arteries or damage to the heart muscle after a heart attack. A heart at rest may show no irregularities, but the ECG of a person under stress may show variances from normal. As the demand for oxygen increases, if the coronary arteries are narrowed, not enough oxygen gets to the heart, and this produces an irregular ECG. Damage from a myocardial infarct (heart attack) may also produce an irregular ECG. Typically, the greater the damage, the greater the irregularity.

Caution

If you have a personal or family history of heart problems, do *not* do this part of the lab exercise. If you exercise regularly and are in good health, the ECG should produce tracings that are different from a resting ECG.

Do vigorous exercise for about 5 minutes. You can run around, climb stairs, do jumping jacks, and so on, or you can stay hooked up to the ECG machine if you have an exercise bike, a treadmill, or any other apparatus in your lab. Record the ECG at lead II for several cycles.

How does the heart rate change with exercise? _____

_____

Is the height of the QRS complex during or after exercise different than at rest? _____

_____

Do you perceive any irregularities in the ECG? _____

_____

Irregularities in the ECG

An excessively high heart rate is termed **tachycardia.** In young adults, a heart rate above 100 beats per minute is considered tachycardia. However, 100 beats per minute in small children may be perfectly normal. An excessively low heart rate is termed **bradycardia.** In young adults, a rate below 60 beats per minute is considered bradycardia unless they are highly trained aerobic athletes. The whole cardiac cycle should take, on average, 0.7–0.8 second.

The **PQ interval** is the time between the beginning of atrial depolarization and the beginning of ventricular depolarization. PQ (PR) intervals should normally be about 0.16 second. If the PQ interval is longer than 0.2 second (5 mm on the chart paper), then this could be indicative of a **heart block.** Heart blocks result from reduced conduction between the atria and the ventricles. This may be caused by damage to the AV node or decreased transmission in the AV bundle. In a **complete heart block,** the atria do not stimulate the ventricular depolarization at all; therefore, the atria fire independently from the ventricles. In a complete heart block, the P waves may be spaced at 0.8 second apart, which is the SA node depolarization rate, but the ventricles fire at a much lower rate (1.5–2.0 seconds apart). Are the times measured on your recording within normal limits? Record the PQ interval in table 28.1.

The **QRS complex** is 0.08 to 0.10 second, on average. If the QRS complex spans longer than 0.12 second, this may indicate a **right** or **left bundle branch block.** In this condition, the two ventricles contract at slightly different times, increasing the length of the QRS complex. Measure the length of the QRS complex and record your findings in table 28.1.

The **QT interval** is 0.3 second, on average. The interval becomes shorter as the heart rate increases, and the interval becomes longer as the heart rate slows down. Record your results in table 28.1.

Exercise 28 Review

Electrical Conductivity of the Heart

Name: _____

Lab time/section: _____

Date: _____

1. The sinoatrial node has a common name. What is it? _____

2. Which two chambers of the heart (atria or ventricles) contract last in a normal cardiac cycle? _____

3. What two chambers are stimulated immediately after the SA node depolarizes? _____

4. After the AV node depolarizes, what structures conduct the impulse to the myocardium of the ventricles? _____
_____

5. What are the main events recorded by an ECG? _____
_____

6. What electrical event in the heart does the QRS complex represent? _____
_____

7. Ventricular repolarization is represented by what part of an ECG? _____
_____

8. What ECG wave represents the atrial depolarization? _____
_____

9. Why is the ECG event indicating atrial repolarization not seen in an ECG? _____
_____

10. What does a heart block do to impulse transmission in the heart? _____
_____
_____

11. Fibrillation is uncoordinated cardiac muscle contraction. Predict what an ECG would look like if there were no uniform conduction of electrical activity in the heart. Draw what it might look like.

12. What consequence does fibrillation have for cardiac muscle contraction and for the pumping efficiency of the heart? Which is more serious—atrial or ventricular fibrillation? _____

_____

_____

13. If a myocardial infarct (heart attack) destroyed a portion of the right or left bundle branches, what potential change would you see in an ECG? _____

_____

_____

14. Tape or paste your ECG in the following space. Label the P wave, the QRS complex, and the T wave.

Exercise 29
Functions of the Heart

INTRODUCTION

In terms of efficiency and overall endurance, the human heart is a remarkable organ. It pumps blood continuously at the average rate of about 72 beats per minute for the lifetime of the individual. The heart increases the volume of blood it pumps when more oxygen is required by increasing the rate of contraction, the volume ejected per contraction, or both. In this exercise, you explore some of the functions of the human heart, with emphasis on how the heart is regulated in terms of speed and contraction strength. The functions of the heart are covered in *Principles of Anatomy and Physiology* in chapter 17, "Heart."

Cardiac Pacemaker

The pacemaker cells of the **sinoatrial (SA) node** of the heart spontaneously depolarize. Sodium and calcium channels in the plasma membrane of pacemaker cells slowly leak their respective ions until threshold is reached. Once this occurs, a spontaneous opening of ion gates leads to depolarization of the pacemaker cells, which subsequently causes the depolarizing of adjacent cells near the pacemaker. This is called a **field effect.** This depolarization spreads throughout the atria and eventually to the ventricles, stimulating the cardiac muscle to contract.

The native sinoatrial depolarization rate is approximately 100 depolarizations per minute. The parasympathetic nervous system in a person at rest slows down this depolarization rate to an average of 72 beats per minute. The **vagus nerve** conveys parasympathetic fibers that innervate the SA node. **Acetylcholine (ACh)** is a neurotransmitter released by the vagus nerve, and it promotes potassium (K^+) leakage from the nodal cells to the interstitial fluid, thus **hyperpolarizing** the cells. By increasing the difference in voltage between the inside and the outside of the cells, it takes longer for the cells to reach threshold and depolarize. This decreases the heart rate.

On the other hand, the heart rate can be elevated by increasing the firing rate of the SA node. From the resting heart rate of 72 beats per minute to about 100 beats per minute (the native nodal firing rate), the initial control of increased firing rate is due to gradual inhibition of the vagus nerve. As the nerve secretes less acetylcholine at the node, the firing rate increases. Above 100 beats per minute, the heart rate is controlled by the sympathetic nervous system. **Cardiac nerves** from the cervical region of the spinal cord release **norepinephrine,** which binds to **beta-adrenergic receptors.** Calcium channels open somewhat and decrease the threshold, thus causing more rapid firing of the SA node.

The electrical activity of the nodes of the heart occurs in specialized muscle cells and not in nervous tissue. The process of electrochemical transfer of impulses by cardiac muscle cells is known as **myogenic conduction.** The muscle tissue may be influenced by the nervous system, but the activity is initiated and generated in specialized muscle fibers. In addition to the influence of the nervous system, certain hormones, some drugs, and ion concentration have an effect on the heart rate.

Because the heart muscle cells are linked together by intercalated disks, they act as one electrical unit. An impulse generated in the nodal tissue spreads to the surrounding muscle, causing it first to depolarize (an electrical event) and subsequently to contract (a mechanical event). Review the conduction pathway of the heart in Laboratory Exercise 28.

Cardiac Muscle Characteristics

Cardiac muscle cells, like neurons, undergo a normal **polarization** process by activation of the **sodium–potassium pump.** The eventual result of polarization is an increase in sodium and potassium ion concentration on the outside of the plasma membrane relative to their concentration on the inside. This produces the **resting membrane potential.** When the pacemaker cells depolarize, the surrounding cardiac muscle cells also depolarize. This is an electrochemical event. After the cells have depolarized, they contract. This is a physical event.

In this exercise, you examine the natural rate of heart contraction in a frog and what occurs when various solutions are added to the heart muscle. The heart of a frog is physically somewhat different from a human heart. The frog has two atria and only a single ventricle; however, the physiologic response of the frog to various cardiac-influencing substances is similar to that of humans.

OBJECTIVES

At the end of this exercise, you should be able to

1. describe the basic contraction characteristics of the heart;
2. list the effects of various substances, such as epinephrine, calcium chloride, and acetylcholine, on the heart rate;
3. outline the mechanisms by which these substances change heart rate;
4. measure the resting pulse rate of the heart;
5. correlate the sounds the heart makes with the action of the heart;
6. explain how the valves of the heart increase the efficiency of the contraction.

MATERIALS
Pulse Rate and Heart Sounds

Clock or watch with second hand
Stethoscope
Alcohol wipes

Frog Heart Rate and Contraction Strength

Ph.I.L.S. Physiology Interactive Lab Simulations 3.0 CD (available through McGraw-Hill)
Compatible computer
Frogs
Clean dissection instruments
    Scissors
    Razor blades or scalpels
    Dissection tray
Small heart hook (fishhook, copper wire, or Z wire)
Pins
Thread
Frog board
Myograph transducer
Physiology recorder (computer, physiograph, or duograph)
Frog Ringer's solution—room temperature, 37°C, and iced
Solutions
    2% calcium chloride solution
    0.1% acetylcholine chloride solution
    0.1% epinephrine solution
    Saturated caffeine solution
Disposal container

Heart Valve Experiment

Dissection tray
Squeeze bottle of water
Scissors or razor blade
Fresh or thawed sheep heart
Clamp (hemostat) or string

PROCEDURE

Virtual Experiment Ph.I.L.S.

Frog Heart Functions 18. Thermal and Chemical Effects

1. Open the program; click on and read the **Objectives and Introduction** tab.
2. Take the pre-lab quiz to make sure that you understand the concepts presented in the virtual experiment. The Wet Lab portion describes the process as it is done on a live frog and the video clips demonstrate the technique.
3. Follow the instructions carefully in the Laboratory Exercise.
4. Turn on the power unit and attach the blue plug to the Recording Input 1. Touch the tip of the plug to the unit and it should insert.
5. Click the pipette once and drag it to the location where the simulated heart is attached to the myograph transducer.
6. Click the bulb of the pipette again to get it to dispense the regular frog Ringer's solution.
7. Let the heart beat for three or four cycles and then add the cold frog Ringer's solution. Once you get three or four more cycles, click on the **Stop** button. In order to record the data you will have to click on the left arrow at the bottom of the chart to get to the first recording.
8. Move the mouse to the top of one of the curves and click it there.
9. Move the mouse to the bottom of one of the curves and click it there. This will set the data points for the amplitude of the contraction.
10. Click on the **Journal** tab to get a record of the amplitude written in the journal.
11. Make sure that you push the **Start** button prior to adding new solutions. You will need to add regular frog Ringer's solution again before you add another solution.
12. Wait for three or four cycles, then add the adrenaline solution via a dropper bottle. As in the first part of the experiment you will need to click on the left arrow at the bottom of the chart in order to begin your tracing. Record the amplitude of the contraction and the interval between contractions as you did in the first recording.

Do the same with the acetylcholine solution as you did with the adrenaline.

Record the amplitude of the various solutions in the following spaces.

    Regular frog Ringer's solution: _____

    Frog Ringer's solution with adrenaline: _____

    Frog Ringer's solution with acetylcholine: _____

Record the interval between contractions here.

    Regular frog Ringer's solution: _____

    Frog Ringer's solution with adrenaline: _____

    Frog Ringer's solution with acetylcholine: _____

Live Frog Experiment

Examination of Frog Heart Rate and Contraction Strength in Real Frogs

Either you have a frog prepared for you or your instructor may wish for you to double pith a frog for this exercise. If you need to pith a frog, follow the instructions outlined in Exercise 11 and then follow the procedure described next.

Measurement of Baseline Data

Once the frog is prepared, cut through the pectoral skin with a scissors and through the sternum. The heart should be beating inside the pericardial sac. Pin the frog to a board and cut open the pericardial sac with a pair of scissors or a scalpel. Count the number of beats per minute and record this number as the resting heart rate of the frog.

    Resting heart rate: _____ bpm

Keep the heart moistened with frog Ringer's solution. Unless directed to do otherwise, use the Ringer's solution that is at room temperature.

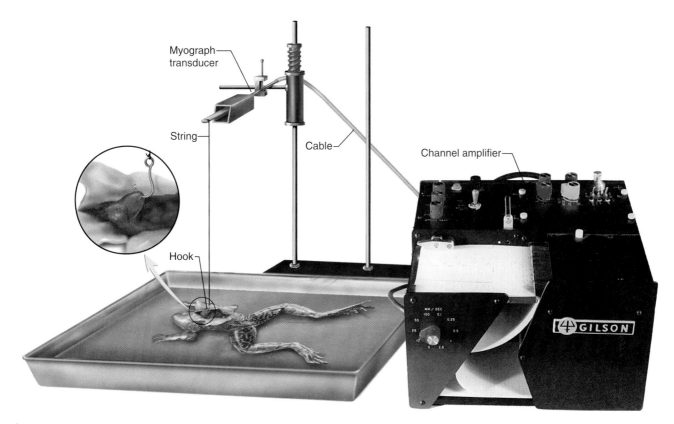

Figure 29.1 Frog Setup

Carefully cut the pericardial membrane of the heart away from the dorsal surface of the frog and attach a metal hook or wire to the membrane attached to the dorsal portion of the heart. You can pierce the apex of the heart with a wire or hook, but you must be very careful not to puncture the ventricle. Attach the hook to a thread and tie the thread to the myograph transducer, adjusting the tension until you get a deflection in the myograph leaf. Your setup should look similar to figure 29.1.

Set the chart speed to 0.5 cm/second and record the contraction rate and strength of contraction of the heart muscle. The force that is generated by the heart is measured by the height of the stylus on the chart paper. The rate can be determined by the distance between the peaks on the chart paper. Make sure you keep the heart moist with frog Ringer's solution during the experiment.

Effect of Various Chemicals on Heart Rate and Contraction Strength

Addition of Calcium
Add several drops of 2% calcium chloride solution to the heart muscle. Wait until you get a change in the heart rate. Turn on the recording device and produce a tracing. Calculate the heart rate and record the number in the following space.

Heart rate: _____ bpm

What happens to the contraction strength? _____

The addition of calcium chloride can produce pacemakers in regions of the heart other than the SA node. These are called ectopic foci. Rinse the heart with room temperature frog Ringer's solution.

Addition of Acetylcholine
Add several drops of 0.1% acetylcholine solution to the outer surface of the heart. You may have to wait a few minutes to observe any changes in heart rate. Record the heart rate for about 10 seconds on the physiology recorder after waiting 2 minutes. Wait 3 minutes and then record the rate for another 10 seconds. Record the rate in the space. When you have determined a change in the heart rate and recorded your results, flood the heart with frog Ringer's solution.

Heart rate after 2 minutes: _____ bpm

Heart rate after 3 minutes: _____ bpm

Acetylcholine released by the vagus nerve causes the potassium channels in the plasma membrane of the nodal cells to become more permeable, thus increasing the flow of potassium ions to the outside of the membrane. This increases the voltage difference across the plasma membrane, hyperpolarizing the cell. It takes more input to depolarize a cell in this condition, and thus the heart rate slows. Acetylcholine is a **parasympathomimetic substance,** one that mimics the effects of innervation by the parasympathetic nervous system.

Addition of Epinephrine
Add several drops of 0.1% epinephrine solution to the outer surface of the heart. Examine the rate over the next 2 to 3 minutes, and, when you notice a change, record the heart rate on the recorder chart paper for about 10 seconds. Write your results in the space provided. After you record your results, flush the heart with room temperature frog Ringer's solution.

Heart rate after 2 minutes: _____ bpm

Heart rate after 3 minutes: _____ bpm

What is the change in the contraction strength? _____

_____

Epinephrine increases the heart rate and strength of contraction by making the calcium slow channels more permeable, thus lowering the threshold. Calcium (a cation) leaks into the cell, thus decreasing the voltage difference across the membrane. Epinephrine is a **sympathomimetic drug,** one that mimics the effect of sympathetic nervous innervation.

Addition of Caffeine
Add several drops of saturated caffeine solution to the outer surface of the heart and note the time when the rate changes. Once the rate changes, record the heart rate and strength of contraction of the heart muscle for about 10 seconds on the physiology recorder. Enter the rate.

Heart rate with caffeine: _____ bpm

Strength of contraction (increase/decrease): _____

Caffeine is a central nervous system stimulant, and it has an effect on heart muscle. It increases heart rate and strength of contraction. High concentrations of caffeine cause **arrhythmias,** or irregular heart rates.

Changes in Heart Rate with Temperature
Run tests by flooding the heart with iced Ringer's solution. Note the change in heart rate.

Heart rate with iced Ringer's solution: _____ bpm

Readjust the rate by adding room temperature Ringer's solution. Record your results.

Heart rate at room temperature: _____ bpm

You can further study the effect of temperature on enzymes by flooding the frog heart with frog Ringer's solution at 37°C. Record the contraction rate.

Heart rate at 37°C: _____ bpm

The **Q10** is an enzymatic relationship that states that, for every 10° rise in temperature (from 0°C to 50°C), there is a doubling of enzyme activity. Does this correlate with your findings on the temperature relationships with the heart?

_____

> **Clean Up**
> When you are finished, dispose of the frog in the appropriate animal waste container. Clean your desk and equipment. Dispose of scalpel blades in the sharps container.

Human Pulse Rate and Heart Sounds
Measure the pulse rate (heart rate) and listen to the heart sounds of your lab partner. Place your fingers on the radial artery or the carotid artery of your lab partner and count the number of beats that occur in 30 seconds. Multiply this number by 2 and record the resting pulse rate.

Resting pulse rate: _____ bpm

You can calculate the average pulse rate for your class by having all of the members of your class record their pulse rate in beats per minute on the chalkboard (or a piece of paper at the front of the class). Add all the pulse rates together and divide by the total number to determine the average. Compare the value you calculate from your class with the average pulse rate.

Average pulse rate for class: _____ bpm

There are differences in pulse rate, depending on whether you are sitting, lying, or standing. The normal resting pulse rate is measured when you are sitting. One of the functions of changes in pulse rate is to alter blood pressure. The carotid sinus is one of the areas of the body that senses changes in blood pressure. After you measure the pulse rate while sitting, lie down on a table or cot and measure the pulse rate. Record any change in the pulse rate in the following space.

Pulse rate while lying down: _____ bpm

Stand up and record the change in the pulse rate in the following space.

Pulse rate while standing: _____ bpm

Can you explain any changes in the rate from the one you recorded when you were sitting?

Clean the earpieces of a stethoscope with alcohol swabs and examine the stethoscope. The earpieces should point in an anterior direction as you place the stethoscope into your ears. Listen for heart sounds generated by your lab partner by placing the diaphragm of the stethoscope inferior to the second rib on the left side of the body. The pulmonary semilunar valves are best heard in this location. On the right side of the body, in the intercostal space between the second and third ribs, is the location for the aortic semilunar valve. This can be found by locating the sternal angle and placing the diaphragm of the stethoscope interior to the ribs at this location. Use figure 29.2 as a guide for listening to heart sounds.

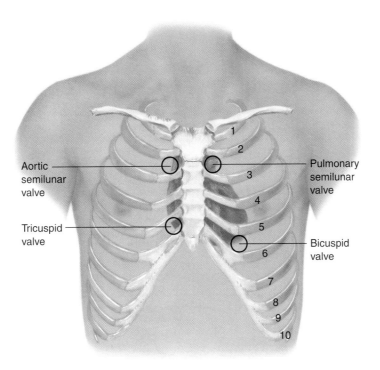

Figure 29.2 Auscultation Areas
Circles are locations where specific valvular sounds can best be heard.

The **lubb/dupp** sounds of the heart reflect the closure of the heart valves. The first heart sound is the lubb sound, and it occurs due to the closing of the atrioventricular valves. The second heart sound is the dupp sound, and it is due to the closing of the semilunar valves.

You may hear an additional whooshing sound as you listen through the stethoscope. This is attributed to the imperfect closure of the valves, which is known as a **heart murmur.**

Mechanical Functions of Sheep Heart Valves

The **semilunar valves** and **atrioventricular valves** prevent blood from flowing in reverse. The semilunar valves keep the backflow of blood from entering the ventricles. The atrioventricular valves keep blood from flowing back into the atria during ventricular contraction. Your instructor may demonstrate the procedure, or you can do it yourself by using a fresh or thawed sheep heart.

Caution
Wear gloves when handling the sheep heart.

Flush any remaining blood from the heart before you locate the valves. Make an incision horizontally into the right atrium, exposing the **tricuspid valve.** Pour water into the right ventricle and notice how the water flows past the tricuspid valve and into the right ventricle. Use a hemostat or string to close off the pulmonary trunk to prevent water from flowing out of the pulmonary trunk. Gently squeeze the right ventricle and notice how the right atrioventricular valve closes, preventing blood from backing up into the right atrium.

What is the adaptive value of the closing of atrioventricular

valves? _____

_____

Cut the **pulmonary trunk** close to the right ventricle and slowly pour water into it, as if you were trying to fill the right ventricle from the pulmonary trunk. The **pulmonary semilunar valve** should fill with water and close the exit of the right ventricle, preventing blood from returning to the right ventricle. This would normally occur as the right ventricle began diastole. Compare these valve closures with figure 29.3.

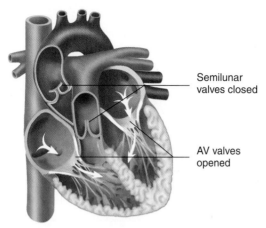 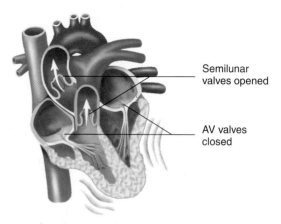

(a) *Diastole: passive ventricular filling.* The AV valves open and blood flows into the relaxed ventricles, accounting for most of the ventricular filling.

(b) *Systole: period of ejection.* Ventricular contraction pushes blood out of the ventricles, causing the semilunar valves to open.

Figure 29.3 Closure of the Heart Valves

(a) Closure of the semilunar valves during filling of the ventricles; (b) closure of the atrioventricular valves and opening of the semilunar valves during ventricular contraction.

Exercise 29 Review

Functions of the Heart

Name:_____

Lab time/section:_____

Date:_____

1. Decreasing heart rate is under the control of what nervous division? _____

2. What is the resting heart rate of the average person? _____

3. Are there more sodium ions inside or outside a cardiac muscle cell during the resting membrane potential? _____
   _____

4. What happens to sodium ions when a membrane depolarizes? _____
   _____

5. What region in the heart depolarizes spontaneously? _____

6. What happens to the heart when an action potential is generated in the SA node? _____
   _____

7. The movement of electrochemical impulses in the myocardium is called _____ conduction.

8. Beta-adrenergic blockers bind to norepinephrine sites, preventing these neurotransmitters from having an effect.

   What effect would the use of "beta blockers" have on heart rate? _____
   _____
   _____

9. How much of a change in the heart rate of the frog did you see after the addition of calcium chloride? _____
   _____

   What was the change in the contraction strength? _____

   What process might account for this in terms of cardiac muscle interactions with calcium? _____
   _____

10. What heart sound is produced by the closure of the atrioventricular valves in the heart? _____

11. A heart murmur is normally caused by what event? _____

12. When would a murmur occur in the lubb/dupp cycle if the AV valves were not closing properly? _____

13. In the Ph.I.L.S. virtual heart experiment, what effect did the addition of the adrenaline solution have on heart rate and strength? _____

14. Did the addition of the adrenaline solution increase or decrease the threshold in the heart muscle tissue? _____

15. In the Ph.I.L.S. virtual heart experiment, what effect did the addition of the acetylcholine solution have on heart rate and strength? _____

16. Did the addition of acetylcholine solution hyperpolarize or depolarize the cardiac muscle cell membrane? _____

Exercise 30
Introduction to Blood Vessels and Arteries of the Upper Body

INTRODUCTION

The heart is the mechanical pump of the cardiovascular system; the blood vessels are the conduits that carry oxygen, nutrients, and other materials to the cells and remove wastes from the extracellular fluid. These vascular conduits consist of numerous vessels, such as the large **arteries** that take blood from the heart. They become progressively smaller and become **arterioles.** Still smaller vessels are **capillaries,** which are the sites of exchange between the blood and cells of the body. **Venules** return blood to the **veins,** which carry it back to the heart.

Arteries are blood vessels that carry blood *away from* the heart. Arteries have thicker walls than veins due to the increased amounts of smooth muscle, and the thickness of arterial walls reflects the higher blood pressure found in them.

Arteries are often named for the region of the body they pass through. The **brachial** artery is in the region of the arm, whereas the **femoral** artery is in the thigh region. In this exercise, you learn the basic structure of blood vessels and locate the arteries of the upper body. The nature of vessels and the descriptions of specific vessels are found in *Principles of Anatomy and Physiology* in chapter 18, "Blood Vessels and Circulation."

OBJECTIVES

At the end of this exercise, you should be able to

1. draw a cross section of the wall of a generalized blood vessel, showing the three layers;
2. distinguish between arteries and veins;
3. distinguish between elastic arteries and muscular arteries;
4. describe the sequence of major arteries that branch from the aortic arch;
5. list the major blood vessels that take blood to the brain;
6. list the arteries of the upper extremity;
7. trace the arteries that supply blood to the head;
8. describe the major organs that receive blood from the upper arteries.

MATERIALS

Microscopes
Prepared slides of arteries and veins
Models of the blood vessels of the body
Charts and illustrations of the arterial system
Microscope slides of arteries and veins in cross section
Cats
Dissection equipment
    Dissection trays
    Scalpels
    Pins
    Blunt (mall) probes
    Gloves
    Sharp scissors or bone cutter
    Forceps
    Animal waste disposal container
    Cat wetting solution

PROCEDURE

Overview of Blood Vessels

Obtain a prepared microscope slide of a cross section of artery and vein. Both arteries and veins have walls that consist of three layers. The outer layer is known as the **tunica adventitia** and consists of a connective tissue sheath. The middle layer, or **tunica media,** is composed of smooth muscle in both arteries and veins. The tunica media is thicker in arteries, and there are pronounced elastic fibers in the wall of the tunica media in arteries. The innermost layer (near the blood) is the **tunica intima** and consists of a thin layer of connective tissue and a thin layer of simple squamous epithelium, known as **endothelium.** Endothelium is the layer closest to the blood. In arteries, there is an **inner elastic membrane,** which is a thin layer of elastic fibers. These layers are seen in figure 30.1. Examine a prepared slide of an artery and a vein and locate these layers.

Arteries do not have valves, but some veins do, and they are described in Exercise 32. You can easily distinguish veins from arteries in a prepared slide where the vessels are cut in cross section. Veins have a large lumen (although the vein is often collapsed in prepared sections) and a thin tunica media relative to their overall size. You may also find nerves in the prepared slide. These are also circular structures, but they are solid, not hollow. Compare your slide with figure 30.2.

Types of Arteries

Elastic arteries are large arteries close to the heart. Their histological appearance is made distinctive by the presence of significant amounts of elastic fibers in the tunica media. **Muscular (distributing) arteries** are farther away from the heart and have more smooth muscle in the tunica media.

Specific Arteries

Examine the models, charts, and illustrations of the cardiovascular system in the lab and locate the major arteries of the upper body. Read the following descriptions of the arteries. Name the vessels that take blood to an artery and those that receive blood from the artery in question. An overview of the major arteries of the body is shown in figure 30.3. You should refer to this illustration for this exercise and for the vessels in Exercises 31 and 32.

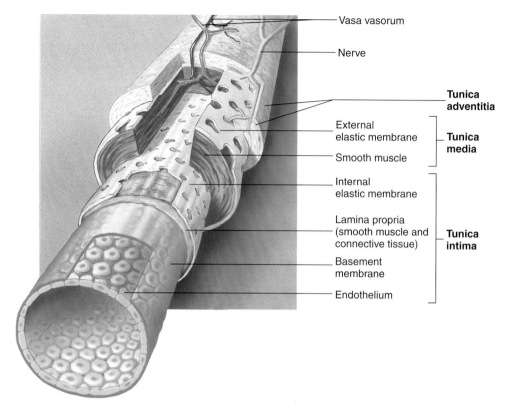

Figure 30.1 Layers of a Generalized Blood Vessel

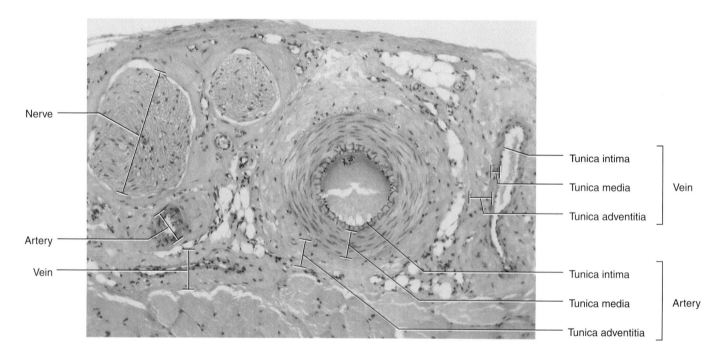

Figure 30.2 Photomicrograph of an Artery, a Vein, and a Nerve (40×)

Aortic Arch Arteries

Locate the heart and find the large **aorta** that exits from the left ventricle. This is the **ascending aorta,** and it is a large vessel about the size of a garden hose. The ascending aorta is relatively thick-walled due to the high pressure of the blood coming from the heart. The first two arteries that arise from the ascending aorta are the **coronary arteries,** which were covered in detail in Exercise 27. The ascending aorta curves to the left side of the body and forms the **aortic arch.** In humans, there are three main arteries that receive blood from the aortic arch. The first major artery to receive blood is on the right side of the body and is called the **brachiocephalic artery.** This artery shortly divides into arteries that feed

Exercise 30 Introduction to Blood Vessels and Arteries of the Upper Body

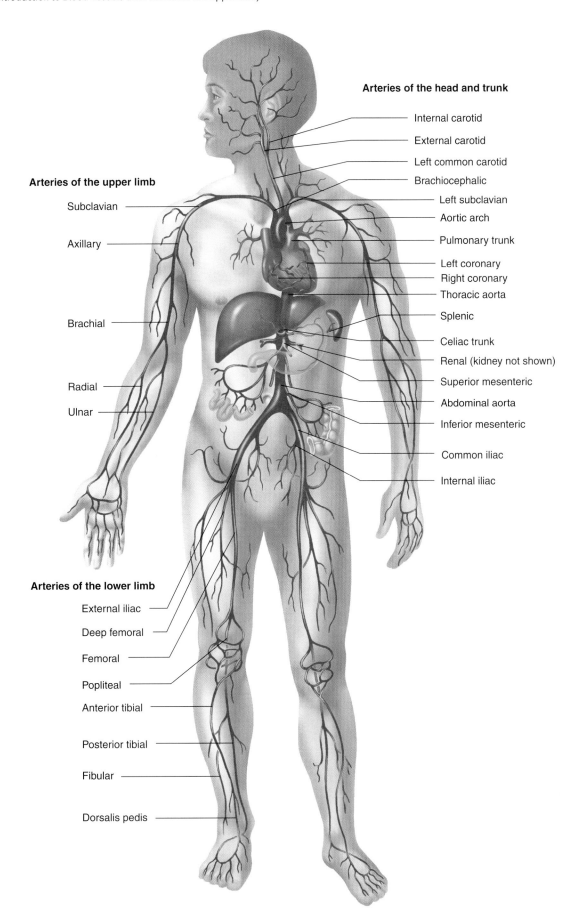

Figure 30.3 Major Arteries of the Body

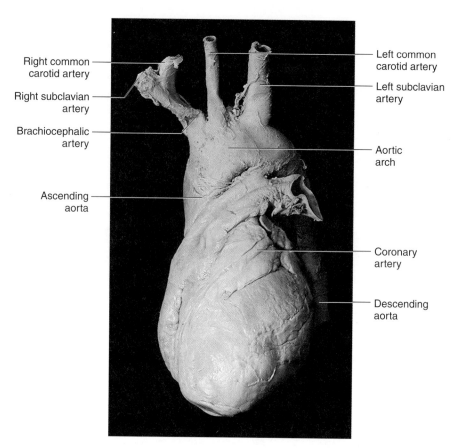

Figure 30.4 Photograph of the Anterior Aspect of the Human Heart and Aortic Arch Arteries

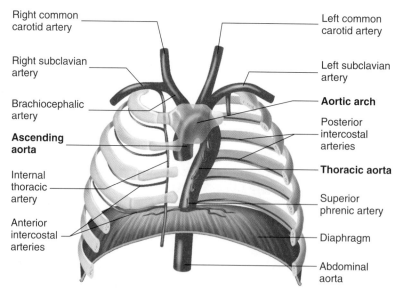

Figure 30.5 Intercostal Arteries

the right side of the head (the **right common carotid artery**) and the right upper extremity (the **right subclavian artery**), as seen in figure 30.4. The aortic arch has two other arteries. On the left side is the **left common carotid artery,** which takes blood to the left side of the head, and the **left subclavian artery,** which takes blood to the left upper extremity.

As the aortic arch turns inferiorly behind the posterior part of the heart, it becomes the **descending aorta,** which is composed of two segments. Above the diaphragm, the descending aorta is known as the **thoracic aorta;** below the diaphragm, it is known as the **abdominal aorta** (figure 30.3). The thoracic aorta has many branches, which run between the ribs, called **intercostal arteries** (figure 30.5).

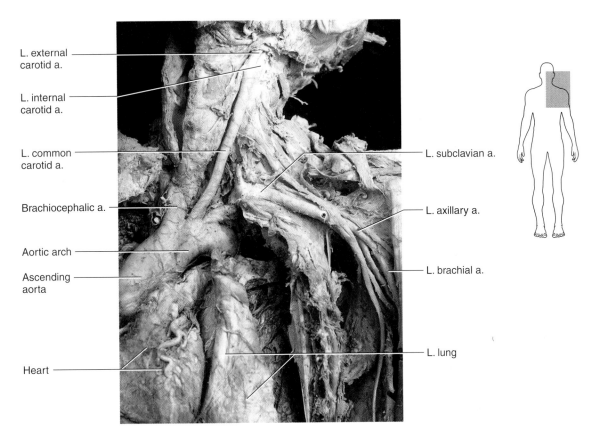

Figure 30.6 Branches of the Subclavian Artery
The abbreviation for artery is a.

Arteries That Feed the Upper Limbs

If you return to the brachiocephalic artery, you can see that it is a short tube that takes blood from the aortic arch and then branches into the **right common carotid artery** and the **right subclavian artery.** The right subclavian artery has one branch that takes blood to the brain, whereas another branch becomes the **axillary artery.** The **left subclavian artery** also has a branch that takes blood to the brain and another branch that becomes the **left axillary artery** (figures 30.3 and 30.6).

The axillary artery on each side of the body turns into **brachial artery,** which has a pulse that can be palpated by finding a groove on the medial, distal region of the arm between the biceps brachii and brachialis muscles. This location is the site for the placement of the diaphragm of a stethoscope during blood pressure measurement. The brachial artery supplies blood to the triceps brachii muscle and the humerus. The brachial artery bifurcates (splits in two) to form the **radial artery** on the lateral side of the forearm and the **ulnar artery** on the medial side of the forearm. These arteries supply blood to the forearm and part of the hand. The radial artery can be palpated just lateral to the flexor carpi radialis muscle at the wrist, which is a common site for the measurement of the pulse. The radial artery and ulnar artery become united again as the **palmar arch arteries.** There is a **superficial palmar arch artery** and a **deep palmar arch artery.** They join, or anastomose, and send out the small **digital arteries** that supply blood to the fingers (figure 30.7).

Arteries of the Head and Neck

The **vertebral arteries** receive blood from the **subclavian arteries** and take it to the brain by traveling through the transverse foramina of the cervical vertebrae and then into the foramen magnum of the skull. The other main arteries that take blood to the brain are the **common carotid arteries** on the anterior sides of the neck. Each common carotid artery branches just below the angle of the mandible to form the **external carotid artery,** which takes blood to the face, and the **internal carotid artery,** which passes through the carotid canal and takes blood to the brain. The external carotid artery on each side has several branches, including the **facial artery,** the **superficial temporal artery,** the **maxillary artery,** and the **occipital artery** (figure 30.8).

Blood Supply to the Brain

The blood vessels that supply nutrients and oxygen to the brain do not actually penetrate the brain tissue itself but, rather, form a meshwork around the brain and into the open spaces in the interior. The brain receives blood from four major arteries. These are the **left** and **right vertebral arteries** and the **left** and **right internal carotid arteries.**

The vertebral arteries pass through the foramen magnum and unite to form the **basilar artery** at the base of the brain (figure 30.9). The two internal carotid arteries pass through the carotid canal and take blood to the base of the brain. The blood from the basilar and the two internal carotid arteries unite again in a vessel called the

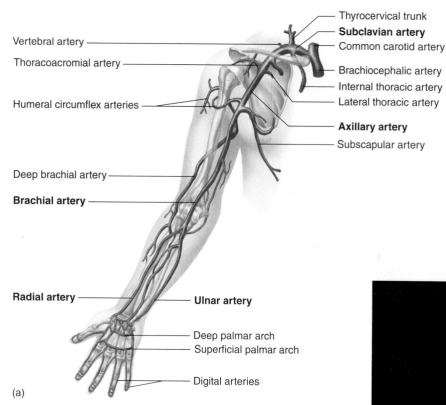
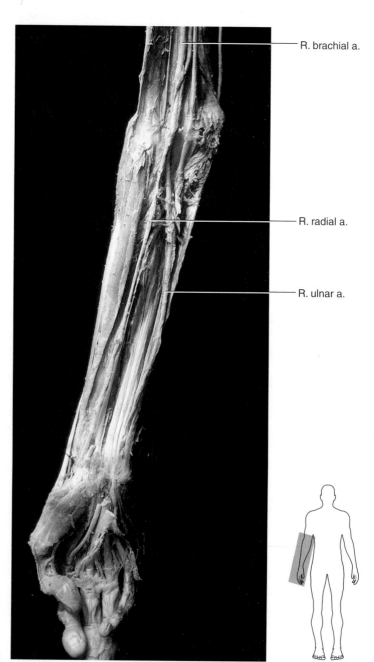

Figure 30.7 Arteries of the Right Upper Extremity
(a) Diagram; (b) photograph.

Exercise 30　Introduction to Blood Vessels and Arteries of the Upper Body

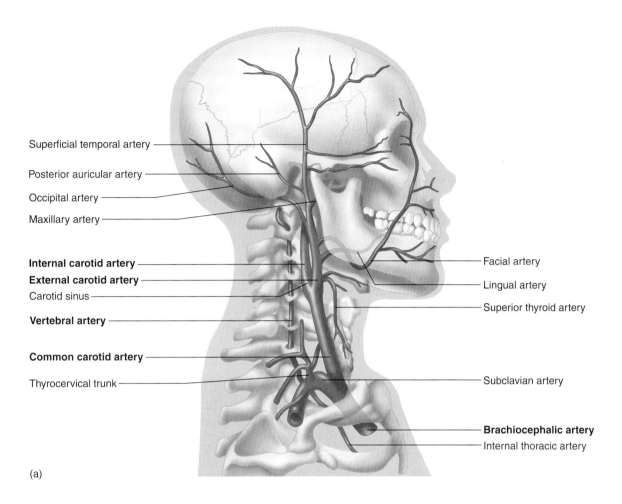

(a)

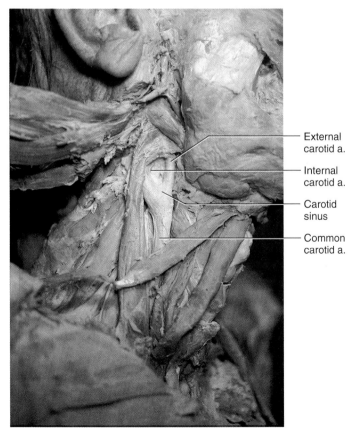

(b)

Figure 30.8 Arteries of the Head and Neck
(a) Diagram; (b) photograph.

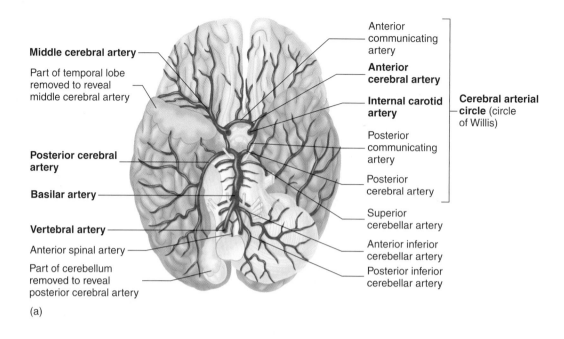

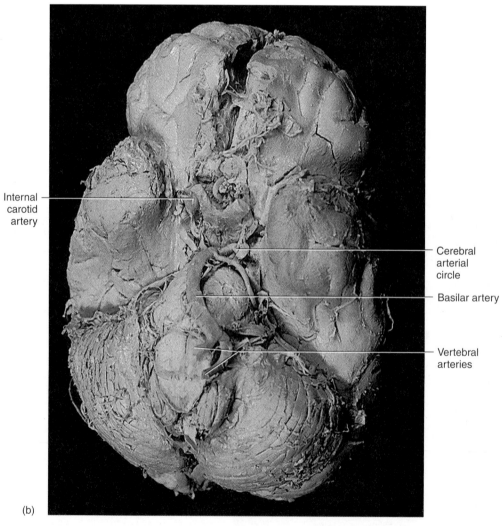

Figure 30.9 Arteries of the Base of the Brain
(a) Diagram; (b) photograph.

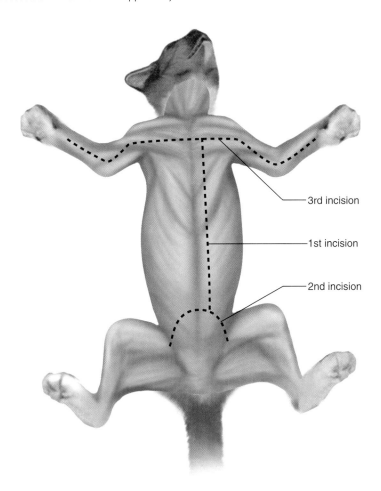

Figure 30.10 Incisions to Open the Thoracic and Abdominal Cavities of the Cat

cerebral arterial circle (circle of Willis). The cerebral arterial circle forms a loop around the pituitary and is composed of the **anterior** and **posterior communicating arteries.** This reuniting of arteries is called **collateral circulation,** which is important in that, if one artery becomes occluded or partially occluded, the region of the brain fed by that artery can be supplied by blood from other collateral arteries. The major arteries that branch off the cerebral arterial circle and take blood directly to the brain are the **anterior, middle,** and **posterior cerebral arteries.** The basilar artery gives rise to the **cerebellar arteries,** which take blood to the cerebellum.

Cat Dissection

Prepare for the cat dissection by taking a dissection tray back to your table, along with a scalpel, scissors or a bone cutter, string, forceps, and, of course, a cat. Remember to place all excess tissue in the appropriate waste container and not in a standard wastebasket or down the sink!

Be careful as you make an incision into the body cavity of the cat. Cut with the scalpel blade facing to one side or, even better, use a lifting motion and cut upward through the tissue. If you use a downward sawing motion, you will damage the internal organs of the cat.

Open the thoracic and abdominal regions of the cat if you have not done so already. Make an incision into the body wall of the cat in the belly, slightly lateral to midline. Cut into the muscle; then grasp the tissue with a forceps and lift the body wall away from the viscera. Continue cutting anteriorly through the belly. Stop at the level of the diaphragm, and then continue cutting a little off center as you cut through the costal cartilages. Use a scalpel or scissors to cut through the thoracic region.

You can now make a transverse incision in the lower region of the belly, so that you have an inverted T-shaped incision (figure 30.10). Lift the tissue near the ribs and expose the diaphragm. Gently and carefully snip or cut the diaphragm away from the ventral region by cutting as close to the ribs as possible.

Next make an incision by cutting transversely across the upper part of the chest. Be careful not to cut the blood vessels of the neck that were exposed during the original skinning process. Open the body cavity by pulling the two flaps laterally, exposing the thoracic and abdominal cavities. You may want to cut through the ribs at their most lateral point to facilitate the opening of the thoracic cavity.

Carefully expose the thoracic region of the cat and locate the two lungs and the heart. Note that the heart is enclosed in the pericardial sac. Look for the fatty material, called the **greater omentum,** that covers the **intestines** in the abdominal region. As you dissect the arteries of the cat, pay careful attention to the veins and nerves that travel along with the arteries. *Do not dissect other organs.* You will study these systems later (for example, respiratory, digestive, and urogenital systems).

As you open the abdominal cavity, you will see the space that contains the internal organs known as the **coelom,** or **body cavity.** The lining on the inside of the body wall is the **parietal peritoneum,** and the lining that wraps around the outside of the intestines is the **visceral peritoneum.**

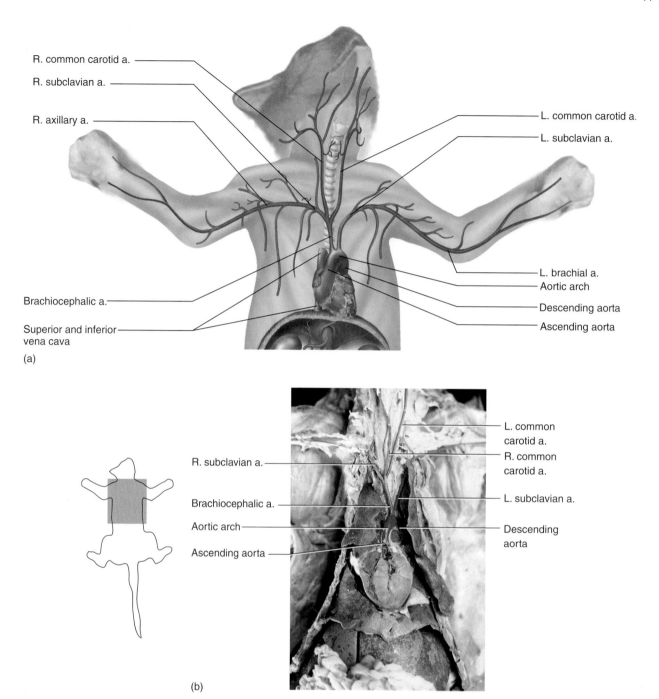

Figure 30.11 Anterior View of the Heart and Aortic Arches of the Cat
(a) Diagram; (b) photograph.

Arteries of the Cat

The blood vessels in the cat have been injected with colored latex. If the cat is doubly injected, the **arteries are red** and the **veins are blue.** If the cat has been triply injected, the hepatic portal vein is injected usually with yellow latex. Be careful locating the arteries in the cat. You can tease away some of the connective tissue with a dissection needle to see the vessel more clearly. Note the difference in the aortic arch arteries in the cat compared with those of the human. Look for the **coronary arteries,** which take blood from the ascending aorta to the heart tissue.

You can begin your study of the arteries of the cat by examining the ascending aorta. Note that the **aorta** is composed of the **ascending aorta,** which bends and leads to the **aortic arch** and then moves posteriorly to form the **descending aorta.** The descending aorta is composed of the **thoracic aorta** above the diaphragm and the **abdominal aorta** below the diaphragm. The pattern of the arteries that leave the aortic arch is somewhat different in cats than in humans. In the cat, typically only two large vessels leave the aortic arch, the large **brachiocephalic artery** and the **left subclavian artery.** The brachiocephalic artery further divides into the **left common carotid artery,** the **right common carotid artery,** and the **right subclavian artery.** Examine figure 30.11 for the pattern in the cat. How is the pattern in the cat different from that in the human?

Several arteries arise from the **subclavian artery.** The **vertebral artery** takes blood to the head, and the **subscapular artery** takes blood to the pectoral girdle muscles, along with the **ventral thoracic artery** and **long thoracic artery.** The ventral thoracic and long thoracic arteries supply the latissimus dorsi and pectoralis muscles with blood. The **left** and **right subclavian arteries** become the **left** and **right axillary arteries,** which in turn take blood to the brachial arteries. The **brachial artery** takes blood to the **radial** and **ulnar arteries.** Additional arteries that branch from the subclavian artery are the **internal mammary artery,** which takes blood to the ventral body wall, and the **thyrocervical artery,** which takes blood to the region of the shoulder and neck. The **costocervical artery** reaches the deep muscles of the back and neck (figure 30.12).

The **common carotid artery** takes blood along the ventral surface of the neck to feed the small **internal carotid artery,** which supplies the brain. The common carotid artery also empties into the **external carotid artery,** which takes blood to the external portions of the head. The **occipital artery** receives blood from the common carotid artery and supplies the muscles of the neck.

The thoracic aorta not only takes blood to the abdominal aorta but also supplies the rib cage with blood from the **intercostal arteries. Esophageal arteries** also branch from the thoracic aorta, supplying the esophagus with blood. The thoracic aorta continues as the **abdominal aorta** as it passes through the diaphragm.

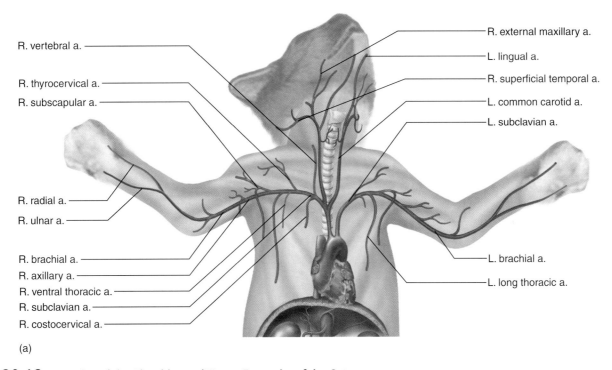

Figure 30.12 Arteries of the Shoulder and Upper Extremity of the Cat
(a) Diagram; (b) photograph.

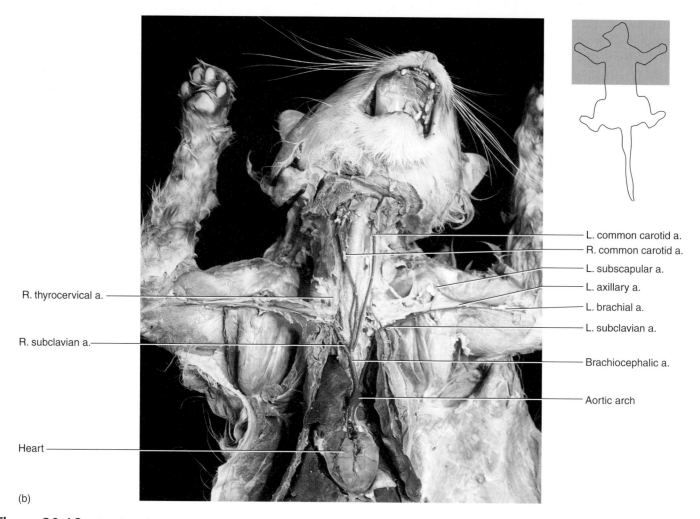

Figure 30.12—*Continued.*

Exercise 30 Review

Introduction to Blood Vessels
and Arteries of the Upper Body

Name: _____

Lab time/section: _____

Date: _____

1. Label the following illustration with the major arteries of the body using the terms provided. Try to complete the illustration first and then review the material in this exercise to determine your accuracy.

   abdominal aorta
   external carotid artery
   brachiocephalic trunk
   thoracic aorta
   common carotid artery
   internal carotid artery
   aortic arch
   subclavian artery

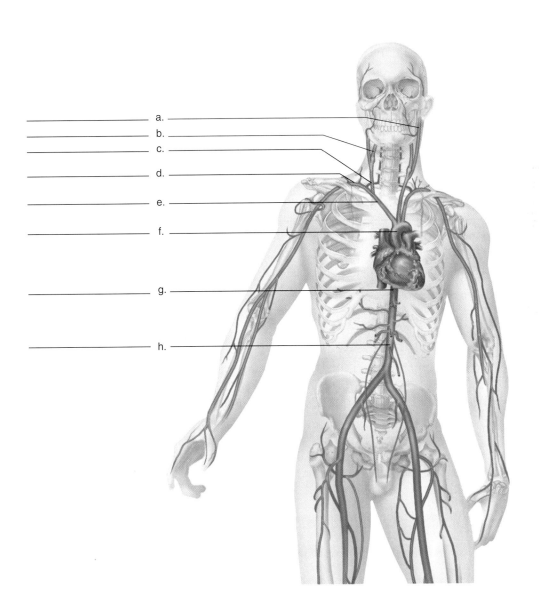

_____ a.
_____ b.
_____ c.
_____ d.
_____ e.
_____ f.
_____ g.
_____ h.

2. Blood from the left subclavian artery flows into what vessels as it moves toward the left arm? _____

_____

_____

3. Blood in the radial artery comes from what blood vessel? _____

391

4. An aneurysm is a weakened, expanded portion of an artery. Ruptured aneurysms can lead to rapid blood loss. Describe the significance of an aortic aneurysm versus a digital artery aneurysm. _____

_____

_____

_____

5. The pulmonary arteries carry deoxygenated blood from the heart to the lungs. Umbilical arteries carry a mixture of oxygenated and deoxygenated blood. Why are these blood vessels called arteries? _____

_____

_____

_____

6. What is the name of the outermost layer of a blood vessel? _____

7. What kind of blood vessel has valves? _____

8. Blood from the common carotid artery next travels to what two vessels? _____

_____

9. Blood from the right brachial artery travels to what two vessels? _____

_____

10. Where does blood in the right subclavian artery come from? _____

_____

11. The internal carotid artery takes blood to what organ? _____

12. The descending aorta receives blood from what vessel? _____

13. What is the general name of a large vessel that takes blood away from the heart? _____

_____

14. Blood in the left common carotid artery receives blood from what vessel? _____

_____

15. Name three blood vessels that exit from the aortic arch. _____

_____

Exercise 30 Introduction to Blood Vessels and Arteries of the Upper Body

16. How do the aortic arch arteries of a cat differ from those of a human? _____

17. Working in pairs, have your lab partner select an artery for you to name. Quiz each other on the material learned in this exercise.

18. Match the letters in the photograph with the terms on the left.

    _____ tunica adventitia
    _____ artery
    _____ vein
    _____ tunica media
    _____ tunica intima

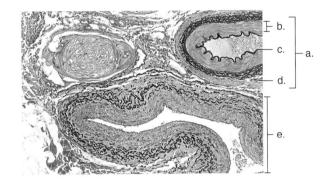

19. Name the specific vessels using the terms provided.

    superficial palmar arch artery
    radial artery
    axillary artery
    digital artery
    subclavian artery
    brachial artery
    ulnar artery

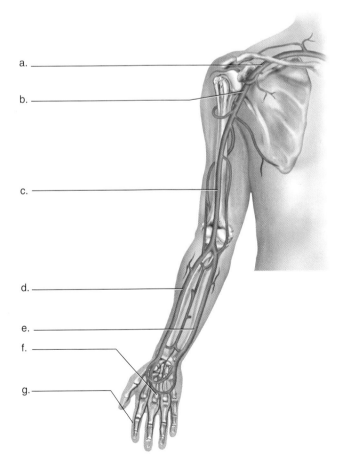

Exercise 31
Arteries of the Lower Body

INTRODUCTION

The arteries of the lower body receive blood directly or indirectly from the descending aorta. As with the arteries of the upper extremity, these vessels are often named for their locations, such as the *iliac* artery, the *femoral* artery, and the *mesenteric* arteries, or they are named for the organs they serve, such as the common *hepatic* (hepatic = liver) artery and the *splenic* artery. These arteries are presented in *Principles of Anatomy and Physiology* in chapter 18, "Blood Vessels and Circulation."

In this exercise, you continue your study of arteries. It is very important that you are careful when dissecting the arteries of the abdominal region of the cat. Do as little damage as possible to the other structures as you examine the arteries in this region.

OBJECTIVES

At the end of this exercise, you should be able to

1. describe the sequence of major arteries that originate from the abdominal aorta or its derivatives;
2. list the arteries of the lower extremities;
3. describe the major organs that receive blood from the lower arteries;
4. name the vessels that take blood to a particular artery and the vessels that receive blood from a particular artery.

MATERIALS

Microscope
Prepared slide of arteriosclerosis
Models of the blood vessels of the body
Charts and illustrations of arterial system
Cadaver (if available)
Cats
Materials for cat dissection
    Dissection trays
    Scalpel
    Gloves (latex or vinyl)
    Blunt (mall) probe
    Pins
    Forceps and sharp scissors
    First aid kit in lab or prep area
    Sharps container
    Animal waste disposal container

PROCEDURE

Examine the models and charts in the lab and the accompanying illustrations to locate the major arteries in the abdomen and lower body. Read the following descriptions of the vessels and locate the individual arteries. You should refer to figure 30.3 for an overview of the major arteries of the body.

Abdominal Arteries

The **abdominal aorta** is the portion of the **descending aorta** inferior to the diaphragm. The first major branch of the abdominal aorta is the **celiac trunk (celiac artery).** The celiac trunk splits into three separate arteries: the **splenic artery,** which takes blood to the spleen, pancreas, and part of the stomach; the **left gastric artery,** which takes blood to the stomach and esophagus; and the **common hepatic artery,** which takes blood to the liver. Locate the celiac trunk and the branches of the celiac in figure 31.1.

Just below the celiac trunk is the **superior mesenteric artery.** This vessel takes blood from the abdominal aorta and continues through the mesentery until it reaches the small intestine and proximal portions of the large intestine, including the cecum, ascending colon, and part of the transverse colon. The major branches of the superior mesenteric artery are the intestinal arteries, the ileocolic arteries, and the right and middle colic arteries.

The next vessels to branch from the aorta are the paired **suprarenal arteries,** which take blood to the adrenal glands. Below the suprarenal arteries are the **left** and **right renal arteries.** These arteries take blood to the kidneys. Locate these vessels in the lab and in figures 31.1 and 31.2.

The **gonadal arteries** branch inferior to the renal arteries and descend to either the testes or the ovaries. The **inferior mesenteric artery** is the next vessel to branch from the aorta, and it takes blood to the lower portion of the large intestine, including part of the transverse colon, the descending colon, and the rectum. These can be located in figures 31.1 and 31.2.

The abdominal aorta terminates by dividing into the two **common iliac arteries.** The common iliac arteries take blood to the internal and **external iliac arteries.** The **internal iliac artery** takes blood to the pelvic region, including branches that feed the rectum, pelvic floor, external genitalia, groin muscles, and hip muscles, as well as the uterus, ovary, and vagina in females. These arteries are illustrated in figure 31.2.

Arteries of the Lower Limb

Each external iliac artery branches from a common iliac artery and exits the body wall near the **inguinal canal.** Once the external iliac

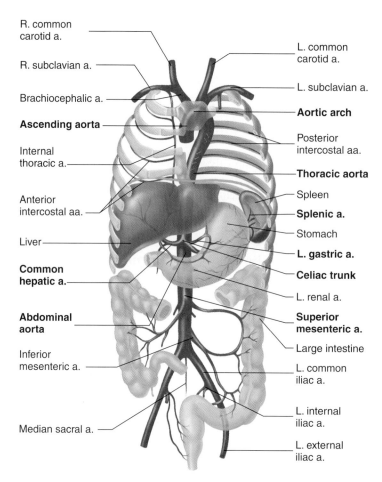

Figure 31.1 Arteries of the Thoracic and Abdominal Regions

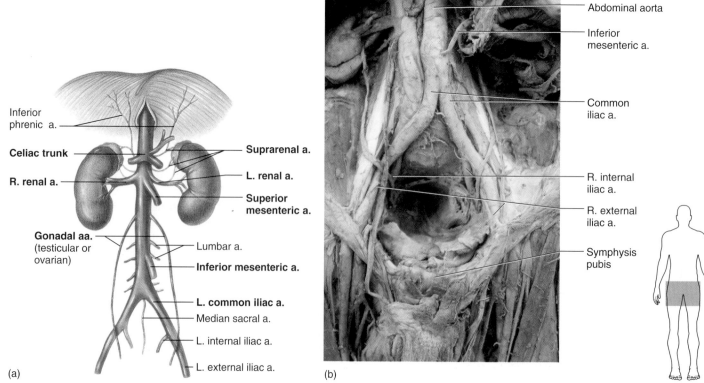

Figure 31.2 Deep Abdominal and Pelvic Arteries
(a) Diagram; (b) photograph.

Exercise 31 Arteries of the Lower Body

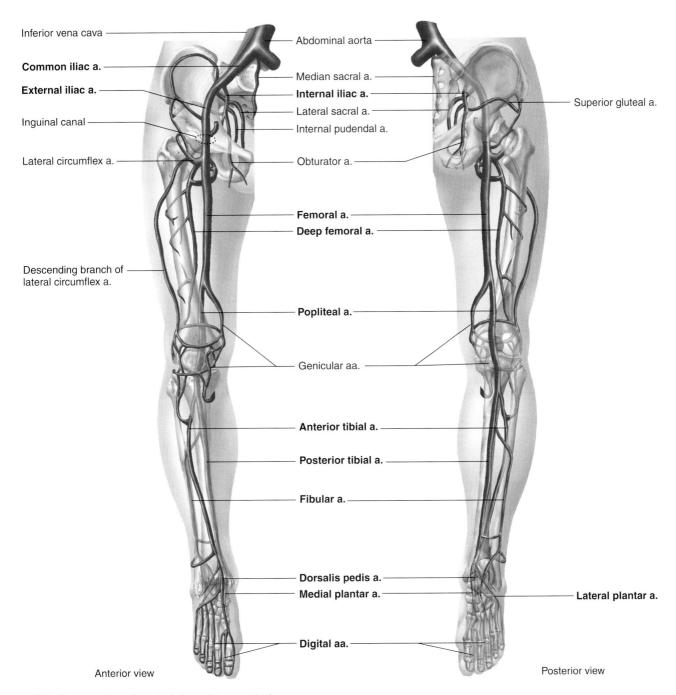

Figure 31.3 Arteries of the Pelvis and Lower Limb

artery leaves the body cavity, it continues into the thigh as the **femoral artery.** The femoral artery has a superficial branch that feeds the thigh and a deeper branch, called the **deep femoral artery,** that takes blood to the muscles of the thigh, knee, and femur. The femoral artery continues posterior to the knee as the **popliteal artery** and then divides into the **posterior tibial artery** and **anterior tibial artery,** which take blood to the knee and leg. Another branch of the popliteal artery is the **fibular artery,** which supplies the muscles on the lateral side of the leg. The foot is supplied with blood from several arteries, including the **plantar arteries,** the **dorsalis pedis artery,** and the **digital arteries.** Locate these arteries in figure 31.3.

Arteriosclerosis

Examine a microscope slide or an illustration of arteriosclerosis and note the development of **cholesterol plaque** under the endothelial layer. **Arteriosclerosis** is a condition known commonly as hardening of the arteries and is caused by the development of cholesterol plaque. Draw what you see in the space provided and compare it with the illustration in figure 31.4.

Drawing of an artery with arteriosclerosis:

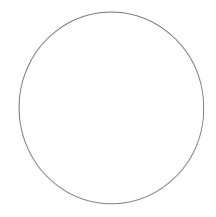

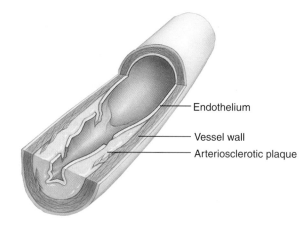

Figure 31.4 Arteriosclerotic Plaque
Note how the plaque develops deep to the endothelial layer.

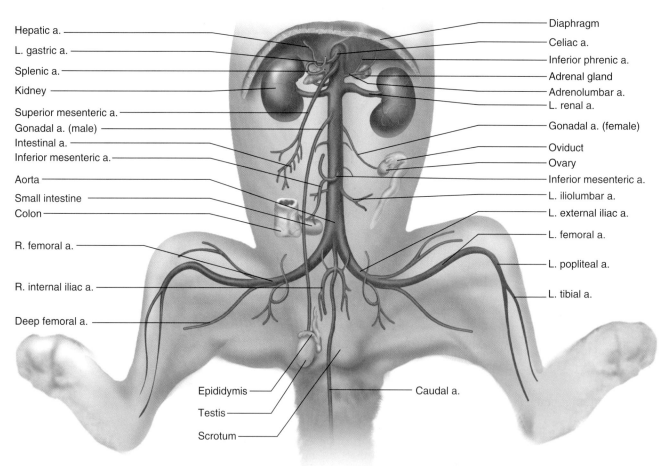

Figure 31.5 Diagram of the Caudal Arteries of the Cat

Cat Dissection

Bring a cat and your dissection material to your lab table. Follow the dissection procedures discussed in Exercise 30. Make sure you do not cut the blood vessels away from the organs that will be studied later. Note the major arteries of the caudal region of the cat in figure 31.5. Refer to this figure and the photographs for the remainder of this exercise.

Abdominal Arteries

The determination of the abdominal arteries is best done by looking for where they leave the aorta and where these arteries enter the organs they supply. The abdominal cavity of the cat should already be opened. If not, make an incision in the lower abdominal wall and carefully cut toward the sternum as described in Exercise 30. Do not cut deeply into the body cavity or you will puncture or cut into the viscera. Locate the major organs of the abdominal region, such as the **stomach, spleen, liver,** and **small** and **large intestines.** Examine the specimen from the left side, gently lifting the stomach, spleen, and intestines toward the ventral right side as you look for the abdominal arteries. The **descending aorta** is located on the left side of the body, whereas the **inferior vena cava** (we will use the human terms for cat vessels) is located on the right side. Locate the short **celiac trunk** (celiac artery), which quickly divides into three major vessels, the **splenic artery,** the **left gastric artery,** and the **hepatic artery.** The splenic artery takes blood to the spleen, the hepatic artery reaches the liver, and the left gastric artery supplies the stomach, as seen in figure 31.6.

The **superior mesenteric artery** is the next major vessel to take blood from the abdominal aorta. The superior mesenteric artery travels to the small intestine and the proximal portion of the large intestine. It branches into the jejunal arteries, the ileal arteries, and the colic arteries. Compare your dissection with figures 31.5 to 31.7.

The paired **adrenolumbar arteries** branch on each side of the abdominal aorta and take blood to the adrenal glands, parts of the body wall, and the diaphragm. Just caudal to the adrenolumbar arteries are the paired **renal arteries,** which take blood to the kidneys. The **gonadal arteries** are the next set of paired arteries, and these take blood to the testes in male cats, or ovaries in the female cat (figure 31.5).

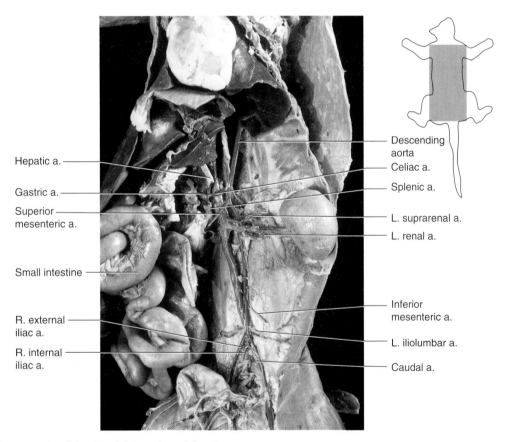

Figure 31.6 Photograph of the Caudal Arteries of the Cat

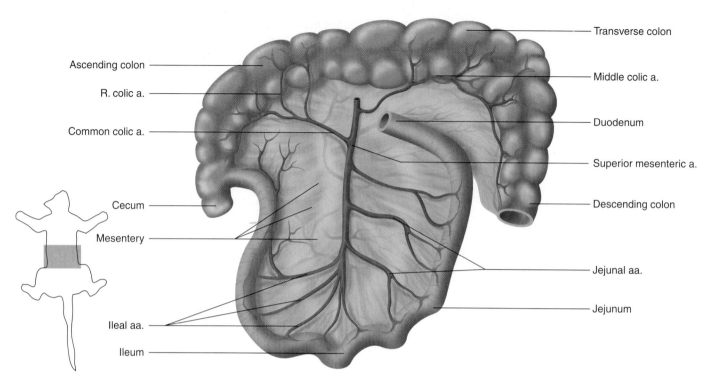

Figure 31.7 Branches of the Superior Mesenteric Artery of the Cat

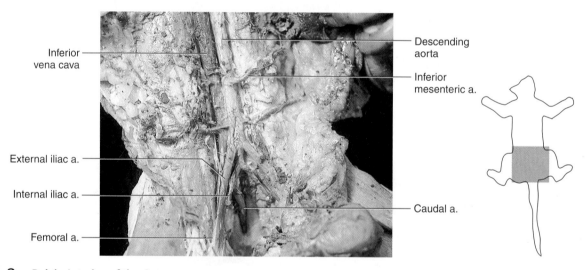

Figure 31.8 Pelvic Arteries of the Cat

A single **inferior mesenteric artery** takes blood from the abdominal aorta to the terminal portion of the large intestine. The inferior mesenteric artery can be found caudal to the gonadal arteries. Locate the lower abdominal arteries in your specimen and compare them with figures 31.5, 31.6, and 31.8.

The pelvic arteries are somewhat different in cats than in humans. The abdominal aorta splits caudally into the **external iliac arteries,** and a short section of the **aorta** continues on and then divides to form the **two internal iliac arteries** and the **caudal artery.** There is no common iliac artery in cats as there is in humans. In cats, the caudal artery takes blood to the tail. The internal iliac artery takes blood to the urinary bladder, rectum, external genital organs, uterus, and some thigh muscles. As the external iliac artery exits from the pelvic region and enters the thigh, it becomes the **femoral artery,** which supplies blood to the thigh, leg, and foot. The femoral artery becomes the **popliteal artery** behind the knee, which further divides to form the **tibial arteries** (figures 31.5 and 31.8).

Clean Up

When you are done, place your cat back in the plastic bag. Remember to place all excess tissue in the appropriate waste container and not in a standard classroom wastebasket or down the sink!

Exercise 31 Review

Arteries of the Lower Body

Name: _____
Lab time/section: _____
Date: _____

Match the organs or blood vessels in the left column with the vessels in the right column that take blood to them. An answer may be used more than once.

1. _____ Kidney

2. _____ Ovaries or testes   a. superior mesenteric artery

3. _____ Small intestine   b. inferior mesenteric artery

4. _____ Distal large intestine   c. renal artery

5. _____ External iliac artery   d. gonadal artery

6. _____ Splenic artery   e. common iliac artery

7. _____ Left gastric artery   f. celiac trunk

8. What artery takes blood directly to the femoral artery? _____

9. Blood from the popliteal artery comes directly from what artery? _____

10. What is arteriosclerosis? _____

11. In what part of the arterial wall does cholesterol plaque develop? _____

12. What is the vessel that takes blood to the adrenal glands in the cat called? _____

13. How do the lower pelvic arteries in humans differ from those in cats? _____

14. Label the following illustration using the terms provided.

celiac trunk
tibial artery
internal iliac artery
renal artery
gonadal artery
popliteal artery
femoral artery
common iliac artery
inferior mesenteric artery

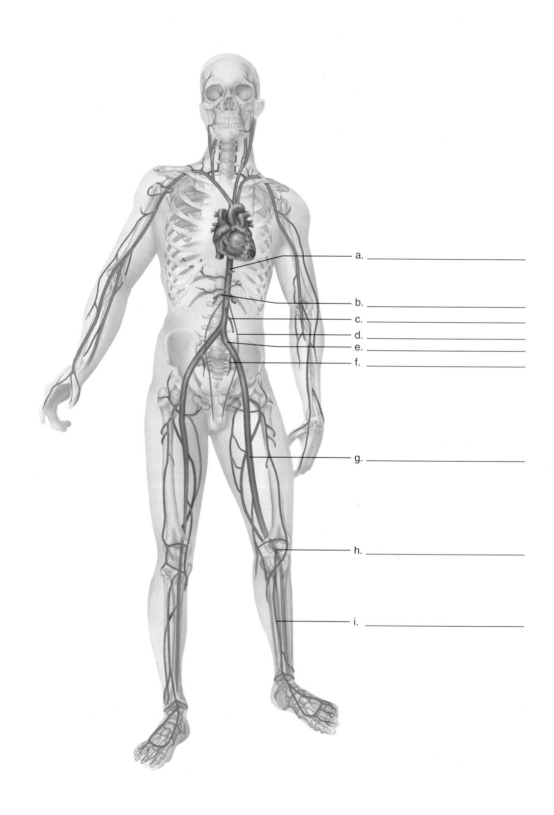

a. _____
b. _____
c. _____
d. _____
e. _____
f. _____
g. _____
h. _____
i. _____

Exercise 32
Veins and Special Circulations

INTRODUCTION

Veins are blood vessels that carry blood toward the heart. They can be superficial (underneath the skin) or deep. The deep veins of the body often travel alongside the major arteries and take on the arterial names, such as the femoral vein, brachial vein, and subclavian vein. The superficial veins have names of their own, such as the cephalic vein, basilic vein, and great saphenous vein. Veins have thinner walls than arteries and contain valves, which maintain a one-way flow of blood to the heart. They are covered in *Principles of Anatomy and Physiology* in chapter 18, "Blood Vessels and Circulation."

A description of the differences between artery walls and vein walls was presented in Exercise 30. In this exercise, you begin the study of veins by examining charts or models of the entire venous system and then study the veins of specific regions of the human body and the cat.

OBJECTIVES

At the end of this exercise, you should be able to

1. describe the sequence of major veins that receive blood from the upper limbs;
2. describe the flow of blood through the veins of the lower limbs;
3. trace the blood flow from the brain to the heart;
4. distinguish a portal system from normal venous return flow;
5. describe the major digestive organs that supply blood to the hepatic portal system;
6. identify major veins in models, a cadaver, or a cat;
7. describe fetal circulation.

MATERIALS

Models of the blood vessels of the body
Charts and illustrations of venous system
Cadaver (if available)
Cats
Materials for cat dissection
    Dissection trays
    Scalpel and two or three extra blades
    Gloves (household latex gloves work well for repeated use)
    Blunt (mall) probe
    Pins
    Forceps and sharp scissors
    First aid kit in lab or prep area
    Sharps container
    Animal waste disposal container

PROCEDURE

Examine the models, charts, and illustrations in the lab and locate the major veins of the body. Read the following descriptions of the veins and find them as they are represented in lab. As you locate a specific vein, name the vessel that takes blood to the vein and those that receive blood from the vein. Veins in this exercise are studied in the direction of their flow from the cells of the body to the heart. In terms of their flow, veins resemble tributaries of rivers, as smaller veins flow into larger veins.

An overview of the major veins of the body is shown in figure 32.1. Compare that figure with the following list of some of the major veins of the body.

    _____ Internal jugular vein
    _____ External jugular vein
    _____ Brachiocephalic vein
    _____ Superior vena cava
    _____ Axillary vein
    _____ Cephalic vein
    _____ Basilic vein
    _____ Inferior vena cava
    _____ Common iliac vein
    _____ External iliac vein
    _____ Internal iliac vein
    _____ Femoral vein
    _____ Great saphenous vein
    _____ Tibial veins

Veins of the Head and Neck

Examine the models and charts in the lab and locate the various veins that drain blood from the head. The drainage of the brain occurs as veins take blood from the brain and pass through the subarachnoid membrane to the venous sinuses in the subdural spaces. To review these structures refer to Exercise 17. The blood from the brain flows into the **internal jugular veins.**

Another vessel that takes blood from the brain is the **vertebral vein,** which, like the vertebral arteries, travels through the transverse foramina of the cervical vertebrae. The vertebral veins take blood to the **subclavian veins,** which in turn flow to the **brachiocephalic veins.** The veins of the head are illustrated in figures 32.2 and 32.3.

Each internal jugular vein passes through a jugular foramen of the skull and then passes along the lateral aspect of the neck as it moves toward the brachiocephalic vein. The brachiocephalic vein is formed by the union of the internal jugular vein and the subclavian vein. The superficial regions of the posterior head (musculature and

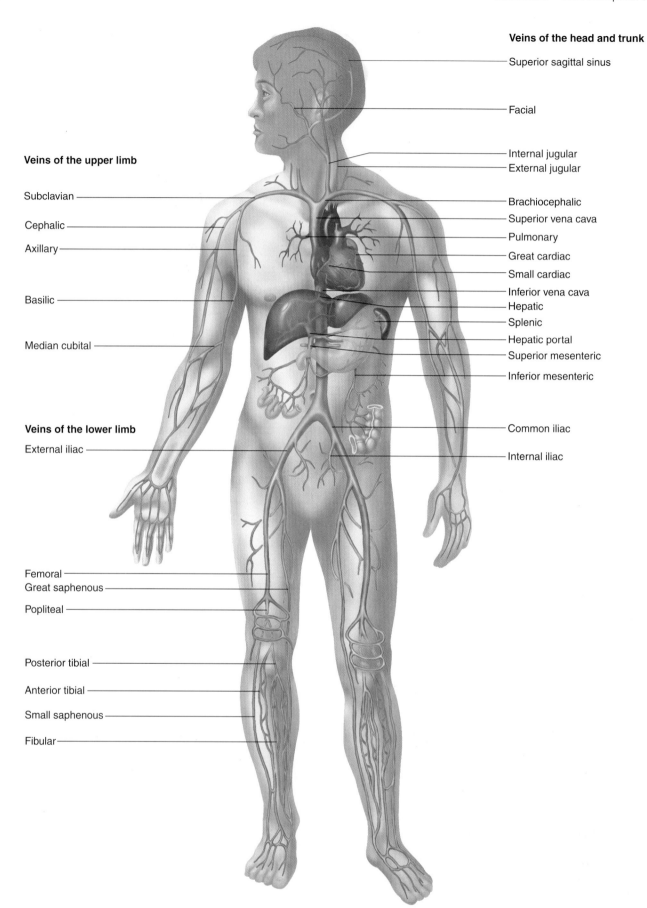

Figure 32.1 Major Systemic Veins

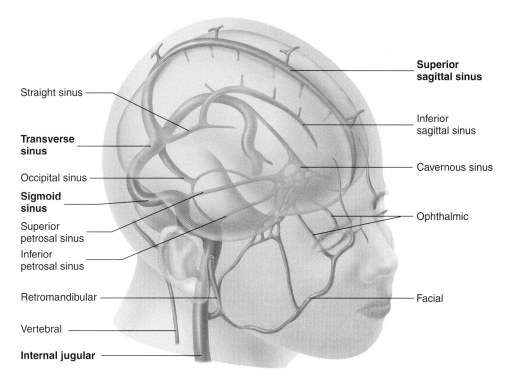

Figure 32.2 Venous Drainage of the Brain

skin of the scalp and face) are drained by the **external jugular vein.** The external jugular veins join with the subclavian veins before reaching the brachiocephalic veins. The left and right brachiocephalic veins take blood to the **superior vena cava.** Locate these major vessels of the head and neck in figure 32.3.

Veins of the Upper Limbs

Examine the models and charts in the lab and locate the veins of the upper limbs. The fingers are drained by the small **digital veins,** which lead to the **palmar arch veins.** The major superficial veins of each upper limb are the **basilic veins,** which are on the anterior, medial side of the forearms and arms, and the **cephalic veins,** which are on the anterior, lateral side of the forearms and arms. The two vessels have many anastomosing branches (cross-connections) between them. One of the significant anastomosing veins is the **median cubital vein,** which crosses the anterior cubital fossa and is a common site for the withdrawal of blood. Locate these superficial veins in figure 32.4.

The deep veins of the forearm are the **radial vein** and the **ulnar vein,** each of which can be found traveling near the artery of the same name. The deep veins of the arm are the **brachial veins,** which are next to the brachial artery in the proximal portion of the arm. The brachial veins are formed by the union of the radial and ulnar veins and merge superiorly with the **basilic vein** to form the short **axillary vein.** The axillary vein connects with the cephalic vein to form the subclavian vein. The drainage of the upper limbs is carried to the heart by the left and right subclavian veins, which take the blood via the brachiocephalic veins to the superior vena cava and finally to the right atrium of the heart. Locate the deep veins of the upper limb in figure 32.4.

Veins of the Lower Limb

The blood vessels that drain the lower limbs operate under relatively low pressure and must take blood back to the heart against gravity. The anterior and posterior tibial veins along with the fibular vein and the small saphenous vein receive blood from the foot and leg and eventually take blood to the femoral vein.

The longest vessel in the human body is the **great saphenous vein,** which can be found just underneath the skin on the medial aspect of the lower limb beginning near the medial malleolus and traversing the lower limb to the proximal thigh. This vessel often is embedded in adipose tissue below the skin in humans, yet it is considered a superficial vein. Two other vessels in the thigh are the **femoral vein** and the **deep femoral vein.** The femoral vein travels alongside the femoral artery. As the great saphenous vein reaches the inguinal region, it joins with the femoral vein, which takes blood from the thigh region and enters the abdominopelvic cavity. The femoral vein then becomes the external iliac vein. Examine these vessels in the lab and compare them with figure 32.5.

Veins of the Pelvis

The **external iliac vein** joins with another vein, the **internal iliac vein,** which drains the region of the pelvis, and they form the **common iliac vein.** The common iliac veins unite and form the **inferior vena cava,** which travels superiorly along the right side of the vertebrae, taking blood to the right atrium of the heart. Locate these veins and compare them with figure 32.5.

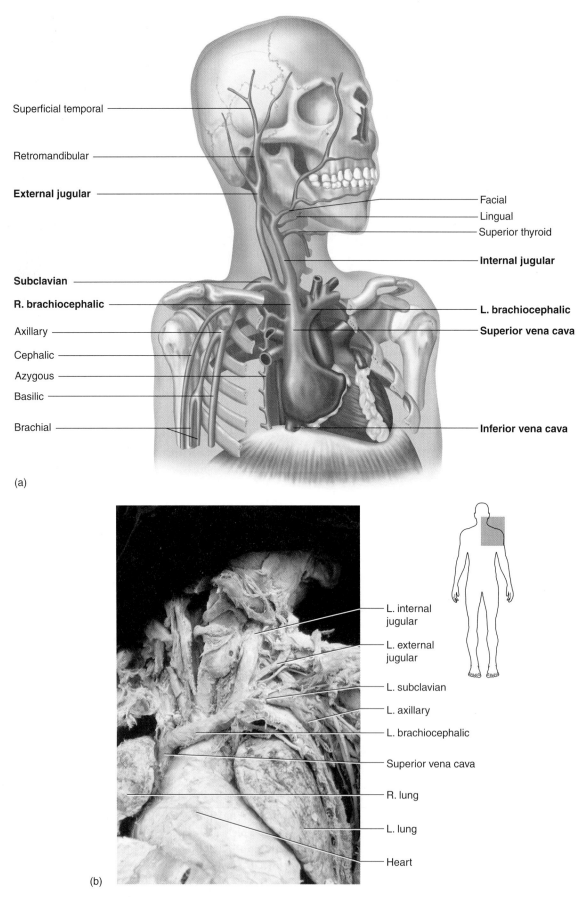

Figure 32.3 Veins of the Head and Neck

(a) Diagram; (b) photograph of cadaver.

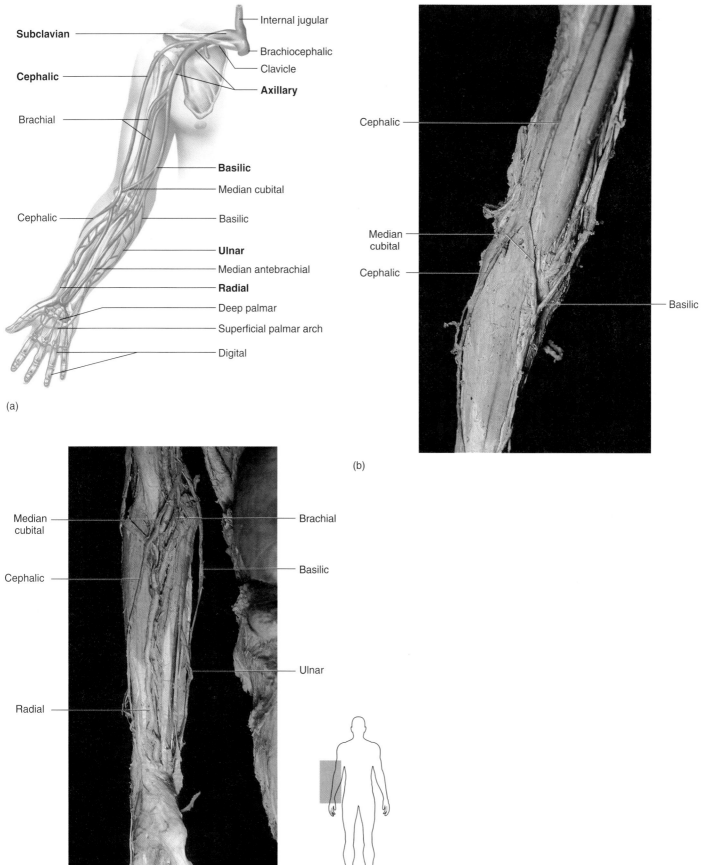

Figure 32.4 Veins of the Upper Limb
(a) Diagram; (b) photograph of superficial veins; (c) photograph of deep veins.

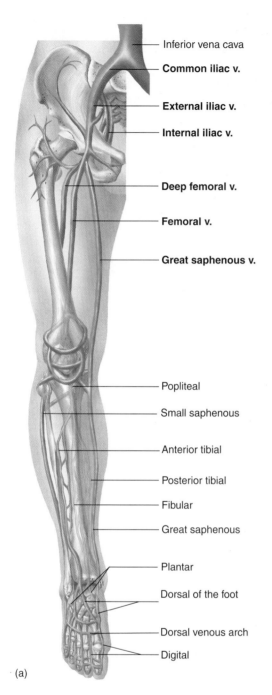
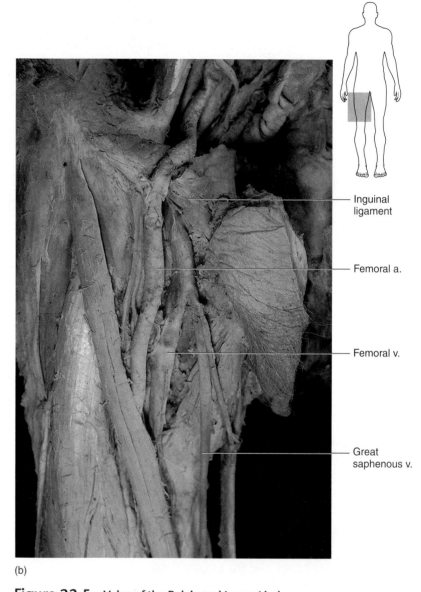

Figure 32.5 Veins of the Pelvis and Lower Limb
(a) Diagram; (b) photograph of the femoral region; (c) photograph of the pelvic region.
v. = vein; a. = artery

Veins of the Trunk

Veins of the abdominal region drain the major organs of the digestive tract and other abdominal organs, such as the spleen. Many of these veins are named for the organs from which they receive blood (for example, the splenic vein takes blood from the spleen). The abdominal veins are a unique group of blood vessels, some of which constitute the hepatic portal system. The portal system pattern is different from the normal venous blood flow. Generally, blood from an organ begins in the capillary bed of that organ and then travels through venules to veins to the heart.

In a portal system, a series of vessels takes blood from the capillary beds of an organ through a series of veins and then to another capillary bed. In the case of the **hepatic portal system,** the blood flows from the capillary beds of the abdominal organs through numerous veins to the capillary bed of the liver. The liver receives the blood from the digestive organs and processes it before sending it through the hepatic vein to the inferior vena cava and toward the heart. Examine figure 32.6 for the main vessels of the hepatic portal system and identify the following veins:

_____ Inferior mesenteric vein
_____ Superior mesenteric vein
_____ Splenic vein
_____ Gastroomental vein
_____ Hepatic portal vein

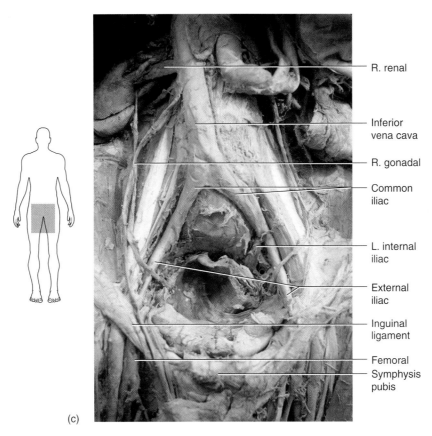

Figure 32.5—*Continued.*

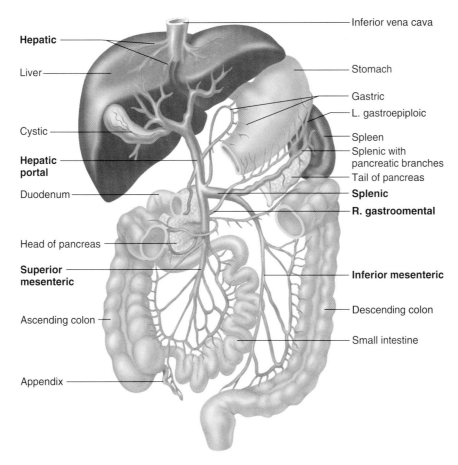

Figure 32.6 Hepatic Portal System

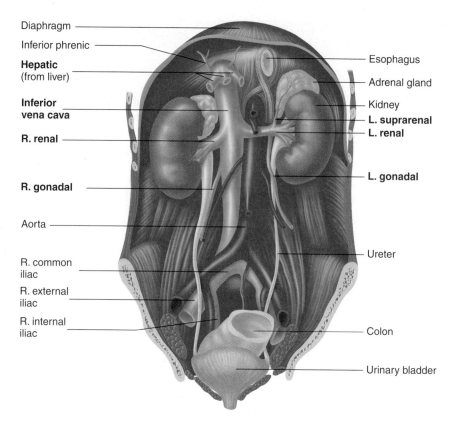

Figure 32.7 Inferior Veins of the Abdomen

Veins of the Abdomen and Pelvis

Some abdominal veins take blood directly to the inferior vena cava, and some pass through the liver before reaching the inferior vena cava. Those that take blood directly to the inferior vena cava are the **renal veins,** the **suprarenal veins,** and the **lumbar veins.** The **right gonadal vein** takes blood directly to the inferior vena cava, but the **left gonadal vein** takes blood to the renal vein before flowing into the inferior vena cava. These vessels can be seen in figures 32.7 and 32.8.

Numerous **intercostal veins** drain the intercostal muscles, and the **azygos vein** and **hemiazygos vein** drain blood from the thoracic region. These are seen in figure 32.8.

Fetal Circulation

The pathway of human fetal blood is somewhat different from that of the adult in that the lungs are nonfunctional in the fetus. Oxygen and nutrients move from the maternal side of the **placenta** to the fetal bloodstream, whereas carbon dioxide and metabolic wastes move from the fetal bloodstream to the placenta. From the placenta, the blood flows through the **umbilical vein,** which is located in the umbilical cord. The blood from the umbilical vein travels through the **ductus venosus,** which is a shunt to the inferior vena cava of the fetus. The maternal blood, which is relatively high in oxygen and nutrients, mixes with the deoxygenated, nutrient-poor fetal blood, and thus the fetus receives a mixture of blood. Examine figure 32.9 for an overview of fetal circulation.

The blood from the inferior vena cava travels to the right atrium of the heart. While in the right atrium, the blood can travel either to the right ventricle (which pumps blood to the lungs) or through a hole in the right atrium called the **foramen ovale,** which leads to the left atrium. Since the lungs do not oxygenate blood in the fetus, the foramen ovale is a bypass route away from the lungs and to the chambers of the heart that will pump blood to the body. Blood in the right ventricle is pumped to the pulmonary trunk, where another shunt vessel, the **ductus arteriosus,** carries blood to the aortic arch, bypassing the lungs. The lungs do receive some blood, but it is for the nourishment of the lung tissue, as opposed to gas exchange. Locate the structures of the fetal heart in figure 32.9.

Blood from the heart exits the left ventricle and passes through the aorta to the systemic arteries. Blood travels down the internal iliac arteries to the lower limbs, and some moves into the **umbilical arteries,** which carry blood to the placenta. At birth, the pressure changes in the newborn's heart cause the closing of a flap of tissue over the foramen ovale, leaving a thin spot in the interatrial septum known as the **fossa ovalis.** Lack of closure of this foramen can lead to a condition known as "blue baby."

Cat Dissection

Bring a cat and dissection equipment back to your lab table. Begin your dissection by reviewing the arterial system previously studied in Exercises 30 and 31. As stated in previous exercises, the blood

Exercise 32 Veins and Special Circulations

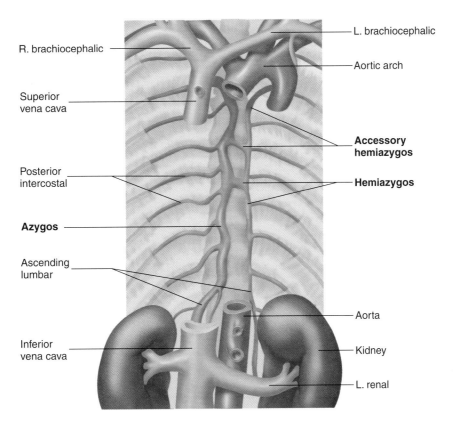

Figure 32.8 Veins of the Thorax

vessels in the cat have been injected with colored latex. If the cat has been doubly injected, the arteries are red and the veins are blue. If the cat has been triply injected, the hepatic portal vein is also injected, typically with yellow latex.

Veins of the Upper Limb

Begin your dissection by examining the veins of the upper limb. The basilic vein is not found in cats, but you should be able to find the **ulnar vein** and note that it joins the **radial vein** to form the **brachial vein**. The brachial vein is next to the brachial artery. Locate the **median cubital vein** and the **cephalic vein**. The **axillary vein** and **subscapular vein** (from the shoulder) join to form the **subclavian vein**, which takes blood to the **brachiocephalic vein**. The cephalic vein continues on into the **transverse scapular vein**, which takes blood to the **external jugular vein**. Locate these veins in figure 32.10.

Veins of the Head and Neck

The veins in this region of the cat have a few variations from those in the human. In the cat, the external jugular vein is larger than the **internal jugular vein**. The reverse is true in humans. What anatomical difference between the cat and human might explain the difference in the volume of blood carried by these two vessels?

The **transverse jugular vein,** which is present in cats but absent in humans, connects the two external jugular veins. The jugular veins, along with the **costocervical veins** and the subclavian veins, unite to form the brachiocephalic veins (figure 32.10). The **vertebral veins** take blood from the head of the cat and empty into the subclavian veins.

Veins of the Lower Limb

Locate the superficial **great saphenous vein** in the cat on the medial side of the lower limb. The great saphenous vein joins the **femoral vein** just above the knee. Note the long **tibial veins** of the leg that join and form the **popliteal veins.** Find the femoral vein, which is found with the femoral artery. The vessels of the distal part of the lower limb take blood to the femoral vein, which turns into the **external iliac vein** as it passes into the body cavity at the level of the inguinal ligament. Find these veins in the cat and compare them with figure 32.11.

Veins of the Pelvis

The external iliac vein and the **internal iliac vein** join to form the **common iliac vein.** The two common iliac veins lead to the **inferior vena cava** along with the **caudal vein,** which takes blood from the tail. Be careful as you dissect these structures in the cat. There are many ducts that cross the external iliac veins, such as the ureter and, in males, the ductus deferens. Look for these structures in figure 32.11, and do not cut them. You will study them later.

1. Blood bypasses the lungs by flowing from the pulmonary trunk through the ductus arteriosus to the aorta.

2. Blood also bypasses the lungs by flowing from the right to the left atrium through the foramen ovale.

3. Oxygen-rich blood is returned to the fetus from the placenta by the umbilical vein.

4. Blood bypasses the liver sinusoids by flowing through the ductus venosus.

5. Oxygen-poor blood is carried from the fetus to the placenta through the umbilical arteries.

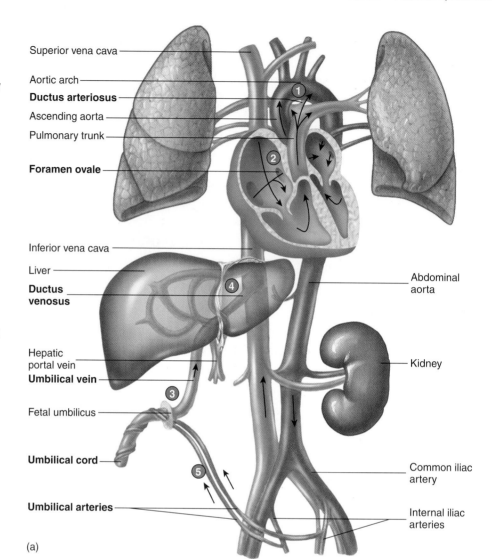

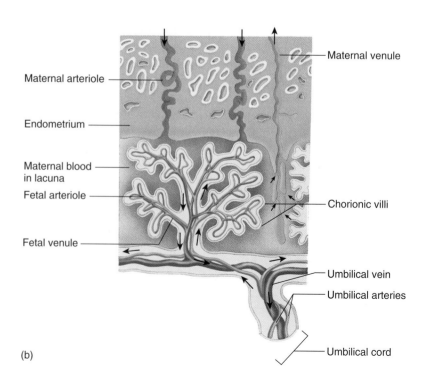

Figure 32.9 **Fetal Circulation in the Human**
(a) Overview of circulation; (b) details of the placenta.

Exercise 32 Veins and Special Circulations

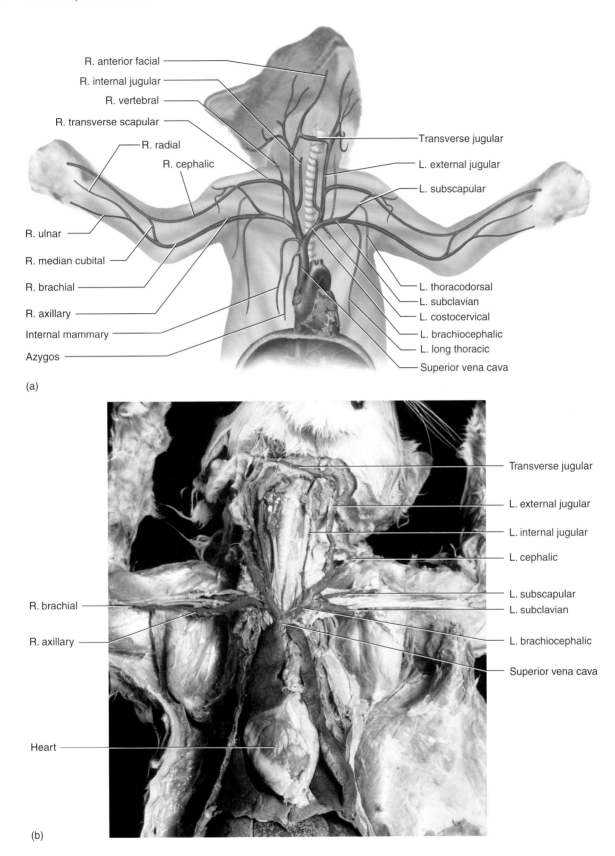

Figure 32.10 Anterior Veins of the Cat
(a) Diagram; (b) photograph.

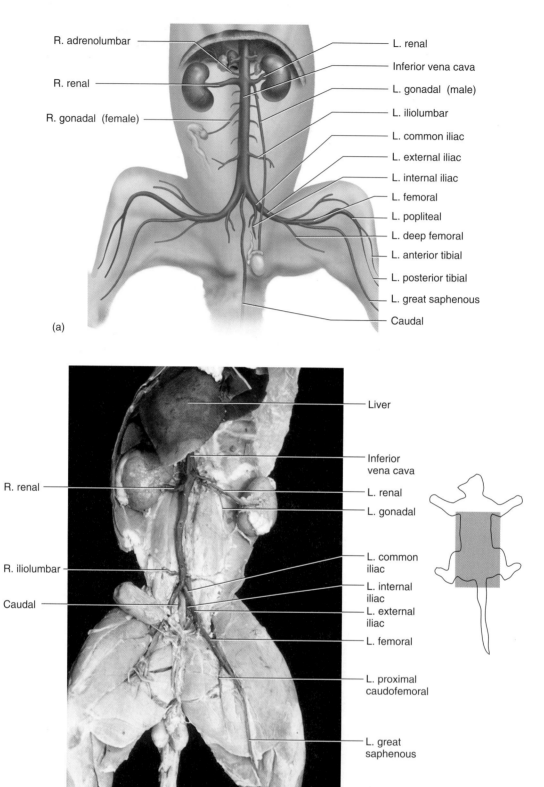

Figure 32.11 Posterior Veins of the Cat
(a) Diagram; (b) photograph.

Abdominal Veins

Identify the major organs in the digestive tract before you proceed with the study of the veins of this region. Pay particular attention to the stomach, liver, spleen, kidneys, adrenal glands, small intestine, and large intestine. The inferior vena cava is a large vessel that travels up the right side of the body in the cat. You may have to gently move some of the digestive organs to the side to find the inferior vena cava. Examine where the common iliac veins join to form the inferior vena cava. The caudal vein (absent in humans) joins the inferior vena cava at this point. The gonadal veins receive blood from the testes or ovaries. The right gonadal vein takes blood to the inferior vena cava, and the left gonadal vein takes blood to the left renal vein, which subsequently leads to the inferior vena cava. The paired **renal veins** receive blood from each kidney and empty into the inferior vena cava. The **iliolumbar veins** take blood from the body wall and return it via the inferior vena cava. The body wall is also drained by the **adrenolumbar veins,** which also receive blood from the adrenal glands. Compare the veins in your cat with figure 32.11.

Hepatic Portal System

The veins that flow into the liver before returning to the heart belong to the **hepatic portal system.** These vessels take blood from a number of abdominal organs and transfer it to the liver where metabolic processing occurs. As you examine the lower portion of the digestive tract (transverse colon and descending colon), locate the **inferior mesenteric vein.** The **superior mesenteric vein** receives blood from the small intestines and part of the large intestines, as well as from the inferior mesenteric vein. The **gastrosplenic vein** joins the superior mesenteric vein and takes blood to the **hepatic portal vein** along with other digestive veins. The gastrosplenic vein receives blood from the spleen and the stomach. The hepatic portal vein thus receives blood from the spleen and the digestive organs and transports that blood to the liver. After the blood reaches the liver, it is transferred to the inferior vena cava by the short **hepatic vein.** This vein is usually difficult to dissect, since it is at the dorsal aspect of the liver and joins the inferior vena cava at about the junction of the diaphragm. Examine the hepatic portal system of the cat while trying to maintain the integrity of the digestive organs. Compare your dissection with figure 32.12.

Thoracic Veins

Anterior to the diaphragm are the veins of the thorax. The internal mammary vein (sternal vein) may be difficult to locate, since it is frequently cut while opening the thoracic cavity. The internal mammary vein receives blood from the ventral body wall. Another vein of the thoracic region is the **azygos vein.** This vessel, located on the right side of the body, takes blood from the esophageal, intercostal, and bronchial veins. Compare the veins in your cat with figure 32.10.

Clean Up

When you are done, place your cat back in the plastic bag. Remember to place all excess tissue in the appropriate waste container and not in a standard classroom wastebasket or down the sink!

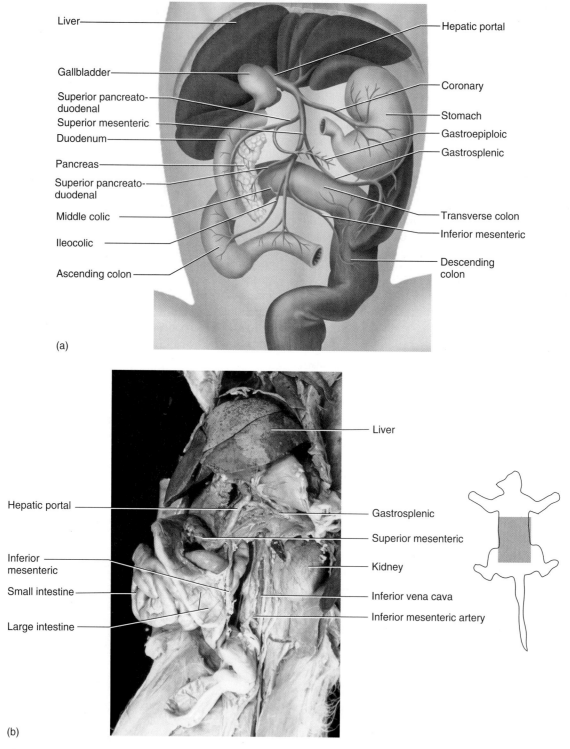

Figure 32.12 Hepatic Portal System of the Cat
(a) Diagram; (b) photograph.

Exercise 32 Review

Veins and Special Circulations

Name: _____

Lab time/section: _____

Date: _____

1. Label the following illustration using the terms provided.

   brachial vein
   brachiocephalic vein
   basilic vein
   superior vena cava
   great saphenous vein
   cephalic vein
   femoral vein
   internal iliac vein
   internal jugular vein
   inferior vena cava
   common iliac vein

   a. _____
   b. _____
   c. _____
   d. _____
   e. _____
   f. _____
   g. _____
   h. _____
   i. _____
   j. _____
   k. _____

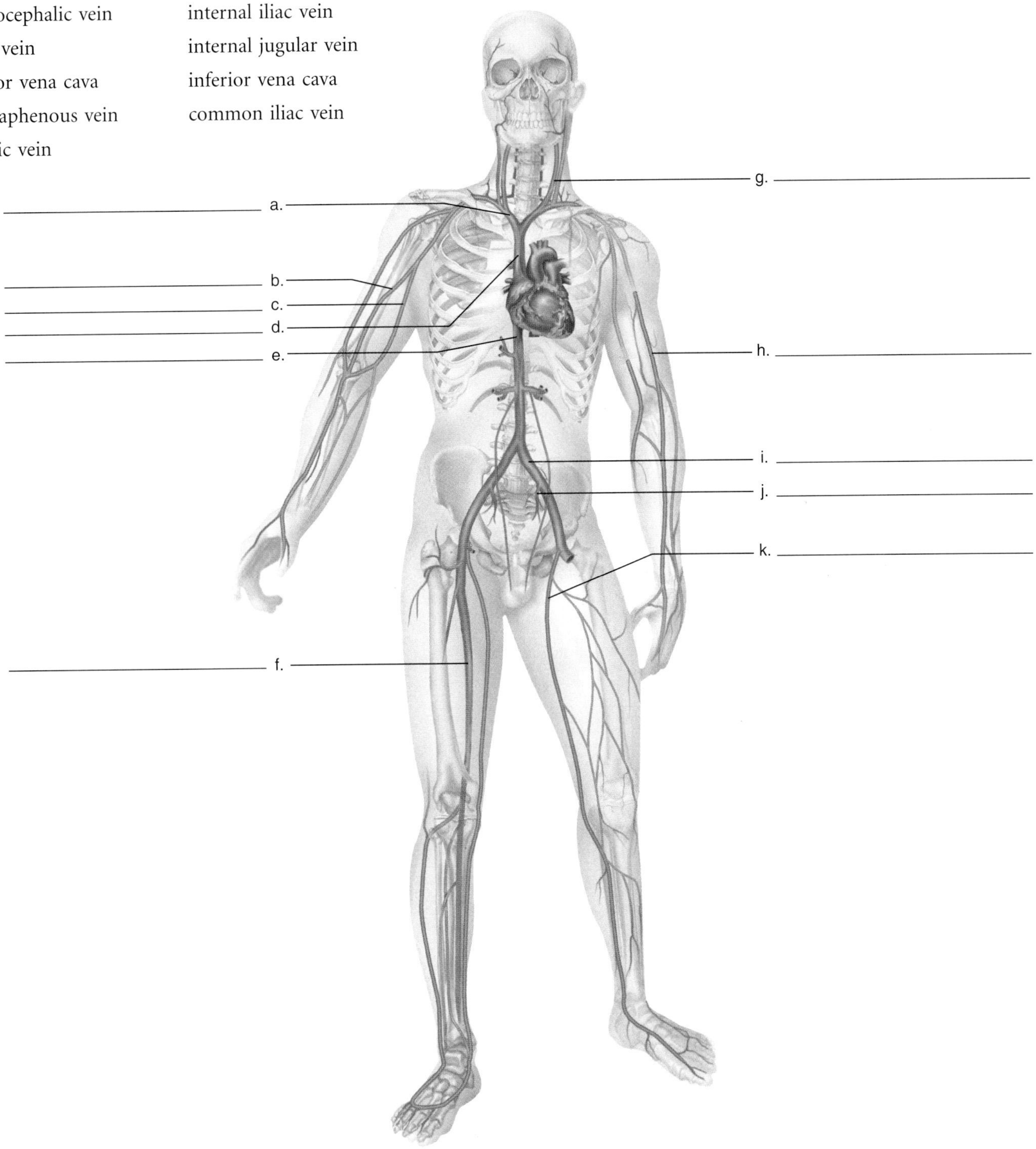

Veins of the Human

417

2. Which veins (superficial/deep) have names that do not correlate to arteries? _____
_____
_____

3. The internal jugular vein takes blood from what area? _____
_____

4. What veins pass through the transverse vertebral foramina? _____
_____

5. What area do the right and left external jugular veins drain? _____
_____

6. The brachiocephalic veins take blood to what vessel? _____

7. Is the radial vein a superficial or deep vein? _____

8. Where is the median cubital vein found? _____

9. What vessel receives blood from the ulnar vein? _____

10. What region of the body houses the cephalic vein? _____

11. Blood from the right axillary vein next travels to what vessel? _____

12. What vessels take blood to the left femoral vein? _____
_____

13. The great saphenous vein is in what region of the body? _____

14. Where does blood flow after it leaves the femoral vein? _____

15. The common iliac vein receives blood from two vessels. What are these two vessels? _____
_____

16. What is the functional nature of a "portal system," and how does it differ from normal venous return flow? _____
_____
_____
_____

Exercise 32   Veins and Special Circulations

17. What major vessels take blood to the hepatic portal vein? _____

18. Blood in the small intestine travels to the hepatic portal vein by what vessel? _____

19. Name five unique ways in which fetal blood flow differs from adult blood flow and relate them to their functions.

Exercise 33
Functions of Vessels and the Lymphatic System

INTRODUCTION

Blood vessels and the lymphatic system are intimately associated in the circulatory pattern. Fluid in the circulatory route may travel by several different pathways. As described in Exercise 30, blood is pumped from the **heart** through **arteries** and distributed through the **arterioles** to the **capillaries,** where nutrients, water, and oxygen are exchanged with the cells of the body. The blood returns via **venules** to **veins** and back to the heart under relatively low pressure. The valves in veins keep blood flowing in the direction of the heart. Their role will be examined in this exercise.

The arteries are thicker than veins, which reflects the greater amount of pressure to which they are subjected. Arteries and veins transport blood cells and plasma. Blood cells and plasma protein typically stay in the vessels, yet a significant amount of fluid from the plasma leaks from the capillaries and bathes the cells of the body. This fluid flows among the cells of the body and is known as **interstitial fluid.** It provides nutrients to the cells and receives dissolved wastes from the cells, along with cellular debris. Most of the interstitial fluid returns to the capillaries, but some is picked up by **lymph capillaries,** which return the fluid, now known as **lymph,** to the **lymphatic vessels.** Lymphatic vessels take the lymph back to the venous system, returning it to the cardiovascular system. This process is illustrated in figure 33.1.

In addition to the functions just described, the lymphatic system is also instrumental in absorbing lipids from the digestive system. Lipids in the cells that line the small intestine are converted to **chylomicrons** (phospholipids and other molecules). These chylomicrons are conducted into the lymphatic system, which then transports the material through the thoracic duct and into the subclavian vein, where it enters the cardiovascular system.

As the fluid flows through the lymphatic vessels, lymph nodes clean the lymph of cellular debris and foreign material (such as bacteria and viruses) that may have entered the lymphatic system. These topics are covered in *Principles of Anatomy and Physiology* in chapter 18, "Blood Vessels and Circulation," and in chapter 19, "Lymphatic System and Immunity." In this exercise, you examine the nature of the lymphatic system.

OBJECTIVES

At the end of this exercise, you should be able to

1. describe the function of venous valves;
2. identify the valves in lymph vessels;
3. locate the major histological features of a lymph node;
4. demonstrate on a model the location of the tonsils and other lymph organs;
5. describe the functions of the spleen and thymus.

MATERIALS

Live frog
Dissection scopes
Paper towels
Small squeeze bottle of water
Microscopes
Microscope slides of lymphatic vessels with valves
Charts, diagrams, and models of blood vessels
Torso models
Cadaver (if available)

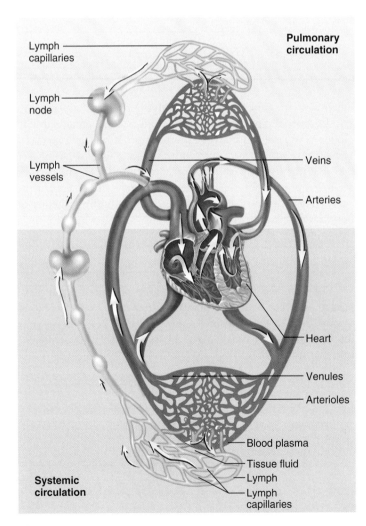

Figure 33.1 Lymph Flow and Its Relationship to the Cardiovascular System

Interstitial fluid is returned to the vascular system by lymphatic vessels and valves.

PROCEDURE

Movement Through Capillaries

Blood capillaries have a relatively simple structure in that they are composed of endothelium. The **endothelium** consists of a single layer of **simple squamous epithelium** that forms a tube slightly larger than the erythrocytes that pass through it. You can observe capillaries in living tissue by examining the blood flow through the webbed foot of a live frog. Obtain a living frog and wrap its body in a wet paper towel. Leave the head and the foot exposed. *Do not let the frog dry out while you are examining it.* Hold the foot of the frog under a dissecting microscope and fan out the webbing of the foot. Make sure to hold the frog securely. Observe the flow of the blood through the capillaries, keeping the foot moist with water. Describe the flow of the blood through the capillaries of the foot in terms of speed and in terms of the size of the capillary relative to the diameter of an erythrocyte.

Your description: _____

_____

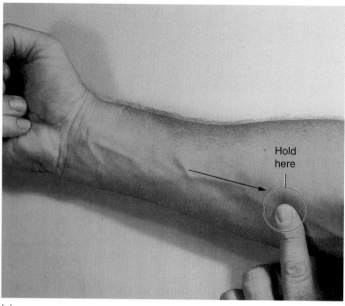

(a)

Demonstration of Valves in Veins

One way to examine the nature of **valves** in veins is to place your arm in anatomic position until the veins become engorged with blood. As you hold your arm in this position, stroke the superficial veins from distal to proximal with the index finger of your free hand, applying uniform and constant pressure (figure 33.2a). Maintain pressure on the vein at its proximal end and see if the vein refills with blood. Record your results, indicating whether or not the vein fills with blood.

Results: _____

_____

Try the experiment again, but keep pressure on the distal part of the vein with your index finger as you push the blood toward the heart with your thumb (figure 33.2b). Release your thumb from the proximal area of the vein and see if blood fills the vein. Record your results.

Results: _____

_____

Do the valves prevent the blood from flowing in a proximal

or distal direction? _____

_____

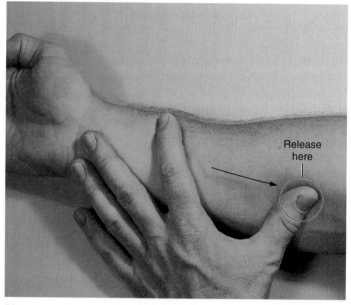

(b)

Figure 33.2 Testing for Valves in Veins
(a) Stroke the vein and keep pressure on the vein; (b) press on the distal part of the vein, stroke proximally, and then release.

Lymphatic System

Lymph originates as interstitial fluid that bathes the cells of the body. This fluid originates from blood capillaries and carries oxygen, nutrients, and other dissolved materials to the cells. The fluid enters the lymph capillaries by way of small, valvelike slits in the lymph capillary wall. These slits act as one-way valves, preventing lymph from returning to the interstices (spaces) between the cells. This process is illustrated in figure 33.3.

The lymphatic system is difficult to study in preserved specimens because it collapses at death. The lymphatic system is composed

Exercise 33 Functions of Vessels and the Lymphatic System

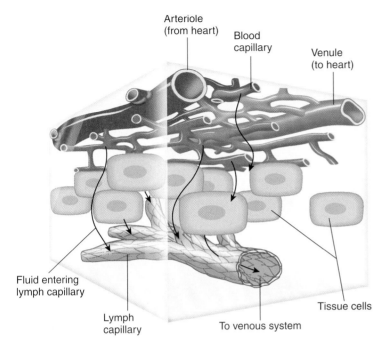

Figure 33.3 Fluid Movement from Blood Capillaries to Interstitial Fluid to Lymph Capillaries

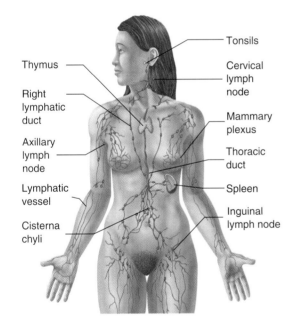

Figure 33.4 Overview of the Lymph System

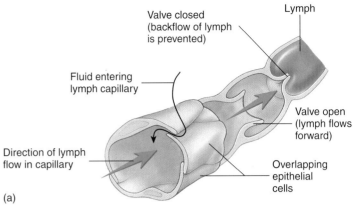

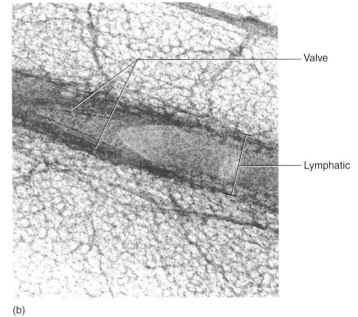

Figure 33.5 Lymphatic Vessel with Valve
(a) Diagram; (b) photomicrograph (100×).

of **lymph capillaries, lymphatic vessels, lymph nodes, lymph organs,** and **lymph tissue.** An overview of the system is illustrated in figure 33.4. Examine charts and models in lab and compare them with this figure.

Lymphatic Vessels

Once the lymph is in the lymph capillaries, it travels through the **lymphatic vessels (lymphatics).** As with the lymph capillaries, a one-way flow occurs in the lymphatics due to the presence of valves. Examine a prepared slide of a lymphatic and note the presence of the **valve.** Compare your slide with figure 33.5.

Use figure 33.4 to examine the drainage pattern of lymph in the body. The **thoracic duct** receives lymph from most of the body, taking lymph to the left subclavian vein where the fluid is returned to the cardiovascular system. Identify the **cisterna chyli,** which is an enlarged portion of the thoracic duct in the abdominal region. The **right lymphatic duct** receives lymph from the right side of the head and neck, the right thoracic region, and the right upper extremity. The right lymphatic duct returns lymph to the right subclavian vein.

Note that the lymphatic vessels lead to regions of the body where lymph nodes are found. Nodes are clustered in the groin (inguinal region), axilla, antecubital fossa, popliteal region, neck, thorax, and abdomen.

Lymph Nodes

Look at a slide of a **lymph node.** The lymph node is enclosed by a sheath of tissue called the **capsule.** Locate the dark purple **lymph nodules** in the lymph node. The cortex is the outer region of the

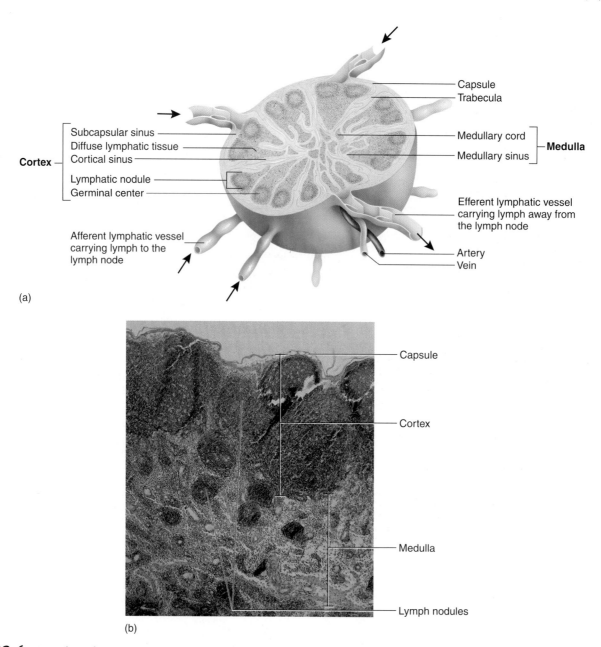

Figure 33.6 Lymph Node
(a) Diagram showing lymph flow by arrows; (b) photomicrograph (40×).

node that has numerous lymph nodules. In the medulla, medullary cords and sinuses cleanse the lymph of foreign particles and cellular debris. **Afferent lymphatic vessels** take lymph to the node, and **efferent lymphatic vessels** take lymph away from the node. Compare your slide with figure 33.6.

Lymph Organs

Examine models and charts in the lab and compare the lymph organs with those represented in figure 33.4. The major lymph organs and tissue of the lymph system consist of the tonsils, thymus, spleen, lymph nodes along the lymphatics, and Peyer's patches in the intestine.

Tonsils are lymph organs in the oral cavity and nasopharynx. Look at illustrations in your text and models in the lab and locate the tonsils. These are the **palatine tonsils,** located along the sides of the oral cavity near the oropharynx, the **lingual tonsils** at the posterior portion of the tongue, and the singular **pharyngeal tonsil,** located in the nasopharynx. **Adenoids** are enlarged pharyngeal tonsils. Compare these with figure 33.7. The tonsils have **crypts** lined with **lymph nodules,** which are the first line of defense against foreign matter that is inhaled or swallowed.

The **spleen** is an important lymph organ that filters blood, removing aging erythrocytes and foreign particles. The spleen is

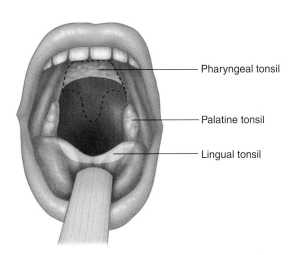

Figure 33.7 Tonsils

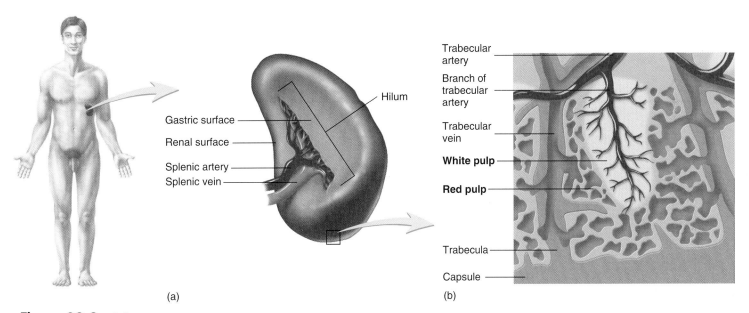

Figure 33.8 Spleen
(a) Whole organ; (b) section.

located on the left hypochondriac region lateral to the stomach. It is a highly vascular organ that filters blood and produces lymphocytes. The spleen contains **red pulp,** which filters blood, and **white pulp,** which is involved in producing lymphocytes. Locate the spleen in models and charts in the lab, or in cadaver specimens, and compare it with figures 33.4 and 33.8.

The **thymus** overlays the vessels superior to the heart. It is an important site in determining **immune competence.** Lymphocytes travel from the bone marrow to the thymus, where they undergo the maturation essential to immune responses. Once these cells mature, they are known as **T cells.** Examine the charts and models, or cadaver specimens in the lab, and compare them with figure 33.4. You will not see a thymus on adult models because it decreases in size with age.

Cat Dissection

Examine small lumps of tissue in the axilla. These are the lymph nodes of the region. Use figure 33.9 as a guide to find other lymph nodes. Note the large **spleen,** which is an approximately 6-inch elongated organ on the left side of the abdominal cavity, and the **thymus,** which is anterior to the heart.

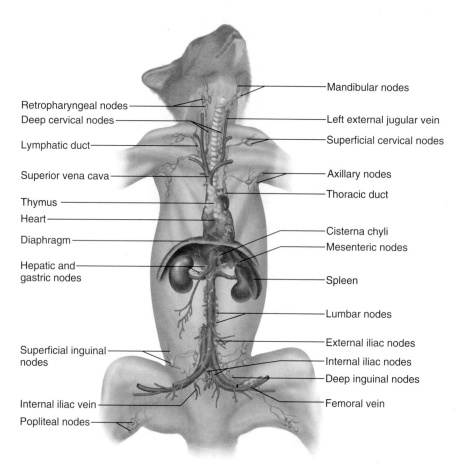

Figure 33.9 Lymphatic System of the Cat

Exercise 33 Review

Functions of Vessels and the Lymphatic System

Name: _____

Lab time/section: _____

Date: _____

1. What cell type makes up the endothelium of capillaries? _____

2. What is the name of the vessels that carry lymph from the lymph capillaries to the veins? _____

3. Once tissue fluid enters the lymphatic vessels, what is it called? _____

4. What is the name of the inner region of a lymph node? _____

5. What kind of vessel takes lymph away from a lymph node? _____

6. The adenoids are enlarged _____ tonsils.

7. Which tonsils are found on the sides of the oral cavity? _____

8. What tonsils are located at the back of the tongue? _____

9. Blood is filtered by which lymph organ in the adult? _____

10. What part of the spleen is involved in producing lymphocytes? _____

11. Where do T cells mature? _____

12. Lymphatic vessels have a one-way flow from the extremities to the heart. Damage to the lymphatic system can lead to edema, or an increase in tissue fluid. From the standpoint of reducing edema, how does the use of medical leeches (segmented worms that drain tissue fluid) work for a region that has suffered trauma? _____
_____
_____
_____

13. From what you know of the functions of lymph nodes, predict the difference between lymph entering a node and lymph leaving a node. What materials may be missing from the lymph leaving the node? _____
_____
_____
_____

14. Elephantiasis is a disease that is caused, in some cases, by a parasitic worm blocking the lymphatic vessels. Examine the following illustration and predict where the lymphatic vessel blockage occurs. _____

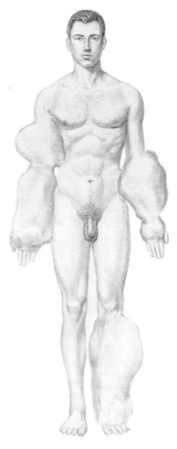

Damage to the Lymphatic System by Elephantiasis

15. In an analysis of breast cancer, lymph nodes of the axillary region are removed and a biopsy is performed. The removal of the nodes is done to determine if cancer has spread from the breast to other regions of the body. What effect would the removal of lymph nodes have on the drainage of the pectoral region? _____

16. Superficial veins contain valves, yet deep veins do not. The deep veins are surrounded by muscles. People who are inactive may have problems with their veins. Can you propose a mechanism by which blood from the deep veins can be returned to the heart (other than by standing on your head)? _____

Exercise 34
Blood Vessels and Blood Pressure

INTRODUCTION

The maintenance of blood pressure is very important for the health of the heart and for the proper functioning of the other organs. **Hypertension** (high blood pressure) increases the workload of the heart by increasing the force with which the heart must pump to provide blood to the body. High blood pressure is also a concern for proper kidney function, as it can damage the filtering membrane of the kidneys. On the other hand, low blood pressure does not provide adequate blood flow to the tissues, including the brain. A decrease in pressure may lead to fainting or dizziness. These topics are covered in *Principles of Anatomy and Physiology* in chapter 18, "Blood Vessels and Circulation."

In this exercise, you learn how to determine blood pressure and some factors affecting blood pressure. You also determine the pulse rate of the heart.

OBJECTIVES

At the end of this exercise, you should be able to

1. determine a person's pulse rate;
2. define *hypertension* in terms of millimeters of mercury pressure;
3. distinguish between systolic pressure and diastolic pressure;
4. properly take and record blood pressure.

MATERIALS

Stethoscope
Alcohol wipes or isopropyl alcohol and sterile cotton swabs
Washable felt pen
Sphygmomanometer
Watch or clock with accuracy in seconds

PROCEDURE

Measurement of Pulse Rate

Pulse rate is important in determining overall fitness and the condition of the heart. You do this by measuring a baseline pulse rate. The pulse rate is measured in beats per minute (bpm), which is done by locating regions of the body where the pulse can be palpated. These areas include the **radial artery, the carotid artery,** and other sites, shown in figure 34.1. If you are recording your lab partner's pulse, make sure you use the tips of your second and third digits and not your thumb. If you use your thumb to measure the pulse rate, you may feel the pulse in your own thumb and not the pulse of the person you wish to record. You can determine beats per minute by counting the pulse for 30 seconds and multiplying it by 2.

Pulse rate of lab partner:  bpm

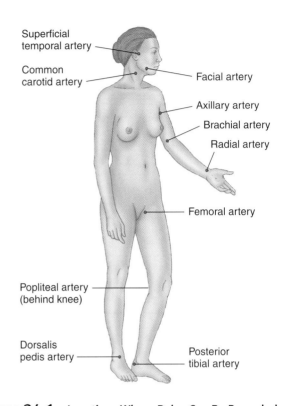

Figure 34.1 Locations Where Pulse Can Be Recorded

Measurement of Blood Pressure

The accuracy of a blood pressure measurement is important, because extremely low blood pressure **(hypotension)** and extremely high blood pressure **(hypertension)** are health risks. Hypotension can lead to dizziness or fainting, and hypertension can lead to cardiovascular disease or strokes. There are many types of **sphygmomanometers** (blood pressure cuffs), which were developed to measure blood pressure indirectly. The original instruments used mercury (Hg) rising in a glass column, and the standard measurement for blood pressure is recorded as pressure in units of millimeters of mercury. There are digital and aneroid (typically using air pressure) models commonly in use. Readings involve examining the level of the mercury column, a needle that moves on a dial, or a digital recording. The greater the pressure, the higher the rise of mercury.

Blood pressure is not uniform throughout the body but is influenced by gravity. Therefore, the pressure in the arteries of the head and neck is less than the blood pressure from the heart. The pressure in the arteries of the leg is greater than the blood pressure in the heart. Blood pressure is measured in the **brachial artery,** which is at the level of the heart and has approximately the same pressure as in the heart. Auscultatory measurement of blood pressure involves listening to sounds as blood passes through the brachial artery. Normally, no sound is heard through a stethoscope

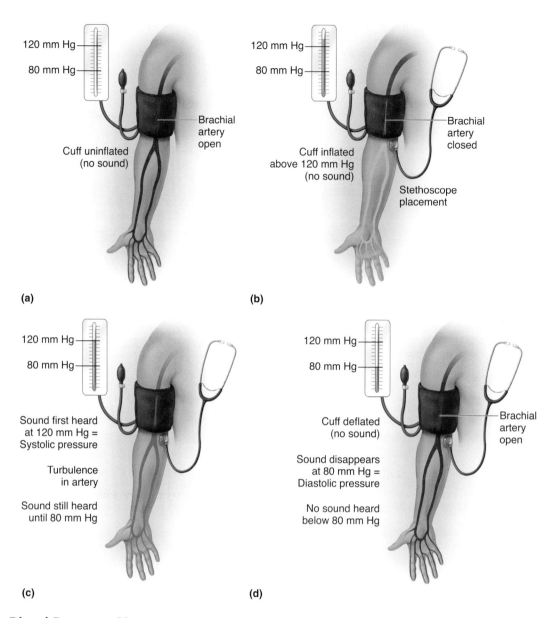

Figure 34.2 Blood Pressure Measurement
Assuming a blood pressure of 120/80. (a) Normal arterial flow with no pressure on the brachial artery. No sound is heard in the stethoscope. (b) Cuff inflated to beyond systolic pressure. Brachial artery is closed and no sound is heard. (c) Release of pressure from the sphygmomanometer just less than blood pressure. Blood rushes into brachial artery and makes arrhythmic "whoosing" sound. Sound increases as pressure is lessened. (d) Sound disappears at 80 mmHg because blood is no longer constricted. Pressure when sound disappears is known as diastolic pressure.

as blood passes through the brachial artery. The blood passes smoothly through the vessel, like water through a garden hose. But, in a garden hose, when pressure is applied by partially pinching it off in one place, the resulting turbulence creates sound. When pressure is applied to the arm due to the constriction from the blood pressure cuff, the turbulence of the blood passing through the vessel also creates sound. The significant difference between the flow of blood in the body and the flow of water in a hose is that the blood pulses rhythmically through the arteries as the heart contracts during ventricular systole. The sounds made by the flow of blood in a constricted artery are known as **Korotkoff sounds,** as seen in figure 34.2.

1. To determine blood pressure, take a sphygmomanometer to your table and clean the earpieces of the stethoscope with alcohol. Locate the pulse of the brachial artery on your lab partner. This can be done by placing two fingers on the medial side of the biceps brachii muscle near the antecubital fossa. You can place a small *X* with a felt pen on this location on the arm.
2. Place the blood pressure cuff around the arm of your lab partner at the level of the heart. Make sure the inflatable portion of the cuff is on the anterior medial side. Some cuffs have a metal bar that has a loop through which a part of the cuff goes. This bar should *not* be located on the medial side of the arm because if the bar is located on the medial side of the

arm, it may not close off the brachial artery effectively as the cuff is inflated.

3. Your lab partner should rest his or her forearm on the lab counter while you place the diaphragm (the flat membrane) of the stethoscope on the X where you previously located the brachial pulse. If you listen through the earpieces of the stethoscope, you should not hear any sound at this time. This is illustrated in figure 34.2.
4. Hold on to the rubber squeeze bulb with the attached rubber tubing leading away from you. Turn the metal dial on the cuff clockwise until it is closed. You can now begin to inflate the cuff.
5. Pump the cuff up to about 90 mm Hg. Look at the mercury in the glass tube, the digital readout, or the needle on the dial. If it bounces up and down a little, then listen closely for the sounds in the stethoscope. (The earpieces are inserted into the ears facing forward.) If you do not hear any sound and the mercury or needle is still, then inflate the cuff to 100 or 120 mm Hg. If you see pulsing, listen again for the sound. Make sure the diaphragm of the stethoscope is in the right place.
6. Once you are sure you have heard the sound of the heart, remove the cuff and place it on the other arm. Excessive constriction of the arm may elevate your lab partner's blood pressure.
7. Inflate the cuff on the other arm, but make sure you exceed the level where the mercury or needle on the dial pulses up and down (usually around 150 mm Hg). Do *not* leave the cuff inflated for a long period of time.
8. Release the knob slowly.
9. Record the mm Hg to the nearest even number when the sound is first heard. This sound represents the **systolic pressure** of the ventricles, which is the pressure the heart generates that exceeds the pressure of the cuff. The sound may muffle a bit and then come in very strongly.
10. Continue to let air out of the cuff slowly until the sound starts to muffle. Listen very carefully until the sound *completely* disappears. The exact level where the sound disappears indicates the **diastolic pressure.**
11. Record the level of blood pressure as the systolic pressure over the diastolic pressure.

$$\frac{\text{Systolic pressure}}{\text{Diastolic pressure}} = \underline{}$$

If you are unsure exactly where the pressure is, you can fine-tune your readings by measuring the cuff pressure at each approximate end to obtain an accurate reading. Do not subject your lab partner to more than two or three attempts. Ask your instructor for help if you have difficulty determining blood pressure.

Hypertension is elevated blood pressure. Normal blood pressure is from 90/60 to 130/80 mm Hg. Hypertension may be due to a number of factors; however, the majority of hypertensive cases occur for reasons as yet unknown. Hypertension is typically a blood pressure in excess of 140/90 mm Hg, although the greatest concern for young adults is in the diastolic reading. The diastolic pressure should be less than 90 mm Hg.

Pulse Pressure

You can now determine the pulse pressure of your lab partner. Low pulse pressure (less than 25 mm Hg) indicates potential blood loss. The pulse pressure is determined by subtracting the diastolic pressure from the systolic pressure. It is normally around 50 mm Hg. The pulse pressure is the pressure exerted by the heart to move the blood through the vessels.

Pulse pressure: _____

Factors Affecting Blood Pressure

Body Position

Measure the blood pressure of your lab partner while he or she lies down. Record this value.

Measurement lying down: _____

Have your lab partner stand suddenly, and quickly take a new measurement. *Be careful NOT to drop the sphygmomanometer!* Record the results.

Blood pressure on immediately standing: _____

How can you account for these two readings? _____

_____

_____

Exercise

Caution
Do not do strenuous exercise if you are not feeling well, have a history of heart trouble or asthma, or have been advised by a physician to avoid exercise.

Have your lab partner do strenuous physical activity (jumping jacks) for a couple of minutes. Measure the blood pressure immediately after the cessation of exercise and record your results.

Blood pressure after exercise: _____

Exercise 34 Review
Blood Vessels and Blood Pressure

Name: _____

Lab time/section: _____

Date: _____

1. The letters *bpm* stand for what phrase in cardiac measurement? _____

2. What does a sphygmomanometer measure? _____

3. If you were to measure blood pressure, what artery would you most commonly use? _____

4. If you have a blood pressure of 140/80, what does the 80 represent? _____
   _____

5. What is the clinical threshold for high blood pressure? _____

6. When the first sound is heard during measurement with a blood pressure cuff, what is measured, systolic or diastolic pressure?
   _____

7. Emotions have an effect on blood pressure. Predict the blood pressure of a person who recently had a heated argument with a roommate about rent money. _____
   _____

8. Illness can affect blood pressure. Illness tends to increase stress responses. Predict the blood pressure of a person with a sinus headache and postnasal drip. _____
   _____

9. Both nicotine and caffeine elevate blood pressure. Explain how an increase in blood pressure can have a negative effect on the pumping efficiency of the heart. _____
   _____
   _____
   _____

10. Record your blood pressure. _____

11. You record a blood pressure of 130/80. What is the pulse pressure? _____

    Is this a normal pulse pressure? (yes/no)

Exercise 35
Structure of the Respiratory System

INTRODUCTION

The body exchanges oxygen and carbon dioxide with the atmosphere. This vital exchange occurs due to the respiratory system, which is discussed in *Principles of Anatomy and Physiology* in chapter 20, "Respiratory System." Atmospheric oxygen moves into the lungs and diffuses into the circulatory system. It subsequently reaches the individual cells of the body while the metabolic waste product, carbon dioxide, is released from the intercellular environment and travels via the blood to the lungs, where it is released by exhalation. In this exercise, you examine the anatomy of the respiratory system. As you study the anatomy of the system, be aware of the role that other systems, such as the cardiovascular system, play in respiration.

OBJECTIVES

At the end of this exercise, you should be able to

1. list the organs and significant structures of the respiratory system;
2. define the role of the respiratory system in terms of the overall function of the body;
3. explain the physical reason for the tremendous surface area of the lungs;
4. identify the cartilages of the larynx;
5. distinguish among a bronchus, a bronchiole, and a respiratory bronchiole;
6. recognize a bronchiole and an alveolus in a prepared slide of lung.

MATERIALS

Lung models or detailed torso model, including midsagittal section of head
Model of larynx
Microscopes
Prepared microscope slides of lung tissue
Prepared microscope slides of "smoker's lung"
Charts and illustrations of the respiratory system
Cats
Materials for cat dissection
- Dissection trays
- Scalpel and two or three extra blades
- Gloves (household latex gloves work well for repeated use)
- Blunt (mall) probe
- Forceps and sharp scissors
- First aid kit in lab or prep area
- Sharps container
- Animal waste disposal container

PROCEDURE

Overview

Look at the charts and models of the respiratory system and compare them with figure 35.1. Locate the following structures:

External nose
Nasal cavity
Pharynx
Larynx
Trachea
Bronchi
Lungs

Nose and Nasal Cartilages

Examine a midsagittal section of a model or chart of the head and look for the **nasus** or **nose, nasal cartilages, external nares (nostrils),** and **nasal septum.** The nasal septum separates the nasal cavities and is composed of the **perpendicular plate of the ethmoid bone,** the **vomer,** and the **septal cartilage.** Examine these features in figure 35.2.

The entrance of the external nares is protected by guard hairs. The region of the nose just posterior to the external nares is the **nasal vestibule,** which is lined with stratified squamous epithelium. Behind the vestibule is the **nasal cavity.** The nasal cavity is lined with a **mucous membrane** that moistens air entering the respiratory system, trapping dust particles. This mucous membrane overlays a superficial venous plexus that warms the air. The membrane consists of connective tissue and **respiratory**

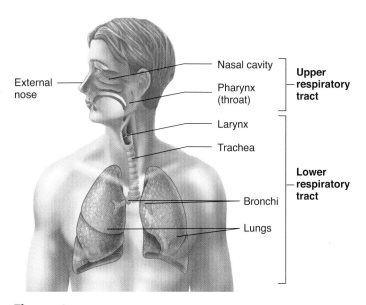

Figure 35.1 Structures of the Respiratory System

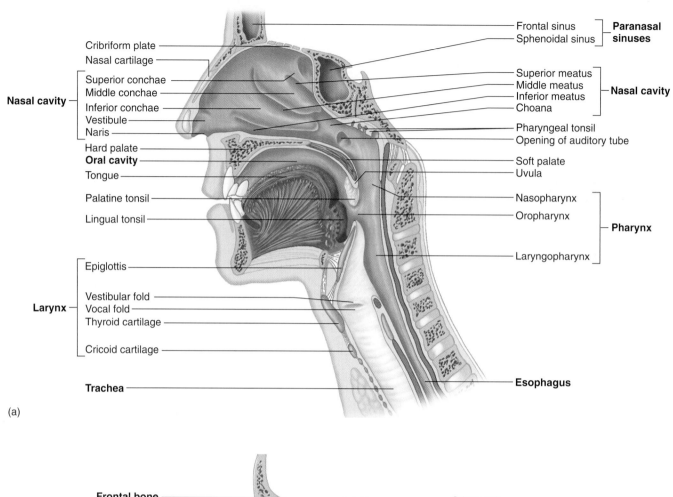

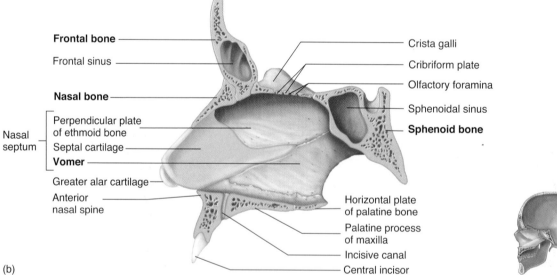

Figure 35.2 Nose, Nasal Cavity, and Nasal Septum
(a) Midsagittal section; (b) nasal septum.

epithelium, which is composed of **pseudostratified ciliated columnar epithelium** with **goblet cells.** The lateral walls of the cavity have three protrusions that push into the nasal cavity. These are the **nasal conchae.** They cause the air to swirl in the nasal cavity and come into contact with the mucous membrane. Locate the **superior, middle,** and **inferior conchae** in the nasal cavity. The nasal cavity terminates where two openings, the **internal nares** (sing., **naris**), or **choanae,** lead to the **pharynx.**

Pharynx

The pharynx can be divided into three regions based on location. The uppermost area is the **nasopharynx**, which is directly posterior to the nasal cavity. The nasopharynx has two openings on the lateral walls, which are the openings of the **auditory,** or **eustachian, tubes.** As the pharynx descends behind the oral cavity, it becomes the **oropharynx.** The oropharynx is a common passageway for food, liquid, and air. The **uvula** is a small, pendulous structure that partially separates the **oral cavity** from the oropharynx. The uvula flips upward during swallowing, helping prevent fluids from entering the nasopharynx. The most inferior portion of the pharynx is the **laryngopharynx,** which is located superior to the larynx. Find the regions of the pharynx in figure 35.2.

Larynx

The **larynx** is commonly known as the "voice box" because it is an important organ for sound production in humans. It is important in the production of the pitch of the voice, while the shape of the oral cavity and the placement and size of the paranasal sinuses are responsible for the sonority (sound quality) of the voice. The larynx is located at about the level of the fourth through sixth cervical vertebrae and consists of a number of cartilages. The most prominent cartilage in the larynx is the **thyroid cartilage,** which is a shield-shaped structure made of hyaline cartilage. Examine models and charts in the lab and compare the structures of the larynx with figure 35.3.

The thyroid cartilage is more prominent in males because of the influence of testosterone. Inferior to the thyroid cartilage is the **cricoid cartilage.** The cricoid cartilage is also composed of hyaline cartilage, and it is relatively narrow when seen from the anterior but increases in size at its posterior surface. Superior to the cricoid cartilage in the posterior wall of the larynx are the paired **arytenoid cartilages.** These cartilages attach to the posterior end of the **vocal folds (true vocal cords).** Muscles pull the arytenoid cartilages, stretching the true vocal folds, increasing the pitch of the voice. This occurs by the contraction of **intrinsic muscles** attached to the arytenoid cartilages from the back, while the vocal cords are held stationary by the thyroid cartilage in the front. Above the true vocal folds are the **vestibular folds (false vocal cords)** (figure 35.3).

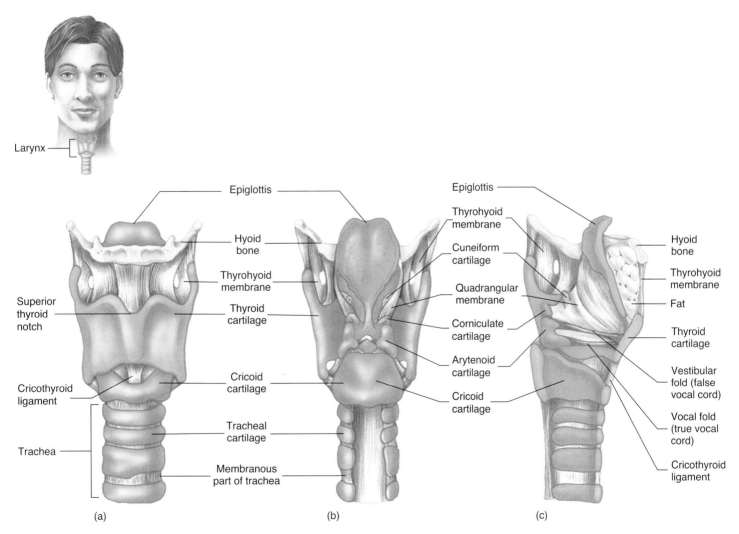

Figure 35.3 Larynx
(a) Anterior view; (b) posterior view; (c) midsagittal section.

At the very posterior, superior edge of the larynx are the **corniculate** and **cuneiform cartilages.** These are also made of hyaline cartilage. The most superior cartilage of the larynx is the **epiglottis,** which is composed of elastic cartilage. During swallowing, the epiglottis covers the opening of the larynx, which is known as the **glottis.** This is not a perfect system, as anyone knows who has started swallowing a liquid and responded by laughing at a joke. Inhalation at the beginning of the laugh causes fluid to move into the larynx and trachea, irritating the respiratory lining. This causes another reflex called the **cough reflex,** which propels the liquid out of the respiratory system (frequently, through the nose). Locate the structures of the larynx on models or charts in the lab and in figure 35.3.

Trachea and Bronchi

The trachea is commonly known as the "windpipe" because it conducts air from the larynx to the lungs. The **trachea** is a straight tube whose lumen is kept open by **tracheal cartilages.** These are C-shaped cartilages. Examine these cartilages by running your fingers gently down the outside of your throat. Palpate the cartilage rings below the larynx. The tracheal cartilages are composed of hyaline cartilage. At the most inferior portion of the trachea is a center point known as the **carina** (*carina* = keel). Locate the features of the trachea in figures 35.4 and 35.5.

The trachea is also lined with respiratory epithelium. Obtain a prepared slide of the trachea and find the tracheal cartilage, respiratory epithelium, and **posterior tracheal membrane** (trachealis muscle; figure 35.5).

The trachea splits into two tubes, which enter the lungs. These tubes are the **main** or **primary bronchi.** Each lung receives air from a primary bronchus. These primary bronchi contain hyaline cartilage and are lined with respiratory epithelium. The primary bronchi of the lung divide into the **lobar** or **secondary bronchi,** and these further divide to form **segmental** or **tertiary bronchi.** The extensive branching of the bronchi produces a structure called the **tracheobronchial tree** (figures 35.4 and 35.6).

Lungs

There are two **lungs** in humans; the right lung has **three lobes** and the left lung has **two lobes.** The **right lung** consists of **superior, middle,** and **inferior lobes,** and the **left lung** has **superior** and **inferior lobes,** as well as an indentation occupied by the heart. This indentation is known as the **cardiac notch.** Look at models or charts in the lab and identify the major features, as shown in figure 35.4.

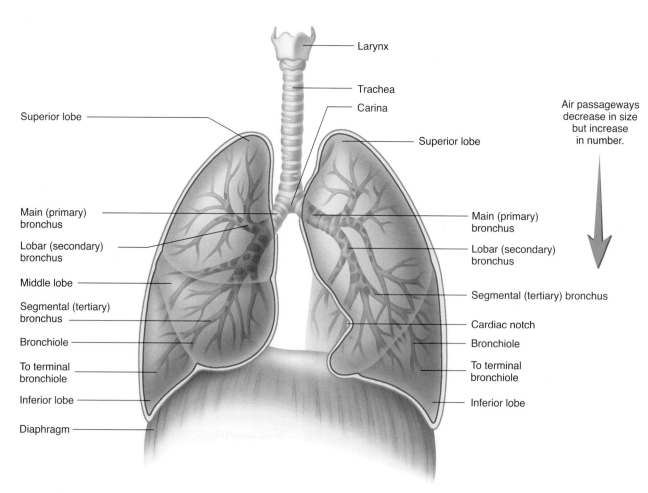

Figure 35.4 Trachea and Tracheobronchial Tree, Anterior View

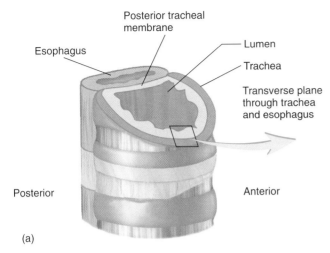
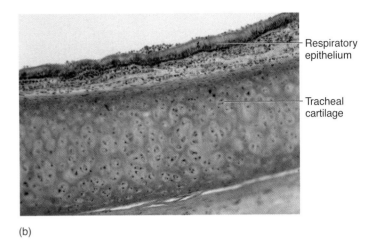

Figure 35.5 Trachea, Cross Section
(a) Diagram; (b) photomicrograph (100×).

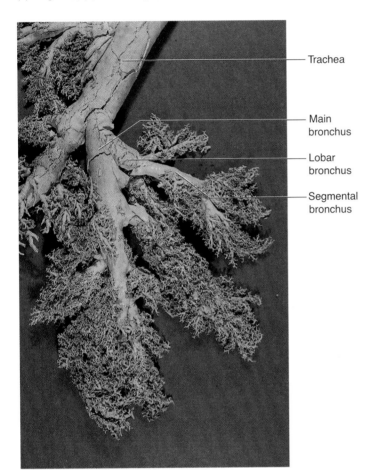

Figure 35.6 Tracheobronchial Tree

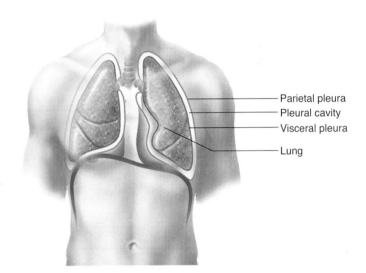

Figure 35.7 Pleural Membranes and Cavity

Histology of the Lung

The bronchi continue to divide until they become **bronchioles,** which are small respiratory tubules with smooth muscle in their walls. The bronchioles in the lung further divide into **respiratory bronchioles,** which are so named because of the small structures, called alveoli, attached to their walls. The respiratory bronchioles lead to passageways known as **alveolar ducts,** which branch into alveoli. **Alveoli** are air sacs in the lung that exchange oxygen and carbon dioxide with the blood capillaries of the lungs. Obtain a prepared slide of lung and scan first under low power and then under higher powers. Examine your slide and locate the bronchi, bronchioles, respiratory bronchioles, alveolar ducts, and alveoli. Alveoli that are clustered around an alveolar duct are collectively known as **alveolar sacs** (figures 35.8 and 35.9).

Type I pneumocytes make up about 90% of the alveoli and they are composed of simple squamous epithelium, as are the capillaries that surround the alveoli. The division of the lung into many small sacs tremendously increases the surface area of the lung. This increase is vital for the rapid and extensive diffusion of

The lungs are located in the **pleural cavities** on each side of the mediastinum. The **parietal pleura** is the outer membrane on the chest cavity wall, and the membrane on the surface of the lungs is the **visceral pleura.** The space between the membranes is known as the pleural cavity. These are shown in figure 35.7.

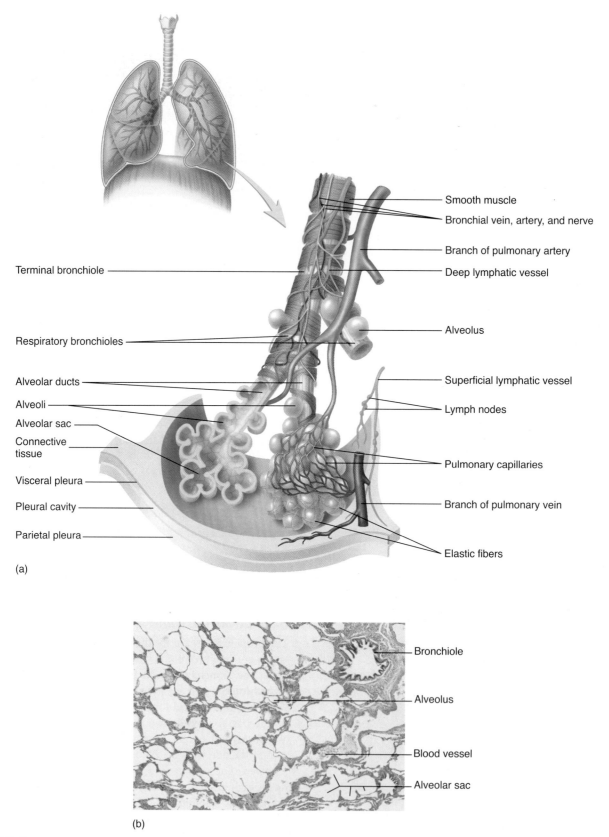

Figure 35.8 Histology of the Lung
(a) Diagram; (b) photomicrograph of lung (40×); (c) photomicrograph of alveolar duct (100×); (d) photomicrograph of alveolus (400×).

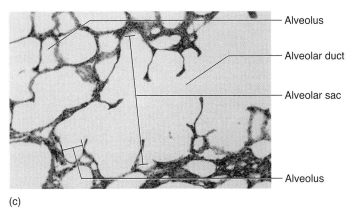
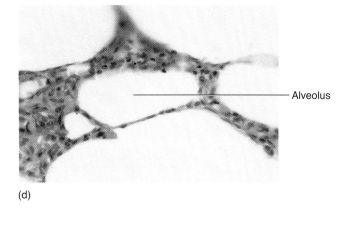

(c)   (d)

Figure 35.8—*Continued.*

oxygen across the respiratory membranes. Oxygen moves across the **respiratory membrane,** which consists of the alveolus, the capillary epithelium, and the basement membrane between the two.

You may see other cell shapes in the prepared lung sections. Some of these cells are **type II pneumocytes (septal cells);** they decrease the surface tension of the lung by secreting **surfactant.** These are seen in figure 35.9.

Examine a prepared slide of smoker's lung. Note the dark material in the lung tissue and the general destruction of the alveoli. Breakdown of the alveoli leads to a disease known as **emphysema.**

Cat Dissection

1. Prepare the cat for dissection, and remember to place all excess tissue in the appropriate waste container and not in a standard wastebasket or down the sink!
2. If you have not opened the chest cavity in your study of the cat, then you should do so now. Removal of the skin is discussed in Exercise 12, and the procedure for opening the thoracic cavity is described in Exercise 30.
3. Locate the **larynx** of the cat above the **trachea.** Notice the broad, wedge-shaped hyaline cartilage structure in the front. This is the **thyroid cartilage.**
4. Make a midsagittal incision through the thyroid cartilage and continue carefully until you have cut completely through the larynx. To examine the larynx more completely, you may wish to continue your midsagittal cut partway down through the trachea.
5. Open the larynx and find the **epiglottis** of the cat. Notice how the elastic cartilage is lighter in color than the thyroid cartilage.
6. Find the **cricoid cartilage,** the **arytenoid cartilages, vestibular folds,** and **vocal folds** (figures 35.3 and 35.10).
7. Examine the trachea as it passes from the larynx into the thoracic cavity. Ask your instructor for permission before you cut the trachea in cross section. If you do, you should see the **tracheal cartilages** and the **posterior tracheal membrane** (trachealis muscle). The posterior portion of the trachea is located ventrally to the esophagus.
8. Examine how the trachea splits into the two **bronchi,** which continue into the lungs. The lobes of the lungs are different in the cat than in the human. The right lung in the cat has four lobes, whereas the left lung has three lobes. How does this compare with the pattern in humans? Examine the structure of the lungs in your specimen and compare it with figure 35.11.
9. Look at the lungs and see how they are covered with a thin serous membrane called the **visceral pleura,** whereas the outer covering (on the deep surface of the ribs and intercostal muscles) is called the **parietal pleura.** The space between these two membranes is the **pleural cavity.** Cut into one of the lungs of the cat and examine the lung tissue. Note how the lungs appear as a very fine mesh sponge. The alveoli of the lungs are microscopic.

At the inferior portion of the thoracic cavity is the diaphragm. As it contracts, the pressure in the thoracic cavity decreases and air fills the lungs.

10. Examine the diaphragm in your specimen and compare it with figure 35.11.

Clean Up
When you are done, place your cat back in the plastic bag. Remember to place all excess tissue in the appropriate waste container and not in a standard classroom wastebasket or down the sink!

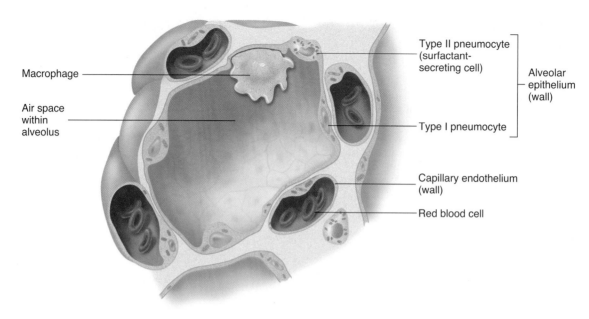

Figure 35.9 Details of a Single Alveolus

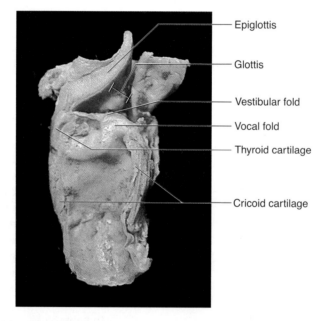

Figure 35.10 Larynx of the Cat, Midsagittal Section

Exercise 35 Structure of the Respiratory System

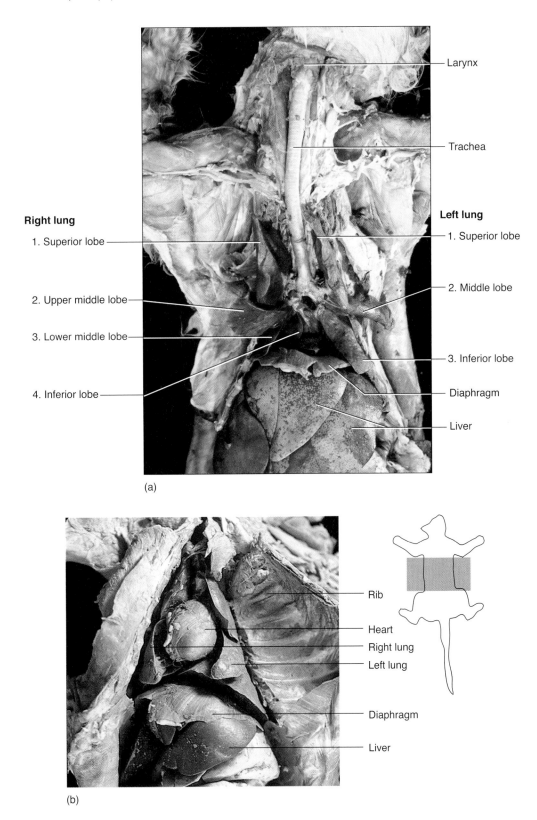

Figure 35.11 Respiratory System of the Cat
(a) Overview; (b) lungs in thoracic cavity.

Exercise 35 Review

Structure of the Respiratory System

Name: _____
Lab time/section: _____
Date: _____

1. What is the common name for the external nares? _____

2. The nasal cartilages are made of hyaline cartilage. What functional adaptation does cartilage have over bone in making up the external framework of the nose? _____

3. The nasal cavities are separated from each other by what structure? _____

4. Three structures make up the nasal septum. What are they? _____

5. What is the function of respiratory epithelium and the superficial blood vessels in the nasal cavity? _____

6. Name the openings between the nasal cavity and the pharynx. _____

7. What is the name of the space behind the oral cavity and above the laryngopharynx? _____

8. What is the name of the structure that prevents fluid from entering the nasopharynx during swallowing? _____

9. What is the name of the large cartilage of the anterior larynx? _____

10. What is the structure that protects the glottis from fluid entering the larynx? _____

11. Which lung has just two lobes in the human? _____

12. What membrane attaches directly to the lungs? _____

13. The trachea branches into two tubes that go to the lungs. What are these tubes called? _____

14. Where is the tracheobronchial tree located? _____

15. What small structure in the lung is the site of oxygen exchange with the blood capillaries? _____

16. The surface area of the lungs in humans is about 70 square meters. How can this be so if the lungs are located in the small space of the thoracic cavity? What role do alveoli play in the nature of surface area? _____

17. Emphysema is a destruction of the alveoli of the lungs. What effect does this have on the surface area of the lungs? _____

18. Label the following illustration using the terms provided.

    main bronchus   internal nares
    nasal cavity   middle lobe
    epiglottis   trachea
    superior lobe   inferior lobe
    cricoid cartilage

    a. _____
    b. _____
    c. _____
    d. _____
    e. _____
    f. _____
    g. _____
    h. _____
    i. _____

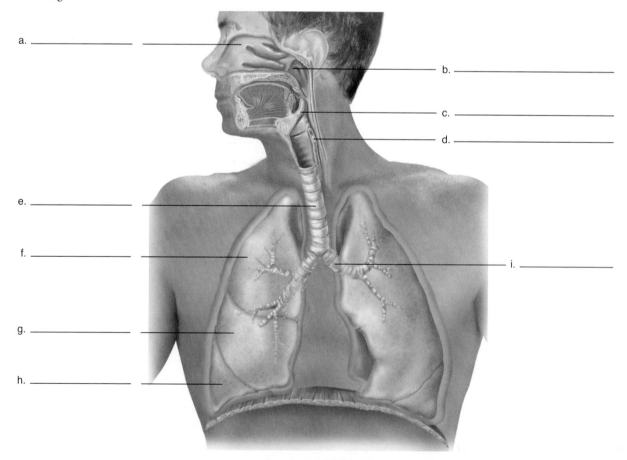

Respiratory System

Exercise 36
Respiratory Function, Breathing, and Respiration

INTRODUCTION

Respiration involves **pulmonary ventilation,** or movement of air into and out of the lungs, and the **exchange of gases** (oxygen and carbon dioxide). The exchange of gases occurs across the respiratory membrane and also at the level of the capillary and the cells of the body.

The amount of air that enters the lungs can be altered by the **ventilation rate** (number of breaths per minute), the change in the **pulmonary volume** (amount of air inhaled and exhaled with each breath) or both. Reduction in the pulmonary volume may be reflective of health conditions such as fibrosis in the lungs, which limits the ability of the lungs to fully expand.

Adequate pulmonary ventilation is critical in clinical settings. The measurement of arterial blood gases is important when patients are artificially ventilated as it allows the respiratory therapist to determine if adequate oxygen is being delivered to the patient. Reduction in pulmonary ventilation can lead to a buildup of carbon dioxide (hypercapnia) in the blood, producing acidosis. Increased pulmonary ventilation (such as being at high altitudes, having severe anemia, and ingestion of certain drugs) can lead to a reduction in carbon dioxide levels (hypocapnia).

Just as changes in the ventilation rate or pulmonary volume can alter the pH of the blood, the reverse is also true. Excess carbon dioxide in the blood stimulates an increase in pulmonary ventilation, which subsequently reduces the carbon dioxide level in the blood. This negative feedback mechanism is important in maintaining homeostasis in the blood. These topics are discussed in *Principles of Anatomy and Physiology* in chapter 20, "Respiratory System."

In this exercise, you measure breathing rates, lung volumes and capacities, and the effects of carbon dioxide on the acid–base balance of a solution.

OBJECTIVES

At the end of this exercise, you should be able to

1. measure the pulmonary volumes and calculate the pulmonary capacities;
2. describe the relationship between the tidal volume, vital capacity, inspiratory reserve volume, and expiratory reserve volume;
3. describe the mechanical process of breathing;
4. identify the tidal volume, expiratory reserve volume, and vital capacity on a spirogram;
5. demonstrate the use of the stethoscope in obtaining respiratory sounds;
6. determine whether a person will inhale or exhale based on the differences in air pressure between the lungs and the external air;
7. explain how resistance in the airways changes the flow of air into or out of the lungs;
8. describe how carbon dioxide in solution changes the pH of the solution.

MATERIALS

Respiration Model
Bell jar respiration model

Pulmonary Volume Setup
Biopac air transducer, wet spirometer, or hand-held spirometers
Disposable mouthpieces to fit respirometer or spirometers
Watch or clock with accuracy in seconds
Biohazard bag
AFT6 – 600 mL calibration syringe
AFT1 – Disposable Bacteriological Filter
AFT2 – Disposable Mouthpiece
SS11LA – Airflow Transducer
Nose clips
Stethoscope
Alcohol wipes

Acid–Base Setup
Litmus solution (2 g litmus powder in 600 mL water)
Sodium bicarbonate powder (baking soda)
Straws
Tape or Parafilm®
Small spatula
100 mL Erlenmeyer flasks
Safety glasses

PROCEDURE

Mechanics of Breathing

Air moves from regions of higher pressure to regions of lower pressure. The lungs fill with air or deflate due to changes in air pressure. If you have a constant amount of air and the pressure increases, then the volume decreases. This relationship is known as **Boyle's law.** In terms of breathing, if the diaphragm contracts, the volume in the chest cavity increases. This causes a decrease in pressure. Air flows from regions of high pressure to low pressure, so inhalation results. Normal atmospheric air pressure is measured in millimeters of mercury (mm Hg), and standard air pressure at sea level is 760 mm Hg. As you move to higher elevations, the atmospheric air pressure decreases. No matter what the elevation, the barometric air pressure is defined as zero, as described in your text. When there is no movement of air into or out of the lungs, the alveolar pressure in the lungs is equal to that of the surrounding air (barometric air pressure), as illustrated in figure 36.1*a*. During inspiration, the diaphragm contracts and/or the intercostal muscles expand the rib cage. The volume in the thoracic cavity increases, and the alveolar pressure becomes

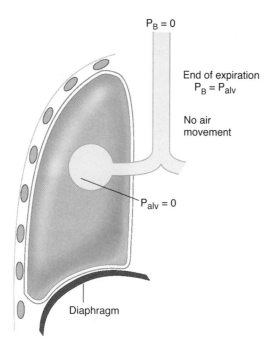

(a) Barometric air pressure (P_B) is equal to alveolar pressure (P_{alv}), and there is no air movement.

less than that of the surrounding environment. Air rushes into the thoracic cavity, thereby inflating the lungs. This is illustrated in figure 36.1b. As the diaphragm relaxes, the volume of air in the thoracic cavity decreases, the alveolar pressure increases, and air moves from the region of higher pressure inside the lungs to the region of lower pressure in the atmosphere, as illustrated in figure 36.1c. Negative intrapulmonary pressure is the pressure between the lung and the parietal pleura. This also has an important role in breathing.

You can use a bell jar model to demonstrate this in lab. This model illustrates the movement of air into or out of the lungs, depending on air pressure differentials between the lungs and the external environment. The glass housing represents the thoracic cage, the balloons represent the lungs, the latex sheeting represents the diaphragm, and the rubber stopper represents the nose. Pull *gently* on the latex diaphragm. As you do this, the volume in the pleural cavity (the space between the balloons and the glass jar) increases, causing a decrease in the pressure of the pleural cavity.

Because the pressure in the pleural cavity decreases, air in the external environment (which is at a higher pressure than that in the jar) moves into the bell jar, filling the balloon lungs. As you gently release the diaphragm, the volume decreases and the pressure increases in the pleural cavity bell jar. This pressure exceeds the external air pressure, and air moves out of the lungs and through the nose (rubber stopper). Although the mechanics of human breathing differ somewhat from this model, the bell jar model demonstrates the basic principle of air moving from a region of high pressure to a region of low pressure.

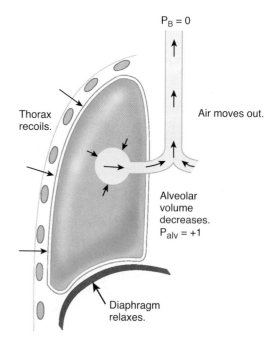

(b) Increased thoracic volume results in increased alveolar volume and decreased alveolar pressure. Barometric air pressure is greater than alveolar pressure, and air moves into the lungs.

(c) Decreased thoracic volume results in decreased alveolar volume and increased alveolar pressure. Alveolar pressure is greater than barometric air pressure, and air moves out of the lungs.

Figure 36.1 Mechanics of Breathing
(a) No movement of air; (b) inspiration; (c) expiration. The blue circles represent the sum of the alveolar pressures and are not represented as anatomically accurate.

Measurement of Breathing Rate

For this experiment, it is important that your lab partner be distracted from thinking about breathing. If you are asked to "breathe normally," typically you become self-conscious about breathing. Therefore, your lab partner should read the remainder of the laboratory exercise while you count the number of breaths he or she takes for a total of 2 minutes. Watch your lab partner's chest rise and fall, or place your hand on his or her back. Divide this number by 2 in order to calculate the average number of breaths per minute. Record that number for your lab partner.

Breaths per minute: _____

Breathing rate varies with oxygen demand and/or carbon dioxide levels in the blood. Before doing any exercise, estimate the breaths per minute your lab partner might take after completing 2 minutes of strenuous exercise. Record your estimation.

Estimation: _____

Caution
If you have a heart condition, a family history of heart failure, asthma, or any other medical condition that prevents you from doing strenuous exercise, then do not do the exercise sections of this lab.

Have your lab partner do strenuous exercise for 2 minutes (doing jumping jacks, running in place or exercising outside of lab, running up and down stairs, etc.). As soon as your lab partner is finished, count the number of breaths per minute for the *first minute* after exercise and record this number.

Number of breaths per minute after 2 minutes of

exercise: _____

How does this compare with your estimation? _____

Measurement of Pulmonary Volumes and Capacities

Pulmonary volumes are the amount of air that flows into or out of the lungs during a particular event. Capacities are the summation of volumes. There are many ways to record pulmonary volumes. An affordable way to measure pulmonary volume is to use a hand-held spirometer in which the exhaled air spins the vanes of the spirometer, estimating the volume of air exhaled. This is like the anemometer used by weather stations to measure wind speed. These are relatively inexpensive pieces of equipment and are good for measuring vital capacity, but they are not as accurate as other equipment. The wet spirometer measures the volume of air exhaled into a chamber, so it measures the actual amount of air exhaled. Some digital recorders, such as those made by Biopac®, use a flow meter to estimate volume.

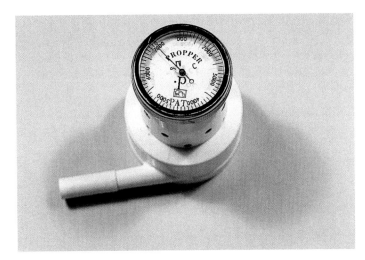

Figure 36.2 Hand-Held Spirometer

Caution
You should *never* inhale using a wet spirometer or a hand-held spirometer.

Spirometry
Tidal Volume with Hand-Held Spirometer

Place a disposable mouthpiece on the spirometer tube. Make sure you set the indicator dial to zero by twisting the knurled ring on the top of the spirometer (zero may be the same as the maximum volume, 7000 in some spirometers; figure 36.2).

As you exhale, estimate what a normal breath volume will be and exhale this amount rather forcefully. If you breathe gently into a hand-held spirometer, you may not cause the vanes to spin enough to get a significant recording. Exhale five breaths and record your results. Do not reset the spirometer after each breath.

Total of five breaths recorded by spirometer:

Divide the total number by 5 and enter the average tidal volume.

Average tidal volume: _____ mL

You may find that the tidal volume is variable among members of your class. This is due to the difficulty of measuring tidal volumes with a standard lab apparatus. The average tidal volume is about 500 mL.

Tidal Volume with a Wet Spirometer

The tidal volume can be measured if your wet spirometer is accurate. Place a disposable mouthpiece into the flexible hose and breathe a normal breath into the mouthpiece. Record your data in the following space.

Tidal volume: _____

Expiratory Reserve Volume

You can determine your **expiratory reserve volume** with the wet spirometer or the hand-held spirometer. The expiratory reserve volume is the maximal amount of air you can exhale after a normal exhalation.

1. Make sure to close off your nostrils and, after a normal exhalation, forcibly expel the remainder of your breath through the mouthpiece into the spirometer.
2. Repeat this two more times for a total of three exhalations, and calculate the average expiratory reserve by dividing the sum of the volumes by 3.
3. Record your results.

   Trial 1: _____

   Trial 2: _____

   Trial 3: _____

   Average expiratory reserve volume: _____ mL

Vital Capacity (VC)

The vital capacity (VC) is the total volume of air that can be forcefully expelled from the lungs after a maximum inhalation.

1. Measure the vital capacity with the use of a wet spirometer, or hand-held spirometer by first breathing in as deeply as you can.
2. Close your nostrils, and then exhale through the mouthpiece completely until you cannot exhale anymore.
3. Force as much air from your lungs as you can.
4. Record your vital capacity.

   Vital capacity: _____ mL

Simple Spirometry

If your lab is not equipped with a commercial spirometer you can make an inexpensive spirometer with a new 13-quart plastic bag (one usually used for kitchen trash cans), a plastic dishpan (11.5 quart), 1 foot of 3/8-inch or 1/2-inch plastic aquarium tubing, a 1-L graduated cylinder, and a sink!

Place the dishpan into the lab sink and fill the dishpan with water, repeatedly, using a 1-L graduated cylinder until you just begin to overflow the pan. Standard dishpans are usually a little less than 11 L. Record the exact volume of the pan in the following space.

   Dishpan volume: _____

Vital Capacity

1. Insert the plastic tubing into the opening of the 13-quart plastic bag. Make sure that there is no air in the bag! Either tighten the bag around the tube with a rubber band or hold tightly onto the bag with your hand to make an airtight seal.
2. Inhale maximally while keeping the plastic tubing sealed in the bag and breathe into the tube, filling the bag, until you can exhale no further.
3. Slide your hand down the bag and force the air into the bottom of the bag until you have the entire volume of exhaled air pushed into the bottom of a bag. Make sure that you do not let any air escape as you do this.
4. While holding tightly onto the bag, slowly push the bag into the water until the volume of air that you exhaled just displaces the water in the pan. Your hand should just be touching the surface of the water. Make sure that the entire volume of enclosed air is submersed.
5. Remove the bag and carefully pour the water from the dishpan into the graduated cylinder and determine how much water is left in the pan. Record this volume in the following space.

   Remaining water left in the dishpan: _____

6. Subtract the remaining volume in the pan from the original volume. That gives you an estimation of your vital capacity. Record your vital capacity.

   Vital capacity: _____

You can determine tidal volume and forced expiratory volume with the following modifications of the procedure.

Tidal Volume

1. Insert the plastic tube into a flattened bag (there should be no air in the bag) or squeeze all the air out of the bag from the first experiment.
2. *After a normal inhalation,* **exhale normally** into the tube that is firmly attached to the collapsed bag.
3. Do this for a total of 5 times, filling the bag with 5 breaths.
4. Take that volume of air in the bag and displace the water in the full dishpan.
5. Pour the water repeated into a graduated cylinder and record the volume in the following space.

   Remaining water left in the dishpan: _____

6. Subtract the remaining volume in the pan from the original volume. Divide that number by 5 (since you took 5 breaths) and record your tidal volume.

   Tidal volume: _____

Expiratory Reserve Volume

1. The expiratory reserve volume is measured by *exhaling normally* and then forcing the remaining air in your lungs into the plastic tube and plastic bag.
2. Displace the water in the dishpan as done previously, measure the remaining water, and subtract that volume from the original volume in the pan to obtain the expiratory reserve volume. Record your expiratory reserve volume in the following space.

   Expiratory reserve volume: _____

Percent of Normal Vital Capacity

You can compare your vital capacity with those of others of your sex, height, and age by using one of the two charts (one for females, one for males) in table 36.1. As you examine the chart for a person of your height, notice that the vital capacity decreases with age.

Calculate your percent of normal vital capacity by dividing your data by the normal data for a person of your age, sex, and height and multiplying the result by 100.

Table 36.1 Predicted Vital Capacities for Females and Males

Females

Height in Centimeters and Inches

Age	cm in.	152 59.8	154 60.6	156 61.4	158 62.2	160 63.0	162 63.7	164 64.6	166 65.4	168 66.1	170 66.9	172 67.7	174 68.5	176 69.3	178 70.1	180 70.9	182 71.7	184 72.4	186 73.2	188 74.0
16		3070	3110	3150	3190	3230	3270	3310	3350	3390	3430	3470	3510	3550	3590	3630	3670	3715	3755	3800
17		3055	3095	3135	3175	3215	3255	3295	3335	3375	3415	3455	3495	3535	3575	3615	3655	3695	3740	3780
18		3040	3080	3120	3160	3200	3240	3280	3320	3360	3400	3440	3480	3520	3560	3600	3640	3680	3720	3760
20		3010	3050	3090	3130	3170	3210	3250	3290	3330	3370	3410	3450	3490	3525	3565	3605	3645	3695	3720
22		2980	3020	3060	3095	3135	3175	3215	3255	3290	3335	3370	3410	3450	3490	3530	3570	3610	3650	3685
24		2950	2985	3025	3065	3100	3140	3180	3220	3260	3300	3335	3375	3415	3455	3490	3530	3570	3610	3650
26		2920	2960	3000	3035	3070	3110	3150	3190	3230	3265	3300	3340	3380	3420	3455	3495	3530	3570	3610
28		2890	2930	2965	3000	3040	3070	3115	3155	3190	3230	3270	3305	3345	3380	3420	3460	3495	3535	3570
30		2860	2895	2935	2970	3010	3045	3085	3120	3160	3195	3235	3270	3310	3345	3385	3420	3460	3495	3535
32		2825	2865	2900	2940	2975	3015	3050	3090	3125	3160	3200	3235	3275	3310	3350	3385	3425	3460	3495
34		2795	2835	2870	2910	2945	2980	3020	3055	3090	3130	3165	3200	3240	3275	3310	3350	3385	3425	3460
36		2765	2805	2840	2875	2910	2950	2985	3020	3060	3095	3130	3165	3205	3240	3275	3310	3350	3385	3420
38		2735	2770	2810	2845	2880	2915	2950	2990	3025	3060	3095	3130	3170	3205	3240	3275	3310	3350	3385
40		2705	2740	2775	2810	2850	2885	2920	2955	2990	3025	3060	3095	3135	3170	3205	3240	3275	3310	3345
42		2675	2710	2745	2780	2815	2850	2885	2920	2955	2990	3025	3060	3100	3135	3170	3205	3240	3275	3310
44		2645	2680	2715	2750	2785	2820	2855	2890	2925	2960	2995	3030	3060	3095	3130	3165	3200	3235	3270
46		2615	2650	2685	2715	2750	2785	2820	2855	2890	2925	2960	2995	3030	3060	3095	3130	3165	3200	3235
48		2585	2620	2650	2685	2715	2750	2785	2820	2855	2890	2925	2960	2995	3030	3060	3095	3130	3160	3195
50		2555	2590	2625	2655	2690	2720	2755	2785	2820	2855	2890	2925	2955	2990	3025	3060	3090	3125	3155
52		2525	2555	2590	2625	2655	2690	2720	2755	2785	2820	2855	2890	2925	2955	2990	3020	3055	3090	3125
54		2495	2530	2560	2590	2625	2655	2690	2720	2755	2790	2820	2855	2885	2920	2950	2985	3020	3050	3085
56		2460	2495	2525	2560	2590	2625	2655	2690	2720	2755	2790	2820	2855	2885	2920	2950	2980	3015	3045
58		2430	2460	2495	2525	2560	2590	2625	2655	2690	2720	2750	2785	2815	2850	2880	2920	2945	2975	3010
60		2400	2430	2460	2495	2525	2560	2590	2625	2655	2685	2720	2750	2780	2810	2845	2875	2915	2940	2970
62		2370	2405	2435	2465	2495	2525	2560	2590	2620	2655	2685	2715	2745	2775	2810	2840	2870	2900	2935
64		2340	2370	2400	2430	2465	2495	2525	2555	2585	2620	2650	2680	2710	2740	2770	2805	2835	2865	2895
66		2310	2340	2370	2400	2430	2460	2495	2525	2555	2585	2615	2645	2675	2705	2735	2765	2800	2825	2860
68		2280	2310	2340	2370	2400	2430	2460	2490	2520	2550	2580	2610	2640	2670	2700	2730	2760	2795	2820
70		2250	2280	2310	2340	2370	2400	2425	2455	2485	2515	2545	2575	2605	2635	2665	2695	2725	2755	2780
72		2220	2250	2280	2310	2335	2365	2395	2425	2455	2480	2510	2540	2570	2600	2630	2660	2685	2715	2745
74		2190	2220	2245	2275	2305	2335	2360	2390	2420	2450	2475	2505	2535	2565	2590	2620	2650	2680	2710

"Predicted Vital Capacities for Females and Males (tables)" by E. A. Gaensler, M.D. and G. W. Wright, M.D., AEH, Vol. 12, pp. 146–189, February 1966. Reprinted with permission of the Helen Dwight Reid Educational Foundation. Published by Heldref Publications, 1319 18th Street NW, Washington, DC 20036–1802. Copyright © 1966.

Table 36.1 Predicted Vital Capacities for Females and Males continued.

Males

Height in Centimeters and Inches

Age	cm in.	152 59.8	154 60.6	156 61.4	158 62.2	160 63.0	162 63.7	164 64.6	166 65.4	168 66.1	170 66.9	172 67.7	174 68.5	176 69.3	178 70.1	180 70.9	182 71.7	184 72.4	186 73.2	188 74.0
16		3920	3975	4025	4075	4130	4180	4230	4285	4335	4385	4440	4490	4540	4590	4645	4695	4745	4800	4850
18		3890	3940	3995	4045	4095	4145	4200	4250	4300	4350	4405	4455	4505	4555	4610	4660	4710	4760	4815
20		3860	3910	3960	4015	4065	4115	4165	4215	4265	4320	4370	4420	4470	4520	4570	4625	4675	4725	4775
22		3830	3880	3930	3980	4030	4080	4135	4185	4235	4285	4335	4385	4435	4485	4535	4585	4635	4685	4735
24		3785	3835	3885	3935	3985	4035	4085	4135	4185	4235	4285	4330	4380	4430	4480	4530	4580	4630	4680
26		3755	3805	3855	3905	3955	4000	4050	4100	4150	4200	4250	4300	4350	4395	4445	4495	4545	4595	4645
28		3725	3775	3820	3870	3920	3970	4020	4070	4115	4165	4215	4265	4310	4360	4410	4460	4510	4555	4605
30		3695	3740	3790	3840	3890	3935	3985	4035	4080	4130	4180	4230	4275	4325	4375	4425	4470	4520	4570
32		3665	3710	3760	3810	3855	3905	3950	4000	4050	4095	4145	4195	4240	4290	4340	4385	4435	4485	4530
34		3620	3665	3715	3760	3810	3855	3905	3950	4000	4045	4095	4140	4190	4245	4285	4330	4380	4425	4475
36		3585	3635	3680	3730	3775	3825	3870	3920	3965	4010	4060	4105	4155	4200	4250	4295	4340	4390	4435
38		3555	3605	3650	3695	3745	3790	3840	3885	3930	3980	4025	4070	4120	4165	4210	4260	4305	4350	4400
40		3525	3575	3620	3665	3710	3760	3805	3850	3900	3945	3990	4035	4085	4130	4175	4220	4270	4315	4360
42		3495	3540	3590	3635	3680	3725	3770	3820	3865	3910	3955	4000	4050	4095	4140	4185	4230	4280	4325
44		3450	3495	3540	3585	3630	3675	3725	3770	3815	3860	3905	3950	3995	4040	4085	4130	4175	4220	4270
46		3420	3465	3510	3555	3600	3645	3690	3735	3780	3825	3870	3915	3960	4005	4050	4095	4140	4185	4230
48		3390	3435	3480	3525	3570	3615	3655	3700	3745	3790	3835	3880	3925	3970	4005	4060	4105	4150	4190
50		3345	3390	3430	3475	3520	3565	3610	3650	3695	3740	3785	3830	3870	3915	3960	4005	4050	4090	4135
52		3315	3355	3400	3445	3490	3530	3575	3620	3660	3705	3750	3795	3835	3880	3925	3970	4010	4055	4100
54		3285	3325	3370	3415	3455	3500	3540	3585	3630	3670	3715	3760	3800	3845	3890	3930	3975	4020	4060
56		3255	3295	3340	3380	3425	3465	3510	3550	3595	3640	3680	3725	3765	3810	3850	3895	3940	3980	4025
58		3210	3250	3290	3335	3375	3420	3460	3500	3545	3585	3630	3670	3715	3755	3800	3840	3880	3925	3965
60		3175	3220	3260	3300	3345	3385	3430	3470	3500	3555	3595	3635	3680	3720	3760	3805	3845	3885	3930
62		3150	3190	3230	3270	3310	3350	3390	3440	3480	3520	3560	3600	3640	3680	3730	3770	3810	3850	3890
64		3120	3160	3200	3240	3280	3320	3360	3400	3440	3490	3530	3570	3610	3650	3690	3730	3770	3810	3850
66		3070	3110	3150	3190	3230	3270	3310	3350	3390	3430	3470	3510	3550	3600	3640	3680	3720	3760	3800
68		3040	3080	3120	3160	3200	3240	3280	3320	3360	3400	3440	3480	3520	3560	3600	3640	3680	3720	3760
70		3010	3050	3090	3130	3170	3210	3250	3290	3330	3370	3410	3450	3480	3520	3560	3600	3640	3680	3720
72		2980	3020	3060	3100	3140	3180	3210	3250	3290	3330	3370	3410	3450	3490	3530	3570	3610	3650	3680
74		2930	2970	3010	3050	3090	3130	3170	3200	3240	3280	3320	3360	3400	3440	3470	3510	3550	3590	3630

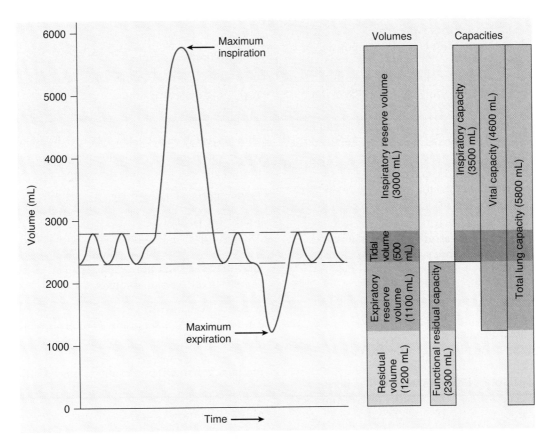

Figure 36.3 Pulmonary Volumes and Capacities

$$\frac{\text{Your vital capacity}}{\text{Normal vital capacity}} \times 100 = \text{Percent of normal vital capacity}$$

Your percent of normal vital capacity: _____

If you come out at 100% of normal, then you have an average vital capacity for a person in your bracket. A percent vital capacity of 80% of average, or better, is in the normal range. If your value is less than 100%, then you have a smaller vital capacity than the average person for your bracket; if your value is larger than 100%, then you have a larger vital capacity than average.

Calculation of the Inspiratory Reserve Volume

The vital capacity consists of the expiratory reserve volume, the tidal volume, and the inspiratory reserve volume. You can indirectly determine the **inspiratory reserve volume (IRV)** by subtracting the **expiratory reserve volume (ERV)** and the **tidal volume (TV)** from the **vital capacity (VC)**. This relationship is indicated in figure 36.3. You cannot measure the inspiratory reserve volume directly with the use of a hand-held spirometer (it records exhalations only). Calculate your inspiratory reserve.

$$IRV = VC - (ERV + TV)$$

Your IRV: _____

The IRV is the amount of air you inhale after a normal inhalation.

Biopac® Respiration Lab

Pulmonary volumes and capacities can be measured directly with the use of a respirometer or spirometer or they can be measured indirectly with the use of a flow meter. One such device is an airflow transducer made by Biopac®. You should have all of the necessary material for measuring pulmonary flow including an MP30 unit, transducer hardware, mouthpieces, bacteriological filters, a compatible computer, and software (Biopac Pro® or Biopac® Student Lessons). Every student in lab should record a spirogram. Make sure that everyone is able to have access to the equipment prior to doing any data analysis.

Recording Your Volumes and Capacities with Biopac®

Caution
When you change subjects always use a new disposable bacteriological filter.

Calibrating the Machine

1. Make sure that the computer is turned on, that the MP30 unit is connected to the computer and turned on, and that the transducer is plugged into the MP30 unit.

2. Select **Lesson 12: Pulmonary 1-Volumes & Capacities (L012-Lung-1)** and type your name where appropriate. If you have a folder with your name already on the computer, then select **Use it** if prompted.
3. Place a new bacteriological filter into the end with the calibration syringe. Insert the syringe into the side of the transducer labeled **Inlet**.
4. Pull the plunger of the calibration syringe all the way out by holding the syringe by the body. Do not hold the unit by the transducer.
5. Click **Calibrate**, read all directions and click **OK**.
6. Push the plunger in and out for a total of 5 cycles (5 in and 5 out). This should take about 30 seconds.
7. Click on **End Calibration**.
8. If the calibration looks abnormal on the screen, then click **Redo**. If it looks okay, then begin the recording of the data.

Recording Data

1. Place a new mouthpiece into the transducer where the calibration syringe was.
2. Pinch off your nose with a nose clip.
3. Click on **Record** and begin breathing easily into the transducer.
4. Take three normal breaths (**tidal volume**) and then inhale maximally and exhale normally (**inspiratory reserve volume**) as seen in figure 36.3.
5. Take a few normal breaths again and then exhale maximally (**expiratory reserve volume**).
6. Take a few normal breaths and then inhale maximally and exhale maximally (**vital capacity**).
7. Click on **Stop**.

If you are not satisfied with your recording (compare what you see with figure 36.3), click on **Redo**. If you are happy with your recording, then click on **Done** and remove the bacteriological filter from the apparatus.

You can select **Review Saved Data** from the menu.

Data Analysis

Select **CH2** (which is the volume) and select **p-p** (peak to peak) which selects the range of minimum and maximum values. Select CH2 and click on "max" then select CH2 and click on "min." Select CH2 again and click on "delta." Compare your data with figure 36.3.

Select the I-beam and highlight one of the tidal volume measurements from the peak to the depth of a breath. Save this data as "Tidal Volume."

Select the area from the peak of a normal inhalation to the maximum air inhaled and save the data as "Inspiratory Reserve Volume."

Find the point of a normal exhalation (after a tidal volume) to the complete exhalation and save the data as "Expiratory Reserve Volume."

Find the point of the maximum inhalation to the maximum exhalation and save the data as "Vital Capacity."

Enter your data from the Biopac® Respiration Lab in the following section.

Tidal Volume   _____
Inspiratory Reserve Volume   _____
Expiratory Reserve Volume   _____
Vital Capacity   _____

Your percent normal vital capacity can be calculated as it was using wet spirometers or hand-held spirometers. Record your percent normal vital capacity in the following space.

Percent normal vital capacity: _____

Residual Volume

There is still air left in the lungs after a maximal exhalation. This is known as the **residual volume (RV)**, and it is approximately 1000 mL. The **total lung capacity (TLC)** is the sum of the vital capacity and the residual volume.

Minute Ventilation

The total amount of air inspired and expired in 1 minute of normal (tidal volume) breathing is known as the **minute ventilation.** This is calculated by multiplying the number of breaths per minute by the tidal volume. Record your minute volume in the following space.

Minute volume: _____

Respiratory Sounds

In this section, you listen to breathing sounds, which can be heard with a stethoscope.

1. Before you begin the experiment, clean the earpieces of the stethoscope with an alcohol wipe and let them dry. The stethoscope earpieces should point toward the anterior as you insert them into your ears.
2. Locate the larynx of your lab partner and place the diaphragm of the stethoscope just inferior to it.
3. Listen for the sound as your lab partner inhales and exhales. The sound should be smooth with no gurgling, rasping, or rattling. These are the tracheal and bronchial sounds.
4. Locate the triangle of auscultation (figure 36.4), which is just medial to the inferior angle of the scapula. This is an ideal area for listening to lung sounds because the thoracic cage is not covered by muscles in this location.
5. Have your lab partner inhale and exhale deeply several times.
6. Listen for a smooth flow of air into and out of the lungs. Wheezing or rattling indicates congestion in the lungs.
7. Record the sounds you hear in the space provided and indicate the condition of your lab partner.

Breathing sounds and condition: _____
_____
_____

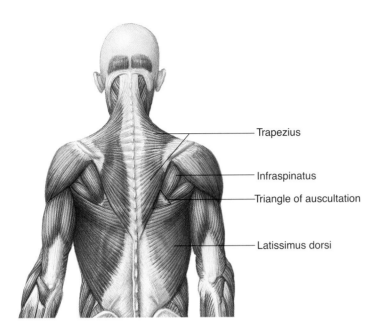

Figure 36.4 Triangle of Auscultation
Find the area between the trapezius, the medial border of the scapula, and the latissimus dorsi to listen to respiratory sounds.

Flow and Resistance

The flow of air is proportional to the pressure of the air and inversely proportional to the resistance in the airways $\left(\text{Flow} = \dfrac{\text{Pressure}}{\text{Resistance}}\right)$.

As the pressure difference between the outside air and the air inside the lungs increases, the flow of air increases. If the resistance in the air passageways increases, then the flow rate decreases. You can see the effects of changing the resistance of the airways in a simple demonstration. Do NOT do this experiment if you have a respiratory condition such as asthma, a cold, or severe sinus allergies.

1. Using a stopwatch or a watch with a second hand, record how long it takes to forcibly inhale the maximum amount of air that your lungs can hold. Do this by breathing through both your mouth and your nose. Record the time in the following space.

   Time for maximum inhalation: _____ seconds

2. Close your mouth and one nostril and try the experiment again. Breathe through only one nostril, and record the time that it takes to maximally inhale. Record this in the following space.

   Time for inhalation through one nostril:
   _____ seconds

   Closing off the respiratory passages (mouth and one nostril) increases the resistance in the respiratory system. How does this change the rate of airflow? _____

Asthma occurs due to a restriction of the bronchioles in the lung. If this occurs, is there a change in the pressure between the alveoli and the external air or an increase in the resistance to airflow into the lungs? _____

Acid–Base Effects of the Respiratory Gases

The respiratory system is involved in helping to maintain an adequate acid–base balance in the blood. If the blood is too acidic (has too many hydrogen ions), an increase in breathing rate can decrease that level. This is due to the interaction of water and carbon dioxide in aqueous solutions.

Carbon dioxide combines with water to form carbonic acid, which subsequently dissociates into bicarbonate and hydrogen ions, as represented in the following equation:

$$\underset{\text{Carbon dioxide}}{CO_2} + \underset{\text{Water}}{H_2O} \rightarrow \underset{\text{Carbonic acid}}{H_2CO_3} \rightarrow \underset{\text{Bicarbonate ion}}{HCO_3^-} + \underset{\text{Hydrogen ion}}{H^+}$$

The chemical equation can proceed from left to right where carbon dioxide and water form bicarbonate ion and hydrogen ions, or it can proceed from right to left where an excess of hydrogen ions and bicarbonate can produce carbon dioxide and water. If you exhale carbon dioxide, the respiratory system assists in decreasing the hydrogen ion level in the blood, preventing acidosis.

Carbon dioxide is a waste product of respiration. Increased carbon dioxide levels in the blood change the pH of the blood, and this can cause metabolic problems not only in the blood but elsewhere in the body such as in the brain, producing states of disorientation or confusion. In this part of the exercise you examine the impact of elevated carbon dioxide levels in solution.

1. Add about 30 mL of litmus solution to a small (100 mL) Erlenmeyer flask. If the solution is red, add small amounts of sodium bicarbonate until the color just begins to turn blue.
2. Cover most of the opening with tape or parafilm. Insert a drinking straw into the flask and *gently* blow air into the flask.
3. Bubble your exhaled breath into the flask and look for a color change. Blue indicates an alkaline condition, and red indicates an acid condition.
4. Using the preceding equation, determine how the exhalation into the flask alters the acid–base conditions.

   Observations: _____

   How does the exhalation of air into the litmus solution reflect the carbon dioxide in the breath? _____

Exercise 36 Review

Respiratory Function, Breathing, and Respiration

Name: _____

Lab time/section: _____

Date: _____

1. What is pulmonary ventilation? _____

2. Would rigorous exercise immediately increase, decrease, or have no effect on:

   Tidal volume _____

   Inspiratory reserve volume _____

   Expiratory reserve volume _____

   Vital capacity _____

3. What is the tidal volume in liters for an average adult? _____

4. Define *tidal volume*. _____

5. What instrument do you use to measure breathing volumes? _____

6. If you inhale maximally, what is the name of the volume of air that you completely exhale? _____

7. Examine the predicted vital capacity chart. What is the approximate percent decrease of vital capacity in the same person from age 25 to age 75? _____

8. How does the decrease in vital capacity potentially influence a person's athletic performance or aerobic condition as he or she ages? _____

9. What is the pressure difference between the external air and the pleural cavity when inhalation just begins? _____

10. Calculate the IRV of a person with a vital capacity of 4360 mL, an expiratory reserve volume of 1300 mL, and a tidal volume of 500 mL. _____

11. How does carbon dioxide change the acid–base condition of a solution when present in excess? _____

Exercise 37
Physiology of Exercise and Pulmonary Health

INTRODUCTION

An understanding of exercise physiology is important not only for athletic training but also for wellness and general health. Significant changes occur during exercise that affect many organ systems of the body. The changes vary, depending on the type of exercise. Exercise is broadly classified as either **aerobic exercise** or **anaerobic exercise.** Aerobic exercises increase heart rate and breathing rates at moderate levels for extended periods of time. Aerobic exercises do not deplete the level of oxygen consumption. Anaerobic exercises result in the consumption of available oxygen faster than it can be supplied to the muscle tissue. The tissue uses glucose anaerobically and produces lactic acid.

Exercise has a profound effect on the muscular, skeletal, cardiovascular, and respiratory systems. Consider the increased metabolic demands placed on skeletal muscles during repeated contractions. Skeletal muscle is more metabolically active when contracting than when at rest, and it uses additional oxygen and nutrients. Oxygen diffuses from the blood to the muscle tissue due to the concentration gradient of oxygen between these two areas. In addition to the diffusion of oxygen, the precapillary sphincters of the circulatory system open and the arterioles dilate, providing greater blood flow to the muscles. This additional blood flow requires an increase in the volume of blood passing through the heart with each contraction. Heart size increases and pulse rate decreases with long-term rigorous exercise.

The respiratory system responds to this greater oxygen demand with increased volume per breath and a greater number of breaths per minute. This increases the minute volume. The lung capillaries expand as well, and a greater diffusion of oxygen occurs between the alveoli and the blood capillaries. This respiratory response is covered in the *Principles of Anatomy and Physiology* text in chapter 20, "Respiratory System."

In this exercise, you examine respiratory health, the effects of exercise on the body, and the body's comparative responses to aerobic exercise.

OBJECTIVES

At the end of this exercise, you should be able to

1. list the major organ systems directly involved in fitness;
2. describe basic physiologic differences between a person who is physically active and one who is physically inactive;
3. determine the forced expiratory volume exhaled in 1 second;
4. calculate the personal fitness index of a subject who performs the Harvard step test or Cooper's 12-minute run test.

MATERIALS

Biopac® setup
AFT6 – 6000 mL calibration syringe
AFT1 – Disposable Bacteriological Filter
AFT2 – Disposable Mouthpiece
SS11L or SS11LA – Airflow Transducer
MP30 – Data Acquisition Unit
Compatible computer
Noseclips
16-inch step
20-inch step
Metronome or clock with second hand
Treadmill

PROCEDURE

Pulmonary Health

Forced Expiratory Vital Capacity (FEV)

Indications of health can be roughly correlated with the amount of air expelled from the lungs in 1 second. This is usually expressed as a percent when compared to the person's vital capacity (VC) as $FEV_1/VC\%$. This should be approximately 75% of the VC in healthy adults. In this exercise you will use the Biopac setup, which is similar to the one used in Exercise 36. Decreases in forced expiratory vital capacity may be caused by asthma, emphysema, or other pulmonary conditions.

Biopac Lesson 13 - Pulmonary Function II

1. Setup

Make sure that the computer is on but the MP30 unit is off. Plug in the airflow transducer to the MP30 unit and turn the unit on. Open the Biopac Student Lab (BSL) Software and select Biopac Lesson 13 - Pulmonary Function II. You will collect data for your respiratory volumes by breathing into a flow meter that measures the force of air passing through it, which it then will convert to a volume. The resulting "spirogram" is displayed on the computer monitor and saved as a file in your Biopac folder. These are the steps you will take:

Click on the **Biopac** icon on the computer desktop to start Biopac. A menu of lessons will appear. Select **Biopac Lesson 13 - Pulmonary Function II.** Type in your (folder) **name.** If you have a folder on this computer station a window should appear with the message "A folder with this name already exists. Would you like to use it or create a new folder?" Choose **Use it.**

Place the bacterial filter onto the end of the calibration syringe. Insert the calibration syringe/filter assembly into the side of the airflow transducer labeled **"Inlet."**

2. Calibration

Pull the calibration syringe plunger all the way out and hold the calibration syringe horizontal so that the airflow transducer is upright. It must remain vertical for the calibration and experimentation as the membrane is sensitive to changes in deflection due to gravity. Hold the calibration syringe by the barrel (body), not by the airflow transducer (connecting duct). Click on **Calibrate.** After the first stage of the calibration is recorded a second dialog box will appear prompting you to read all of the directions. Do so and click on **Yes.** Cycle the syringe plunger in and out 5 times (10 strokes). Use a rhythm of about 1 second per stroke with 2 seconds between each stroke. Click on **End Calibration.** If the calibration looks good, then detach the calibration syringe and proceed to Step 3 (**Record**). (You should be able to see 5 downward and 5 upward deflections on the screen.) If the calibration peaks are uneven or significantly vary from one another, click **Redo.**

3. Record Data

Insert a clean mouthpiece (where the calibration syringe had been attached). Place a clean noseclip on your nose. Make sure that you hold the transducer upright at all times. Click on **Record** and breathe as follows:

Be sure to start the recording before taking the first normal breath.

1. Take 3 normal breaths (3 inhales and 3 exhales).
2. Inhale as deeply as you can (take in as much air as you can). Hold it for an instant.
3. Exhale as fast and as much as you can.
4. Click on **Stop.**

If the recording is not good, click on **Redo** to repeat it. If the recording looks good, click on **Done.** Select **Record from Another Subject** or, if you have recorded everyone's data, select **Analyze Current Data File.**

4. Data Analysis for FEV

Open your folder in the **Data Files** folder (probably listed as "*your name FEV-L13*"), and then open the data file. Select **Display Preferences** from the **File** menu, choose **Grids**, click on **Show Grids** and then **OK.**

Use the **I-beam** to highlight your graph. The **p-p** (peak to peak) will represent your vital capacity.

Use the **I-beam** to select the volume of air you exhaled for the first second. This will be the FEV_1.

Save and print the entire spirogram you record to include in your lab report.

5. Interpretation

Spirometry is an important measurement tool to assist in diagnosing asthma, chronic obstructive pulmonary disease (COPD), and cystic fibrosis.

Pulmonary obstruction occurs when the respiratory passages are narrowed. This can be due to conditions such as asthma, excess mucus, and inflammation such as bronchitis. In these cases the vital capacity of the individual is normal but the $FEV_1/VC\%$ is low. It takes the subject longer to exhale completely.

Pulmonary restriction occurs when the lungs cannot fully inspire or expire the full volume of air. In these cases the vital capacity of the individual is reduced. This can be due to fibrosis of the lungs (cystic fibrosis, or fibrosis due to asbestos, or silica), scarring of the lung tissue as when a subject has had chronic lung infections, adhesions of the lung to the chest wall due to extreme emphysema, or a subject who has had a section of lung removed. It may also be due to damage to the phrenic nerve which stimulates the diaphragm. In these cases the vital capacity of the individual is low but the $FEV_1/VC\%$ is normal.

Most COPD is caused by smoking. Clinical values for FEV_1 compared to vital capacity are listed in Table 37.1.

TABLE 37.1

$FEV_1/VC\%$	Status
100–75	Normal
74–60	Mild COPD
59–50	Moderate COPD
<50	Severe COPD

Measurement of Heart Rate

The measurement of fitness in this part of the exercise is on a voluntary basis. It is best if the class can obtain data from a person who regularly participates in aerobic exercises (three to six times per week) and from a person who does not exercise. The Harvard step test and the Cooper's 12-minute run test are two reliable measurements of fitness.

Caution

Do not do these exercises if you are at risk for heart disease or have a family history of heart disease or another condition, such as asthma, for which exercise is harmful. Stop if you feel exhausted or faint; if you develop chest pain or pain radiating down the left arm, seek medical attention immediately.

A correlation exists between heart rate and fitness level. The resting heart rate of people involved in regular, active aerobic exercise is lower than the rate of sedentary people. In addition, the heart recovers faster in people who have a regular exercise program than in those who do not exercise. In this experiment, record the measurements of at least two volunteers from the class. The greater the number of students who participate, the better the data. Read the entire exercise before beginning.

Harvard Step Test

The Harvard step test was developed during World War II at Harvard University to determine a person's physical fitness. In this exercise, students should select either the 20-inch step for people 5′8″ or taller or the 16-inch step for people under 5′8″. With one person acting as an observer, the subject should step up on the step in 1 second and down on the floor in another second, thus completing 30 complete cycles in 1 minute. The subject should keep the body upright and keep pace with the observer's count or with a metronome set at 60 beats per minute. The subject should exercise for at least 3 minutes but no more than 5 minutes. If the subject stops due to exhaustion, then the observer should note the time of exercise. After the period of exercise, the subject should sit and rest for 1 minute. The number of beats should be taken from 1 minute to 1 minute 30 seconds and recorded in the following space. (See "A" on figure 37.1.)

Pulse from 1 minute to 1 minute 30 seconds: _____

The subject should rest for another 30 seconds before the pulse is taken from 2 minutes to 2 minutes 30 seconds after exercise and recorded here. (See "B" on figure 37.1.)

Pulse from 2 minutes to 2 minutes 30 seconds: _____

The subject should remain seated for another 30 seconds before the pulse is taken from 3 minutes to 3 minutes 30 seconds and recorded here. (See "C" on figure 37.1.)

Pulse from 3 minutes to 3 minutes 30 seconds: _____

Add the three pulse counts recorded.

Sum of three pulse counts: _____

To determine the personal fitness index (PFI), use the following formula:

$$\text{PFI} = \frac{\text{Number of seconds of exercise}}{2 \text{ (sum of three pulse counts)}} \times 100$$

Therefore, if you exercised for 4 minutes and your pulse counts were 50, 48, and 45, then the PFI was

$$\text{PFI} = \frac{(4 \times 60) \times 100}{2(50 + 48 + 45)} = \frac{240 \times 100}{2(143)} = \frac{24{,}000}{286} = 84$$

Determine the fitness evaluation of the subject using the following data:

Below 55: poor physical condition
55–64: low average physical condition
65–79: high average physical condition
80–89: good physical condition
90 and above: excellent physical condition

Record the PFI of the subject: _____

What is the fitness evaluation for this person: _____

Cooper's 12-Minute Run Test

Another fitness test involves determining how far a person can run and/or walk in 12 minutes.

Caution
As with the Harvard step test, do not do these exercises if you are at risk for heart disease or have a family history of heart disease or another condition, such as asthma, for which exercise is harmful.

Procedure

Warm up for 15 minutes prior to doing this test. Find a suitable running track or use a treadmill that is set to no incline and run or walk as fast and as far as you can in 12 minutes. You will not be able to sprint for the entire time, so pace yourself. If you cannot run anymore during the test, you should finish the test by walking or jogging or a combination of running and walking.

Record the distance you traveled in 12 minutes: _____

The results of your test and general fitness can be judged by the following table.

Under Age 40

Distance in 12 minutes	Physical Condition
Over 2700 m	Excellent
2300 m to 2700 m	Good
1900 m to 2299 m	Average
1500 m to 1899 m	Below average
Less than 1500 m	Poor

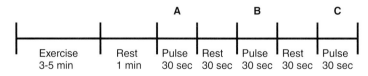

Figure 37.1 Harvard Step Test

Over Age 40

Distance in 12 minutes	Physical Condition
Over 2500 m	Excellent
2100 m to 2500 m	Good
1700 m to 2099 m	Average
1500 m to 1699 m	Below average
Less than 1400 m	Poor

Note:

During the 2005 Tour de France, Lance Armstrong had a resting heart rate of 32–34 bpm. This is less than half the rate of the average resting heart rate of 72 bpm. He had 4% body fat, whereas the average person watching the race had body fat of 15–20%. The maximum amount of oxygen he could use (VO_2 max) was 84 mL/kg/min (6 liters of oxygen per minute). A physically active person has a VO_2 max of 40–50 mL/kg/min, whereas a physically inactive person has a VO_2 max of 30–40 mL/kg/min. The normal oxygen use at rest is about 3.5 mL/kg/min.

Exercise 37 Review

Physiology of Exercise and Pulmonary Health

Name: _____

Lab time/section: _____

Date: _____

1. Define *FEV*. _____

   _____

   _____

2. Record your forced expiratory vital capacity or the one you measured in lab. _____

3. Does your forced expiratory vital capacity value fall within normal limits? _____

4. How do the FEV_1 and the VC compare in a person with a pulmonary obstructive condition such as asthma? Why? _____

   _____

5. How do the FEV_1 and the VC compare in a person with a pulmonary restrictive condition such as asbestosis? Why? _____

   _____

6. If a person had a smaller body and therefore a smaller vital capacity, would the $FEV_1/VC\%$ necessarily change?

   _____

7. What is your personal fitness index or the one measured in lab? _____

8. If the heart rate after 5 minutes of exercise was 70, 68, and 66 beats in the consecutive 30-second trials, what was the personal

   fitness index and what condition does that represent? _____

   _____

9. Record the results of the Harvard step test or the Cooper's 12-minute run test for the members in the class who performed them. Next to the Personal Fitness Index (Harvard) or Physical Condition (Cooper's) for your own results, write an *S* if you smoke cigarettes or an *N* if you are a nonsmoker.

   PFI (Personal Fitness Index)
   Physical Condition

1	2	3
4	5	6
7	8	9
10	11	12
13	14	15
16	17	18
19	20	21
22	23	24

Exercise 38
Anatomy of the Digestive System

INTRODUCTION

The digestive system can be divided into two major parts, the **digestive tract** and the **accessory organs.** The digestive tract is a long tube that runs from the mouth to the anus and comes into contact with food and the breakdown products of digestion. The digestive tract is also known as the **alimentary canal** and is commonly called the "food tube." Some of the organs of the digestive tract are the esophagus, stomach, intestines, colon, rectum, and anus. The accessory organs are important in that they secrete many substances necessary for digestion, yet these organs do not come into direct contact with food. Examples of accessory organs are the salivary glands, liver, gallbladder, and pancreas.

The functions of the digestive system are many. They include the ingestion of food, the physical breakdown of food, the chemical breakdown of food, food storage, absorption of water and products of digestion, vitamin synthesis, and the elimination of indigestible material. The anatomy and physiology of the digestive system are covered in the *Principles of Anatomy and Physiology* text in chapter 21, "Digestive System."

In this exercise, you examine the anatomy of the digestive system in both human and cat, correlating the structure of the digestive organs with their functions.

OBJECTIVES

At the end of this exercise, you should be able to

1. list, in sequence, the major organs of the digestive tract;
2. describe the basic functions of the accessory digestive organs;
3. note the specific anatomical features of each major digestive organ;
4. describe the layers of the wall of the digestive tract;
5. describe the major functions of the stomach and the small and large intestines.

MATERIALS

Models, charts, or illustrations of the digestive system
Hand mirror
Cats
Dissection trays
Scalpels
Latex gloves
Animal waste disposal container
Skull, human teeth, or cast of teeth
Cadaver (if available)
Microscopes
Microscope slides
    Esophagus   Large intestine
    Stomach   Liver
    Small intestine

PROCEDURE
Overview of the Digestive System

Begin this exercise by examining a torso model or charts in the lab, and compare them with figure 38.1. Locate the major digestive organs and place a check mark in the appropriate space.

_____ Oral cavity   _____ Liver
_____ Pharynx   _____ Pancreas
_____ Esophagus   _____ Anus
_____ Stomach   _____ Gallbladder
_____ Small intestine   _____ Salivary glands
_____ Large intestine

Organs of the Upper Digestive Tract

Begin your study of the digestive tract with the mouth and the oral cavity. Examine a midsagittal section of the head, as represented in figure 38.2, and locate the major anatomical features.

At the beginning of the digestive tract is the **mouth,** which is the opening surrounded by lips, or **labia.** The **labial frenulum** is a membranous structure that keeps the lip adhered to the gums, or **gingivae.** Behind the mouth is the **oral cavity,** which is a space bordered in front by the mouth, behind by the oropharynx, and on the

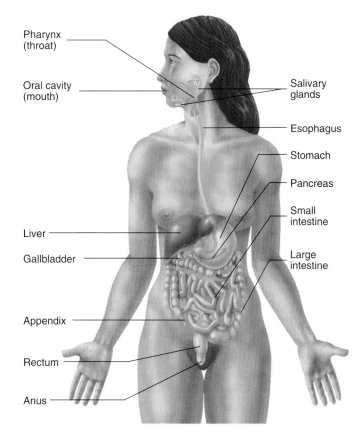

Figure 38.1 Overview of the Digestive System

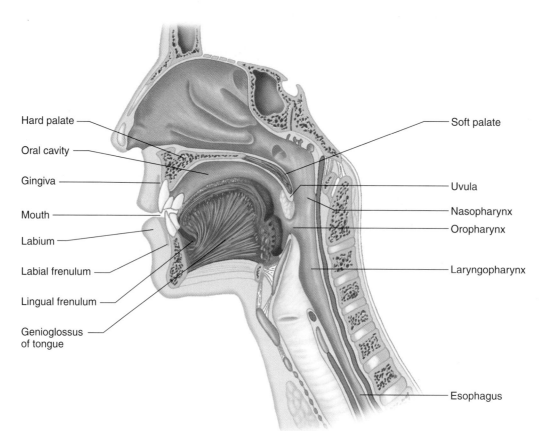

Figure 38.2 Oral Cavity, Midsagittal Section

sides by the inner wall of the cheeks. The hard and soft palates form the roof of the oral cavity, and the floor of the chin is the inferior border. The **hard palate** is composed of the palatine processes of the maxillae and palatine bones. The **soft palate** is composed of connective tissue and a mucous membrane. At the posterior portion of the oral cavity is the **uvula,** a small, grapelike structure suspended from the posterior edge of the soft palate. The uvula helps prevent food and liquid from moving into the nasal cavity during swallowing. The oral cavity is lined with **nonkeratinized stratified squamous epithelium,** which protects the underlying tissue from abrasion. Using a mirror, examine your tongue, which is made of skeletal muscle. One of the major muscles of the tongue is the **genioglossus.** The tongue is important in speech, taste, the movement of food toward the teeth for chewing, and swallowing. The tongue acts as a piston to propel food to the **oropharynx,** the space behind the oral cavity. The tongue is held down to the floor of the mouth by a thin mucous membrane called the **lingual frenulum.**

There are four types of **papillae,** or raised areas, on the tongue: the **foliate, fungiform, filiform,** and **circumvallate** papillae (figure 38.3). Foliate papillae are leaf-shaped, fungiform are mushroom-shaped, filiform are threadlike, and circumvallate are large papillae. Papillae increase the frictional surface of the tongue. The tongue also has taste receptors located in **taste buds.** Taste buds are located on the tongue along the sides of some of the papillae. The sense of taste is covered in Exercise 21.

The oral cavity is important for the physical breakdown of food. This process is driven by powerful muscles called the **muscles of mastication.** The **masseter** and the **temporalis muscles** are involved in the closing of the jaws, and the **pterygoid muscles** are important in the sideways grinding action of the molar and premolar teeth. The **digastric** and **platysma** muscles open the mandible.

Structure of Teeth

1. Examine models of teeth, dental casts, or real teeth on display in the lab and compare them with figure 38.4. A tooth consists of a **crown, neck,** and **root.** The crown is the exposed part of the tooth, the neck is a constricted portion of the tooth normally located at the surface of the gingivae, and the root is embedded in the jaw.
2. Examine a model or an illustration of a longitudinal section of a tooth and find the outer **enamel,** an extremely hard material. Inside this layer is the **dentin,** which is made of bonelike material. The innermost portion of the tooth consists of the **pulp cavity,** which leads to the **root canal,** a passageway for nerves and blood vessels into the tooth. The nerves and blood vessels enter the tooth through the **apical foramen** at the tip of the root of the tooth. The teeth are in depressions in the mandible or maxilla called **alveolar sockets,** and they are anchored to the bone by periodontal ligaments.

There are four types of teeth in the adult mouth:

- **Incisors** are flat, bladelike front teeth that nip food. There are eight incisors in the adult mouth.
- **Canines,** or **cuspids,** are the pointed teeth just lateral to the incisors that shear food. There are four cuspids in the adult, and they can be identified as the teeth that have just one cusp, or point.

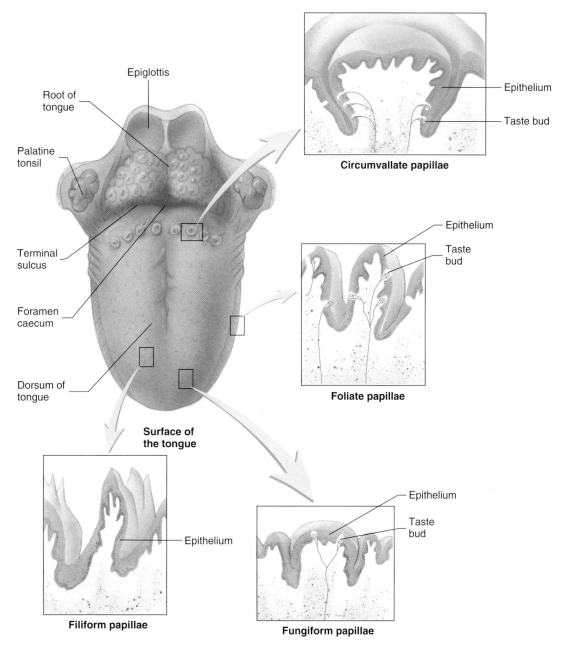

Figure 38.3 Tongue, Location of the Papillae on the Tongue, and Types of Papillae

- **Premolars,** or **bicuspids,** are posterior to the cuspids and grind food. There are typically eight premolars in adults, and they can be identified by their two cusps.
- **Molars** are the most posterior. Like the premolars, these teeth grind food. There are 12 molar teeth in the adult mouth (including the third molars, or **wisdom teeth**). Molars typically have three to five cusps.

3. Examine a human skull, isolated real teeth, or dental cast and identify the characteristics of the four types of teeth (figure 38.5).

    Humans have two sets of teeth. The **primary,** or **deciduous, teeth** (milk teeth) appear first, and these are replaced by the **secondary,** or **permanent, teeth.** There are 20 deciduous teeth. There are no deciduous premolar teeth, and there are only 8 deciduous molar teeth. In adults, there are 8 premolar teeth and 12 molar teeth.

4. Compare figure 38.5, the adult pattern, with figure 38.6, the deciduous teeth.

    Often, the pattern of tooth structure is represented by a dental formula, which describes the teeth by quadrants. The dental formula for the primary teeth is illustrated here: I = incisor, C = cuspid, P = premolar, and M = molar.

Deciduous Teeth (20 Total) Dental Formula

I	C	P	M	One side (quadrant)
2	1	0	2	Top
2	1	0	2	Bottom

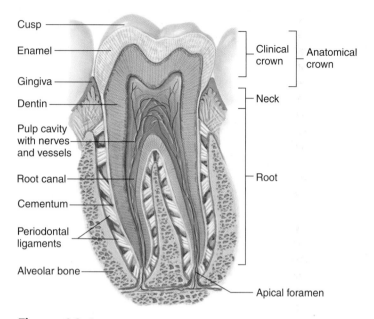

Figure 38.4 Molar Tooth, Longitudinal Section

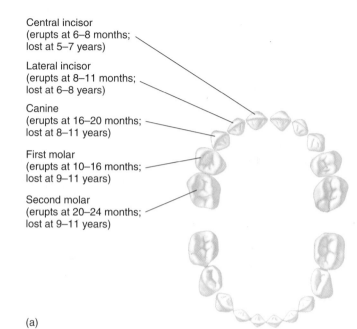

(a)

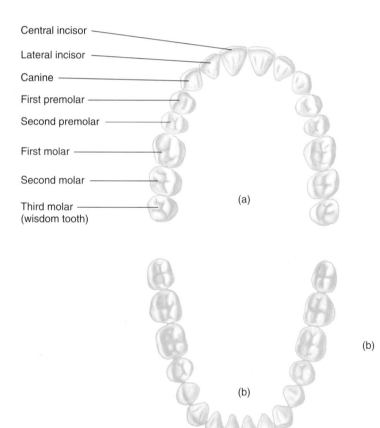

Figure 38.5 Adult (Permanent) Teeth
(a) Upper teeth; (b) lower teeth.

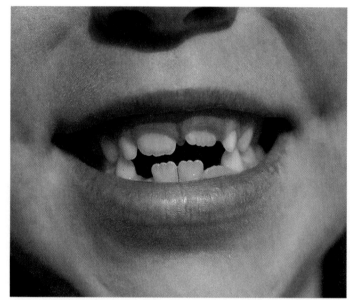

Figure 38.6 Deciduous Teeth
(a) Diagram; (b) photograph of replacement of teeth in a child.

The upper numbers refer to the number of teeth in the maxilla on one side of the head. The lower numbers refer to the number of teeth in the mandible on one side of the head. The adult dental formula is as follows:

Adult (Permanent) Teeth (32 Total)

	I	C	P	M
	2	1	2	3
	2	1	2	3

Pharynx

The space behind the oral cavity is the **oropharynx.** Above the oropharynx is the **nasopharynx,** which leads to the nasal cavity, and below the oropharynx is the **laryngopharynx,** which leads to the larynx and the esophagus. The nasopharynx begins at the internal nares and ends at the soft palate. The oropharynx begins there and continues to the larynx. Behind the larynx is the laryngopharynx. The oropharynx is composed of nonkeratinized stratified squamous epithelium and is a common passageway for food, liquids, and air. Muscles around the wall of the oropharynx are the **pharyngeal constrictor muscles** and are involved in swallowing. Food is moved by the tongue to the region of the pharynx, where it is propelled into the **esophagus.** Locate the oropharynx and the esophagus in figure 38.2.

Esophagus

The esophagus conducts food from the oropharynx, through the diaphragm, and into the stomach. Normally, the esophagus is a closed tube that begins at about the level of the sixth cervical vertebra. As a lump of food, or **bolus,** enters the esophagus, **skeletal muscle** begins to move it toward the stomach. The middle portion of the esophagus is composed of both skeletal and smooth muscle, whereas the lower portion of the esophagus is made of **smooth muscle.** In the esophagus, muscle contracts, moving the bolus by a process known as **peristalsis.** Locate the esophagus in charts and models in lab and compare it with figure 38.1.

The esophagus has an inner epithelial lining of stratified squamous epithelium and an outer connective tissue layer called the **adventitia.** The space in the esophagus that the food passes through is called the **lumen,** which continues through the digestive tract. The inferior portion of the esophagus has a **lower esophageal sphincter,** which prevents the backflow of stomach acids. Heartburn or gastroesophageal reflux disease (**GERD**) occurs if the stomach contents pass through the lower esophageal sphincter and irritate the esophageal lining.

Abdominal Portions of the Digestive Tract

The inner structure of the body has been referred to as a "tube within a tube." The body wall forms the outer tube, and the digestive tract, including the stomach, small intestine, and large intestine, forms the inner tube. Specialized serous membranes cover the various organs and line the inner wall of the **coelom** (the **body cavity**). The membrane lining the outer surface of the digestive tract is called the **visceral peritoneum (serosa).** It continues as a double-folded membrane called the **mesentery,** which attaches the tract to the back of the body wall. The mesentery is continuous with the membrane on the inner side of the body wall, where it is called the **parietal peritoneum.**

1. Locate these three membranes in figure 38.7.

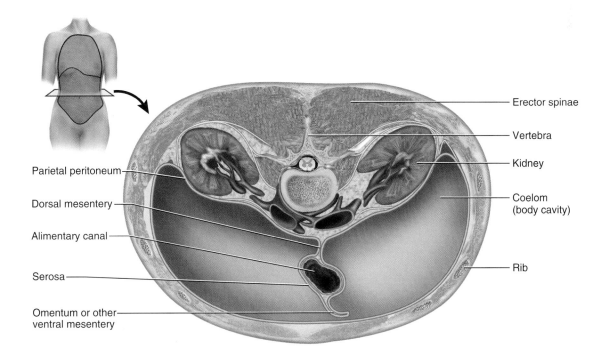

Figure 38.7 Membranes of the Digestive Tract

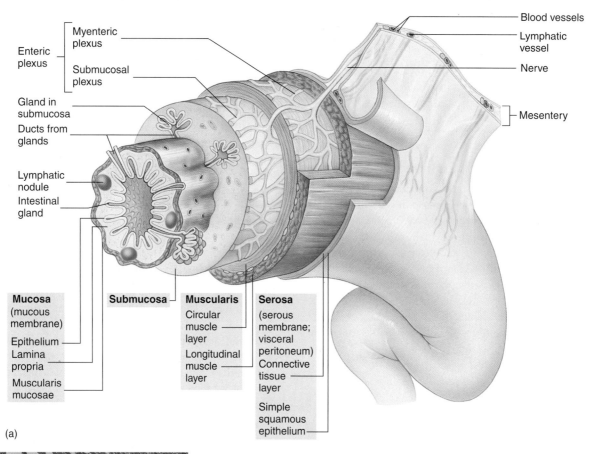

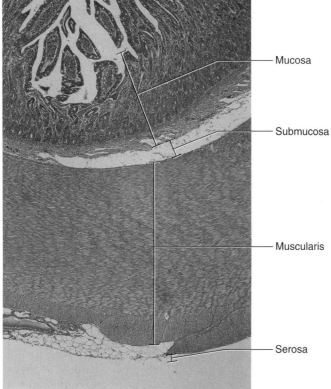

Figure 38.8 Gastrointestinal Tract (Small Intestine), Cross Section

(a) Diagram; (b) photomicrograph (100×).

2. Examine a microscope slide of the digestive tract (small intestine) under the microscope. In general, the stomach, small intestine, and large intestine have the same layers from the lumen to the coelom. The innermost layer is the **mucosa,** which consists of **mucous epithelium** closest to the lumen, a connective tissue layer called the **lamina propria,** and an outer muscular layer called the **muscularis mucosa.** These three layers are called the **mucous membrane.**

   The next layer is called the **submucosa,** which is mostly made of **connective tissue** and contains many blood vessels. The next layer is the **muscularis (muscularis externa),** which is typically made of two or three layers of smooth muscle. The muscularis propels material through the digestive tract and mixes ingested material with digestive juices. The outermost layer is called the **serosa,** or **visceral peritoneum,** and this layer is closest to the coelom.

3. Locate the layers, as represented in figure 38.8.

Stomach

The **stomach** is located on the left side of the body; it receives its contents from the esophagus. The food that enters the stomach is stored and mixed with enzymes and hydrochloric acid to form a soupy material called **chyme.** The stomach can have a pH as low as 1 or 2. Chyme remains in the stomach as the acids denature proteins and enzymes reduce proteins to shorter fragments.

Examine a model of the stomach or charts in the lab and compare them with figure 38.9. Locate the upper portion of the stomach, called the **cardiac part,** or **cardia.** A portion of the cardiac

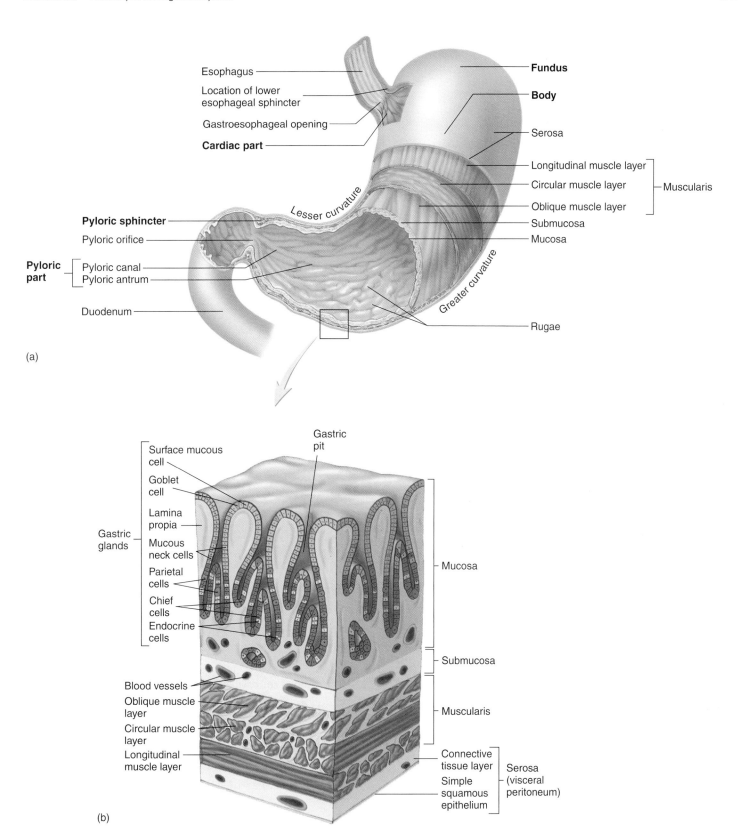

Figure 38.9 Anatomy of the Stomach
(a) Gross anatomy; (b) diagram of histology; (c) photomicrograph (40×); (d) mucosa (100×).

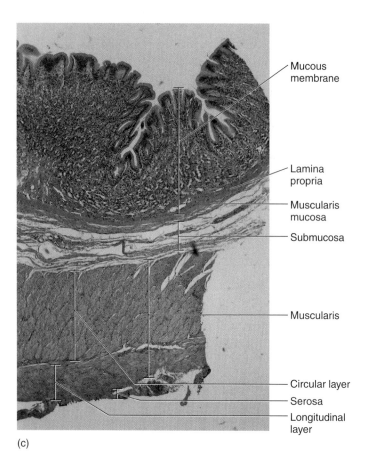

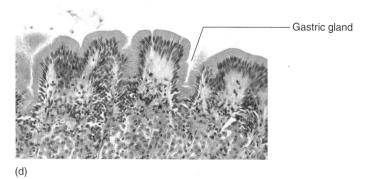

Figure 38.9—Continued.

part extends superiorly as a domed section called the **fundus.** The main part of the stomach is called the **body,** and the terminal portion of the stomach, closest to the small intestine, is called the **pyloric part.**

The pyloric part has an expanded area called the antrum and a narrowed region called the pyloric canal. The pylorus leads to the duodenum, and this opening is controlled by the **pyloric sphincter,** which prevents premature release of stomach contents into the small intestine.

The left side of the stomach is arched and forms the **greater curvature,** whereas the right side of the stomach is a smaller arch, forming the **lesser curvature.** The inner surface of the stomach has a series of folds called **rugae,** which allow for significant expansion of the stomach. The contents are held in the stomach by two sphinc-

ters. One is the **lower esophageal sphincter** and it prevents stomach contents from moving into the esophagus. The other sphincter is called the **pyloric sphincter.**

Stomach Histology

Examine a prepared slide of stomach tissue. Identify the four primary layers: **mucosa, submucosa, muscularis,** and **serosa.** Notice how the mucosa in the prepared slide has a series of indentations. These depressions are **gastric glands,** which are in the inner lining of the stomach. Find the three layers of the mucosa, including the **epithelial layer,** the **lamina propria,** and the **muscularis mucosa.** The epithelial layer consists of **simple columnar epithelium,** as well as specialized cells, such as **surface mucous cells, chief cells,** and **parietal cells.** Surface mucous cells secrete mucus, which protects the stomach lining from erosion by stomach acid and proteolytic (protein-digesting) enzymes. Chief cells secrete **pepsinogen** (the inactive state of a proteolytic enzyme). In some prepared slides, chief cells can be recognized by their blue-staining granules. Parietal cells secrete **HCl.** They typically contain orange-staining granules. When pepsinogen comes into contract with HCl, it is activated as **pepsin.** Deep to the epithelial layer is the lamina propria, a connective tissue layer. The deepest layer of the mucosa is the muscularis mucosa, a smooth muscle layer that moves the mucosa. The submucosa is the next layer and is typically lighter in color in prepared slides.

The next layer of the stomach is the **muscularis.** In some parts of the stomach, there are three layers of the muscularis—an inner **oblique layer,** a middle **circular layer,** and an outer **longitudinal layer.** The muscularis moves chyme from the stomach through the pyloric sphincter and into the small intestine.

The outermost layer is the **serosa,** and it is composed of a thin layer of connective tissue and **simple squamous epithelium.** Locate these structures and compare them with figure 38.9.

Small Intestine

The **small intestine** is approximately 5 m (17 feet) long in living humans, yet it can be longer in cadaveric specimens due to the relaxing of the smooth muscle. Movement through the small intestine occurs by peristalsis, which is smooth muscle contraction. It is called the small intestine because it is small in diameter. The small intestine is typically 3 to 4 cm (1.5 inches) in diameter when empty. The primary function of the small intestine is nutrient absorption.

1. Locate the three major regions of the small intestine on a model or chart and compare them with figure 38.10. The first part of the small intestine is the **duodenum,** which is a C-shaped structure attached to the pyloric region of the stomach. The duodenum begins in the body cavity and then moves behind the parietal peritoneum and returns to the body cavity. The duodenum is approximately 25 cm (10 inches) long. It receives fluid from both the **pancreas** and the **gallbladder.** The junction between the duodenum and the jejunum is known as the duodenojejunal flexure.

The second portion of the small intestine is the **jejunum,** which is approximately 2 m (6.5 feet) long. The terminal portion of the small intestine is the **ileum,** which is approximately 3 m (10 feet) long. A closure between the small intestine and large intestine is called the **ileocecal valve.** This

Exercise 38 Anatomy of the Digestive System

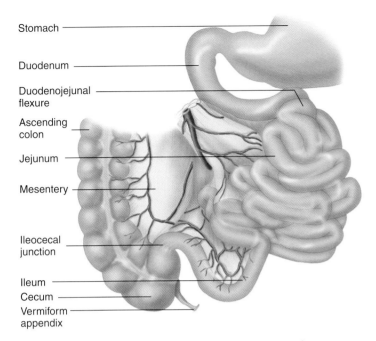

Figure 38.10 Gross Anatomy of the Small Intestine

valve keeps material in the large intestine from reentering the small intestine.

2. Locate the small intestine and associated structures in figure 38.10.

Histology of the Small Intestine

The small intestine, due to the presence of **villi,** can be distinguished from both the stomach and the large intestine. Villi are fingerlike projections that increase the surface area of the mucosa. Each villus contains **blood vessels** that transport sugars and amino acids from the intestine to the liver. In addition to this, the villi also contain **lacteals,** which transport fatty acids via lymphatics to the venous system. The villi give the lining of the small intestine a velvety appearance to the naked eye. As with the stomach, the inner lining of the small intestine consists of **simple columnar epithelium** with **goblet cells.** You may distinguish the three sections of the small intestine by noting that the duodenum has **duodenal glands** in the portion of the mucosa away from the lumen. The jejunum and the ileum lack these glands. The ileum is distinguished by the presence of **aggregated lymph nodules,** or **Peyer's patches,** in the mucosa. These lymph nodules produce lymphocytes, which protect the body from the bacterial flora in the lumen of the small intestine. Compare the prepared slides of the small intestine with figure 38.11.

Large Intestine

The **large intestine** is so named because it is large in diameter. The large intestine is approximately 7 cm (3 inches) in diameter and 1.4 m (4.5 feet) long. The function of the large intestine is the absorption of water and the formation of feces. The mucosa of the large intestine is made of **simple columnar epithelium** with a large number of **goblet cells.** Look at a model or chart of the large intestine and note the major regions. Compare them with figure 38.12.

- **Cecum:** first part of the large intestine. The cecum is a pouchlike area that articulates with the small intestine at the ileocecal valve.
- **Ascending colon:** on the right side of the body. It becomes the transverse colon at the **hepatic (right colic) flexure.**
- **Transverse colon:** traverses the body from right to left. It leads to the descending colon at the **splenic (left colic) flexure.**
- **Descending colon:** passes inferiorly on the left side of the body and joins with the sigmoid colon.
- **Sigmoid colon:** an S-shaped segment of the large intestine in the left inguinal region.
- **Rectum:** a straight section of colon in the pelvic cavity. The rectum has superficial veins in its wall called **hemorrhoidal veins.** They may enlarge and cause the uncomfortable condition known as **hemorrhoids.**

The large intestine has some unique structures. The longitudinal muscle of the large intestine is not a continuous sheet but is located along the length of the large intestine as three bands called **teniae coli.** These muscles contract and form pouches or puckers in the intestinal tract called **haustra** (sing. **haustrum**). Another unique feature of the outer wall of the large intestine is the fat lobules called **epiploic appendages.** Locate these structures in figure 38.12.

Fecal material passes through the large intestine by peristalsis and is stored in the rectum and sigmoid colon. Defecation occurs as mass peristalsis causes a bowel movement.

Histology of the Large Intestine

Examine a prepared slide of the large intestine and compare it with figure 38.13. The large intestine is distinguished from the small intestine by the absence of villi and from the stomach by the presence of large numbers of goblet cells. There are no villi or aggregated lymph nodules present, yet the wall of the large intestine has solitary lymph nodules. Examine your slide for these characteristics.

Anal Canal

The **anal canal** is not part of the large intestine but is a short tube that leads to an external opening, the **anus.** Locate the anal canal in figure 38.12.

Accessory Structures

Salivary Glands

The accessory structures are associated with segments of the digestive tract. The first of these structures, the **salivary glands,** are located in the head and secrete saliva into the oral cavity. Saliva is a watery secretion that contains **mucus,** a protein lubricant, and **salivary amylase,** a starch-digesting enzyme. The average adult secretes about 1.5 liters of saliva per day.

There are three pairs of salivary glands. The most superior of these, the **parotid glands,** are located just anterior to the ears. Each gland secretes saliva through a **parotid duct,** a tube that traverses the buccal (cheek) region and enters the oral cavity just posterior to the upper second molar. The **submandibular glands** are located medial to the mandible on each side of the face. The submandibular glands secrete saliva into the oral cavity by a single

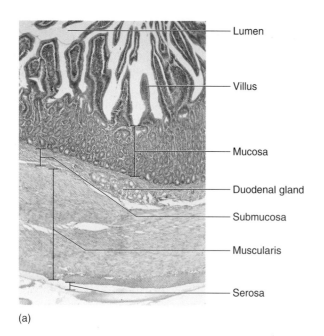
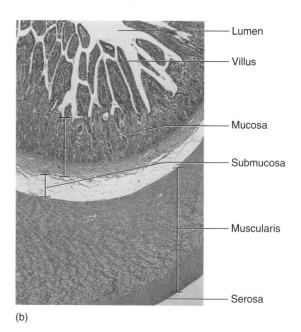
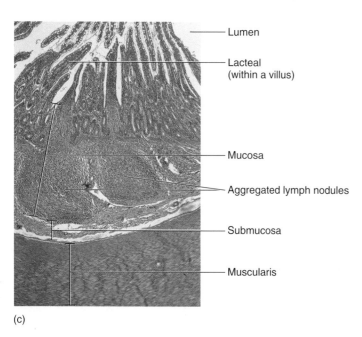
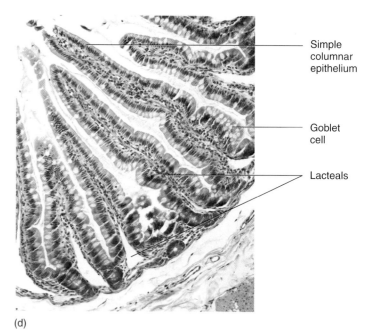

Figure 38.11 Sections of Small Intestine (40×)
(a) Duodenum; (b) jejunum; (c) ileum; (d) close-up of villus (100×).

duct on each side of the oral cavity inferior to the tongue. The **sublingual glands** are located inferior to the tongue and open into the oral cavity by several ducts. Locate these salivary glands on a model of the head and in figure 38.14.

Vermiform Appendix

The **vermiform appendix** is about the size of your little finger and is located near the junction of the small and large intestines (at the region of the ileocecal valve). Locate the appendix on a torso model or chart in the lab and compare it with figures 38.12 and 38.15.

Omenta

The **lesser omentum** is an extension of the peritoneum that forms a double fold of tissue between the stomach and the liver. The **greater omentum** is a section of peritoneum on the transverse colon and drapes over the intestines as a fatty curtain. Locate the omentum in figure 38.7.

Liver

The **liver** is a complex organ with many functions, some of which are digestive but many of which are not. The liver processes digestive

Exercise 38   Anatomy of the Digestive System

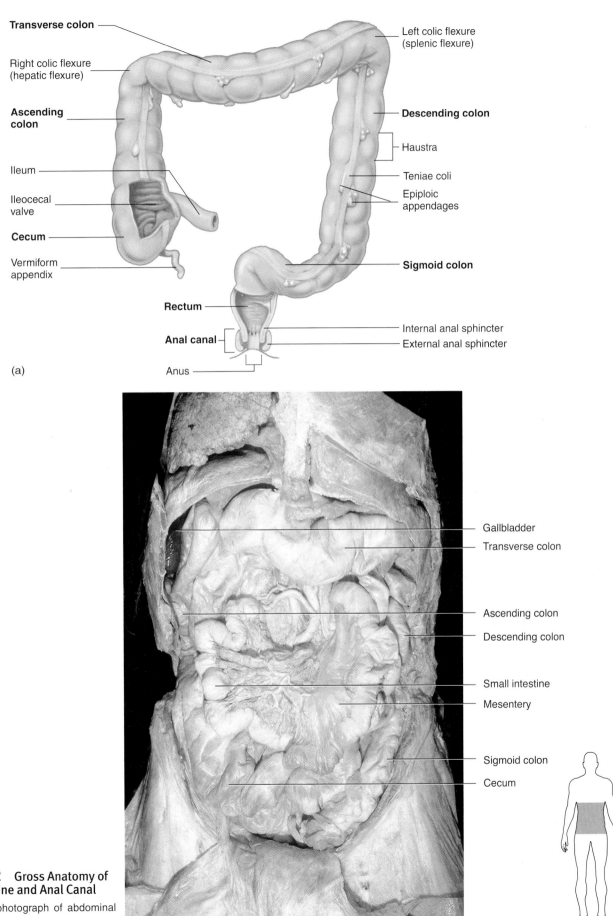

Figure 38.12 Gross Anatomy of the Large Intestine and Anal Canal

(a) Diagram; (b) photograph of abdominal cavity of the cadaver.

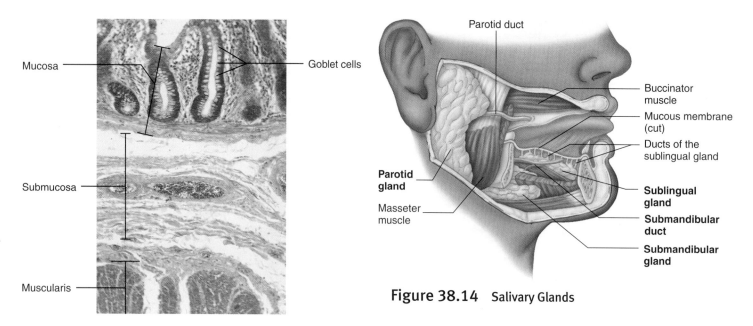

Figure 38.13 Histology of the Large Intestine (100×)

Figure 38.14 Salivary Glands

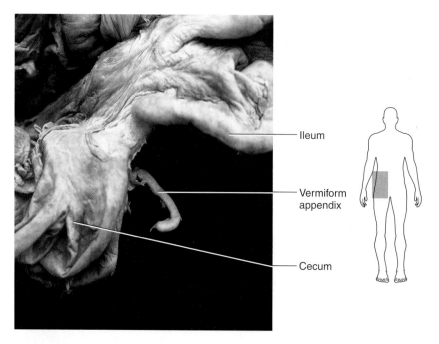

Figure 38.15 Vermiform Appendix

material from the vessels that return blood from the intestines and has a role in both moving nutrients into the bloodstream and storing them in the liver tissue. The liver also produces blood plasma proteins, detoxifies harmful material that has been produced by the body or introduced into the body, and produces bile.

Examine a model or chart of the liver and note its relatively large size. The liver is located on the right side of the body and is divided into four lobes: the **right, left, quadrate,** and **caudate lobes.** The right lobe of the liver is the largest lobe; the left is also fairly large. The quadrate lobe is located in the middle portion of the liver ventral to the caudate lobe. Traversing through the liver is the **falciform ligament,** which attaches the liver to the inferior side of the diaphragm. Locate the liver, and compare it with figure 38.16.

Liver Histology

Examine a prepared slide of liver tissue. Note the hexagonal structures in the specimen. These are the **liver lobules.** Each lobule has a blood vessel in the middle called the **central vein.** Vessels that carry blood to the central vein are the **liver sinusoids,** and these are lined with a double row of cells called **hepatocytes.** Hepatocytes carry out the various functions of the liver as just described. The **hepatic arteries,**

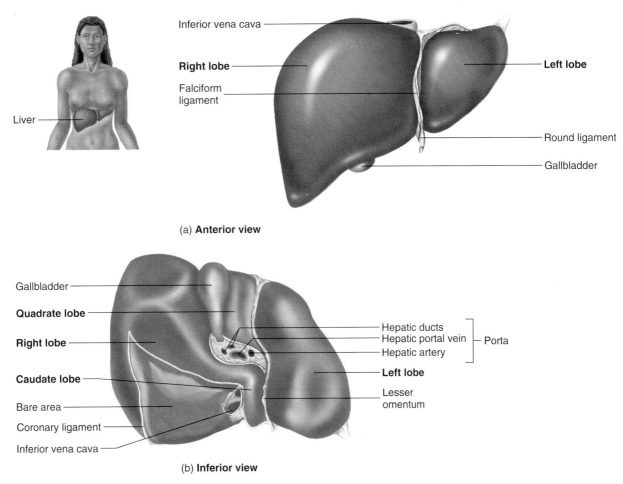

Figure 38.16 Liver
(a) Anterior view; (b) inferior view.

hepatic portal veins, and **hepatic ducts** frequently are located together, forming the **portal (hepatic) triad** at the corners of the liver lobules. The tissue of the liver is extremely vascular and functions as a sponge. Fresh, oxygenated blood from the hepatic artery and deoxygenated blood from the hepatic portal vein mix in the liver. **Kupffer cells,** found throughout the liver, are phagocytic cells. Locate the lobules, central vein, sinusoids, and hepatocytes in a prepared slide, using figure 38.17 to aid you.

Pancreas

The **pancreas** is located inferior to the stomach and on the left side of the body. It has both endocrine and exocrine functions. The hormonal function of the pancreas is covered in Exercise 24. The exocrine function of the pancreas is studied in Exercise 39. The pancreas consists of a **tail** near the spleen; an elongated **body;** and a rounded **head** near the duodenum. Enzymes and buffers pass from the tissue of the pancreas into the **pancreatic duct** and then into the duodenum. Locate the pancreatic structures, as represented in figure 38.18.

Gallbladder

The **gallbladder** releases bile, which emulsifies lipids, into the duodenum. The lipids break into smaller droplets, which increase the surface area for digestion. The gallbladder is located just inferior to the liver. The liver is the site of bile production, and the bile flows from the liver through **left** and **right hepatic ducts** (figure 38.18, step ①) to enter the **common hepatic duct.** Once in the common hepatic duct, the sphincter ④ closes. Bile fills the ducts and flows into the **cystic duct** ② and then is stored in the gallbladder. As the stomach begins to empty its contents into the duodenum, the gallbladder constricts, and bile flows from the gallbladder back into the cystic duct and into the **common bile duct,** which empties into the duodenum ④. Locate these structures on a model and compare them with figure 38.18.

The bile may be transported to the duodenum by the common bile duct. The pancreas secretes many digestive enzymes and buffers, which neutralize the stomach acids. These are secreted into the duodenum by the **pancreatic duct.** In some cases, the pancreatic duct joins with the **common bile duct** to form the **hepatopancreatic ampulla** (ampulla of Vater).

Cat Dissection

1. Prepare for the cat dissection by obtaining a dissection tray, a scalpel, scissors or bone cutter, string, forceps, and a plastic bag with a label. Remember to place all excess tissue in the appropriate waste container and not in a standard wastebasket or down the sink!

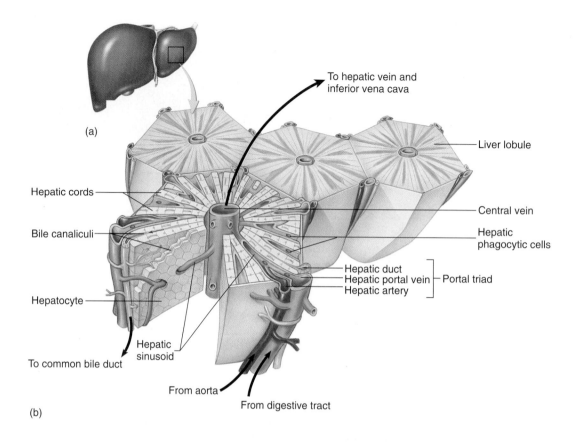
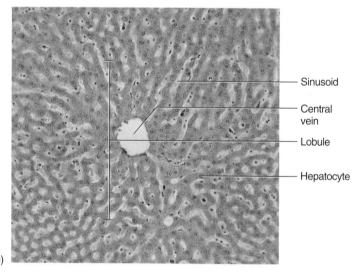

Figure 38.17 Histology of the Liver
(a) Anterior view of liver; (b) diagram of liver lobule; (c) photomicrograph of liver lobule (100×).

2. Begin your study of the digestive system of the cat by locating the large **salivary glands** around the face. You must first remove the skin from in front of the ear. If you have dissected the face for the musculature, you may have already removed the **parotid gland,** which is a spongy, cream-colored pad anterior to the ear. You can also locate the **submandibular gland** as a pad of tissue slightly anterior to the angle of the mandible and lateral to the digastric muscle. The **sublingual gland** is just anterior to the submandibular gland and is an elongated gland that parallels the mandible. Use figure 38.19 to locate these glands.

3. Examine the **tongue** of the cat and note the numerous papillae on the dorsal surface. These are **filiform papillae,** and they have both a digestive and a grooming function. You can lift up the tongue and examine the **lingual frenulum** on the ventral surface.

4. Dissection of the head of the cat should be done only if your instructor gives you permission to do so. To dissect the head,

Exercise 38 Anatomy of the Digestive System   479

1. The hepatic ducts, which carry bile from the liver lobes, combine to form the common hepatic duct.

2. The common hepatic duct combines with the cystic duct from the gallbladder to form the common bile duct.

3. The common bile duct and the pancreatic duct combine to form the hepatopancreatic ampulla.

4. The hepatopancreatic ampulla empties bile and pancreatic secretions into the duodenum at the major duodenal papilla.

5. The accessory pancreatic duct empties pancreatic secretions into the duodenum at the minor duodenal papilla.

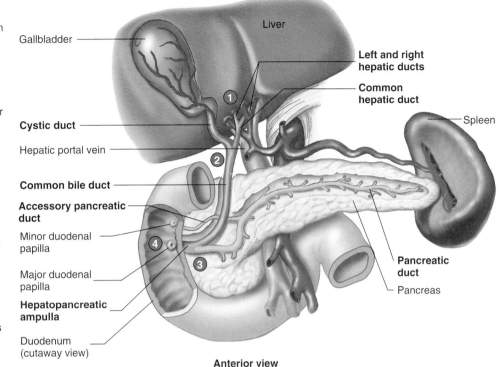

Figure 38.18 Liver, Gallbladder, and Pancreas
Bile flow moves from the liver through the hepatic ducts (1), to the cystic duct (2). Bile is stored in the gallbladder then flows back out the cystic duct to the common bile duct (3) and then into the duodenum (4).

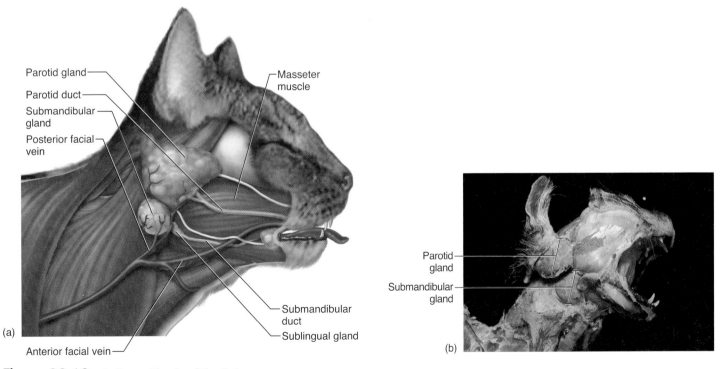

Figure 38.19 Salivary Glands of the Cat
(a) Diagram; (b) photograph.

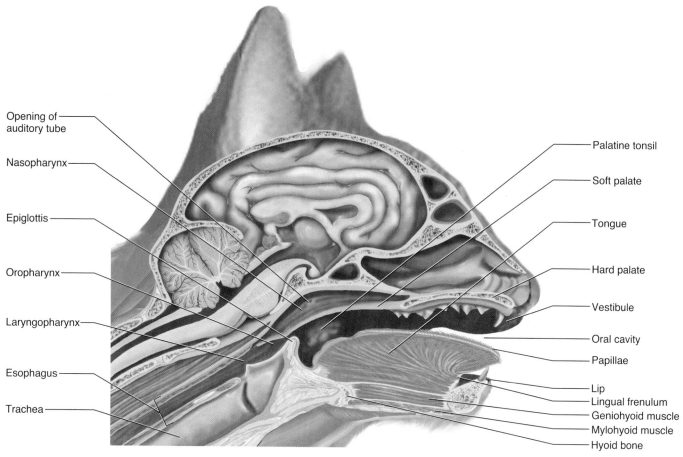

Figure 38.20 Midsagittal Section of the Head and Neck of the Cat

you should first use a scalpel and make a cut in the midsagittal plane from the forehead of the cat to the region of the occipital bone.

5. Cut through any overlying muscle in the cat. With a small saw, gently cut through the cranial region of the skull, being careful to stay in the midsagittal plane.
6. Once you make an initial cut through the dorsal side of the head, cut the mental symphysis of the cat, separating the mandible.
7. You can then use a long knife (or scalpel) to cut through the softer regions of the skull, brain, and tongue. Use a scalpel to cut through the floor of the oral cavity to the hyoid bone. It is best to use a scissors or small bone cutter to cut through the hyoid bone. This should allow you to open the head and examine the structures seen in a midsagittal section, as in figure 38.20.
8. Locate the hard palate, tongue, oral cavity, and oropharynx. Note how the larynx in the cat is closer to the tongue than in humans.
9. Look for the muscular tube of the **esophagus** by carefully lifting the trachea ventrally away from the neck.
10. Insert your blunt dissection probe into the oral cavity and gently wiggle it into the esophagus. The tongue and esophagus are represented in figure 38.20.

Abdominal Organs

11. To get a better view of the abdominal organs, it is best to cut the **diaphragm** away from the body wall. Carefully cut the lateral edges of the diaphragm from the ribs. You should see a fatty curtain of material covering the intestines. This is the **greater omentum.** Just caudal to the diaphragm, note the dark brown, multilobed **liver.** In the middle of the liver is the green **gallbladder.** To the left of the liver (in reference to the cat) is the J-shaped **stomach.**
12. Lift the liver and locate the **lesser omentum,** which is a fold of tissue that connects the stomach to the liver. Locate these structures in figure 38.21.
13. If you make an incision into the stomach, you should be able to see the folds known as **rugae.** Place a blunt probe inside the stomach and move it anteriorly. Notice how the **esophageal sphincter** makes it difficult to pierce the diaphragm. If you do move the probe into the esophagus, you should be able to see it as you look anterior to the diaphragm.
14. The stomach has an anterior **cardiac part,** a domed **fundus, greater** and **lesser curvatures,** and a **pyloric part.** Cut into the pyloric part of the stomach and then through the duodenum. Locate a sphincter muscle between the stomach and the duodenum. This is the **pyloric sphincter.** As you move into the duodenum, you may have to scrape some of

Exercise 38 Anatomy of the Digestive System

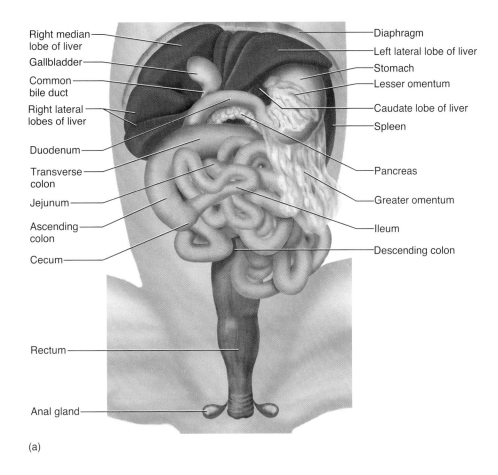

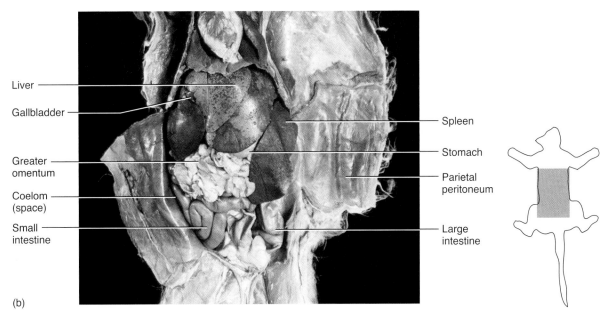

Figure 38.21 Abdominal Organs of the Cat
(a) Diagram; (b) photograph.

the chyme away from the wall of the small intestine to be able to see the fuzzy texture of the intestinal wall. This texture is due to the presence of **villi.** As in the human, the small intestine is composed of three regions: the **duodenum,** the **jejunum,** and a terminal **ileum.**

15. Compare the stomach and small intestine with figure 38.21.

16. Elevate the greater omentum, and locate the **pancreas,** dorsal to the stomach. The pancreas appears granular and brown. The **tail** of the pancreas is near the **spleen,** which is a brown, elongated organ on the left side of the body.

    The small intestine is an elongated, coiled tube about the diameter of a wooden pencil. Note the **mesentery,** which holds the small intestine to the posterior body wall near the vertebrae. The small intestine is extensive in the cat and rapidly expands into the large intestine. At the junction of the small and large intestines, note that there is no appendix in the cat.

17. The **large intestine** in the cat is a fairly short tube with a diameter slightly larger than your thumb. The first part of the large intestine is a pouch called the **cecum.** The remainder of the large intestine can be further divided into the **ascending, transverse,** and **descending colon** and the **rectum.** Compare these with figure 38.21.

18. Examine the **parietal peritoneum** along the inner surface of the body wall and the **visceral peritoneum** that envelopes the intestines.

Clean Up
When you are done with your dissection, carefully place your cat back in the plastic bag. Place all waste material in the appropriate container.

Exercise 38 Review
Anatomy of the Digestive System

Name: _____
Lab time/section: _____
Date: _____

1. The pancreas belongs to what part of the digestive system (alimentary canal/accessory organ)? _____

2. The descending colon belongs to what part of the digestive system (alimentary canal/accessory organ)? _____

3. Name the layer of material on the outer surface of the stomach and small intestine. _____

4. In the stomach, what is partially digested food called? _____

5. What is the name of the portion of the stomach closest to the small intestine? _____

6. What cell type comprises the inner lining of the mucosa (near the lumen) of the small intestine? _____

7. What is the middle part of the small intestine called? _____

8. Where are lacteals located in the digestive tract? _____

9. What membrane holds the tongue to the floor of the oral cavity? _____

10. What part of the tooth is located above the neck? _____

11. What is the layer of a tooth superficial to the dentin? _____

12. What are the adult teeth that are directly posterior to the canine teeth called? _____

13. What muscle cell type lines the small intestine? _____

14. The segments, or pouches, of the large intestine have what name? _____

15. What are the names of the salivary glands located anterior to the ear? _____

16. Where is the lesser omentum found? _____

17. Where does the cystic duct take bile for storage? _____

18. Trace the flow of bile from the liver to the duodenum, listing all the structures that come into contact with the bile on its journey. _____

_____

_____

_____

19. How does the large intestine differ from the small intestine in terms of length? _____

_____

20. How does the large intestine differ from the small intestine in terms of diameter? _____

_____

21. Name two functions of the pancreas. _____

_____

_____

Exercise 38 Anatomy of the Digestive System

22. Label the following illustration using the terms provided.

ascending colon vermiform appendix sigmoid colon
rectum tongue small intestine
oral cavity duodenum esophagus
stomach parotid gland liver
submandibular gland transverse colon cecum
descending colon

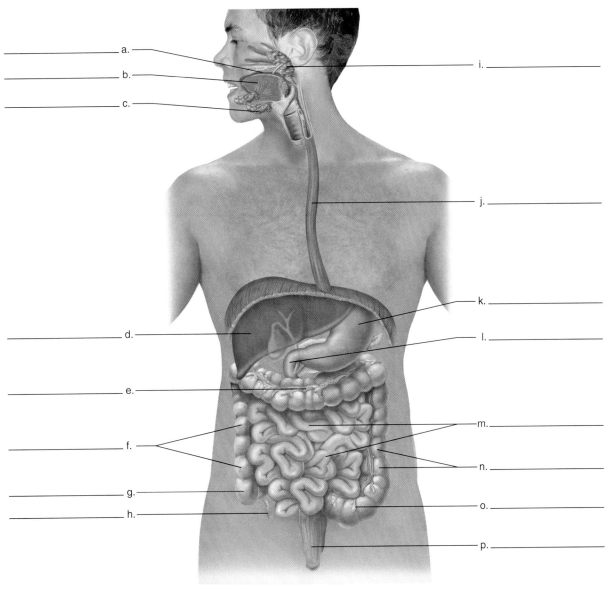

Digestive Organs

Exercise 39
Digestive Physiology

INTRODUCTION

An understanding of the digestive process is fundamental to the study of human physiology. Digestive physiology is covered in the *Principles of Anatomy and Physiology* text in chapter 21, "Digestive System," and chapter 22, "Nutrition, Metabolism, and Temperature Regulation."

We obtain almost all the raw materials the body uses for growth, development, life-sustaining energy, and other metabolic functions from digestion. Digestion and nutrition have significant health implications. Poor diet has been linked to inadequate body functions and to conditions such as heart disease and colon cancer. Digestive physiology also involves the physical and chemical breakdown of material, the absorption of nutrients, coordination between sections of the digestive tract and accessory organs (liver secretions and processing of absorbed foods), and the elimination of fecal material. In this exercise, you look at the physical (mechanical) and chemical digestion of food.

To develop some understanding of the digestive process, you examine the digestive functions of the enzymes. In this experiment, you analyze **pancreatin** (an extract of the pancreas containing several enzymes important in digestion) and amylase. You determine the effectiveness of these enzymes in digesting four materials commonly found in plant and animal tissue: starch, lipid, cellulose, and protein. The goal of the experiment is to determine if these enzymes can break down each of the four substances into smaller products. If the enzymes are effective, then starch should be broken down into sugars, lipids broken down into glycerol and fatty acids, cellulose reduced to sugars, and proteins split into amino acids. Important biomolecules are illustrated in your text (in the *Principles* book, they are in chapter 2, "The Chemical Basis of Life" in the "Organic Chemistry" section).

Furthermore, you examine the mechanical breakdown of food that precedes most of the chemical processes. By taking solid food and increasing its surface area, you determine if the rate of digestion increases. This is an example of the role of mechanical breakdown in the digestion of food.

OBJECTIVES

At the end of this exercise, you should be able to

1. discuss the importance of catabolism in the digestive process;
2. describe how an enzyme is important in chemical digestion;
3. list possible human digestive enzymes as determined by experiment;
4. demonstrate the effectiveness of enzymes in food digestion;
5. describe the mechanisms involved in the mechanical digestion of food;
6. state one of the limitations of human digestive enzymes.

MATERIALS

Microscopes

General Supplies

80 test tubes (15 mL each)
12 test tube holders
6 test tube racks
12 test tube brushes and soap
Warm water bath, set at 37°C, with test tube racks and thermometer
Hot water bath or hot plates and 400 mL beakers for hot bath (100°C) and thermometer
6 permanent markers
18 pipettes of 5 mL each
6 pipette pumps
7 100 mL beakers to pour from stock bottle for distribution
1 250 mL bottle of tap water
6 small spatulas
100 mL 2% pancreatin solution, fresh
300 mL 1% alpha amylase solution, fresh
100 mL 0.1% alpha amylase solution, fresh
300 mL Benedict's reagent (6 50 mL bottles)
Parafilm squares (10 per table)
Biohazard container or container with 10% bleach solution

Starch Digestion

6 dropper bottles of iodine solution
250 mL 0.5% potato starch solution
6 boxes of microscope slides (1 per table)
6 boxes of coverslips (1 per table)
6 eyedroppers

Sugar Test

100 mL 1% maltose solution
6 10 mL graduated cylinders

Cellulose Digestion

Cellulose—2 cotton balls cut into small pieces
6 small dropper bottles of water

Lipid Digestion

200 mL litmus cream
6 dropper bottles of "acid solution" (lemon juice or vinegar)
6 dropper bottles of 1% sodium hydroxide solution (NaOH)
12 pairs of goggles (during use of NaOH)
12 pairs of latex gloves

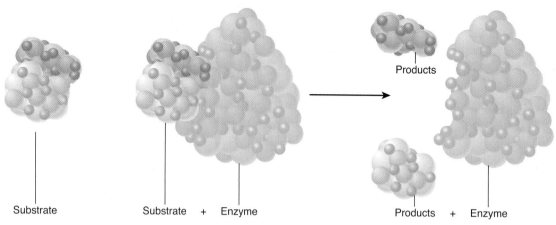

Figure 39.1 Catabolic Reactions with Enzymes

Protein Digestion

25 mL 1% BAPNA solution; BAPNA (benzoyl-DL-arginine-p-nitroanilide), available from Sigma B4875 or Aldrich 85,711-4

Surface Area and Digestion

2 large potatoes per class
2 breadboards
2 number 4 cork borers
2 small rods (applicator sticks) to push through the borers
12 rulers (15 cm)
6 razor blades
2 mortars and pestles

PROCEDURE

General Procedures

Read all of each experiment before beginning. Your instructor may want you to do these experiments in student pairs or as members of a larger group. If you do these experiments as part of a larger group, make sure that you *witness* and *record* the results of your group. **Do not discard any material until everyone in your group has seen the result** or until your instructor directs you to do so. While you are waiting for one set of materials to incubate, you can begin the next set of experiments.

Label all test tubes with your group name and the test tube number. This is very important, because removing the wrong tube will give you erroneous results, and removing a tube belonging to another group will produce the wrong result not only for your group but for the other group as well.

Tests in this exercise are divided into **reagent tests** and **experimental tests.** The reagent tests allow you to see the results of an experiment when using known samples. The experimental tests determine if a digestive reaction has taken place.

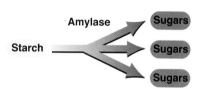

Figure 39.2 Starch Digestion

The Nature of Enzymes

The function of a digestive enzyme is to break down large **macromolecules** in the stomach and small intestine into smaller biomolecules in a process known as **catabolism.** This enzymatic process is shown in figure 39.1.

Starch Digestion

Starch digestion occurs by the breakdown of a large molecule of starch into smaller molecules of sugar by enzymes called **amylases.** Amylases catabolize starches as represented by figure 39.2. One way to analyze the effectiveness of amylases is to see if the enzyme removes starch from solution. The decrease or absence of starch in solution indicates that amylases are present.

Experiment 1: Reagent Test—Iodine

Use **iodine** as a reagent test to determine the presence of starch. If starch is present, iodine produces a blue-black color, which is a **positive result.** If starch is absent, then the solution remains yellow, and this is a **negative result.**

1. Label a test tube with your group name and the number 1.
2. Pour a small amount of starch solution from the stock bottle into a small labeled beaker. *Never return excess solution to the stock bottle!*
3. Take a pipette pump and pipette and place 1 to 2 mL of starch solution from the small beaker into the test tube labeled number 1.

Caution
Never pipette material by mouth; use a pipette pump or bulb.

4. Add four or five drops of iodine solution to the starch solution in test tube 1. Cover with parafilm. The solution should turn blue or black if starch is present. This is a positive test. Save this test result for future comparison and record the outcome of your test.

   Color of reaction (blue/black or yellow): _____

   Test result (positive/negative for starch): _____

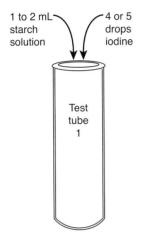

Experiment 2: Determination of Starch Digestion by Amylase

1. Using a pipette, remove 5 mL of starch solution from the beaker that received the stock solution and place it in a small test tube that you have labeled with your group name and the number 2a.
2. Pour a small amount of 1% amylase from the stock bottle into a small labeled beaker. With a clean pipette, withdraw 2 mL of amylase solution from the beaker and add it to the starch solution. Discard any remaining solution in your pipette. Do not return it to the beaker. You can substitute 2 mL of saliva for the amylase if directed by your instructor. Do not use your saliva if you recently had sugar in your mouth. Remember to be careful with bodily fluids. Stir the solutions in the test tube by flicking the bottom of the tube gently with your finger.
3. Incubate the mixture in test tube 2a in a **warm** water bath (37°C) for 60 minutes.

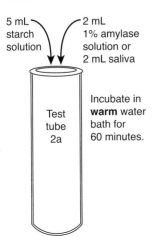

4. After 60 minutes, take 1 to 2 mL of the solution from test tube 2a and place it in a new test tube labeled 2b.
5. Test the solution in test tube 2b with four or five drops of iodine solution and record your results.
6. Compare the results of your experiment with test tube 1.

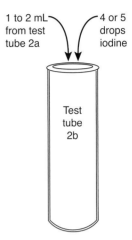

   Color of the solution with iodine: _____

   Test results (positive/negative): _____

7. From test tube 2b, withdraw a small sample with an eyedropper and place a drop on a microscope slide.
8. Examine the slide under low power and look for starch grains. Compare this slide with one that you made from a drop of starch from the *undigested* starch solution (from test tube 1). Is there a difference? Record your observations in the following space. Save this reaction mixture.

   Observations: _____

   _____

   _____

   _____

Sugar Production

The preceding experiment tested for the *decrease* in starch as a way to determine if a catabolic reaction has taken place. Another way to look at the effectiveness of amylase is to determine the *presence* of sugar after the exposure of starch to the enzyme. In this experiment, instead of looking for the absence of the **substrate** (starch) to determine enzyme effectiveness, you can test for the formation of the **product** (sugar). If amylase is effective, then the presence of sugar in solution, after exposure to the enzyme, indicates the conversion of starch to sugar.

Experiment 3: Reagent Test—Benedict's Test

The first test in this section involves the detection of sugar in a solution. This is done with the Benedict's test.

Caution
Benedict's reagent is poisonous. Use caution when mixing the solutions in the test tube.

1. Label a test tube with the number 3 and your group name and then pipette 2 mL of 1% maltose solution into the test tube. Maltose is a sugar.
2. Using a pipette and a pipette pump (never by mouth) add 1 mL of Benedict's reagent to the maltose solution in the test tube. Hold the test tube by the top and gently flick the bottom of the tube with the pad of your index finger to mix the contents.
3. Place this mixture in a **hot** (100°C) water bath for about 10 minutes. If the mixture turns green, yellow, orange, or brick red, then sugar is present.

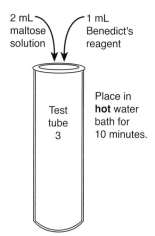

The sequence of colors indicates the presence of sugar in increasing amounts.

Presence of sugar:

Blue = 0
Green = trace
Yellow = moderate
Orange, red = large amounts

If there is no color change (if the solution stays blue) after 10 minutes, then remove the tube from the bath and note the absence of sugar (a negative result). Record your results.

Color of test solution: _____

Test result (positive/negative for sugar): _____

Experiment 4: Determination of Sugar (Maltose) Production by Amylase

1. Take 4 mL of the solution from test tube 2a (the starch and amylase reaction) and put it in a test tube labeled with your group name and the number 4.
2. To the solution in test tube 4, add 4 mL of Benedict's reagent.
3. Place the tube in a **hot** water bath (100°C) for 10 minutes. Record your results.

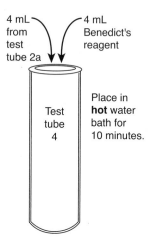

Color of solution with Benedict's reagent: _____

Compare these results with the sequence of colors in experiment 3.

Presence of sugar: _____

Effectiveness of digestion: _____

Cellulose Digestion

Cellulose is composed of repeating **glucose** units. Starch is also composed of repeating glucose units. If amylase converts cellulose to sugar, then the presence of sugar in solution after incubation indicates that the enzyme amylase is effective in breaking down cellulose into glucose. This is the same process as in experiment 4, except that the substrate is changed from starch to cellulose. To test for the effectiveness of amylase in digesting cellulose, you incubate a cellulose mixture with amylase in a warm water bath. A positive reaction indicates the presence of sugar after incubation.

Experiment 5: Determination of Sugar Production from Cellulose

1. Place a small pinch of cellulose fibers (the size of a green pea) in a test tube and add 2 mL of water and either 3 mL of 1% amylase solution or 2 mL of saliva. You should have a total volume of 4–5 mL. Make sure the majority of cellulose fibers are not stuck to the sides of the test tube. You can use a small spatula to push the fibers into the liquid, if necessary. The test tube should be labeled with your group name and the number 5.

2. Incubate for 60 minutes in a **warm** (37°C) water bath.

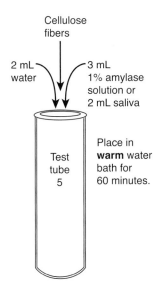

3. After 60 minutes, remove the solution from the bath and add 4 mL of Benedict's reagent.
4. Place the tube in a **hot** water bath (100°C) for 10 minutes. Record the presence or absence of sugar. A positive reaction to the Benedict's test indicates the presence of sugar after incubation.

   Color of Benedict's test: _____

   Presence of sugar: _____

   Effectiveness of digestion: _____

Lipid Digestion

In the digestive process, some **lipids (triglycerides)** are broken down into **monoglycerides** and **free fatty acids.** The fatty acids make the solution more **acidic,** thus lowering the **pH.** The decrease in pH can be used as a measure of digestion. The greater the acidity of the solution, the greater the digestion.

Experiment 6: Reagent Test—Litmus Test

1. To determine the effectiveness of the litmus reagent, pour 3 mL of **litmus cream** into a test tube labeled with your group name and the number 6. Litmus cream consists of dairy cream (containing fat) and litmus powder (a pH indicator).
2. Add an acid solution (lemon juice or vinegar) drop by drop while flicking the bottom of the test tube until the color changes from blue to pink. Do not add too much acid but just enough to change the color. A pink color indicates an acidic condition.
3. Now add 1% sodium hydroxide solution (NaOH) a drop at a time until you reverse the color. This indicates that the solution is alkaline.

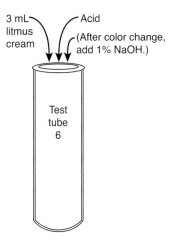

Caution

Sodium hydroxide is a caustic solution that can dissolve your skin, so be careful! Wear protective goggles and gloves as you add the sodium hydroxide solution.

Experiment 7: Determination of Lipid Digestion by Pancreatin

1. Pipette 3 mL of litmus cream into a small test tube labeled with your group name and the number 7.
2. Add 1 mL of 2% pancreatin solution (pancreatin contains many digestive enzymes) to the litmus cream and stir well with a spatula.
3. Incubate for 60 minutes in a **warm** (37°C) water bath.

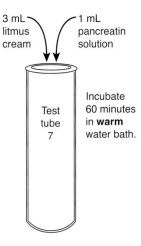

4. After 60 minutes, look for a color change. If the cream turns pink, then digestion has occurred. If the color remains blue, then no digestion has occurred. Record your results.

   Color of solution after incubation (pink/blue): _____

   Condition of the litmus cream (acidic/alkaline): _____

   Effectiveness of digestion: _____

   Change in smell from normal cream: _____

Protein Digestion

Proteins are made of **amino acids.** These amino acids are linked by **peptide bonds** to form **polypeptide chains.** If the chains are long enough, they form **proteins.** Proteins are split into amino acids by a number of digestive enzymes called proteases. The effectiveness of proteases can be studied with the use of a chromogenic (color-producing) substance known as BAPNA (benzoyl-DL-arginine-p-nitroanilide). If proteases are present, they release a yellow aniline dye from the larger BAPNA molecule.

Experiment 8: Determination of Protein Digestion by Pancreatin

1. Label a test tube with your group name and the number 8.
2. Pipette 1 mL of BAPNA solution into the tube. To this, add 1 mL of pancreatin solution and stir by flicking the bottom of the test tube.
3. Place the tube in a **warm** (37°C) water bath for 15 minutes.

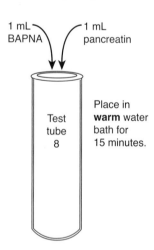

4. After 15 minutes, remove the tube from the bath and examine the test tube to see if it has turned yellow. Yellow indicates protein-digesting abilities of the enzyme. Record your results.

   Color of the tube (yellow/clear): _____

   Digestion capability (yes/no): _____

Experiment 9: Surface Area and Digestion

As food is broken down physically, the surface area increases relative to the volume of the food. This experiment examines the effectiveness of an increase in surface area for digestion.

1. Label two test tubes, each having your group name and the number 9. Write "mashed" on one tube and "entire" on the other.
2. Carefully plunge a number 4 cork borer into the center of a raw potato. *Make sure your hand is not on the receiving end of the cork borer! Use a breadboard.*
3. Remove the cork borer from the potato and, using a small rod, push the potato cylinder from the borer.
4. Using a ruler and a razor blade, cut the potato cylinder into two pieces, each 1 cm in length. *It is important that the two pieces are the same length!*
5. Rinse off both pieces with running water, and place one piece of potato in the test tube labeled "entire."
6. Chop the other piece into several little pieces and mash these with a mortar and pestle until they are well pulverized.
7. Carefully place all of the mashed potato in the other test tube with the "mashed" label.
8. Pipette exactly 4 mL of **0.1%** amylase solution* (or 0.5 mL saliva and 4 mL water) into each test tube and incubate them for **20** minutes in a **warm** water bath.

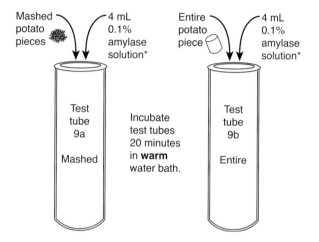

9. After incubation, place exactly 2 mL of Benedict's reagent into each tube and place the two tubes in the hot water bath for 10 minutes.

   The amount of digestion can be estimated by the color of the test. As described in experiment 3, blue indicates no sugar, with green, yellow, orange, and brick red indicating increasing amounts of sugar. In which tube did the greatest amount of digestion take place? In which tube did the least amount of digestion take place? Record your results.

   Color of solution with intact potato piece: _____

   Color of solution with mashed potato piece: _____

   Relative amount of digestion of intact piece: _____

   Relative amount of digestion of mashed piece: _____

*You can use a solution of 0.1% alpha amylase or dilute the 1% stock solution by adding 1 mL of 1% amylase solution to 9 mL of water.

Clean Up

Place all test tubes, pipettes, and other materials in either a biohazard container or a 10% bleach solution.

Exercise 39 Review
Digestive Physiology

Name: _____

Lab time/section: _____

Date: _____

1. Why is pancreatin used in a digestion experiment? _____
   _____

2. Substrates are converted into what substances by enzymes? _____
   _____

3. Name the group of enzymes that digest starch. _____
   _____

4. Starch is reduced by enzymes into smaller molecules, known as _____ .

5. What is catabolism? _____
   _____
   _____

6. What are the bonds that hold amino acids together? _____
   _____

7. Explain why a negative iodine test for starch indicates a positive result for the enzymatic degradation of starch by amylase.
   _____
   _____
   _____
   _____

8. What would a positive iodine test indicate in the reaction in question 7? _____
   _____
   _____

9. Record all your negative results. Determine if these negative results indicate digestion or no digestion. _____
   _____
   _____
   _____

10. Explain why cellulose could or could not be digested by amylase. (Hint: Look in your text in the organic chemistry section.)

11. What effect does chewing food have on digestion? What experiment did you perform to evaluate the effectiveness of chewing for digestion? _____

12. Lipase is an enzyme that converts some lipids to monoglycerides and free fatty acids. Which substance used in lab probably contained lipase? _____

13. What does a negative Benedict's test (a blue color) indicate? _____

14. What does a pink color in the lipid digestion experiment indicate? _____

15. Pancreatin (which has amylase) and amylase are often prepared with lactose or other sugars as extenders. Determine which experiments would give erroneous results if pancreatin or amylase were contaminated with sugar. _____

Exercise 40
Anatomy of the Urinary System

INTRODUCTION

The organs of the urinary system consist of two kidneys, two ureters, a single urinary bladder, and a single urethra. The urinary system filters dissolved material from the blood, regulates electrolytes and fluid volume, concentrates and stores waste products, and reabsorbs metabolically important substances, returning them to the circulatory system. **Filtration** occurs when one or more substances pass through a selectively permeable membrane, whereas others do not. Filtration in the kidney involves both metabolic waste products (urea) and material beneficial to the body. It is not desirable to have all filtered material removed from the body. Glucose and other materials, such as sodium and potassium ions, are **reabsorbed** from the kidney back into the circulatory system. The kidney **secretes** urea; some drugs; hydrogen and hydroxyl ions. The kidneys excrete metabolic wastes, hydrogen ions, toxins, water, and salts. The kidney also produces the enzyme renin and the hormone erythropoietin. These topics are covered in the *Principles of Anatomy and Physiology* text in chapter 23, "Urinary System and Body Fluids."

In this exercise, you examine the gross and microscopic anatomy of the urinary system, study the major organs as represented in humans, and dissect a mammal kidney, if available.

OBJECTIVES

At the end of this exercise, you should be able to

1. list the major organs of the urinary system;
2. describe the blood flow through the kidney;
3. describe the flow of filtrate through the kidney;
4. name the major parts of the nephron;
5. trace the flow of urine from the kidney to the exterior of the body;
6. distinguish among the parts of the nephron in histological sections;
7. compare the male and female urinary anatomy.

MATERIALS

Models and charts of urinary system
Models and illustrations of kidney and nephron system
Microscopes
Microscope slides of kidney and bladder
Samples of renal calculi (if available)
Preserved specimens of sheep or other mammal kidney
Dissection trays and materials
Scalpels
Forceps
Blunt (mall) probes
Latex gloves
Waste container

PROCEDURE

You can begin the study of the urinary system by locating its principal organs. Compare figure 40.1 with charts and models available in the lab. Find the **kidneys, ureters, urinary bladder,** and **urethra.**

Kidneys

The kidneys are **retroperitoneal** (posterior to the parietal peritoneum) and are embedded in **renal fat pads (perirenal fat).** These adipose pads cushion the kidneys, which are located mostly below the protection of the rib cage (figure 40.1). The kidneys are located adjacent to the vertebral column about at the level of T12 to L3. The right kidney is slightly more inferior than the left.

1. Examine a model of the kidney and compare it with figure 40.2.
2. Locate the outer **renal capsule,** a tough connective tissue layer, the outer **cortex,** and the inner **medulla** of the kidney. The kidney has a depression on the medial side where the **renal artery** enters the kidney and the **renal vein** and the **ureter** exit the kidney. This depression is called the **hilum.**
3. Examine a coronal section of the kidney and locate the **renal pyramids,** which are separated by the **renal columns** in the renal medulla. Each renal pyramid ends in a blunt point called the **renal papilla** (figure 40.2). Urine drips from many papillae toward the middle of the kidney.

   The urine drips into the **minor calyces** (sing. **calyx**), which enclose the renal papillae. Minor calyces are somewhat like funnels that collect fluid. Minor calyces lead to the **major calyces** and these, in turn, conduct urine into the large **renal pelvis.** The renal pelvis is located in a space known as the **renal sinus.** The renal pelvis is like a glove in a coat pocket. The pocket is the renal sinus, and the membranous glove that occupies the space is the major calyces, some of the minor calyces, and the renal pelvis.
4. Locate these structures in figure 40.2. The renal pelvis is connected to the ureter at the medial side of the kidney.

Blood Flow Through the Kidney

The kidney filters material from the blood and returns important material such as water, glucose, and sodium back to the blood. It is not a perfect system as some urea is also returned to the cardiovascular system. There is an arterial system that takes blood to the cortex of the kidney and several capillary beds, which is somewhat unusual compared to other organs. The venous system returns blood to the inferior vena cava.

Examine models or charts in lab and look at the vascular system in figure 40.3. The arterial system begins with the **renal artery,** which exits the **abdominal aorta** and enters the kidney. The arterial

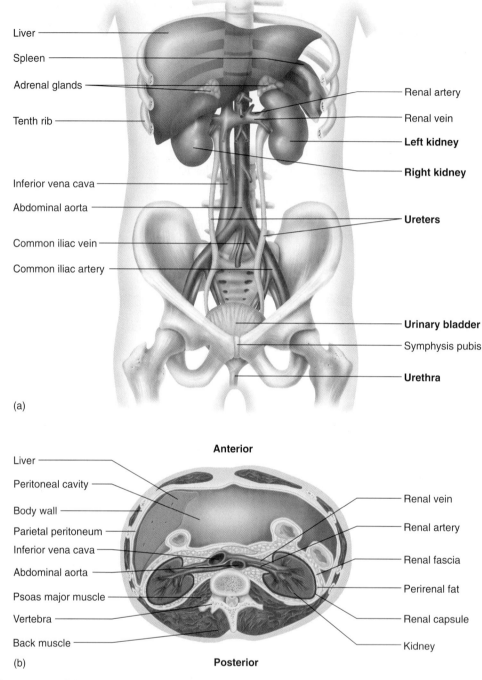

Figure 40.1 Major Organs of the Urinary System
(a) Anterior view; (b) transverse section of the system, showing kidneys in retroperitoneal position.

system branches extensively throughout the kidney. The first branches are the **segmental arteries,** which take blood from the renal artery. The segmental arteries are found inside the renal sinus. The kidney is divided into lobes, and there are **interlobar arteries** that take blood from the segmental arteries and pass through the renal columns. The interlobar arteries are relatively large and make a sharp bend, becoming the **arcuate arteries.** The arcuate arteries form arcs between the cortex and the medulla and they obtain their name from these arcs. Branching from the arcuate arteries are the **interlobular arteries** that move into the cortex of the kidney. The names *interlobar* and *interlobular* can be confusing at first, but the interlobar arteries are found in the medulla and are larger and the interlobular arteries are found in the cortex and are smaller.

In the cortex of the kidney, the interlobular arteries branch and form the **afferent arterioles,** which take blood to the **glomerulus,** a cluster of capillaries where filtration occurs. Blood then travels through the **efferent arterioles** and then to the **peritubular capillaries,** where reabsorption and secretion take place. In regions of the cortex near the medulla there are other vessels that branch from the efferent arteriole. These are the **vasa recta.**

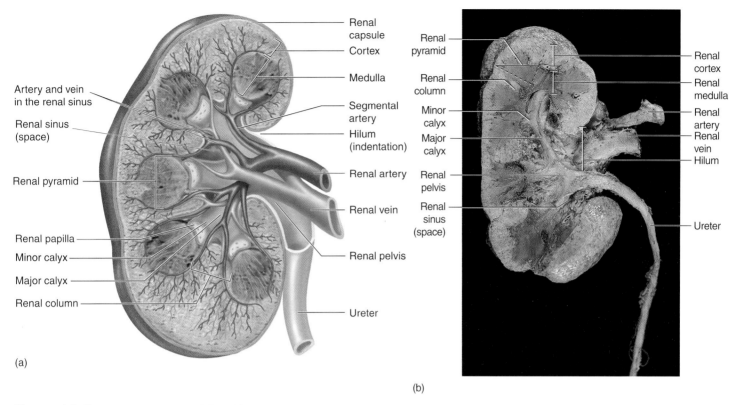

Figure 40.2 Coronal Section of the Kidney
(a) Diagram; (b) photograph.

They represent only a small number of capillaries in the kidney, but they are important in producing a concentrated urine by reabsorption and secretion. The vasa recta are found in association with special nephrons called **juxtamedullary nephrons.**

Thus there are three capillary beds in the kidney, the **glomeruli** (plural of glomerulus), the peritubular capillaries, and the vasa recta. The blood flow in the kidney forms a **portal system,** which is defined as a group of blood vessels in which blood flows from one capillary bed (the glomerulus) to another capillary bed (the peritubular capillaries or vasa recta) with an arteriole or venule between them prior to returning to the heart.

The return flow to the heart occurs as blood returns via the **interlobular veins,** to the **arcuate veins,** to the **interlobar veins,** and to the **renal vein,** which leads to the inferior vena cava.

Ureters

The **ureters** are long, thin tubes that conduct urine from the kidneys to the urinary bladder. The ureters have **transitional epithelium** as an inner lining and smooth muscle in their outer wall. Urine is expressed by **peristalsis** from the kidney to the urinary bladder. Examine the models in the lab and compare them with figure 40.4.

Urinary Bladder

The **urinary bladder** is located anterior to the parietal peritoneum and is thus described as being **anteperitoneal.** Locate the urinary bladder in the torso model and note that it is found just posterior to the symphysis pubis (figure 40.1). **Transitional epithelium** lines the inner surface of the bladder, whereas layers of smooth muscle known as **detrusor muscles** are located in the wall of the bladder. On the posterior wall of the urinary bladder is a triangular region known as the trigone. The **trigone** is defined by the superior entrances of the ureters and the inferior exit of the urethra, as illustrated in figure 40.4.

Histology of the Urinary Bladder

Transitional epithelium has a special role in the urinary bladder. This epithelium can distend (stretch) when the bladder fills with urine.

1. Examine a slide of transitional epithelium under the microscope and compare it with figure 40.5.
2. Look at the inner surface of the prepared section for the epithelial layer. Note that the cells are shaped somewhat like teardrops. Transitional epithelium can be distinguished from stratified squamous epithelium in that the cells of transitional epithelium from an empty bladder do not flatten at the surface of the tissue.

Urethra

The terminal organ of the urinary system is the **urethra.** The urethra is approximately 3–4 cm long in females. It passes from the urinary bladder to the **external urethral orifice,** located anterior to the vagina and posterior to the clitoris (figure 40.6). The urethra is about 20 cm long in males. It begins at the urinary

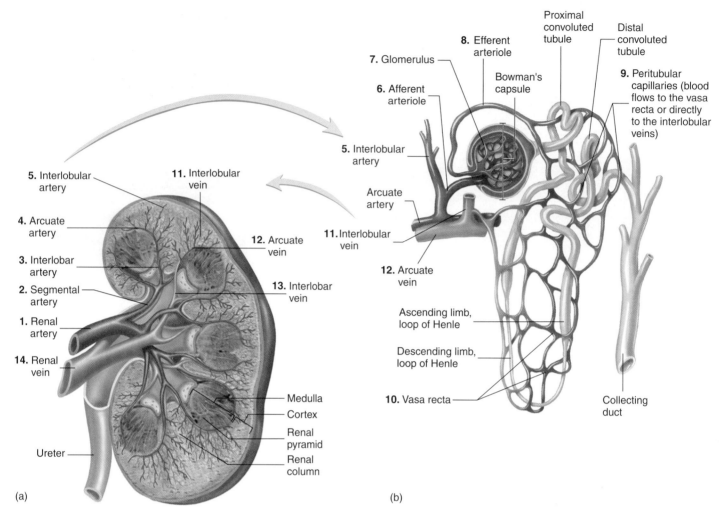

Figure 40.3 Blood Vessels and Ultrastructure of the Kidney
(a) Major vessels; (b) nephron and associated vessels. The blood flow can be followed by the numbers on the illustration.

bladder, passes through the body wall, and then exits through the penis to the external urethral orifice at the tip of the glans penis. Urinary bladder infections are more common in females than in males, because of the difference in length of the urethra in males and females.

Microscopic Examination of the Kidney

Before you examine the sections of kidney under the microscope, first become familiar with the structure of the **nephron.** Look at figures 40.3, 40.7, and 40.8, and locate the **renal corpuscle (Bowman's capsule** and **glomerulus), proximal convoluted tubule, loop of Henle,** and **distal convoluted tubule.** These structures make up the nephron.

Blood travels to the nephron via the **afferent arteriole.** When the blood reaches the **glomerulus,** a cluster of capillaries in the renal cortex, the plasma is filtered by blood pressure forcing fluid across the capillary membranes. This fluid, which is present in the nephron, is called filtrate.

As the filtrate flows through the nephron, water, glucose, and many electrolytes are returned to the blood. The urea is concentrated as it passes through the entire nephron and **collecting duct,** a tube that receives the end product of the nephrons. Locate the glomerulus, Bowman's capsule, proximal convoluted tubule, loop of Henle, distal convoluted tubule, and collecting duct in figure 40.3 and on models or charts in the lab.

1. Examine a kidney slide under low power. You should see the **cortex** of the kidney, which has a number of round structures scattered throughout. These are the **glomeruli,** and they are composed of capillary tufts. The other part of the kidney slide should have fairly open parallel spaces. These spaces are the **collecting ducts,** and they are located in the **medulla.** Compare the slide with figure 40.7a.
2. Examine the slide under higher magnification and locate the glomerulus and **Bowman's capsule.** The capsule is composed of simple squamous epithelium and specialized cells called podocytes.
3. Examine the outer edge of the capsule around the glomerulus. If you move the slide around in the cortex, you should find the **proximal convoluted tubules** with the **brush border,** or **microvilli,** on the inner edge of the tubule. The inner surface of the tubule appears fuzzy. The microvilli increase the surface area of the proximal convoluted tubule.

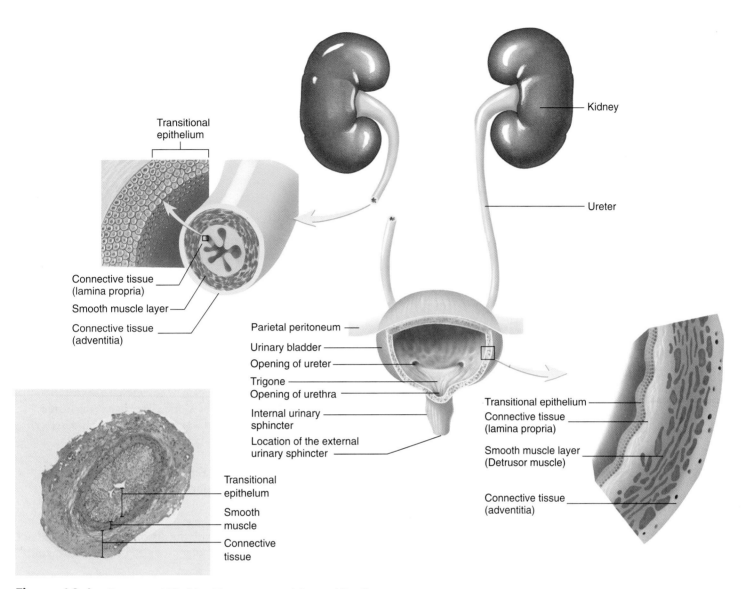

Figure 40.4 Ureter and Bladder, Transverse and Coronal Sections

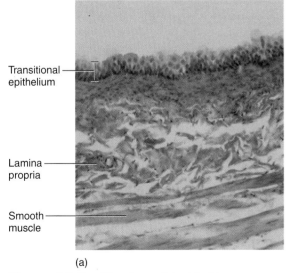

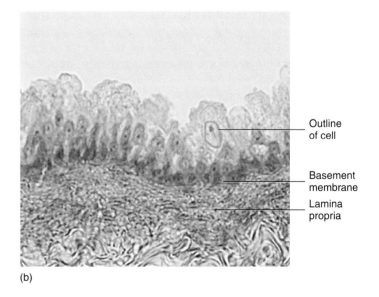

Figure 40.5 Histology of the Bladder
(a) Overview (100×); (b) detail of epithelium.

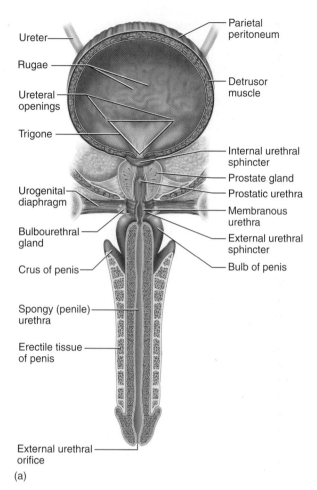
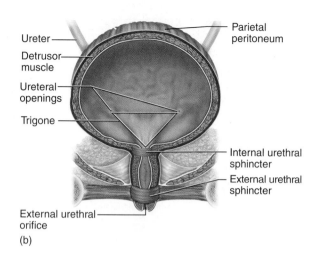

Figure 40.6 Urethra
(a) Male; (b) female.

The **distal convoluted tubules** do not have brush borders; therefore, the inner surface of a tubule does not appear fuzzy. The cells of the distal convoluted tubules generally have darker nuclei and cytoplasm that is relatively clear when compared with the cells of the proximal convoluted tubules (figure 40.7c).

4. Examine the medulla of the kidney under high magnification and locate the thin-walled **loop of Henle** and the larger-diameter **collecting ducts** (figure 40.7d).

The **juxtaglomerular apparatus (JGA)** is located in special nephrons where the distal convoluted tubule connects to the afferent arteriole. The JGA is involved in the control of blood pressure. When blood pressure decreases, the afferent arteriole constricts. This decreases the blood flow through the kidney and allows more blood to remain in the systemic circulation, thereby increasing the blood pressure.

Dissection of the Sheep Kidney

1. Take a sheep kidney and dissection equipment back to your table.
2. Place the kidney on a dissection tray before you and examine the outer **capsule** of the kidney. You may see some tubes coming from a dent in the kidney. The dent is the **hilum**, and the tubes are the **renal artery, renal vein,** and **ureter.**
3. Make an incision in the sheep kidney a little off center in the coronal plane (figure 40.9).
4. Locate the **renal cortex, renal medulla, renal pyramids, papillae, minor** and **major calyces,** and **renal pelvis.**
5. Lift the renal pelvis somewhat to pull it away from the **renal sinus.**
6. When you are finished with the dissection, place the material in the proper waste container provided by your instructor.

Cat Dissection

1. Open the abdominal region of the cat, if you have not done so already. The **kidneys** are on the dorsal body wall of the cat and are located dorsal to the parietal peritoneum.
2. Examine the kidneys and the structures that lead to and from the **hilum** of the kidney.
3. Locate the **renal veins** that take blood from the kidney. The veins are larger in diameter than the **renal arteries,** and they are attached to the inferior vena cava of the cat, which runs along the ventral, right side of the vertebral column.
4. Find the renal arteries that take blood from the abdominal aorta to the kidneys. The aorta lies to the left of the inferior vena cava.
5. Locate the **ureters** as they run inferiorly from the kidney to the **urinary bladder.** Do not dissect the **urethra** at this time. You can locate the urethra during the dissection of the reproductive structures in Exercises 42 and 43. Compare your dissection with figure 40.10.

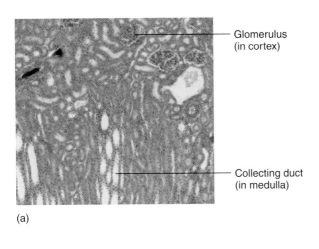
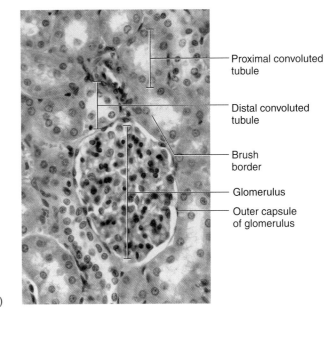
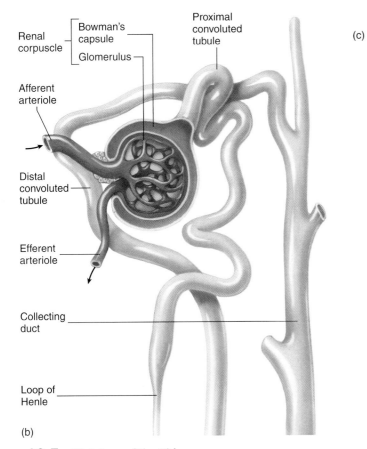
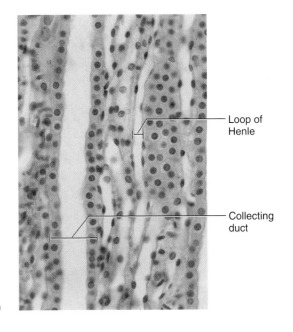

Figure 40.7 Histology of the Kidney

(a) Overview (40×); (b) diagram of nephron; (c) cortex (400×); (d) medulla (400×).

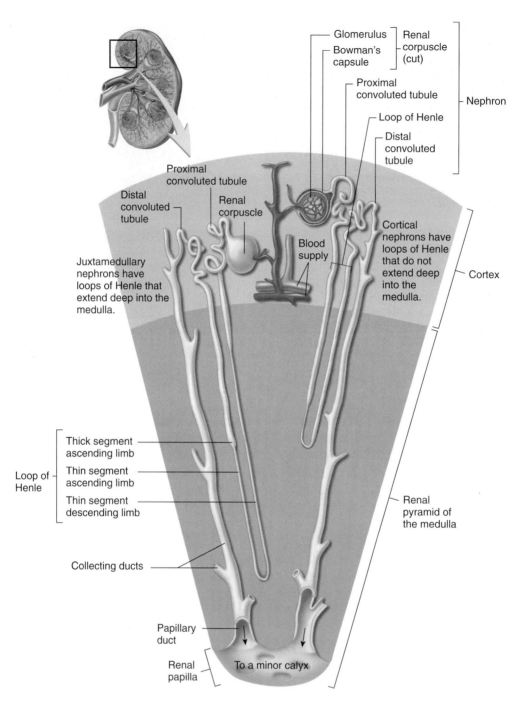

Figure 40.8 The Nephron

Exercise 40 Anatomy of the Urinary System

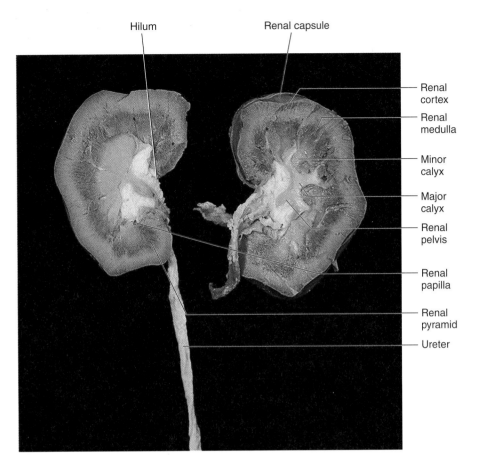

Figure 40.9 Dissection of the Sheep Kidney

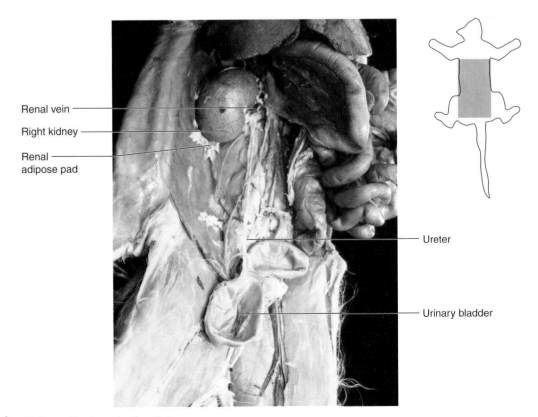

Figure 40.10 Urinary System in the Cat

Exercise 40 Review

Anatomy of the Urinary System

Name: _____

Lab time/section: _____

Date: _____

1. What is the outer region of the kidney called? _____

2. Describe the kidneys with regard to their position in relation to the parietal peritoneum. _____
_____

3. What takes urine directly to the urinary bladder? _____

4. What takes urine from the urinary bladder to the exterior of the body? _____

5. What blood vessel takes blood from the kidney? _____

6. On the posterior urinary bladder, there is a triangular region. What is it called? _____
_____

7. What is the name of the cluster of capillaries in the kidney where filtration occurs? _____
_____

8. Distal convoluted tubules flow directly into what structures? _____

9. What is a renal papilla? _____
_____

10. Name the connective tissue that covers a kidney. _____

11. Urine flows from the tip of a renal pyramid into what structure? _____

12. Blood in the glomerulus next flows to what arteriole? _____

13. Which shows the greatest anatomical difference between the sexes: ureters, urinary bladder, or urethra? _____
_____

14. Blood in an arcuate vein next flows into what structure? _____

15. What histological feature distinguishes a proximal convoluted tubule from a distal convoluted tubule? _____
_____
_____

16. Name the parts of the nephron. _____

17. What type of cell lines the urinary bladder? _____

18. Trace the flow of blood from the renal artery through the kidney to the renal vein. _____

19. What anatomical feature in females is responsible for the higher level of urinary tract infections in females? _____

20. Trace the path of filtrate and urine from the glomerulus to the external urethral opening. _____

21. Label the following illustration using the terms provided.

renal artery	renal vein	renal pelvis	renal medulla
renal papilla	major calyx	ureter	minor calyx
renal capsule	renal pyramid	renal cortex	

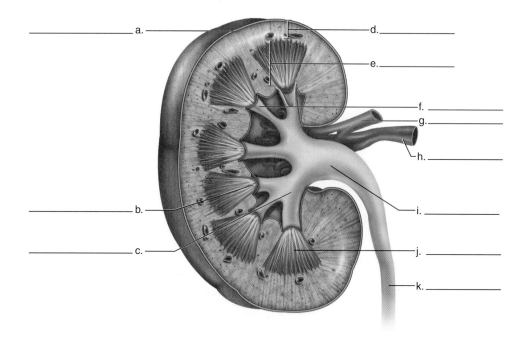

Exercise 41
Urinalysis

INTRODUCTION

Urinalysis is important in the clinical assessment of an individual's physical condition. Traces of blood in the urine can indicate the presence of kidney stones or renal damage; elevated levels of glucose can indicate diabetes mellitus; high bilirubin levels can indicate liver problems; and high bacterial counts can indicate bladder or kidney infections. The analysis of urine provides a broad diagnostic indicator of general health. The analysis involves visual inspection, chemical tests, and microscopic examination. The function of the urinary system is covered in the *Principles of Anatomy and Physiology* text in chapter 23, "Urinary System and Body Fluids." In this exercise, you analyze a sample of urine or synthetic urine for materials dissolved in the urine or suspended in it.

Normal values of urine vary, based on water intake and diet. A person who drinks 3 quarts of water per day has a very different urine composition than one who drinks 3 cups of water per day. Not only does the volume of water in the urine vary but the concentration of other urinary solutes varies as well. Some cations, such as potassium and sodium, are also diet-dependent. A general table of urine values is listed in table 23.1 in the *Principles* text. The process of collecting urine and examining it under the microscope may seem unappealing at first, yet the exercise is usually appreciated after it is completed.

OBJECTIVES

At the end of this exercise, you should be able to

1. use a Multistix or Chemstrip test and determine if the values obtained are within the standard range;
2. list the sediments commonly found in urine;
3. relate urinalysis results such as glycosuria, bilirubinuria, ketonuria, hematuria, albuminuria, pyuria, and calculi, to potential diseases or homeostatic imbalances;
4. distinguish among casts, crystals, and microbes in a urine sample;
5. prepare a stained sediment slide and identify the major components of the sediment.

MATERIALS

Sterile urine collection containers
Permanent marker or wax pencil
Microscope slides
Coverslips
Urine sediment stain (Sedistain, Volusol, etc.)
Chemstrip or Multistix 10SG urine test strips
Tapered centrifuge tubes
Test tube racks
10 mL pipette and pipette pump
Pasteur pipettes and bulbs
Centrifuge
Protective gloves (latex or plastic)
Protective eyewear
Biohazard bag
Microscopes
Urine specimen (yours or synthetic/sterilized urine)
10% bleach solution

Caution

Urine is potentially contaminated with pathogens. Wear latex barrier gloves and protective eyewear during the entire exercise. Place all disposable material that comes into contact with urine in the biohazard bag. Work only with your own urine and avoid any contact between urine and an open wound or cut. If you spill your sample, notify your instructor, wipe up the spill, and swab the countertop with a 10% bleach solution. When you are finished with the exercise, place reusable glassware in a 10% bleach solution. Read all procedures before beginning this exercise.

PROCEDURE

Preliminary Urinalysis

You will use either your own urine for urinalysis or synthetic urine provided for you. If you use synthetic urine, you can test for color, pH, specific gravity, glucose, and protein. Some tests are provided in kit form from biological supply houses. If you use your own urine, follow these directions: Using a marker or wax pencil, write your name on a sterile collection container. Proceed to the restroom and void a little urine prior to collecting a sample. The reason for voiding a little urine first is to flush microbes that normally occur in the urethra. You want a **midstream sample,** which should produce a **clean catch** (one with pure, uncontaminated urine). It is preferable to analyze the urine within 1 hour of the collection. If you collect your urine at home, do so in the morning and store the sample in a collection cup in a sealed plastic bag in the refrigerator. Take the specimen back to the lab and note the color of the sample.

Urine color: _____

The color of urine should normally be a very pale yellow. This is due to the pigment **urochrome,** which is a metabolic product of hemoglobin breakdown. Urine can vary in color due to vitamin B, some foods (such as beets or carrots), certain diseases, or some medicines. A low fluid intake may also cause the urine to be a darker yellow. Typically, freshly voided urine is clear or slightly cloudy. The factors that affect urine turbidity (cloudiness) are increased numbers of red blood cells, white blood cells, epithelial cells, microorganisms, mucus, lipids, or crystals. Crystals make the urine cloudy usually when the urine cools and the crystals precipitate. Examine your sample and determine into which category your urine falls.

Urine Turbidity

_____ Clear
_____ Slightly cloudy
_____ Cloudy
_____ Opaque

If your urine is cloudy or opaque you should be able to see evidence of that in the microscopic examination portion of the lab.

Odor: _____

Urine should normally have a faint but characteristic odor. The consumption of certain foods, such as asparagus, may produce sulfur compounds in urine, which produce stronger odors.

Chemical Urinalysis

Gently swirl the urine before testing and pour a 10 mL sample into a clean test tube. Follow your instructor's directions for determining standard values in urine. You can test for specific compounds in urine individually, or you can use a Chemstrip or Multistix to run a battery of tests in a few minutes. The Multistix procedure is outlined next. Before you proceed with the test you should have the urine sample, the test strip, a pencil or pen, and a container handy. Do not touch the indicator pads on the test strip as this may alter the results. The test covers 10 specific exams in 2 minutes, so you have to be organized to record the results accurately. Consider any result read after 2 minutes invalid. Multistix test strips are composed of sections of paper with test reagents embedded into the fibers. They react with urine components if present. Examine figure 41.1 and note the way the stick is read. Read the test strips first from the region near where you hold the strip and continue away from that area to the end of the strip.

Multistix/Chemstrip Procedure

If you are using a Multistix or Chemstrip, be sure to pay attention to the time limits provided on the container for reading the results. Wear gloves to dip the test strip into the urine sample and leave the strip in the urine for no longer than 1 second. Follow the directions

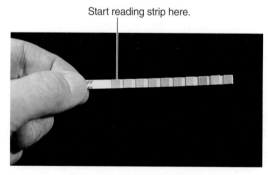

Figure 41.1 **Reading a Urine Test Strip**
Read the strip near the end where you hold on to the strip with your fingers and progress to the other end of the strip.

on the test container precisely. As you read the test strip, record the results in the following spaces. When you finish the test, compare your results with the normal values in each paragraph following a specific test.

Glucose: _____

Although glucose is present in blood plasma, there is normally no glucose in urine. **Glycosuria** is the condition of having glucose in the urine. Levels of glucose should be minimal (less than 40 mg/dL) or absent after fasting. Trace amounts of glucose can be found in the urine after ingesting food that is high in carbohydrates (especially sugar), and significant levels can be found in people with diabetes mellitus. Normally, glucose transporters take glucose out of the nephron and back to the blood. If the blood glucose levels are too high (such as in cases of diabetes mellitus), there are not enough transporters to handle the excess glucose, and the glucose passes into the urine.

Bilirubin: _____

The bilirubin test should be negative, since bilirubin is not normally present in urine. Bilirubin is a metabolic waste product from the destruction of erythrocytes. In the condition of **bilirubinuria**, bile pigments and bilirubin are present in urine. This can be due to erythrocyte destruction (hemolytic anemia), blockage of the bile duct, or liver damage, such as hepatitis or cirrhosis.

Ketones: _____

There should be only trace amounts of (5 mg/mL) or no ketones (acetone) in the urine. Ketone bodies in urine (**ketonuria**) represent the general body condition known as **ketosis**. This is due to the mobilization and use of fat stores in people during periods of starvation, in people with diabetes mellitus, and in people with an abnormally high-fat diet.

Specific gravity: _____

The specific gravity of pure water is 1.000. The normal values for specific gravity are variable between labs, but they generally range from a dilute 1.005 to 1.030 in concentrated urine. When urine has a high specific gravity, there are more dissolved solutes in the urine. A high specific gravity may be due to dehydration or diabetes mellitus (where there is more sugar in the urine) among other things. A low specific gravity may be due to diabetes insipidus (where the body is either secreting too little ADH or the kidneys are not responding to ADH), increased water intake, or renal failure.

Hematuria: _____

There should be no blood present in urine. The presence of erythrocytes or hemoglobin in urine may be due to some forms of anemia, erythrocyte destruction after incompatible blood transfusions, renal disease or infection, kidney stones, or urinary contamination during the menses (bleeding period) of a female's menstrual cycle. If the indicator strip shows a spotted pattern, then the erythrocytes are intact (nonhemolyzed). If the red blood cells have ruptured, then the presence of **hematuria** shows up as uniform color changes on the indicator strip.

pH: _____

The pH of urine can vary from around 4.5 to 8.2. This represents more than a 1000-fold change in hydrogen ion concentration. The dramatic fluctuation in pH occurs because the kidneys regulate blood pH by removing ions from the blood. Urine is normally slightly acidic. In a diet high in protein, the urine is more acidic, whereas a diet high in vegetable material yields a urine that is more alkaline.

Protein: _____

There should be no protein in urine. The most common blood protein is albumin, and its presence in urine is known as **albuminuria.** This may be due to diabetes mellitus, renal damage (kidney disease), extreme physical activity, or hypertension.

Urobilinogen: _____

Normal ranges of urobilinogen are 0.2 to 1 mg/dL. Increases in the secretion of urobilinogen indicate significant **hemolysis** of erythrocytes to the point that the liver cannot process the bilirubin. The bilirubin increases in the plasma, and the formation of urobilinogen in the intestines increases as well. The urobilinogen diffuses into the blood, where it is filtered by the kidneys.

Nitrites: _____

The normal urine value of nitrites should be negative. A positive value for nitrites indicates the presence of large numbers of bacteria in the urinary tract.

Leukocytes: _____

The normal value for leukocytes is negative or present in trace amounts. An elevated level of leukocytes in urine is known as **pyuria.** This indicates a potential urinary tract infection—typically, bladder or kidney infection.

Note that some drugs and vitamins can give false positive results in urinalysis tests. Any test that indicates renal disease should be further validated by professionals before you take any course of action.

After you have finished collecting your results, complete table 41.1.

Sediment Study

1. Obtain a centrifuge tube from the supply area and write your name on the tube for identification.
2. Pipette or pour 10 mL of urine into the tube and place it in the centrifuge opposite another tube containing an equal volume of urine. If there is an uneven number of tubes, make sure to fill a tube with water to the same level as the unpaired tube and place it opposite that tube in the centrifuge. This balances the centrifuge while it is operating. The centrifuge must be balanced before it begins to rotate. An unbalanced centrifuge can cause damage to the rotor and is extremely dangerous.
3. Spin the tubes for about 4 minutes at a slow speed (1500 rpm [revolutions per minute]) and let the rotor of the centrifuge slow down over time. If you brake the spinning, you may resuspend the sediments.
4. Carefully remove your tube and place it in a test tube rack on your desk.
5. Examine the tube. You should have an upper fluid layer and a small amount of urine sediment at the bottom of the tube.
6. Place one drop of urine sediment stain (Sedistain or other urine stain) on a clean microscope slide.
7. Withdraw a small sample from the bottom of the centrifuge tube with a Pasteur pipette and place a drop of the sediment on the slide. *Do not put the urine on the slide before you put on*

TABLE 41.1 Urine Values

Characteristic	Normal Value or Range	Your Results
Appearance	Almost clear to deep amber	
Odor	Faint odor	
Glucose	0–trace (less than 40 mg/dL)	
Bilirubin	None	
Ketones	None–trace (5 mg/dL)	
Specific gravity	1.005–1.030	
Free hemoglobin	None	
pH	4.5–8.2	
Albumin	None–trace	
Urobilinogen	Trace (0.2–1.0 mg/dL)	
Nitrites	None	
Leukocytes	None–trace	

Cells (Figure 41.2)

Epithelial cells
    Cells of urethra (squamous cells)—normal constituent
    Cells of bladder or ureter (transitional cells)—normal constituent
Bacterial cells in urine—indicate bacteriuria
Erythrocytes (RBCs)—menstrual blood in females or blood from the passage of kidney stones
Leukocytes (WBCs)—indicate urinary tract (bladder/kidney) infections
Microorganisms—yeast (*Candida*) and *Trichomonas*, common urinary pathogens

Crystals (Figure 41.3)

Crystals usually indicate reduced water intake.

    Struvite
    Calcium oxalate—small, green crystals
    Calcium carbonate
    Calcium phosphate
    Uric acid
    Ammonium urates
    Cystine—oxidation product of an amino acid

Casts (Figure 41.4)

Casts are conglomerations of cells, usually from the kidneys.

    Leukocytes—indicate pyelonephritis, inflammation of the kidney
    Erythrocytes—indicate glomerulonephritis, inflammation of glomeruli
    Hyaline

Artifacts (Figure 41.4)

    Clothing fibers, skin oil, bath powder

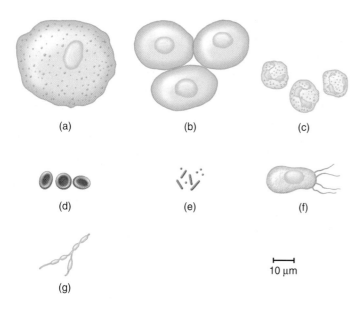

Figure 41.2 Urine Sediments—Cells
(a) Squamous epithelial cells; (b) transitional epithelial cells; (c) white blood cells; (d) red blood cells; (e) bacteria; (f) *Trichomonas*; (g) *Candida*.

the stain, because you may contaminate the stain if the stain bottle comes in contact with the sediment.

8. Place a coverslip on the sample and examine the slide under the microscope. Examine the slide under low power first, and look at the free edge of the coverslip. Red-orange elongated crystals may appear here later as the stain evaporates. Do not confuse the stain crystals with sediment crystals.
9. Now examine the slide under high power and compare your sample with the common urine sediments in figures 41.2 through 41.4.
10. Record the material you find in urine sediment in the review section at the end of this laboratory exercise.
11. Some of the common sediments found in urine are as follows:

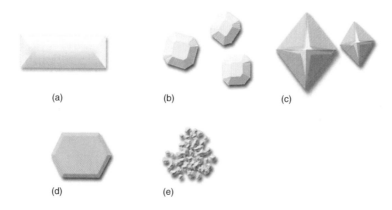

Figure 41.3 Urine Sediments—Crystals
Sizes are variable (a) struvite; (b) calcium carbonate; (c) calcium oxalate; (d) cystine; (e) urate.

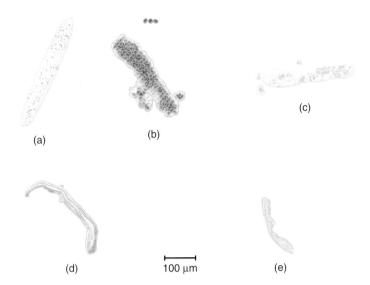

Figure 41.4 Urine Sediments—Casts, Fibers, and Artifacts
(a) Hyaline cast; (b) erythrocyte cast; (c) leukocyte cast; (d) vegetable fiber; (e) mucous thread.

Renal Calculi

Kidney stones, or **renal calculi,** are caused either by mineral buildup in the kidneys or by amino acid or amino acid oxidative product deposition, such as tyrosine or cystine crystals. Kidney stones are commonly composed of calcium oxalate or calcium phosphate and occasionally uric acid or magnesium ammonium phosphate (struvite). A common preventative for mineral stones is simply drinking more water. The stones are masses of crystals that fuse into a larger form; they can block the ureter, causing a retention of fluid in the kidney. Stones can be exceptionally painful. Examine kidney stones in the lab, if available.

Clean Up

Place urine-contaminated material to be disposed of in the biohazard container and all the urine-contaminated material to be reused in the 10% bleach container. Swab the countertops with 10% bleach, which destroys infectious agents. Clean any stain from the objective lenses of the microscopes.

Exercise 41 Review

Urinalysis

Name: _____

Lab time/section: _____

Date: _____

1. Assume that a person did not collect a midstream sample of urine but collected a sample from the beginning of urination. What additional materials might be in greater numbers in this sample? _____

2. What metabolic by-product from hemoglobin colors the urine yellow? _____

3. How can adequate water intake be judged by the color of urine? _____

4. What is the name of the condition of having measurable amounts of sugar in urine? _____

5. What is the normal value for glucose in urine? What might produce higher levels of glucose in urine? _____

6. What substance in urine creates ketonuria? _____

7. What was the specific gravity of your urine sample? _____

8. What is the range of the specific gravity in normal urine values? _____

9. If you collected your own urine, is there a correlation between the amount of water that you drink and the specific gravity of your urine? _____

10. What is hematuria? _____

11. The kidneys are very efficient at balancing blood pH. It is critical that blood acidity remain constant. If excess hydrogen ions are present in the blood and increase blood acidity, the kidneys secrete the hydrogen ions. What effect does this have on the pH of urine? _____

12. There are about 200 grams of protein in the blood plasma and normally none in urine. What mechanism keeps protein out of the urine, and what structure might be damaged (review Exercise 40, if necessary) if protein is found in significant amounts in urine? _____

13. Elevated levels of white blood cells produce what condition in urine? _____

14. If you process 180 liters of water through the kidney each day, yet produce only 1.8 liters of urine, approximately how efficient are your kidneys at reabsorbing the water passing through them? _____
_____

15. List the materials you found in the urine sample. _____
_____
_____

16. What mechanism normally keeps glucose out of urine? _____

17. What cells found in the urine originally come from the walls of the urethra? _____

18. What cells in the urine come from the wall of the urinary bladder? _____
_____

19. If your urine contained large numbers of calcium crystals, what might this tell you about the amount of water you drink?
_____
_____

Exercise 42
Male Reproductive System

INTRODUCTION

The male reproductive system produces **male gametes (sperm cells),** transports the gametes to the female reproductive tract, and secretes the male reproductive hormone, **testosterone.** The gonad, or gamete-producing structure, of the male reproductive system is the testis (plural *testes*). The testes produce sperm cells in the seminiferous tubules. The endocrine function of the testes is secretion of testosterone. In this exercise, you examine the gross anatomy of the male reproductive system, the histology of the system, and the male reproductive system of the cat. The structure and function of the male reproductive system are covered in the *Principles of Anatomy and Physiology* text in chapter 24, "Reproductive System."

OBJECTIVES

At the end of this exercise, you should be able to

1. identify the gamete-producing organ of the male reproductive system;
2. describe the anatomy of the major structures of the male reproductive system;
3. describe the glands that contribute to the formation of semen;
4. list the pathway that sperm cells follow from production to expulsion;
5. describe the anatomy of the spermatic cord;
6. identify the stages of sperm development on a chart or model;
7. name the three cylinders of erectile tissue in the penis;
8. compare and contrast the anatomy of the cat reproductive system with that of the human.

MATERIALS

Charts, models, and illustrations of the male reproductive system
Microscopes
Prepared slides of a cross section of testis and sperm smear
Cats
Materials for cat dissection
    Dissection trays
    Scalpel and two or three extra blades
    Gloves (household latex gloves work well for repeated use)
    Blunt (mall) probe
    Forceps and sharp scissors
    Sharps container
    Animal waste disposal container

PROCEDURE
Overview of the Gross Anatomy of the Male Reproductive System

Examine the models and charts of the male reproductive system available in the lab, and locate the following structures in figure 42.1:

    Testis
    Epididymis
    Scrotum
    Ductus (vas) deferens
    Seminal vesicle
    Prostate gland
    Bulbourethral (Cowper's) gland
    Penis

Testes

The **testes** are paired organs wrapped in a tough connective tissue sheath called the **tunica albuginea** (figure 42.2). They lie outside of the body cavity, where the temperature is somewhat cooler, and are surrounded by the **scrotum,** which envelops the testes. The testes are the site of **spermatozoa (sperm cells)** production, and this process must occur at about 35°C (a few degrees cooler than the core body temperature). The scrotum is lined with a layer of muscle called the **dartos muscle** (figure 42.3). It is composed of smooth muscle fibers that contract when the testes are cold, thus bringing them closer to the body. When the environment around the testes is warm, the dartos muscles relax and the testes descend from the body, becoming cooler. The cremaster muscles in the spermatic cord also regulate the elevation of the testes. Examine charts and models of the testes and compare them with figures 42.1 through 42.3.

Histology of the Testis

The testis, as the gamete-producing organ of the male reproductive system, should be viewed in a prepared slide. Many tubules are seen in cross section. These are the **seminiferous tubules.** The gametes, or sperm cells, are produced in seminiferous tubules in the testis (figures 42.2 and 42.4).

Find the clusters of cells that often appear as triangles between the tubules. These are called **interstitial cells.** They produce the male sex hormone, **testosterone.** Examine the seminiferous tubules under high magnification. You should be able to see the outer row of cells, called the **spermatogonia.** These cells reproduce by mitosis to produce **primary spermatocytes. Sustentacular (Sertoli) cells** assist in the movement of the spermatocytes and isolate the sperm cells with the blood testis barrier (BTB).

The primary spermatocytes undergo meiosis, or reduction division, to eventually produce the sex cells (sperm cells). The

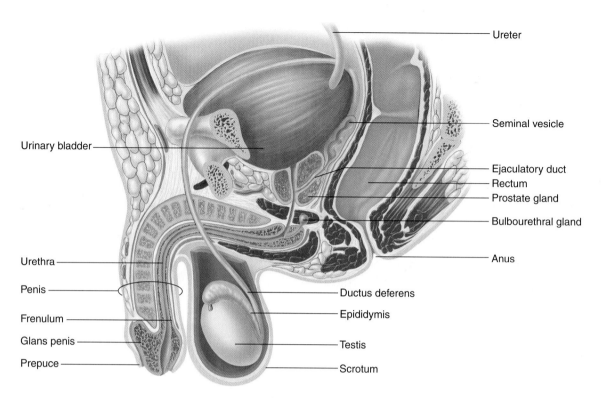

Figure 42.1 Male Reproductive System—Sagittal View

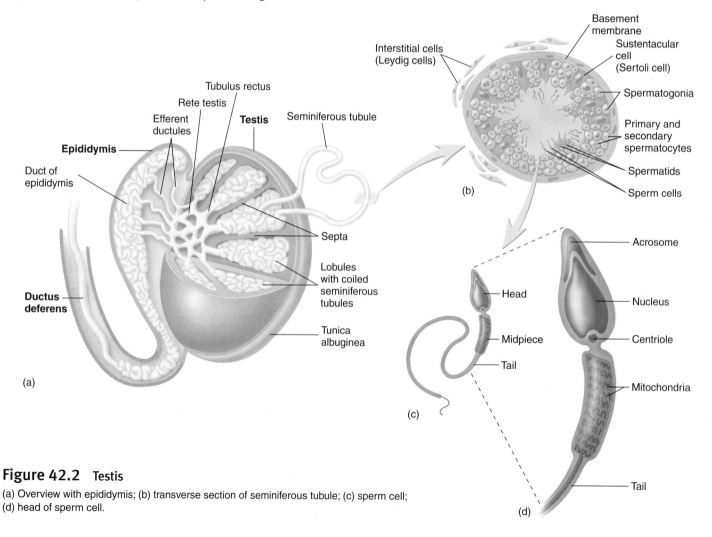

Figure 42.2 Testis
(a) Overview with epididymis; (b) transverse section of seminiferous tubule; (c) sperm cell; (d) head of sperm cell.

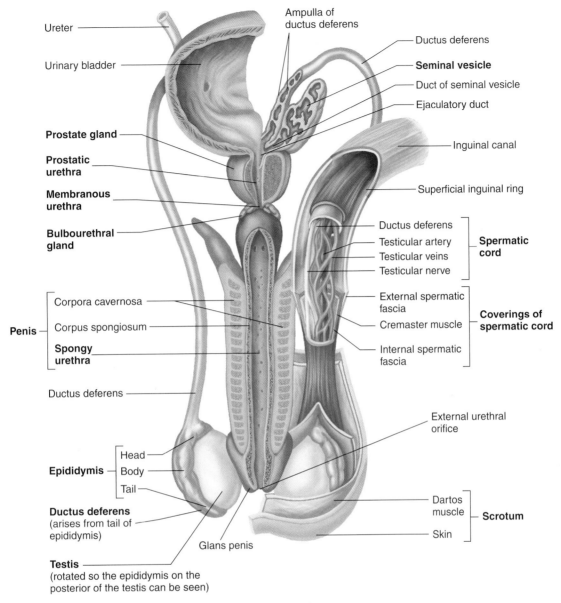

Figure 42.3 Scrotum, Testis, Spermatic Cord, and Penis—Coronal View

primary spermatocytes divide to form **secondary spermatocytes,** which are located closer to the lumen. The secondary spermatocytes become **spermatids.** Spermatids lose their remaining cytoplasm and mature into **sperm cells.** The process of sperm formation from spermatogonia to sperm cells is called **spermatogenesis.** Examine a prepared slide of testis and compare it with figure 42.4. Locate the spermatogonia, primary and secondary spermatocytes, spermatids, and sperm cells.

Sperm

The structure of an individual sperm cell consists of a **head, midpiece,** and **tail.** The head contains the genetic information (DNA), as well as a cap known as the **acrosome.** The acrosome contains digestive enzymes that digest the exterior covering of the female gamete. The midpiece of the sperm contains **mitochondria** that provide ATP to the sperm cell tail. The tail of the sperm is a flagellum that propels the sperm forward. Examine a prepared slide of sperm and compare it with figure 42.2. Draw what you see in the following space.

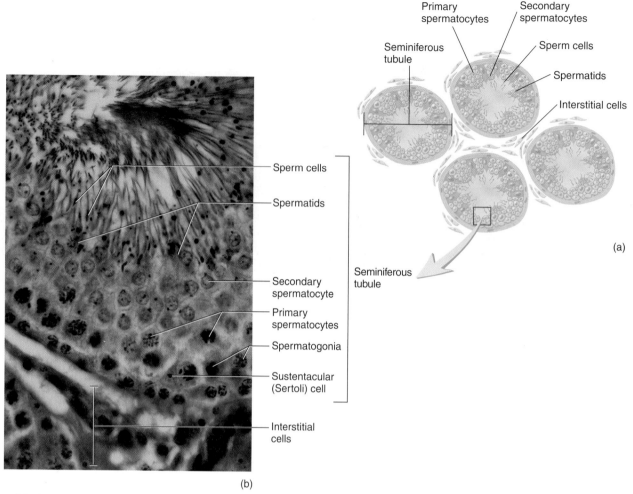

Figure 42.4 Testis
(a) Illustration; (b) histology (400×).

Epididymis

Sperm cells from each testis travel from tubules in the testis to the **rete testis** and into the **epididymis** where they are stored and mature. Each epididymis has a blunt, rounded **head;** an elongated **body;** and a tapering **tail** that leads to the ductus deferens. Sperm maturation, or **capacitation,** occurs in the epididymis. If sperm cells are removed from the testis proper, they are not capable of fertilizing the female oocyte (egg). This fact was discovered in early studies of fertility; for example, when a man had an injury to the spinal cord and ejaculation was not possible. Sperm cells move slowly through coiled tubules of the epididymis taking about 72 days to mature. Examine a model or chart of the longitudinal section of a testis and epididymis and locate the structures by comparing them with figures 42.2 and 42.3.

Spermatic Cord

Sperm cells travel from the epididymis into the **ductus deferens.** The ductus deferens is enclosed in the **spermatic cord,** which is a complex cable consisting of the ductus deferens, the **testicular artery** and **vein,** the **testicular nerves,** and the **cremaster muscle.** The cremaster muscle is a cluster of skeletal muscle fibers. The spermatic cord is longer on the left side than on the right; therefore, the left testis is lower than the right. Locate the structures of the spermatic cord in figure 42.3.

As the spermatic cord reaches the **inguinal canal,** each ductus deferens enters the abdominopelvic cavity and passes posterior to the urinary bladder. You can trace the course of the ductus deferens until it reaches the inferior portion of the bladder. The ductus deferens enlarges somewhat here to form the **ampulla.** Each ductus deferens joins with a **seminal vesicle,** which is a gland that adds fluid to the sperm cells and is composed of fructose, prostaglandins, and proteins that cause temporary clotting of the semen. The union of the ductus deferens and the seminal vesicles produces the **ejaculatory duct.** Locate the seminal vesicle and the ejaculatory duct in figure 42.3. The fluid from the seminal vesicles adds about 60% to the final volume of semen.

From this location, the ejaculatory ducts lead to the inferior portion of the bladder and join with the urethra, which passes through the prostate gland. The **prostate gland** is located just inferior to the urinary bladder, and the urethra that passes through the gland is known as the **prostatic urethra.** The prostate gland adds a buffering fluid to the secretions of the testes and seminal vesicles. The prostate fluid makes up slightly less than 40% of the final semen volume. As the prostatic urethra exits the

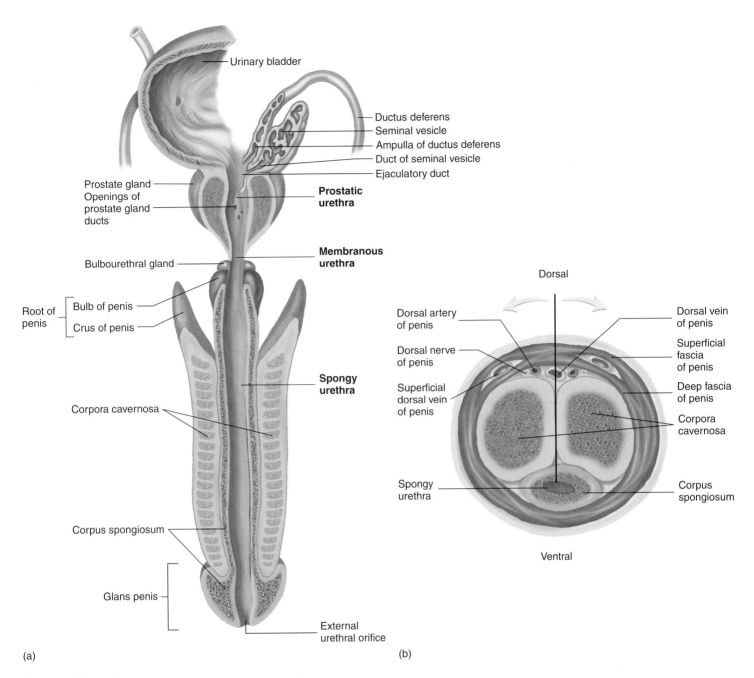

Figure 42.5 Penis
(a) Coronal section; (b) transverse section.

prostate gland, it becomes the **membranous urethra** and passes through the body wall. Here the paired **bulbourethral (Cowper's) glands** are located. They add a lubricant to the seminal fluid. **Seminal fluid** consists of secretions from the seminal vesicles, prostate gland, and bulbourethral glands. **Semen** consists of seminal fluid plus the sperm cells from the testes. Sperm cells make up much less than 1% of the total volume of semen. The urethra passes out of the body cavity and becomes the **spongy,** or **penile, urethra** of the penis. Locate the portions of the urethra and the accessory glands in figures 42.3 and 42.5.

External Genitalia

Penis

The **penis** consists of an elongated shaft and a distally expanded **glans penis.** The glans is covered with the **prepuce,** or **foreskin,** which is removed in some males by a procedure called a **circumcision.** At the inferior portion of the glans is a region richly supplied with nerve endings, called the **frenulum.** The glans penis is an expanded region that stimulates the genitalia of the female and, in turn, is stimulated by the female genitalia. The penis is, on average, about 16 cm in length. These structures can be seen in figures 42.1, 42.3, and 42.5.

The penis contains three cylinders of erectile tissue. The **corpus spongiosum** is the cylinder of erectile tissue that contains the spongy (penile) urethra. The two **corpora cavernosa** are located anterior to the corpus spongiosum. Examine a model or chart of a cross section of penis and compare it with figure 42.5. The proximal parts of the cylinders of erectile tissue are anchored to the body. Locate the **crus,** which is an expansion of the corpora cavernosa. The **bulb** of the penis is an extension of the corpus spongiosum. The crus and the bulb form the **root** of the penis. The corpus spongiosum expands distally to form the glans penis. Note the **dorsal arteries** of the penis. They take blood to the penis. Locate the **dorsal veins** of the penis. When the arteries of the penis dilate, the erectile tissues engorge with blood and the penis becomes erect. The erection subsides as the arteries constrict, decreasing blood flow into the penis. Examine a model of the penis and find the features in figure 42.5.

The floor of the pelvis as seen from the outside is referred to as the **perineum.** It can be divided into a posterior **anal triangle** and an anterior **urogenital triangle.** The anal triangle surrounds the anus, and the urogenital triangle encloses the penis and scrotum (figure 42.6).

Male Contraception

The main goal in male contraception is the prevention of sperm cells from reaching the female gametes. **Abstention,** or refraining from sexual intercourse, is the most effective method. The next most effective method is **male sterilization,** which is most commonly performed by a procedure called a **vasectomy** (the cutting and tying of the two **vas,** or **ductus, deferentes**). In a vasectomy, an incision is made on each side of the scrotum and both ductus deferentes are cut. The free ends of the cut ductus deferentes are tied, preventing sperm from traveling from the testes to the spermatic cords (figure 42.7). The sperm is reabsorbed by the body, and vasectomies do not cause any physical change in erectile function. The proper use of a barrier, such as a **condom,** is relatively effective contraception, since it prevents ejaculated sperm from entering the female reproductive tract. **Coitus interruptus,** or preejaculatory

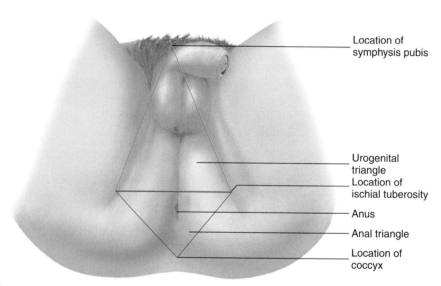

Figure 42.6 Perineum

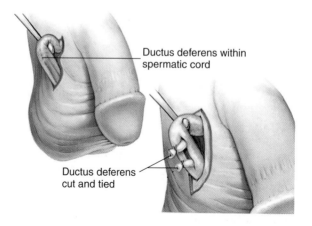

Figure 42.7 Vasectomy

withdrawal, is not a very effective method, since sperm may be present in seminal fluid prior to ejaculation, and pregnancy can result.

Cat Dissection

Prepare for the cat dissection by obtaining a dissection tray, a scalpel, scissors or bone cutter, string, forceps, and a plastic bag with a label. Remember to place all excess tissue in the appropriate waste container and not in a standard wastebasket or down the sink!

Check to see whether the cat is male or female. The **penis** may be retracted in your specimen, so look for the opening of the penile urethra and the scrotum. Team up with a lab partner or group that has a cat of a different sex than your specimen, so you can learn both male and female reproductive systems. Once you are sure you have a male cat, identify the penis, the glans penis, the **scrotum,** and paired **testes.** Make an incision on the lateral side of the scrotum and locate the testis inside the scrotal sac. If the cat was neutered, you will not be able to locate the testis. If the testes are present, cut through the connective tissue of the scrotum (**tunica vaginalis**) and observe both the testis and the **epididymis.** The testis is covered by a tough connective tissue membrane called the **tunica albuginea.** Use figure 42.8 as a guide. Sperm cells move from the testis and into the epididymis, where the sperm cells mature.

Once you have located the epididymis, proceed in an anterior direction and trace the thin **ductus deferens** from the epididymis into the **spermatic cord.** The spermatic cord traverses the body wall on the exterior and enters the body of the cat at the **inguinal canal** (an opening that pierces the inguinal ligament). You may want to gently insert a blunt probe into the inguinal canal so you can locate the ductus deferens as it passes into the **coelom.** Locate the ductus deferens as it enters the body cavity and notice how it arches around the ureter on the dorsal side of the urinary bladder. You may have cut the ductus deferens in an earlier exercise, so, if you cannot find it on one side, look for it on the other side.

You may want to look for the accessory organs of the male reproductive system, but this takes some significant dissection. *Check with your instructor before cutting through the pelvis of the cat.* If your instructor directs you to do so, then begin by cutting through the musculature of the cat at the level of the **symphysis pubis.** Make a midsagittal incision through the groin muscles, and carefully cut through the cartilage of the symphysis pubis. You should now be able to open the pelvic cavity and locate the single **prostate gland** and the paired **bulbourethral glands** (figure 42.8). Much of the anatomy of the cat is similar to that of the human, except that there are no seminal vesicles in the cat. Trace

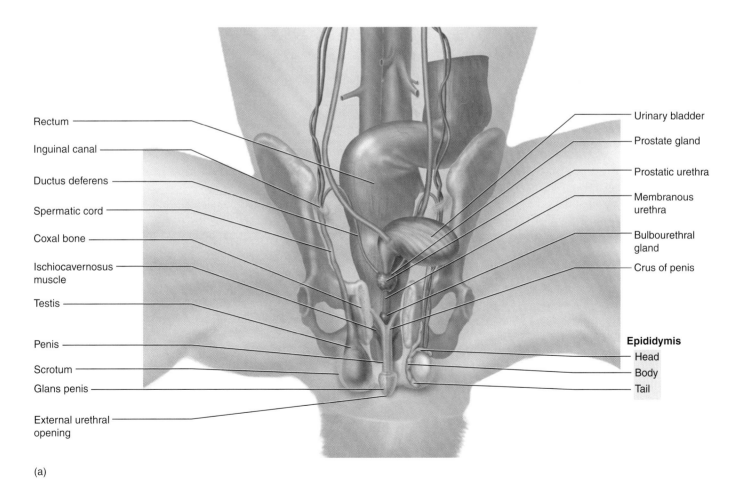

(a)

Figure 42.8 Male Reproductive Organs of the Cat
(a) Diagram; (b) photograph.

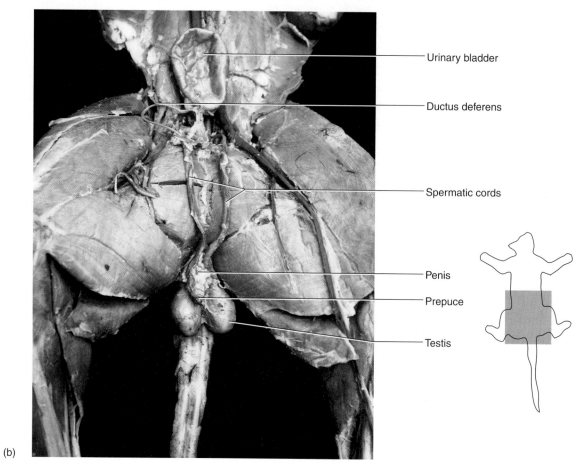

(b)

Figure 42.8—*Continued.*

the ductus deferens from the posterior surface of the bladder through the prostate gland and the penis. You can make either a longitudinal section through the penis to trace the urethra in the erectile tissue or a cross section of the penis to see the three cylinders of erectile tissue—the **corpus spongiosum** and the **two corpora cavernosa.**

Clean Up
Place all excess tissue in the appropriate waste container and not in a standard wastebasket or down the sink!

Exercise 42 Review
Male Reproductive System

Name: _____
Lab time/section: _____
Date: _____

1. What is the gonad in the male reproductive system? _____

2. What cells and what hormone come from the testes? _____
_____
_____

3. Proper sperm production must occur at what temperature? _____

4. What does the condition of cryptorchidism (undescended testicles) have on male fertility? Why? _____

5. Name the lining of the scrotal sac that consists of a smooth muscle layer. _____

6. In what structure are sperm cells produced in the testis? _____

7. What are the cells called that initiate sperm cell production? _____

8. Where do sperm cells move after leaving the epididymis? _____

9. Where is the cremaster muscle found? _____

10. List all the structures involved in producing semen. _____

11. Name the three organs that produce seminal fluid. _____

12. Describe the developmental stages in sperm development. _____

13. A vasectomy is the cutting and tying of the two ductus deferentes at the level of the spermatic cords. Considering the percentage that sperm cells contribute to the total volume of semen, what effect does a vasectomy have on semen volume? _____

14. Which one of the seminal fluid glands is not a paired gland? _____

15. Name the section of urethra that passes through the prostate gland. _____

16. Trace the pathway of sperm from formation to release. List the structures that the sperm pass through and the additions to the sperm along the way. _____

17. What is the cylinder of erectile tissue below the corpora cavernosa? _____

18. What male reproductive gland is missing in the cat but present in the human? _____

19. Label the following illustration using the terms provided.

    seminal vesicle   prostate gland   bulbourethral gland
    glans penis   prepuce   epididymis
    bulb of penis   urinary bladder   corpus cavernosum
    ductus (vas) deferens   testis   scrotum

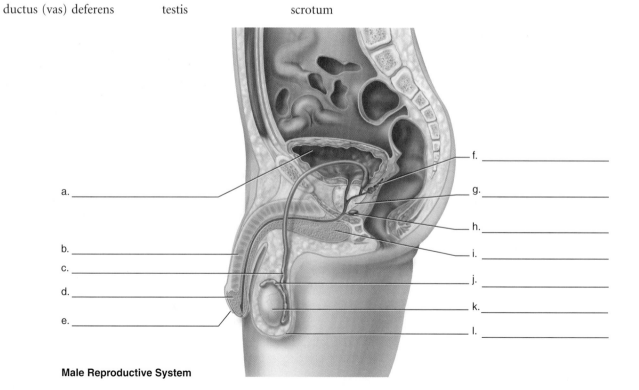

Male Reproductive System

Exercise 43
Female Reproductive System

INTRODUCTION

The female reproductive system is functionally more complex than the male reproductive system. In the male, the reproductive system produces gametes and delivers them to the female reproductive system. The female reproductive system not only produces gametes and receives them from the male but also, under the hormonal influence of HCG from the early cell mass and later from the placenta, provides space and maternal nutrients for the developing **conceptus.** Finally, the female reproductive system delivers the child into the outer environment. The female reproductive system is covered in the *Principles of Anatomy and Physiology* text in chapter 24, "Reproductive System."

The ovaries are the gamete-producing organs of the female reproductive system. They produce oocytes, the female sex hormones, estrogen (estradiol), and progesterone. In this exercise, you learn about the structure and function of the female reproductive system.

OBJECTIVES

At the end of this exercise, you should be able to

1. identify the gamete-producing organ of the female reproductive system;
2. identify the stages of maturation of the oocyte on charts or models;
3. trace the pathway of a gamete from the ovary to the usual site of implantation;
4. list the structures of the vulva;
5. describe the function of each organ in the female reproductive system;
6. name the layers of the uterus from superficial to deep;
7. compare and contrast the anatomy of the cat reproductive system with that of the human.

MATERIALS

Charts, models, and illustrations of the female reproductive system
Microscopes
Prepared slides of ovary and uterus
Cats
Materials for cat dissection
    Dissection trays
    Scalpel and two or three extra blades
    Gloves (household latex gloves work well for repeated use)
    Blunt (mall) probe
    Forceps and sharp scissors
    Sharps container
    Animal waste disposal container

PROCEDURE
Overview of the Gross Anatomy of the Female Reproductive System

Examine a model or chart of the female reproductive system and locate the following major reproductive organs there and in figure 43.1.

    Ovary
    Uterine tube
    Uterus
    Vagina
    Clitoris
    Labia minora (sing. *labium minus*)
    Labia majora (sing. *labium majus*)

Ovaries

Each **ovary** is approximately 3–4 cm long and oblong in form. The ovaries produce **oocytes** (eggs), which are shed from the outer surface of the ovary during **ovulation.** From there, the oocytes (which mature into ova if fertilized) move into the **uterine (fallopian) tube.** The ovaries are not directly attached to the uterine tube, and the oocytes must move from the surface of the ovary into the uterine tube.

Histology of the Ovary

Examine a prepared slide of the ovary under the microscope on low power. Locate the background substance of the ovary, which is known as the **stroma.** Look for circular structures in the ovary. These are the **ovarian follicles.** The smallest of the follicles are the **primordial follicles.** Locate the primordial follicles in your slide and compare them with the follicles in figure 43.2. The process of oogenesis involves the production of the gametes.

Also locate the **primary** and **secondary follicles.** Some of the follicles may contain **oocytes.** Primordial and primary follicles contain **primary oocytes;** secondary follicles contain **secondary oocytes.** The largest follicles in the ovary are the **graafian follicles,** or **mature ovarian follicles,** and you may be able to see one if it is present in your slide. During ovulation in humans, usually one secondary oocyte is shed from the ovary. In cats, many secondary oocytes may be shed. The process of the maturation of the oocyte is known as **oogenesis.**

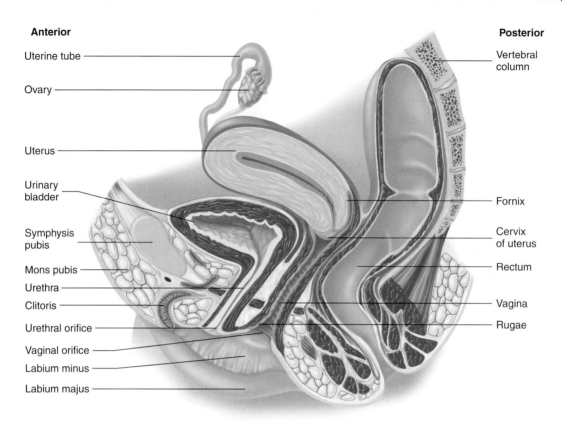

Figure 43.1 Female Reproductive System, Midsagittal View

Examine your slide and compare it with figures 43.2 and 43.3. Draw what you see in your slide in the following space.

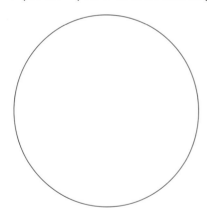

After ovulation, the remains of a mature ovarian follicle become a **corpus luteum,** which primarily secretes progesterone. If pregnancy does not occur, the corpus luteum decreases in size and becomes the **corpus albicans.** This progression is seen in figure 43.4. Examine a prepared slide of the ovary with a corpus luteum or corpus albicans and compare it with figures 43.4 and 43.5.

Uterine Tubes

The uterine tube has a small fringe on the distal region comprising the **fimbriae.** These are small, fingerlike projections attached to an expanded region known as the **infundibulum.** The uterine tube also has an enlarged region known as the **ampulla** and a narrower portion toward the uterus. Examine a model or chart of the female reproductive system and compare it with figure 43.6.

Uterus

The **uterus** is a pear-shaped organ with a domed **fundus,** a **body,** and a circular, inferior end called the **cervix.** The **uterine tubes** enter the uterus at about the junction of the fundus with the uterine body. A constricted portion of the inferior uterus is called the **isthmus.** The uterine wall is composed of three layers. The outer (superficial) surface of the uterus is called the **perimetrium.** This is located near the body cavity. The majority of the uterine wall consists of the **myometrium,** a thick layer of smooth muscle, and the innermost (deepest) layer of the uterus is the **endometrium.** Examine the models or charts in the lab and locate the structures in figure 43.6.

Histology of the Uterus

Examine a prepared slide of the uterus and locate its three layers. Locate the outer **perimetrium,** the smooth muscle of the **myometrium,** and the inner **endometrium.** Compare them with figures 43.6 and 43.7.

Now examine the two layers of the endometrium of the uterus under higher magnification. The **functional layer** is the one that is shed during menstruation. It is composed of **spiral arterioles** and **uterine glands** that appear as wavy lines toward

Exercise 43 Female Reproductive System

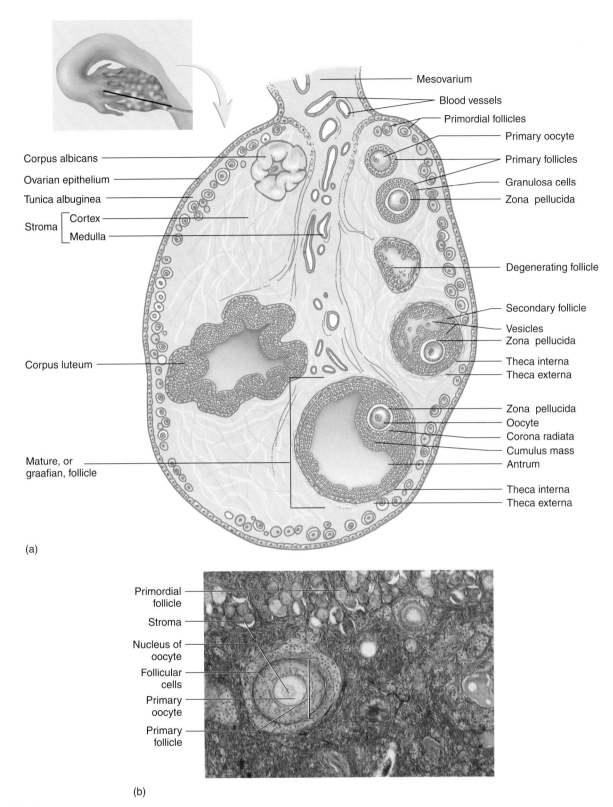

Figure 43.2 Histology of the Ovary
(a) Diagram; (b) photomicrograph (100×).

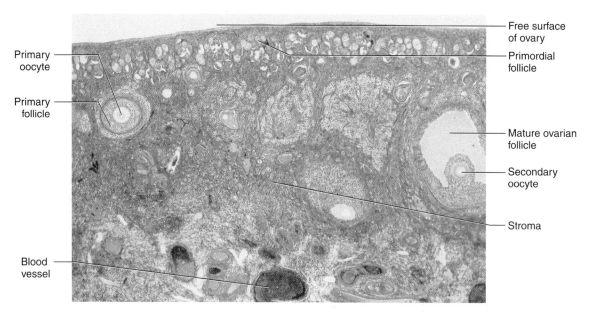

Figure 43.3 Follicles of the Ovary
Photomicrograph (40×).

the edge of the tissue. The **basal layer** is deeper and contains **straight arterioles.** Deep to the endometrium is the myometrium, which is mostly composed of smooth muscle.

Vagina

The **vagina** is a tough, muscular tube that joins the cervix of the uterus at its superior end and terminates at the **vaginal orifice** at its inferior end. It exits the body between the clitoris and the anus. The cervix pokes into the vagina, leaving a recessed area around the cervix known as the **fornix**. The **vaginal canal** is poorly supplied with nerves, is lined with nonkeratinized stratified squamous epithelium, and has cross ridges in the wall known as **rugae**. These allow the vagina to expand greatly during the delivery of a child. Locate these structures in figures 43.1 and 43.6.

External Female Genitalia

As in the male, the floor of the pelvis can be divided into the **anal triangle** and the **urogenital triangle.** The anal triangle consists of the **anus,** and the urogenital triangle consists of the **external female reproductive structures,** or **female genitalia.** The external female reproductive structures are collectively referred to as the **vulva.** The **mons pubis** is the anteriormost structure of the vulva. It is an adipose pad that overlies the symphysis pubis. Posterior to the mons pubis is the **clitoris,** which is a cylinder of erectile tissue embedded in the body wall that terminates as the **glans clitoris.** The body of the clitoris is curved, as illustrated in figure 43.1, and the glans clitoris is the superficial portion. The glans clitoris is anterior to the **external urethral orifice.** The clitoris has the same embryonic origin as the penis in males, and like the penis it is richly supplied with nerve endings. It is a major center for sexual pleasure in females. The anterior edge of the clitoris is enclosed by the **prepuce,** which is an extension of the labia minora.

Posterior to the clitoris is the external urethral orifice, and posterior to this is the **vaginal orifice.** The vaginal orifice is partially enclosed by a mucous membrane structure known as the **hymen.** The hymen is variable anatomically and has historically (and sometimes incorrectly) been used as an indicator of virginity. Lateral to the vaginal orifice are the **labia minora** (sing. **labium minus**). The space between the labia minora is known as the **vestibule,** and located laterally and posteriorly to the vestibule are the **greater vestibular glands (Bartholin's glands).** Lateral to the labia minora are the paired **labia majora** (sing. **labium majus**). Locate these structures on models or charts in the lab and in figure 43.8.

Ligaments

The uterus and ovaries are suspended in the pelvic cavity by a number of connective tissue sheaths called ligaments. The **broad ligament** anchors the uterus to the anterior body wall. The **round ligament** attaches the uterus to the anterior body wall at about the region of the inguinal canal. The **ovarian ligament** directly attaches the ovary to the uterus, and the **suspensory ligament** attaches the ovaries to the lumbar region. Locate these structures in models or charts in the lab and in figure 43.6.

Ovarian and Menstrual Cycles

The endometrium undergoes dynamic changes during the **menstrual cycle.** Gonadotropin-releasing hormone (GnRH) is secreted by the hypothalamus and stimulates the anterior pituitary to release hormones. **Luteinizing hormone (LH)** and **follicle-stimulating hormone (FSH)** from the anterior pituitary, and

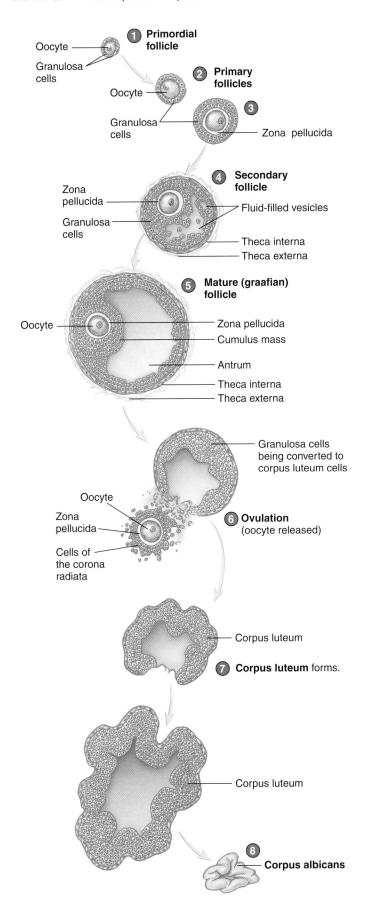

Figure 43.4 Maturation of the Oocyte and the Follicle
Arrows indicate a time line of development from primordial follicles to the corpus albicans.

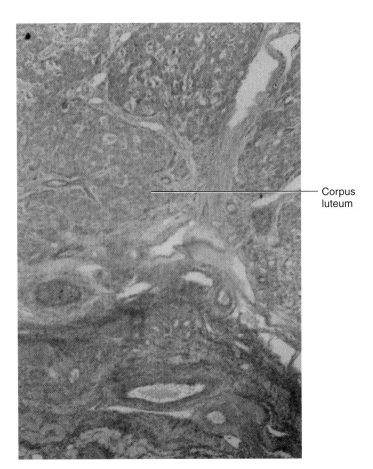

Figure 43.5 Postovulatory Ovary (40×)

subsequently **estrogen** and **progesterone** from the ovary, have significant effects on the endometrium. FSH stimulates the ovarian follicles to secrete estrogens and some progesterone. Elevated levels of these hormones promote the thickening of the endometrium, as indicated in figure 43.9.

After ovulation, progesterone levels increase and the endometrial cells remain intact. Toward the end of a woman's menstrual cycle, estrogen and progesterone levels drop, causing the functional layer of the endometrium to slough off. The loss of the endometrial layer is the beginning of a woman's period, or **menstruation**. These events are briefly outlined in figure 43.9. Examine the figure and note how the endothelial layer begins to decrease when estrogen and progesterone levels fall (about day 25 in figure 43.9).

Female Contraception

As with male contraception, preventing male and female gametes from uniting is the goal in female contraception. Abstention, or refraining from intercourse, is the most effective method of birth control. Other methods involve oral contraceptives or hormonal implants; tubal ligation (cutting and tying the uterine tubes—see figure 43.10); the use of barrier methods, such as the condom or diaphragm; and the use of spermicidal foams. The use of oral contraceptives, which are synthetic estrogens and progesterones, decreases the levels of LH and FSH in the pituitary (by negative-feedback mechanisms), thus preventing ovulation. Progesterone

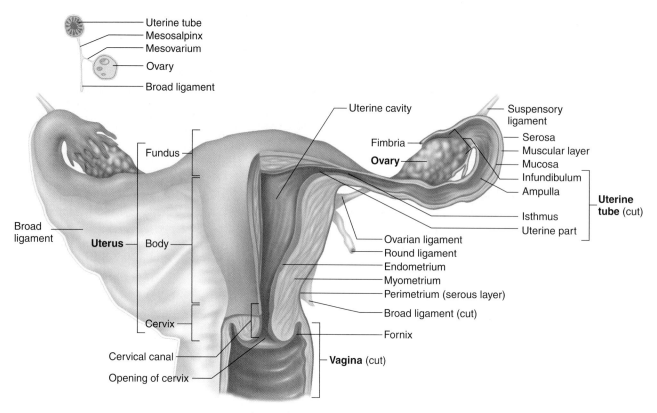

Figure 43.6 Female Reproductive System, Anterior View

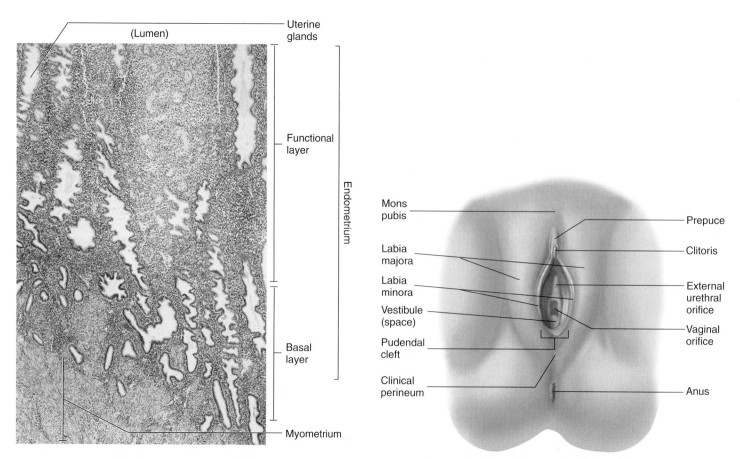

Figure 43.7 Histology of the Uterus (40×)

Figure 43.8 External Female Genitalia

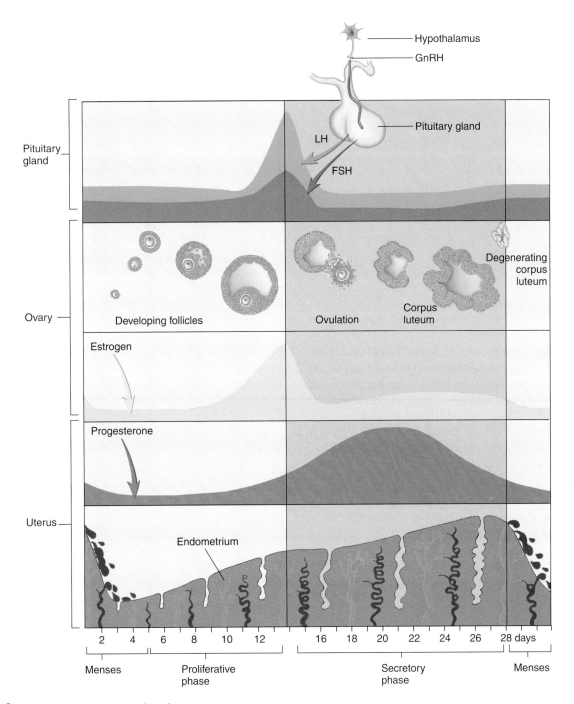

Figure 43.9 Ovarian and Menstrual Cycles

implants elevate hormone levels, which also prevent ovulation. Examine the levels of progesterone in figure 43.9. When these levels are elevated in the later part of the ovarian cycle, note how the levels of FSH and LH are low. An antiprogesterone compound, mifepristone, terminates pregnancy and is known as the "morning after" pill.

Anatomy of the Breast

The structure of the breast derives from the integumentary system, yet the female breast is discussed as a reproductive structure due to its importance as a source of nourishment for offspring. The major structures of the external breast are the pigmented **areola,** the protruding **nipple,** the **body** of the breast, and the **axillary tail (tail of Spence).** The axillary tail is of clinical importance in that breast tumors often occur there. Examine the surface features of the breast in figure 43.11 and locate the structures listed.

Compare models or charts in the lab and locate the internal structures of the breast. Note that the breast is anchored to the pectoralis major muscle with **mammary (suspensory) ligaments.** Much of the breast is composed of **adipose tissue,** and embedded in the adipose tissue are the **mammary glands.** The mammary glands are responsible for the production of milk in

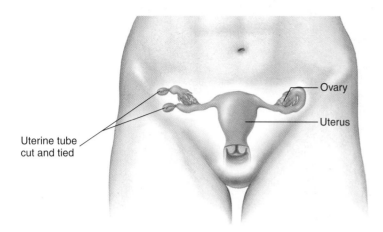

Figure 43.10 Female Contraception

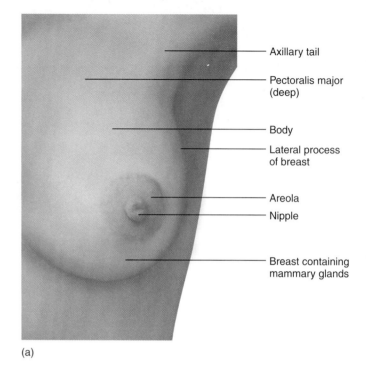

lactating females. The glands are clustered in **lobes**, and there are about 15 to 20 lobes in each breast. The mammary glands increase in size in nursing women and lead to **lactiferous ducts**, which subsequently lead to **lactiferous sinuses (ampullae).** These lead to other lactiferous ducts that exit via the nipple. Humans have several ducts leading to each nipple, whereas in some mammals, such as cows, just one tube exits the nipple. The mammary glands in females begin to undergo changes prior to puberty and become functional glands after the delivery of a child. Note the features listed in figure 43.11.

Cat Dissection

Prepare for the cat dissection by obtaining a dissection tray, a scalpel, scissors or bone cutter, string, forceps, and a plastic bag with a label.

You probably already removed the multiple mammary glands of the female cat during the removal of the skin and the study of the muscles. Examine the external genitalia for the **urogenital orifice.** *Check with your instructor before cutting through the pelvis of the cat.* If your instructor directs you to do so, then begin by cutting through the musculature of the cat at the level of the **symphysis pubis.** Continue to cut in the midsagittal plane cranially through the abdominal muscles, as described in the dissection of the male cat. This should expose the reproductive organs (figure 43.12).

Unlike the human female, the cat has a **horned (bipartite) uterus.** The uterus of the cat has the appearance of a Y, with the upper two branches being the **uterine horns** and the stem of the Y the **body** of the uterus. Humans normally have single births from a pregnancy, whereas the expanded uterus in cats facilitates multiple births. If your cat is pregnant, the uterus will be greatly enlarged, and you may find many fetuses inside.

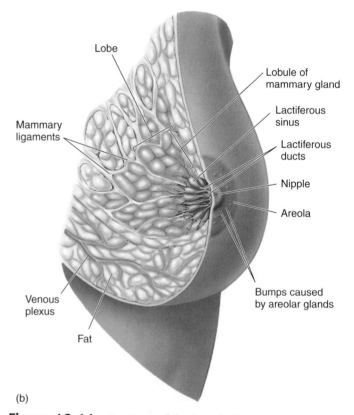

Figure 43.11 Anatomy of the Female Breast
(a) Surface features; (b) interior of the female breast.

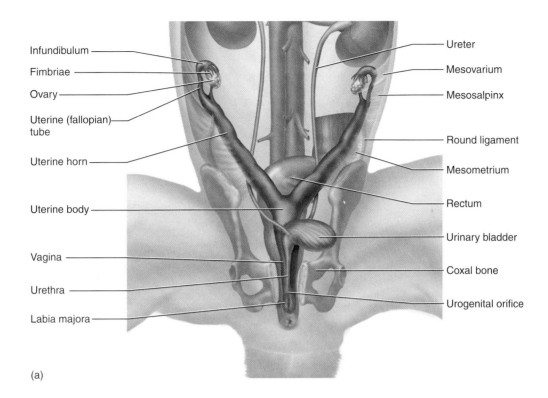

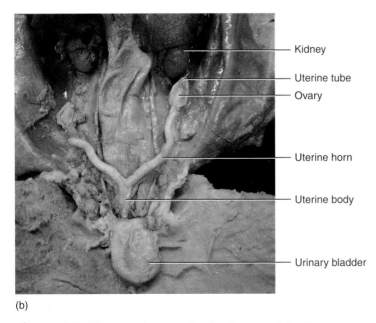

Figure 43.12 Female Reproductive Organs of the Cat
(a) Diagram; (b) photograph.

The **ovaries** in cats are caudal to the kidneys and are relatively small organs. Examine the paired ovaries and the short **uterine tubes (oviducts)** in the cat. If you have difficulty locating them, trace the uterus toward the uterine horns and locate the ovaries. The uterine tubes in the human female are proportionally longer than those in the cat. The opening of the uterine tube near the ovary is called the **ostium,** and it receives the oocytes during ovulation.

At the termination of the uterus is the **cervix,** which leads to the **vaginal canal.** The vagina in cats is different from that in humans in that the **urethral opening** is internally enclosed in the vaginal canal. This region where the vagina and the urethral opening occurs is called the **vaginal vestibule.** Thus, the opening to the external environment is a common urinary and reproductive outlet called the **urogenital orifice.** Locate these structures in figure 43.12.

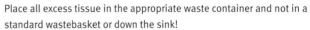

Clean Up
Place all excess tissue in the appropriate waste container and not in a standard wastebasket or down the sink!

Stages of Development
Early Development
The union of the sperm and egg in a process known as **fertilization** initiates a remarkable phenomenon of growth and differentiation from the single-celled **zygote** to the adult human.

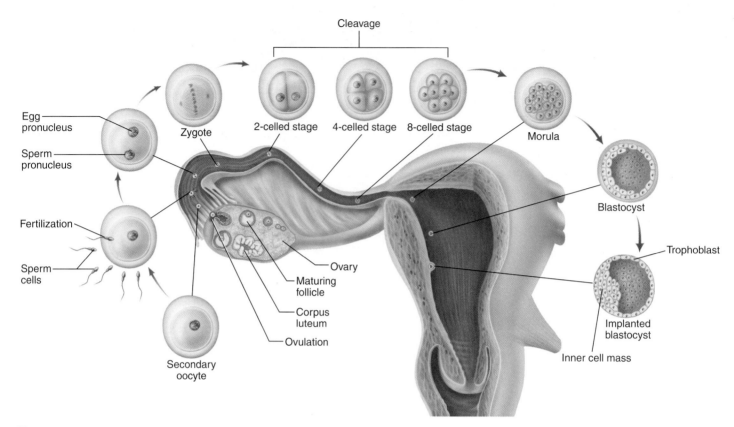

Figure 43.13 Early Stages of Development

Fertilization usually occurs in the uterine tube, and the zygote divides into 2 cells, then 4, 8, 16, and so on, until a solid cluster of cells called a **morula** is formed. The morula continues to divide until it becomes a hollow ball of cells known as the **blastocyst.** The covering of cells on the outside of the blastocyst is called the **trophoblast,** and the cluster of cells on the inside is known as the **inner cell mass.** Review these stages in figure 43.13.

Embryonic Tissues

The early stage of development continues to progress, and the formation of three embryonic tissues occurs. These are the **ectoderm,** the **mesoderm,** and the **endoderm.** The ectoderm gives rise to the outer layer of skin and the nervous tissue, the mesoderm gives rise to bones and muscles, and the endoderm gives rise to many internal organs, such as digestive and respiratory organs. The early development of these layers is illustrated in figure 43.14.

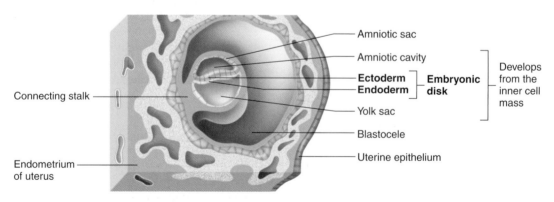

Figure 43.14 Embryonic Development

Exercise 43 Review
Female Reproductive System

Name: _____

Lab time/section: _____

Date: _____

1. What is the gonad in the female reproductive system? _____

2. What is the inner layer of the uterus called? _____

3. What happens to estrogen and progesterone levels just prior to menstruation? _____

4. The ovaries attach to the uterus by what structure? _____

5. What three hormones are at elevated levels just prior to ovulation? _____

6. Where is the fornix in the female reproductive system? _____

7. Which is more anterior, the urethral opening or the clitoris? _____

8. What is the background substance of the ovary called? _____

9. What is the name for the expulsion of the secondary oocyte from the ovary? _____

10. What is the layer of the endometrium closest to the myometrium called? _____

11. Name the part of the breast nearest to the shoulder. _____

12. What are the milk-producing glands of the breast called? _____

13. A zygote is formed from the fusion of what two cells? _____

14. Embryonic tissue consists of three layers. What are these called? _____

15. What is a morula? _____

16. Trace the pathway of milk from the mammary glands to expulsion. _____

_____

_____

17. Ectopic pregnancies are those that occur outside the endometrial layer of the uterus. Explain how pregnancies may occur in the uterine tube (thus, a tubal pregnancy) or in the abdominopelvic cavity. _____

_____

_____

_____

18. How does the uterus of the human differ from that of the cat? _____

_____

19. How does the structure of the uterus in the cat correlate to multiple births from each pregnancy? _____

_____

_____

20. Label the following illustration using the terms provided.

cervix   clitoris   vaginal orifice   fundus
urinary bladder   fornix   labium minus
rugae   vaginal canal

a. _____

b. _____

c. _____

d. _____

e. _____

f. _____

g. _____

h. _____

i. _____
(ridges)

Appendix A
Measurement Conversions

The **metric system** is universally used in science to measure certain **values** or **quantities.** These values are **length, volume, mass** (weight), **time,** and **temperature.** The metric system is based in units of 10, and conversion to higher or lower values is relatively easy when compared with using the U.S. customary system. In the United States, length is typically measured in inches, feet, yards, and miles. The **base unit** for length in the metric system is the meter. As shown in table A.1, 1 meter is equivalent to 1.09 yards or 39.4 inches. When the measurement of great lengths makes the use of meters impractical, then the kilometer (1000 meters) is used. When the measurement of extremely small lengths is needed, then the centimeter (1/100 of a meter), millimeter, micrometer, nanometer, or angstrom is used. The common multiples or fractions for the values are outlined in table A.1.

TABLE A.1 Conversions of Standard and Metric Systems

Value	Base Unit	×1000	1/100	1/1000	1/1,000,000
Length	Meter (m)	Kilometer (km)	Centimeter (cm)	Millimeter	Micrometer
	1 m = 39.4 in.	1 km = 5/8 mi	2.54 cm = 1 in.		1 nanometer = 10^{-9}m
	1 m = 1.09 yd				(1 angstrom = 10^{-10}m)
Mass	Gram (g)	Kilogram (kg)	Centigram	Milligram	Microgram
	454 g = 1 lb	1 kg = 2.2 lb			
	1 g = 0.036 oz				
Volume	Liter (L)	Kiloliter	Centiliter	Milliliter (mL)	Microliter
	1 liter = 1.06 qt			1 mL = 1 cc	
	1 gal = 3.78 L				
Time	Second (s)	Kilosecond	Centisecond	Millisecond	Microsecond
Temperature	Degrees	Celsius			
	0°C = freezing				
	100°C = boiling				
	°F = 9/5 °C + 32				

Appendix B
Preparation of Materials

The following preparations are designed for a lab of 24 students. It is best to estimate how many milliliters of stock solutions are required for all the students in a lab and then double that amount. The preparations are listed alphabetically. The number of the lab exercise follows the solution description for cross-referencing.

Acetylcholine chloride solution (0.1%)
Add 0.1 g of acetylcholine chloride to 100 mL frog Ringer's solution. Pour into a small, labeled dropper bottle. (Exercise 29)

Acid solution
Pour 200 mL lemon juice or vinegar into six dropper bottles labeled "Acid Solution." (Exercise 39)

Agar plates
Add 22 g agar to enough water to make 1 L of solution. Boil and stir the agar until it all dissolves. Pour into petri dishes for three dishes per table. (Exercise 3)

Alpha amylase solution (1%)
Add 3 g of alpha amylase to a graduated cylinder and add water to the 300 mL mark. Pour into a small, labeled bottle or beaker. (Exercise 39)

Alpha amylase solution (0.1%)
Add 0.1 g of alpha amylase to a graduated cylinder and fill to 100 mL with water or make a 0.1 serial dilution by taking 10 mL of the above solution and adding 90 mL of water. Pour into a small, labeled bottle or beaker. (Exercise 39)

Ammonia
Small bottle of ammonia. (Exercise 21)

BAPNA solution (1%)
N-alpha-Benzoyl-DL-arginine-4-nitroanilide hydrochloride 1% BAPNA solution (50 mL) in two bottles of 25 mL each. BAPNA is an expensive material, but you do not need much of it for the lab. Available from Sigma B4875 or Aldrich 85,711–4. (Exercise 39)

Benedict's reagent

A copper sulfate solution that turns color if reducing sugars are present and remains blue in the absence of reducing sugars. Add 35 g sodium citrate and 20 g sodium carbonate (Na_2CO_3) to 160 mL of water. Filter through paper into a glass beaker. Dissolve 3.5 g copper sulfate ($CuSO_4$) in 40 mL of water. Pour the copper sulfate solution into the 160 mL, stirring constantly. (Exercise 39)

Bleach solution (10%)

Mix 100 mL of household bleach (sodium hypochlorite) with 900 mL tap water. (Exercises 25, 26, 41)

Caffeine solution, saturated

Add small amounts of caffeine to 50 mL of water until no more will dissolve. Decant the solution into small dropper bottles. (Exercise 29)

Calcium chloride solution (2%)

Weigh 5 g of calcium chloride and place in a graduated cylinder. Add frog Ringer's solution to make 250 mL. Pour into dropper bottles. (Exercise 29)

Cat wetting solution

Numerous formulations are available for keeping preserved specimens moist. Some commercial preparations are available that reduce the exposure of students to formalin or phenol. You may not need any wetting solution at all if the cats are kept in a plastic bag that is securely tied closed. You can make a wetting solution by putting 75 mL of formalin, 100 mL glycerol, and 825 mL distilled water in a 1 L squeeze bottle. Another mixture consists of equal parts Lysol and water. (Exercises 12–15)

Cellulose

Cut several (3–4) grams of pure cotton (cotton wool, cotton balls) into fine pieces (0.5 cm or less). Label "Cellulose." (Exercise 39)

ELISA test kits such as:
Wards Lyme Disease # 36 V 8907
Carolina Elisa Simulation Kit # 211248
Fisher Scientific HIV/AIDS Test Simulation #S66656 (Exercise 26)

Epinephrine solution (0.1%)

Add 0.1 g of adrenalin chloride in 100 mL of frog Ringer's solution. Label and pour the solution into small dropper bottles. (Exercise 29)

Essential oils preparation

Fill several small, screw-top vials with peppermint, almond, wintergreen, and camphor oils (available from local drug stores). Label "Peppermint," "Almond," "Wintergreen," and "Camphor," respectively, and keep vials in separate wide-mouthed jars to prevent cross-contamination of scent. (Exercise 21)

Fill four small vials halfway to the top with cotton and color them red with food coloring. Label the vials "Wild Cherry" and add benzaldehyde solution until the cotton is moist. Fill four small vials halfway to the top with cotton but do NOT color them red. Label the vials "Almond." You can use the almond vials from the previous preparation. (Exercise 21)

Filtration solution (1% starch, charcoal, and copper sulfate solution)

Place 5 g of starch, 5 g of powdered charcoal, and 5 g of copper sulfate ($CUSO_4$) in a 1 L beaker. Add enough water to make 500 mL. Stir well and pour into a 500 mL bottle. Label "Filtration Solution." (Exercise 3)

Frog Ringer's solution

Weigh and place the following materials in a 1 L graduated cylinder:
6.5 g NaCl (sodium chloride)
0.2 g $NaHCO_3$ (sodium bicarbonate)
0.1 g $CaCl_2$ (calcium chloride)
0.1 g KCl (potassium chloride)
To these, add enough water to make 1000 mL. This solution should be prepared fresh and used within a few weeks. Place in large dropper bottles. (Exercises 11, 19, and 29: In Exercise 29, there should be three solutions prepared—one at room temperature, one at 37°C, and one in an ice bath.)

Hydrochloric acid solution (0.1%)

Add 1 mL of concentrated HCl to 1 L of water. Remember "AAA"—Always Add Acid to water. Pour into a small, labeled dropper bottle. (Exercise 19)

India ink

Dropper bottles of India ink. (Exercise 3)

Iodine solution

See Lugol's iodine solution.

Litmus cream

Use approximately 250 mL of heavy cream. To this, add powdered litmus until the cream is a light blue. Pour into two separate bottles and label "Litmus Cream." (Exercise 39)

Litmus solution

Weigh 1 g litmus powder and dissolve in 300 mL water. Titrate HCl into the solution while stirring until the color begins to turn from blue to red. Add NaOH drop by drop until the solution just turns back to blue. Pour into two bottles labeled "Litmus Solution." (Exercise 36)

Lugol's iodine solution

Prepare by adding 10 g I_2 (iodine) and 20 g KI (potassium iodide) to 1 L of distilled water. Store in small, dark dropper bottles. Label "Lugol's Iodine." (Exercises 3 and 39)

Maltose solution (1%)

Add 2 g of maltose in enough water to make 200 mL. Stir until dissolved and pour into two clean bottles. Label "1% Maltose Solution." (Exercise 39)

Methylene blue (1%)

Add 5 g methylene blue powder in 500 mL distilled water. Pour into labeled dropper bottles. (Exercise 2)

Methylene blue solution (0.01 M)

Add 3.2 g methylene blue (MW 320) to distilled water to make 1 L of solution. Place in dropper bottles. (Exercise 3)

Dark corn syrup or concentrated sucrose solution (20%)

Use undiluted corn syrup or a 20% sugar solution. To make the sugar solution, add 100 g of table sugar (sucrose) to water to make 500 mL of solution. Add food coloring to the sucrose solution. Make sure that the sucrose is completely dissolved. (Exercise 3)

Nitric acid (1 N) for decalcifying bones

Add 64 mL of concentrated nitric acid (70%) slowly to water to make 1 L of solution. Remember "AAA"—Always Add Acid to water. (Exercise 6)

Pancreatin solution (2%)

Place 4 g of pancreatin in a graduated cylinder and add water to make 200 mL. Stir well. Adjust the pH with 0.05 M sodium bicarbonate until neutral (pH 7). Pour into two different stock bottles and label "2% Pancreatin Solution." Pancreatin may be stored frozen (not in a frost-free freezer that regularly cycles between freezing and defrosting). Also note that commercially prepared pancreatin has an optimum pH. If the pH is too low, then the reaction will be slowed or stopped. (Exercise 39)

Perfume, dilute solution

Add 10 mL inexpensive perfume to 100 mL of isopropyl alcohol. (Exercise 21)

Phosphate buffer solution

Add 3.3 g of potassium phosphate (monobasic) and 1.3 g of sodium phosphate (dibasic) to 500 mL of water. Label and pour into squeeze bottles. (Exercise 25)

Potassium dichromate solution (0.01 M)

Add 2.94 g of potassium dichromate (MW 294) crystals to water to make 1 L of solution. Label and pour into dropper bottles. (Exercise 3)

Potassium permanganate solution (0.01 M)

Add 1.58 g of potassium permanganate (MW 158) crystals to water to make 1 L of solution. Label and pour into dark brown dropper bottles. (Exercise 3)

Procaine hydrochloride
Place 1 g of procaine hydrochloride solution in 1 mL water. Add to this 30 mL of pure ethanol. Place in small, screw-capped bottles labeled "Procaine Hydrochloride Solution." (Exercise 19)

Saltwater solution (3%)
Using a clean, food-grade container, add 15 g of NaCl crystals (table salt is acceptable) to water to make 500 mL of solution. Pour into clean, food-grade dropper bottles labeled "Salty" for lab. (Exercise 21)

Sodium bicarbonate (Exercise 36)

Sodium chloride solution (5%)
Add 25 g of NaCl crystals to water to make 500 mL of solution. Label the solution "Sodium Chloride Solution 5%" and place in small dropper bottles. (Exercise 3)

Sodium chloride solution (5%)
Add 2.5 g of NaCl crystals to water to make 50 mL of solution. Label the solution and place in a small beaker. (Exercise 19)

Sodium chloride solution 0.9% (physiologic saline)
Put 9.0 g NaCl in 1000 mL water and pour into small dropper bottles. Label the solution "Sodium Chloride Solution 0.9%." (Exercise 3)

Sodium hydroxide solution (1%)
Add water to 2 g NaOH to make 200 mL of solution. Stir the mixture carefully. The reaction is exothermic and will generate quite a bit of heat. (Exercise 39)

> **Caution**
> NaOH is caustic. Wear gloves and goggles. Pour into dropper bottles and label "1% NaOH Solution—Caustic."

> **Caution**
> The reaction is exothermic and will generate heat. Wear protective gloves and eyewear. If you spill this on your skin, make sure that you flush your skin immediately with cold water.

Starch solution (1%)
Boil 500 mL of distilled water. Remove the water from the heat and add 5 g of cornstarch (or 5 g of potato starch). Filter the mixture through cheesecloth into bottles. (Exercise 3)

Starch solution (0.5%)
A potato starch solution is made by first boiling 500 mL of water. Remove the water from the hot plate and add 2.5 g of potato starch powder. Stir and cool the mixture. Do not boil the starch and water mixture, as this will lead to some hydrolysis of starch to sugar. Test for the presence of sugar by using Benedict's reagent. There should be no sugar present. Place into 250 mL bottles and label "0.5% Starch Solution." (Exercise 39)

Sugar solution (3%)
Using a clean, food-grade container, dissolve 15 g of table sugar in enough water to make 500 mL of solution. Pour into clean, food-grade dropper bottles labeled "Sweet" for lab. (Exercise 21)

Sugar solutions
Four table sugar solutions of 2 L each. Color each solution so that there are four different colored solutions. (Exercise 3)

0%—2 L of water
5%—Dissolve 100 g sugar in enough water to make 2 L of solution.
15%—Dissolve 450 g of sugar in enough water to make 3 L of solution; pour into a 2 L bottle and a 1 L bottle.
30%—Dissolve 600 g of sugar in enough water to make 2 L of solution.

Tonic water
Select a clean, food-grade dropper bottle that holds 100 mL. Label "Bitter." Fill with commercial tonic water. (Exercise 21)

Umami solution
Using a clean, food-grade container, dissolve 15 g of monosodium glutamate in 500 mL water. Pour into clean, food-grade dropper bottles labeled "Umami." (Exercise 21)

Vinegar solution
Select a clean, food-grade container. Use household vinegar or make a 5% acetic acid solution by adding 5 mL of concentrated food-grade acetic acid to about 50 mL of water and then adding water to make 100 mL. Pour into clean, food-grade dropper bottles labeled "Acidic" for lab. (Exercise 21)

Wright's stain
Wright's stain is available as a commercially prepared solution from a number of biological supply houses. Place in labeled dropper bottles. (Exercise 25)

Appendix C
Lab Reports

Part of working in science involves writing lab reports. You should write lab reports in a certain style and follow the basic guidelines that are generally accepted in the field. The general format for lab reports includes four sections: **introduction, materials and methods, results,** and **conclusion.** Each of these parts is important in the write-up, and all these sections must be included in the lab report. Your report should have a title, your name, your instructor's name, the course name and semester, and your lab section.

The purpose of a scientific report is to explain an investigation. Your instructor is your primary audience for your report in this class, so you should communicate that you understand the basic information. Part of a grade in a lab report depends on doing the experiment correctly and demonstrating that you are aware of the outcome of the experiment and its relationship to theory or applications you have learned in lecture.

If your experiment did not come out as you thought it would, you still can do well in your lab report. Results from your experiment may have come out satisfactorily and be perfectly fine even if you think that the data should have been different from what you obtained.

Introduction

The introduction section consists of a description of the problem and the subject of your study. In professional journals, the introduction often includes a history of experimentation or current knowledge in the field, but you will probably not have this in your lab report. You should pose the question, or hypothesis, that your experiment is trying to resolve in the introduction. You may be conducting an experiment to determine whether an enzyme functions on a substrate or whether a particular effect occurs when you perform an action.

Materials and Methods

The materials and methods section is a clear description of what equipment, animals, chemicals, and so on were used and the experimental procedure followed. You must write this section in very clear and precise terms. From this section, you should expect that someone could repeat your experiment and produce the same results.

In one way, this is the "recipe" for your experiment. As in baking a cake, it is not good enough simply to list the materials. You must include quantities, in what sequence the materials were added, and how long things were stirred. You must provide specific details about your procedure. Make sure that you state how much material was added to a sample. For example, you should write "We added 5 mL." This is a known quantity that people can duplicate, whereas "We added a little" is vague. You must make sure that you include all steps in your procedure. If you leave something out in your description, then a person following your directions might get very different results.

Results

In the results section, you state the outcome of your experiment in a clear and defined way. Your data must be clearly presented. This may consist of the tabulation of data that you acquired from your experiment in the form of a line graph, bar graph, or table. Remember that any graph should have a complete description. If you are studying the effect of exercise on heart rate, then exercise is the **independent variable** and is listed on the horizontal axis, whereas heart rate is the **dependent variable** and is listed on the vertical axis, as illustrated in the following graph.

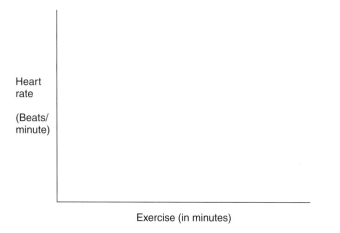

The data for your lab results section may be in raw form (direct measurements from your experiment) or organized by calculating the mean (average) data and range (high and low), as determined from your experiment. You should not try to explain your results at this point but simply state what the outcome was. Save the interpretation of the data for the conclusion.

Conclusion

In the conclusion, you can analyze and interpret your data and determine whether the results of your experiment proved your initial question (make sure you remember to answer your original question). This is the most important part of your lab report in terms of finding out if you understood the lab. Did you prove your hypothesis, did you disprove your hypothesis, or are the results inconclusive? Where should you go from here?

Sometimes experiments go awry and the results do not support the hypothesis. This can be due to many things, such as experimental error, which is a big factor in lab experiments, especially at the undergraduate level. Someone may not have followed the experimental procedure correctly and added too much, too little, or none of the materials that should have been part of the procedure. Sometimes it is a matter of timing—the process did not go on as long as it should have or went on too long. In some cases, the data are recorded incorrectly or read wrong. A 3 in the data might be mistaken for an 8, or the data might be entered in the wrong location. There are hundreds of possibilities as to why error plays a part in experiments, and your instructor is probably very familiar with many of them.

Another factor is faulty equipment, supplies, or setup. Other problems may be due to the variation in living organisms. Humans and other animals have their own idiosyncratic physiologic responses. Not every person and not all experimental animals respond in the same way. For example, almost all product information sheets that come with pharmaceutical drugs list adverse reactions seen in some people. If everyone responded the same way, there would be no need for warnings on drug products.

If you obtain variances from what you expect to get, then you should write this in your report. Sometimes experimental procedures are included that produce results other than what you expect to see. You should be honest in your recording of data and resist the temptation to "fudge" the data so that you can get a better result and therefore a better grade on your report.

Sometimes people in industry change their data to produce more desirable results. If this is discovered by independent investigations, the corporation that allowed this to happen usually pays significant fines. Students who change data and are discovered are also subject to penalty, which is usually far more severe than what they would have received due to the "bad" data they think they have gotten. In addition to being honest in the reporting of data, you also want to make sure that your words are your own. Even though you may have performed an experiment word for word from this lab manual or another source, do not plagiarize the material. You should paraphrase material if you want to have the same meaning but not the same words as someone else.

Your lab report should follow the format described here unless your instructor decides to make significant changes in the write-up. Lab reports tend to take a long time to write, but the analysis in the lab report is where you work toward understanding the experiment.

INDEX

A
Abdominal aorta, 381f, 382, 388, 389, 396f–397f, 412f, 496f
Abdominal arteries, 395, 396f, 398f, 399f
Abdominal cavity, 5, 6f
Abdominal muscles, 221–224, 222t, 223f, 229f
Abdominal oblique muscle(s)
    external, 161f–162f, 167f–168f, 221, 222t, 223f, 228f
    internal, 221, 222t, 223f, 228f
Abdominal organs, 480, 481f
Abdominal regions, 7, 7f–8f
Abdominal veins, 410, 410f, 414f, 415
Abdominopelvic cavity, 6, 6f
Abducent (VI) nerve, 251, 253f, 254t, 255f, 255t, 256f
Abduction, 139, 140f
Abductor digiti minimi muscle, 164t, 174, 175f
Abductor pollicis brevis muscle, 164t, 174, 175f
Abductor pollicis longus muscle, 164t, 174, 176f–177f
ABO blood system, 345–346, 346t–347t
Abstention, 520, 529
Acceleration, 316
Accessory (XI) nerve, 252, 253f, 254t, 255f, 255t, 268f
Accessory hemiazygos vein, 411f
Accessory organs, of digestive system, 465, 473–477
Accessory pancreatic duct, 479f
Accommodation, visual, 304
Acetabular labrum, 134, 137f
Acetabular ligament, transverse, 137f
Acetabular notch, 88f
Acetabulofemoral joint, 134–136, 137f
Acetabulum, 87, 87f–88f
Acetylcholine, 147, 371, 373–374
Acetylcholine chloride solution, A–1
Achilles tendon, 162f, 195, 198f, 201f
Achilles tendon reflex, 282, 283f
Acid solution, 491, A–1
Acid-base effects, of respiratory gases, 455
Acinar cells, 330f
Acromial end, of clavicle, 82, 83f
Acromial region, 7f
Acromioclavicular ligament, 136f
Acromiodeltoid muscle, 181

Acromion process, 81, 82f, 136f, 171f
Acromiotrapezius muscle, 180, 181f
Acrosome, of sperm, 516f, 517
ACTH, 327
Actin, 24, 26f, 147
Action(s)
    at joints, 139–141
    of muscles, 159
Action potential, 237f, 278, 366, 366f
Acuity, visual, 305
Adaptation, 289–291, 296
Adduction, 139, 140f
Adductor brevis muscle, 190t, 191, 192f, 272f
Adductor femoris muscle, 199, 199f
Adductor longus muscle, 161f, 190t, 191, 191f–195f, 199, 199f
Adductor magnus muscle, 162f, 190t, 191, 192f–193f, 199, 272f
Adductor magnus tendon, 138f
Adductor muscles, 190–193, 193f
Adductor pollicis muscle, 175f, 269f
Adductor tubercle, 89, 89f–90f
Adenohypophysis, 326, 327f. See also Pituitary gland
Adenoids, 424
ADH, 327
Adipocytes, 50, 51f
Adipose tissue, 50, 50f, 531
Adrenal glands, 3f, 325f, 329–330, 330f
Adrenocorticotropin (ACTH), 327
Adrenolumbar arteries, 398f, 399
Adrenolumbar vein, 414f, 415
Adventitia, 469f, 499f
Aerobic exercise, 459
Afferent arteriole, 496, 498, 498f, 501f
Afferent division, of peripheral nervous system, 233
Afferent lymphatic vessels, 424f
Afferent neurons, 234, 235f, 281f
Afterimages, 308
Agar, 31
Agar plates, A–1
Agglutinates, 345
Agglutinins, 345
Agglutinogens, 345
Aggregated lymph nodules, 473, 474f
Agranular leukocytes, 339f, 340–341, 341f, 342t
Agranulocytes. See Agranular leukocytes

Air flow, 455
Ala (pl., alae), 101, 102f
Alar cartilage, greater, 120f, 436f
Albumins, 337
Albuminuria, 509, 509t
Aldosterone, 329
Alimentary canal, 465, 469f
All-or-none principle, 147
Alpha amylase solution, A–1
Alpha cells, 329, 330f
Alveolar bone, 468f
Alveolar ducts, 439, 440f–441f
Alveolar epithelium, 442f
Alveolar pressure, 448, 448f
Alveolar processes, 111, 113f–114f
Alveolar sac, 439, 440f–441f
Alveolar sockets, 466
Alveolar volume, 447–448, 448f
Alveoli
    maxillary, 111, 112f
    pulmonary, 439, 440f–442f
Amino acids, 492
Ammonia, A–1
Amniotic cavity, 534f
Amniotic sac, 534f
Amphiarthrotic joints, 133
Ampullae
    of ductus deferens, 517f, 518, 519f
    of mammary glands, 532
    of semicircular canals, 316, 317f
    of uterine tube, 526
    of Vater, 477, 479f
Amygdaloid nucleus, 251f, 257f, 295f
Amylases, 473, 488f, 489–492
Anaerobic exercise, 459
Anal canal, 473, 475f
Anal gland, 481f
Anal sphincter, 475f
Anal triangle, 520, 520f, 528
Anaphase, 28, 29f, 30t
Anatomic position, 3, 4f
Anatomy, 1
Anconeus muscle, 176f, 182, 183f, 269f
Androgens, 330
Anemia, 347
Anesthetics, nerve response to, 280
Angle
    of mandible, 114f, 116f, 117
    of rib, 102, 104f
    of scapula, 81, 82f
    of sternum, 104f, 105
Angular acceleration, 316
Annular ligament, 134, 137f
Ansa cervicalis, 268f

Antagonists, 141
Antebrachial region, 7f
Antebrachial vein, median, 407f
Anteperitoneal, 497
Anterior, 3, 4f, 5t
Anterior arch, cervical, 100f
Anterior chamber of eye, 302f, 308
Anterior commissure, 251f
Anterior compartment, of eye, 300, 302f, 308
Anterior crest, 90, 91f
Anterior horn
    of lateral ventricle, 243f
    of spinal cord, 264, 264f
Anterior median fissure, 264–265, 264f
Anterior nuclei of thalamus, 251f
Anterior surface, of sacrum, 101, 102f
Anti-B agglutinins, 345
Antibodies, 340–341, 345–347
Antibody-mediated immunity, 341
Anti-D, 347
Antidiuretic hormone (ADH), 327
Antigens, 329, 345
Antiserum A, 346, 346f
Antiserum B, 346, 346f
Antrum, 527f, 529f
Anus, 2f, 465f, 473, 475f, 516f, 520f, 528, 530f
Aorta, 3f, 353, 380, 388, 398f, 400f, 410f–411f
    abdominal, 381f, 382, 388, 389, 396f–397f, 412f, 496f
    ascending, 356f–357f, 358, 361, 380, 382f–383f, 388, 388f, 396f, 412f
    descending, 382, 382f, 388, 388f, 395, 399, 399f–398f
    of sheep, 361, 361f
    thoracic, 381f, 382, 382f, 388, 396f
Aortic arch, 355f–357f, 358f–359f, 361, 380, 380f–383f, 388, 388f, 390f, 396f, 411f
Aortic arch arteries, 380–382, 382f
Aortic semilunar valve, 355f, 358f, 359, 359f, 374, 375f–376f
Apex
    of heart, 353, 356f–360f
    of patella, 91f
Apical foramen, 468f
Apical foramen, of tooth, 466
Apocrine sweat glands, 62
Aponeuroses, 179
Appendicular skeleton, 69, 69f–70f, 71t, 81–96

Appendix, 2f, 465f, 473f, 474, 474f–475f
Aqueduct of Sylvius, 243
Aqueous humor, 301, 308
Arachnoid granulations, 249f
Arachnoid mater, 241, 242f, 249f, 252, 265, 265f
Arbor vitae, 249f, 250, 257f
Arch, of foot, 92f
Arcuate arteries, 496, 498f
Arcuate line, 88, 88f
Arcuate veins, 497, 498f
Areola, 531, 532f
Areolar connective tissue, 48–49, 49f, 301f
Areolar glands, 532f
Arm, 6
    muscles of, 163t–166t, 170f–171f
Armstrong, Lance, 462
Arrector pili muscle, 60f, 63, 63f–64f
Arterial circle, 256f, 386f
Arteriles, 421, 421f
Arterioles, 379, 496, 498f, 501f, 526–528
Arteriosclerosis, 397–398, 398f
Artery(ies), 60f, 379, 380f, 387–390, 388f–390f, 395–398, 398f–400f, 421, 421f
    of abdomen, 395, 396f
    of brain, 383–386, 386f
    of head, 381f, 383, 385f
    of heart, 353f–357f, 356, 358f
    of kidney, 495, 498f
    of lower body, 395–402
    of lower limb, 381f, 395–397, 397f
    of neck, 383, 385f
    of pelvis, 397f
    of trunk, 381f
    of upper body, 379–394
    of upper limb, 381f, 383, 384f
    structure of, 379, 380f
    types of, 379
Arthrology, 129
Articular cartilage, 71, 73f, 131, 133f, 139f
Articular disk, 134, 135f
Articular facet
    inferior, 97, 99f
    superior, 97, 99f, 101, 101f, 102f
Articular process
    inferior, 97, 99f
    superior, 97, 99f, 101, 102f
Articulations, 129–146
Artifacts, in urine, 510, 510f
Arytenoid cartilage, 437, 437f

Ascending aorta, 356f, 358, 361, 380, 381f–382f, 388, 388f, 396f, 412f
Ascending colon, 473, 473f, 475f, 481f, 482
Ascending limb, of loop of Henle, 498f
Ascending tract, 264
Assimilation, nervous system and, 233
Association areas, 244
Association neurons, 234, 235f, 280
Asthma, 460
Astigmatism, 305–306, 305f
Astral fibers, 28, 29f
Astrocytes, 236, 238f
Atlas, 98, 99f–100f
Atria. See Atrium (pl., atria)
Atrial depolarization, 366, 366f
Atrial repolarization, 366, 366f
Atrial systole, 365
Atrioventricular bundle, 365, 366f
Atrioventricular canals, 359f
Atrioventricular groove, 358
Atrioventricular node, 365, 366f
Atrioventricular sulcus, 356
Atrioventricular valves, 358, 359f, 375, 376f
Atrium (pl., atria), 353–356
    left, 355f–362f, 356, 360
    right, 356, 356f–357f, 358f–362f, 360
Audiometer test, 316–317
Auditory (acoustic) meatus
    cat, 217f
    external, 112, 116f, 118f, 118t, 125f, 311, 312f
    internal, 118, 118t, 119f–120f, 124, 125f
Auditory association area, 245f
Auditory cortex, 245f, 246, 314, 315f
Auditory tube, 311, 312f, 436f, 437
Auricles, 311, 312f, 356, 356f, 360, 362f
Auricular artery, posterior, 385f
Auricular surface, of sacrum, 88f, 101, 102f
Auricularis anterior muscle, 212f
Auricularis posterior muscle, 212f
Auricularis superior muscle, 212f
Auscultation, 374, 375f, 455
Autonomic nervous system (ANS), 233
Autonomic reflex, 280
Axial skeleton, 69, 69f–70f, 71t, 97–107, 111–128
Axillary arteries, 381f, 383, 383f
    left, 383f–384f, 389, 393f
    pulse measurement at, 429f
    right, 388f, 389, 389f
Axillary lymph node, 2f, 423f, 426f

Axillary nerve, 266, 268f
Axillary region, 7f
Axillary tail, of breast, 531, 532f
Axillary veins, 404f, 405, 406f–407f, 411, 413f
Axis (second cervical vertebra), 98, 99f–100f
Axon, 48, 48f, 147f, 233, 235f–238f
Axon hillock, 48f, 233, 235f
Azygos vein, 406f, 410, 411f, 415

B

B cells, 340–341
Babinski reflex, 283, 283f
Back, muscles of, 221, 222t, 224f, 228f
Bacterial cells, in urine, 509, 510f
Balance, 318–319
Ball-and-socket joints, 134, 135f
BAPNA solution, A–1
Barany's test, 318–319
Barometric air pressure, 447, 448f
Baroreceptors, 287
Bartholin's glands, 528
Basal layer, of uterus, 528, 530f
Basal metabolic rate, 329
Basal nuclei, 250, 250f, 257f
Base
    of heart, 353, 356f
    of metacarpal, 85, 86f
    of patella, 91f
    of phalanx, 85, 86f
Base unit, in measurement, A–1, A–1t
Basement membrane, 42, 44f, 61, 61f, 63f, 380f, 499f, 516f
Basilar artery, 256f, 383, 386f
Basilar membrane, 313, 313f–315f
Basilic veins, 403, 404f, 405, 406f–407f
Basophils, 339f–340f, 340, 342, 342t
Bell jar model, of breathing, 448, 448f
Belly, of muscle, 179
Benedict's reagent, 490, A–2
Benedict's test, 490
Beta cells, 329, 330f
Beta-adrenergic receptors, heart rate and, 371
Biaxial joints, 134
Bicarbonate ion, 455
Biceps brachii muscle, 2f, 161f, 163t, 166f–167f, 169, 171f–172f, 182, 182f
Biceps brachii reflex, 282, 282f
Biceps brachii tendon, 136f–139f, 170f
Biceps femoris muscle, 162f, 179f, 190t, 192, 195f, 199, 200f
Biceps femoris tendon, 138f

Bicipital groove, 83, 84f
Bicuspid valve, 355f, 359, 359f, 361f, 375f–376f
Bicuspids, 467
Bifid spinous process, 98, 99f–100f
Bile canaliculi, 478f
Bile duct, common, 477, 479f, 481f
Bilirubinuria, 508, 509t
Binocular visual field, 304
Biopac®, 453–454, 459–460
Bipartite uterus, 532
Bipolar layer, 302, 303f
Bipolar neurons, 234, 236f, 302
Bladder. See Gallbladder; Urinary bladder
Blastocele, 534f
Blastocyst, 534, 534f
Bleach solution, A–2
Blind spot, 308
Blindness, color, 307
Blink reflex, 283
Blood, 53, 54f, 337–344
    in urine, 509
    withdrawal of, 338
Blood cell(s), 337–342
Blood cell count, 348
Blood flow, through kidneys, 495–497, 498f
Blood pressure, 429–434
Blood smear, 338, 338f–339f
Blood testing, 345–352, 346f
Blood typing, 345–348, 346t
Blood vessels, 59, 379, 380f. See also specific types and vessels
Blood-brain barrier, 236
Blue sensitive cones, 307
Body
    of breast, 531, 532f
    of epididymis, 517f, 518, 521f
    of femur, 89f
    of hyoid, 103f, 105
    of mandible, 113f–114f, 116f, 117
    of pancreas, 477
    of pubis, 88f
    of rib, 104f
    of sternum, 104f, 105
    of stomach, 471f, 472
    of uterus, 526, 530f, 532, 533f
    of vertebrae, 97, 99f–100f
Body cavities, 5–6, 6f, 387, 469, 469f
Body position, and blood pressure, 431
Body regions, 6–7, 7f
Bolus, 469
Bone(s), 53, 69–76, 70f–75f, 72t
Bone cells, 75, 75f
Bone marrow, 51, 51f, 72, 75f
Bone salt, 53
Bony labyrinth, 313, 313f–314f
Bowman's capsule, 498, 498f, 501f–502f

Index

Boyle's law, 447, 448f
Brachial arteries, 3f, 379, 381f, 383, 383f–384f, 389, 389f–390f
   deep, 384f
   left, 383f, 388f
   measurement at, 428f–429f, 429
   right, 384f, 389
Brachial plexus, 266, 267f–268f, 274, 274f
Brachial region, 7f
Brachial veins, 403, 405, 406f–407f, 411, 413f
Brachialis muscle, 163t, 169, 170f–171f, 182, 183f
Brachiocephalic arteries, 380–382, 381f–385f, 388, 388f–390f, 396f
Brachiocephalic veins, 403, 404f, 406f–407f, 411, 411f
Brachioradialis muscle, 161f, 163t, 169, 171f–172f, 176f, 182, 183f, 269f
Bradycardia, 368
Brain, 2f, 234f, 241–251, 242f–251f
   arteries of, 383–386, 386f
   coronal section, 250
   embryonic development of, 241, 242f
   inferior view, 247, 248f
   meninges of, 241, 242f
   midsagittal section, 247–250, 249f
   olfactory transmission to, 295
   parts of, 243t
   sheep, dissection of, 252, 255f–257f
   superior view, 247f
   surface view, 245–247
   veins of, 403–405, 405f
   ventricles of, 241–243, 243f–244f, 249f–250f
Braincase, 111, 115, 124
Brainstem, 250–251, 252f
Breast, 2f, 531–532, 532f
Breast-feeding, hormones and, 328, 328f
Breathing rate, measurement of, 449
Breathing, mechanics of, 447–448, 448f
Broad ligament, 528, 530f
Broca's area, 244, 245f
Bronchi, 2f, 435f, 438, 438f, 441
Bronchial artery, 440f
Bronchial nerve, 440f
Bronchial vein, 440f
Bronchioles, 438f, 439, 440f
Bronchitis, 460
Brown, Robert, 31
Brownian motion, 30–31
Brush border, 498, 501f

Buccinator muscle, 210t, 212, 212f–213f, 214f, 476f
Bulb
   of hair, 62, 63f–64f
   of penis, 500f, 519f, 520
Bulbar conjunctiva, 301f
Bulbourethral glands, 500f, 516f–517f, 519, 519f, 521, 521f
Bundle branch block, 368
Bundle branches, 365, 366f
Bundle of His, 365, 366f
Bursae, 133, 133f–134f, 139f

C

C cells, 328f, 329
Caffeine, 374
Caffeine solution, saturated, A–2
Calcaneal (Achilles) tendon, 162f, 195, 198f, 201f
Calcaneal (Achilles) tendon reflex, 282, 283f
Calcaneal region, 7f
Calcaneus, 90, 92f
Calcitonin, 329
Calcium, 147, 373
Calcium chloride solution, A–2
Calculi, renal, 511
Calyx (pl., calyces), 495, 497f, 500, 503f
Canal, 72t
Canal of Schlemm, 301, 302f
Canaliculi, 74f, 75, 75f
Cancellous bone, 53, 53f, 72, 73f
Canines, 114f, 466, 468f
Capacitation, 518
Capillaries, 379, 421, 421f, 423f
   lymph, 421, 421f, 423, 423f
   movement through, 422, 423f
   peritubular, 496, 498f
   pulmonary, 440f
Capitate bone, 85, 86f
Capitulum, 83, 84f
Capsule
   of kidney, 495, 496f, 500
   of lens, 302f
   of lymph nodes, 423, 424f
   of spleen, 425f
Carbon dioxide, 455
Carbonic acid, 455
Cardia, 470, 471f
Cardiac muscle, 46–47, 47f, 371
Cardiac nerves, 371
Cardiac notch, 438, 438f
Cardiac pacemaker, 365, 371
Cardiac part, of stomach, 470, 471f, 480
Cardiac veins, 353, 356, 358
   great, 356f–358f, 358, 404f
   middle, 357f–358f
   small, 357f, 358, 358f, 404f

Cardiopulmonary resuscitation, 105
Cardiovascular system, 3, 3f
Carina, 438, 438f
Carotid arteries, 3f
   common, 382f–385f, 389, 389f–390f, 429f
   external, 381f, 383, 385f, 389
   internal, 381f, 383, 383f–386f, 389
   left, 381f, 382f–383f, 383, 384f, 388f–390f, 396f
   pulse measurement at, 429, 429f
   right, 381f–383f, 383, 388f–390f, 396f
Carotid canal, 117, 118f–119f, 118t, 123f
Carotid sinus, 385f
Carpal(s), 70f, 85–86, 86f
Carpal region, 7f
Cartilage, 51–53
Cartilaginous joints, 129, 130, 130t
Casts, in urine, 510, 510f–511f
Cat
   abdominal muscles, 224, 226f, 229f
   abdominal organs of, 480, 481f
   arteries of, 387–390, 388f–390f, 398–400, 398f–400f
   back muscles of, 224, 228f, 229f
   digestive system of, 477–482, 479f–481f
   dissection of, 176–184, 177f–184f, 214, 410–416, 425, 520–521
   female reproductive system of, 532–534, 533f
   forelimb muscles of, 182–183, 182f–184f
   head and neck muscles of, 214, 216f–217f
   heart of, 390f
   leg muscles of, 198–204, 199f–204f
   lymphatic system of, 425, 426f
   male reproductive system of, 521–522, 521f–522f
   mammary glands, 178f
   musculature of, 179–184, 179f–184f, 198–204, 214
   pectoral muscles of, 179, 180f
   respiratory system of, 441–443, 442f–443f
   salivary glands of, 478
   scapular muscles of, 182
   shoulder muscles of, 180–181, 181f
   skeletal anatomy of, 75–76, 76f
   spinal nerves and plexuses of, 274, 274f
   thigh muscles of, 198–202, 199f–200f

   urinary system of, 500, 503f
   veins of, 388, 410–416, 413f–416f
Cat wetting solution, A–2
Catabolism, 488, 488f
Cauda equina, 263, 263f, 267f
Caudal, 3, 4f
Caudal arteries, 398f–400f, 400
Caudal veins, 411, 414f
Caudate lobe, of liver, 476, 477f
Caudofemoral vein, left proximal, of cat, 414f
Caudofemoralis muscle, 199, 200f
Cavernous sinus, 405f
Cavities
   body, 5–6, 6f
   in bone, 71, 72t
Cecum, 473, 473f, 474f–475f, 481f, 482
Celiac artery, 395, 396f, 398f, 399, 399f
Celiac trunk, 381f, 395, 396f
Cell(s), 23–35, 25f
   in urine, 510, 510f
Cell connections, 46
Cell cycle, 27–30, 27f, 29f–30f
Cell membrane, 25f–26f, 34–35, 35f
Cell-mediated immunity, 329, 341
Cellulose, 490–491, A–2
Cementum, 468f
Central canal
   of bone, 53, 53f, 73f–74f, 75
   of spinal cord, 243f, 257f, 264–265, 264f
Central incisor, 436f, 468f
Central nervous system (CNS), 2f, 233, 234f, 238
Central sulcus, 244, 245f, 247f
Central vein, 476, 478f
Centrioles, 25f, 27–28, 29f, 516f
Centromere, 27–28, 29f
Centrosomes, 25f, 27
Cephalic, 3, 4f
Cephalic region, 6, 7f
Cephalic veins, 403, 404f, 405, 406f–407f, 411, 413f
Cerebellar arteries, 386f
Cerebellar cortex, 250
Cerebellum, 242f, 245f, 247, 248, 248f–249f, 255f–257f
Cerebral aqueduct, 243, 243f–244f, 249f, 252, 257f
Cerebral arterial circle, 386f, 387
Cerebral arteries, 386f, 387
Cerebral cortex, 250, 250f, 252, 257f
Cerebral hemispheres, 246–247, 247f
Cerebral medulla, in sheep, 257f
Cerebral peduncles, 247, 249f, 252f, 255f
Cerebral vein, 249f
Cerebrospinal fluid (CSF), 241, 265

Cerebrum, 242f, 243, 250f, 252, 255f, 257f
Ceruminous (earwax) glands, 62
Cervical canal, 530f
Cervical curvature, 97, 98f
Cervical enlargement, 263, 263f
Cervical lymph node, 2f, 423f, 426f
Cervical nerves, 265, 267f
Cervical plexus, 265, 267f–268f
Cervical region, 6, 7f, 98f
Cervical vertebrae, 98, 98f–100f
Cervix, 526, 526f, 530f, 533
Chambers, of heart, 353–356, 355f–356f
Chemical stimuli, nerve response to, 280
Chemoreceptors, 287
Chemstrip procedure, for urinalysis, 508–509, 508f
Chest, muscles of, 166f–167f
Chief cells, 329, 471f, 472
Choana, 436, 436f
Cholesterol molecules, 24, 25f
Cholesterol plaque, 397–398
Chondrocytes, 51, 52f–53f
Chorda tympani, 312f
Chordae tendineae, 357f–361f, 358, 361
Chorionic villi, 412f
Choroid, 302, 302f–303f, 307, 307f
Choroid plexuses, 241, 249f
Chromatin, 26–28, 29f
Chromosomes, 27, 28f–29f
Chronic obstructive pulmonary disease (COPD), 460
Chylomicrons, 421
Chyme, 470, 472
Cilia, 25f, 26–27, 314, 316f
Ciliary body, 301, 302f, 307
Ciliary muscle, 301, 302f
Ciliary processes, 302f
Ciliary ring, 302f
Cingulate gyrus, 251f
Circle of Willis, 386f
Circular layer, of muscularis, 470f–472f, 472
Circular muscles, of eye, 301
Circumcision, 519
Circumduction, 140f, 141
Circumferential lamellae, 74f
Circumflex artery, 358, 358f, 397f
Circumvallate papillae, 466, 467f
Cisterna chyli, 423, 423f, 426f
Cisternae, 26
Clavicle, 2f, 70f, 82, 83f, 104f
Clavicular notch, 104f, 105
Clavicular region, 7f
Clavodeltoid muscle, 181, 181f
Clavotrapezius muscle, 180, 181f
Claws, cat, 177
Cleavage furrow, 28, 29f–30f

Cleidomastoid muscle, 214, 216f
Clitoris, 526f, 528, 530f
Coccygeal nerves, 263f, 265, 267f
Coccygeal plexus, 267f
Coccygeal region, 98f
Coccyx, 70f, 89f, 98f, 101, 102f, 520f
Cochlea, 312f, 313–314, 313f–315f
Cochlear duct, 313, 314f–315f
Cochlear ganglion, 314f–315f
Cochlear nerve, 314f
Cochlear nucleus, 315f
Coelom, 387, 469, 469f, 481f, 521
Coitus interruptus, 520–521
Colic arteries
 common, 400f
 middle, 395, 400f
 right, 395, 400f
Colic flexures, left and right, 473, 475f
Colic vein, middle, 416f
Collagen, 49
Collagenous fibers, 48–49, 49f, 51, 52f, 59
Collateral axon, 235f
Collateral circulation, 387
Collateral ligaments
 fibular (lateral), 136, 138f
 tibial (medial), 136, 138f
Collecting ducts, 498, 498f, 500, 501f
Colliculi, inferior/superior, 248, 249f, 252f
Colloid, of thyroid gland, 328, 328f
Colon, 465, 473f, 475f, 481f
Color blindness, 307
Color, of urine, 507
Columnar epithelium, 42–43, 43f, 44, 472, 473, 474f
Columns, of spinal cord, 264, 264f
Common bile duct, 477, 479f, 481f
Common carotid arteries, 381f–383f, 388f–390f, 389, 396f, 429f
Common colic artery, 400f
Common fibular (peroneal) nerve, 267, 271f, 273f, 274, 274f
Common hepatic artery, 395, 396f
Common hepatic duct, 477, 479f
Common iliac arteries, 381f, 395, 396f–397f, 412f
Common iliac veins, 404f, 405, 408f–410f, 411, 414f
Communicating arteries, 386f, 387
Communicating rami, 266f
Communication, nervous system and, 233
Compact bone, 53, 53f, 72, 72t, 73f–75f
Complete heart block, 368
Complete tetanus, 148, 149f, 151
Compound light microscope, 14f

Concentrated sucrose solution, A–3
Concentration gradient, 31, 33–34
Concentric lamellae, 74f
Conceptus, 525
Conclusion, of lab report, A–4
Condoms, 520, 531
Conduction deafness, 317
Conduction fibers, 365, 366f
Conduction problems, 367
Conductivity, 277
Condylar process, 114f, 117
Condyles, 72t, 83. See also specific types
Condyloid joints, 134, 135f
Cones, 302, 303f, 304
Conjunctiva, 299, 300, 301f–302f, 308
Conjunctival fornix, 301f
Connecting stalk, 534f
Connective tissue, 41, 48–53, 49t
Conoid tubercle, 82, 83f
Consensual reflex, 307
Contraception, 520–521, 529–531, 532f
Contractility, of muscles, 147
Contraction phase, of muscle, 148f, 149–151, 153, 153f
Conus medullaris, 263, 263f, 267f
Convoluted tubules, 498, 498f, 500, 501f–502f
Convolutions, 243
Cool receptors, 289
Cooper's 12-minute run test, 459–461
Coordination, nervous system and, 233
Coracoacromial ligament, 136f
Coracobrachialis muscle, 163t, 166f, 169, 170f
Coracoclavicular ligament, 136f
Coracohumeral ligament, 136f
Coracoid process, 82f, 136f
Cornea, 299, 300, 301f–302f, 307–308, 307f
Corneal reflex, 283, 283f
Corniculate cartilage, 437f, 438
Corona radiata, 527f, 529f
Coronal plane, 5, 5f
Coronal suture, 112, 113f, 115f–116f, 131f
Coronary arteries, 358, 380, 382f, 388
 left, 358, 358f, 381f
 right, 356f–357f, 358, 358f, 381f
Coronary ligament, 477f
Coronary sinus, 356, 357f, 358, 358f–359f, 361
Coronary sulcus, 356f–357f
Coronoid fossa, 83, 84f
Coronoid process, 83, 85f, 114f, 116f, 117

Corpora cavernosa, 517f, 519f, 520, 522
Corpora quadrigemina, 243t, 248, 252, 257f
Corpus albicans, 526, 527f, 529f
Corpus callosum, 247, 249f–251f, 252, 257f, 295f
Corpus luteum, 526, 527f, 529f, 531f, 534f
Corpus spongiosum, 517f, 519f, 520, 522
Corrugator supercilii muscle, 210t, 212, 212f–213f
Cortex
 of adrenal glands, 329, 330f
 of hair, 63, 63f
 of kidneys, 495, 497f–498f, 498, 501f–502f
 of lymph nodes, 423–424, 424f
 of ovary, 527f
Corti, organ of, 313, 314f
Cortical sinus, 424f
Corticosteroid hormones, 329
Costal angle, 102, 104f
Costal cartilage, 104f, 130
Costal groove, 102, 104f
Costal notch, 104f, 105
Costocervical artery, 389, 389f
Costocervical veins, 411, 413f
Costochondral joint, 132f
Cough reflex, 438
Cow, eye of, 307–308
Cowper's glands, 500f, 516f–517f, 519, 519f, 521, 521f
Coxal bones, 70f, 86, 87f, 521f, 533f
Coxal region, 7f
Cranial cavity, 5, 117, 119f
Cranial fossa (anterior, middle, and posterior), 117, 119f
Cranial nerves, 234f, 251–252, 253f, 254t, 255f, 255t, 256f
Cranial region, 7f
Cremaster muscle, 517f, 518
Crenation, 35, 35f
Crest, of bones, 72t
Cribriform plate, 118, 119f, 121, 122f, 436f
Cricoid cartilage, 436f, 437, 437f, 442f
Cricothyroid ligament, 437f
Cricothyroid muscle, 211f
Crista ampullaris, 316, 317f
Crista galli, 118, 119f, 121, 122f, 436f
Crown, of tooth, 129f, 466, 468f
Cruciate ligaments, anterior and posterior, 136, 138f
Crural region, 7f
Crus, of penis, 500f, 519f, 520, 521f
Crypts, of tonsils, 424
Crystallines, 301

Crystals, in urine, 510, 510f
Costocervical vein, 413f
Cubital region, 7f
Cubital vein, median, 404f, 405, 407f, 411, 413f
Cuboid, 90, 92f
Cuboidal epithelium, 42, 43f
Cumulus mass, 527f, 529f
Cuneiform bones, 90, 92f
Cuneiform cartilage, 437f, 438
Cupula, 316, 317f
Curvature, spinal, 97, 98f
Cusp, of tooth, 468f
Cuspids, 466
Cusps, of tricuspid valve, 358
Cutaneous maximus, 179
Cuticle
　of hair, 63, 63f
　of nail, 64, 65f
Cystic duct, 477, 479f
Cystic fibrosis, 460
Cytokinesis, 27, 30f
Cytology, 23
Cytoplasm, 24, 25f, 237f
Cytoskeleton, 24, 25f–26f
Cytosol, 24

D

Dartos muscle, 515, 517f
Daughter chromosome, 28
Deafness, 317
Decalcified bone, 75, 75f
Deciduous teeth, 467, 468f
Decussation of the pyramids, 250
Deep, 5t
Deep brachial artery, 384f
Deep fascia, of penis, 519f
Deep femoral artery, 381f, 397, 397f–398f
Deep femoral vein, 405, 408f
Deep fibular (peroneal) nerve, 273f
Deep inguinal nodes, 426f
Deep lymphatic vessel, 440f
Deep palmar arch artery, 384f
Deep palmar arch vein, 407f
Deflection wave, 365, 366f
Dehydroandrostenedione (DHEA), 330
Delta cells, 329
Deltoid muscle, 161f–162f, 163t, 164, 165f–168f, 171f, 179f–181f
Deltoid tuberosity, 83, 84f
Demifacet, 98, 104f
Dendrite, 48, 48f, 233, 235f–238f
Dens, 98, 100f
Dense connective tissue, 49–50
Dentate nucleus, 251f
Denticulate ligament, 265f
Dentin, 466, 468f
Deoxyribonucleic acid (DNA), 26–28

Dependent variables, in experiments, A–4
Depolarization, 365, 366f
Depression (movement), 141, 141f
Depressions, of bones, 71, 72t
Depressor anguli oris muscle, 212f–213f
Depressor labii inferioris muscle, 210t, 212, 212f–213f
Depth of field, 17
Dermal papilla, 62, 63f–64f
Dermal root sheath, 62–63, 63f
Dermis, 42, 59, 60f
Descending aorta, 382, 382f, 388, 388f, 395, 398f–399f, 399
Descending colon, 473, 475f, 481f, 482
Descending limb, of loop of Henle, 498f
Descending tract, 264
Desmosomes, 46, 46f
Determinate hair, 62
Detrusor muscle, 497, 499f
Developmental stages, 533–534, 534f
DHEA, 330
Diabetes mellitus, vision and, 306
Dialysis bags, 33–34, 33f
Diaphragm, 6f, 222, 222t, 224f, 226f, 356f, 438f, 480, 481f
Diaphragm (contraceptive), 531
Diaphysis, 71, 72f–73f, 83
Diarthrotic joints, 133
Diastole, 365, 376f
Diastolic pressure, 430f, 431
Diencephalon, 242f, 243t, 247, 249f, 252f
Differential white blood cell count, 341–342
Diffusion, 23, 30–32
Digastric muscle, 209, 210t, 211f, 214, 217f, 466
Digestion, 487–494
Digestive enzymes, 487–494, 488f
Digestive system, 2f, 3, 465–485, 465f, 479f–481f
　accessory structures of, 465, 473–477
　anatomy of, 465–486
　physiology of, 487–494
Digestive tract, 465
　abdominal portions of, 469–473
　membranes of, 469, 469f
　upper, 465–469
Digital arteries
　foot, 397, 397f
　hand, 383, 384f
Digital region, 7f
Digital veins
　foot, 408f
　hand, 405, 407f

Dilutor papillae, 301
Diopter, of ophthalmoscope, 306, 306f
Diploe, 72
Directional terms, 3–4, 4f, 5t
Dissection, 176–184, 177f–184f, 214, 387f, 410–416, 520–521. See also Cat; Sheep
　of sheep brain, 252, 255f–257f
　of sheep heart, 360–362, 361f–362f
　of sheep or cow eye, 307–308, 307f
Distal, 4f, 5t
Distal convoluted tubule, 498, 498f, 500, 501f–502f
Distal phalanx, 85, 86f, 90–91, 92f
Distributing (muscular) arteries, 379
DNA, 26–28
Dorsal, 3, 4f, 5t
Dorsal artery, of penis, 519f, 520
Dorsal (posterior) column, 264f
Dorsal nerve, of penis, 519f
Dorsal ramus, 265, 266f
Dorsal root, 235f, 264f, 266f
Dorsal root ganglion, 235f, 264f, 265, 265f–266f
Dorsal vein
　of foot, 408f
　of penis, 519f, 520
Dorsal venous arch, 408f
Dorsalis pedis artery, 381f, 397, 397f, 429f
Dorsiflexion, 189
Dorsum sellae, 118, 119f, 123f
Double pith, 151
Ductless secretion, 325
Ductus arteriosus, 410, 412f
Ductus deferens, 516f–517f, 518, 519f, 520, 520f, 521, 521f–522f
Ductus venosus, 410, 412f
Duodenal glands, 473, 474f
Duodenal papilla, minor, 479f
Duodenojejunal flexure, 473f
Duodenum, 471f, 472, 473f, 479f, 481f, 482
Dura mater, 241, 242f, 249f, 252, 265, 265f
Dural venous sinus, 242f
Dynamic balance, 311, 314

E

Ear(s), 311–316, 312f
　external, 311, 312f
　inner, 311–313, 312f, 313f–314f
　middle, 311, 312f
Eardrum, 311, 312f

Earlobe, 311, 312f
Early development, 533, 534f
Earwax glands, 62
ECG. See Electrocardiograph
Ectoderm, 534, 534f
Effector, 280–281
Effector organ, 235f, 281f
Efferent arteriole, 496, 498f, 501f
Efferent division, of peripheral nervous system, 233
Efferent ductules, 516f
Efferent lymphatic vessels, 424f
Efferent neurons, 234, 235f, 280, 281f
Eggs. See Oocytes
Ejaculatory ducts, 516f–517f, 518, 519f
Ejection period, 376f
EKG. See Electrocardiograph
Elastic arteries, 379
Elastic cartilage, 52, 53f
Elastic connective tissue, 50, 50f
Elastic fibers, 48, 49f, 50, 50f, 52, 53f, 440f
Elastic membrane, 379, 380f
Elasticity, of muscles, 147
Elastin, 50
Elbow, 83, 85f
Elbow joint, 134, 137f
Electrical activity, nerve response to, 277–279
Electrical conductivity, of heart, 365–370, 366f
Electrical ground, in electrocardiogram, 366
Electrocardiograph, 365–368, 366f
Electrochemical impulses, 365
Electrode plates, 366
Electron microscopy, 13
Elevation, 141, 141f
ELISA Tests, 349
Ellipsoid joint, 134, 135f
Embryonic development, 534, 534f
Embryonic disk, 534f
Embryonic tissues, 534, 534f
Emphysema, 441, 459–460
Enamel, of tooth, 466, 468f
Endocardium, 353, 355f
Endocrine cells, 471f
Endocrine system, 3, 3f, 325–336
　major organs of, 325–331, 325f
　physiology of, 331–332
Endocytosis, 30
Endoderm, 534, 534f
Endolymph, 313, 313f, 316, 317f
Endometrium, 412f, 526, 530f–531f, 534f
Endoneurium, 265, 265f
Endoplasmic reticulum, 24–26, 25f–26f
Endosteum, 73f, 75, 75f, 313f–314f

Endothelium, 42, 353, 379, 380f, 422
Enteric plexus, 470f
Enzyme-linked immunosorbent assay (ELISA), 349
Enzymes, 487–494, 488f
Eosinophils, 339f–340f, 340, 342, 342t
Ependymal cells, 236, 238f
Epicardium, 353, 355f
Epicondyles, 72t
    femoral, 89, 89f–90f
    humeral, 83, 84f
Epidermis, 42, 59–62, 60f–61f
Epididymis, 511f–516f, 518, 521, 521f
Epidural space, 265
Epigastric region, 7, 8f
Epiglottis, 435f–436f, 438, 442f, 467f, 480f
Epinephrine, 330, 374
Epinephrine solution, A–2
Epineurium, 265, 265f
Epiphyseal line, 71, 73f
Epiphyseal plate, 71, 73f, 130, 132f
Epiphysis, 71, 72f–73f
Epiploic appendages, 473, 475f
Epithelial cells, 326, 327f
    in urine, 510, 510f
Epithelial layer, of mucosa, 472
Epithelial root sheath, 63f
Epithelial tissue, 42–45
Epithelium, mucous, 470
Epitrochlearis muscle, 180f, 182
Eponychium, of nail, 64, 65f
Equilibrium, 31
Erectile tissue, of penis, 500f
Erector spinae muscle, 179f, 221, 222t, 224f, 469f
Erythroblastosis fetalis, 347
Erythrocytes. See Red blood cell(s)
Erythropoiesis, 338
Esophageal arteries, 389
Esophageal sphincter, 469, 471f, 472, 480
Esophagus, 2f, 6f, 226f, 436f, 439f, 464f–465f, 469, 471f, 480, 480f
Essential oils preparation, A–2
Estradiol, 331
Estrogens, 330, 331, 529, 531f
Ethmoid bone, 112, 112f, 113f, 115, 116f, 118, 119f, 120f, 121, 122f, 435, 436f
Ethmoidal labyrinth, 121, 121f
Ethmoidal sinus, 121, 121f, 122f
Eustachian tube, 311, 312f, 436f, 437
Evaporative cooling, 62
Eversion, 141, 141f
Excitability, of muscles, 147
Exercise, 367–368, 431, 459–464
Exocrine glands, 325
Exocytosis, 30

Expiration, 447–448, 448f
Expiratory reserve volume (ERV), 450, 453, 453f
Extensibility, of muscles, 147
Extension, 139, 139f–141f
Extensor carpi radialis brevis muscle, 164t, 174, 176f, 182, 269f
Extensor carpi radialis longus muscle, 164t, 174, 176f, 182, 183f, 184f, 269f
Extensor carpi radialis muscle, 183f
Extensor carpi ulnaris muscle, 164t, 174, 176f, 183, 183f, 184f, 269f
Extensor digiti minimi muscle, 176f, 269f
Extensor digitorum brevis muscle, 197f
Extensor digitorum communis muscle, 183, 184f
Extensor digitorum lateralis muscle, 183, 184f
Extensor digitorum longus muscle, 161f, 190t, 193, 196f, 203, 203f
Extensor digitorum longus tendon, 203f
Extensor digitorum muscle, 164t, 174, 176f–177f
Extensor digitorum tendons, 176f
Extensor hallucis longus muscle, 190t, 193, 196f
Extensor indicis muscle, 176f, 269f
Extensor indicis tendon, 176f
Extensor muscles, 162f, 174, 176f, 269f
Extensor pollicis brevis muscle, 164t, 174, 176f–177f, 183, 184f, 269f
Extensor pollicis longus muscle, 164t, 174, 176f–177f, 269f
Extensor pollicis longus tendon, 176f
Extensor retinaculum, 174, 176f
Extracellular fluid (ECF), 23
Extracellular matrix, 41
Extrinsic muscles, of eye, 299, 300f, 300t, 302, 307
Eye(s), 299–310, 299f, 307f
Eye movements, 302
Eye reflexes, 282–283
Eyebrow, 301f
Eyebrows, 299, 299f
Eyelash, 301f
Eyelashes, 299
Eyelid, 299, 299f, 301f

F

Facet, of bones, 72t
Facial artery, 383, 385f, 429f
Facial bones, 111, 112, 112f

Facial muscles, 161f, 210t
Facial (VII) nerve, 251, 253f, 254t–255t
Facial region, 7f
Facial veins, 404f–406f, 413f
Facilitated diffusion, 30
Falciform ligament, 476, 477f
Fallopian tube, 525, 526f, 530f, 533f
False pelvis, 88, 88f
False ribs, 102, 104f
False vocal cords, 437, 437f
Farsightedness, 306
Fascia, 179
Fasciculi, of spinal cord, 264
Fat cells, 50, 51f
Fat pads, renal, 495
Fat, subcutaneous, 331
Fatigue, muscle, 153–155
Fatty acids, free, 491
Female contraception, 529–531, 532f
Female reproductive system, 2f, 525–536, 526f
Femoral arteries, 3f, 381f, 397f–399f, 400, 400f
    deep, 381f, 396f–397f, 397
    left, 398f
    pulse measurement at, 429f
    right, 398f
Femoral cutaneous nerve, 271f
Femoral nerve, 267, 271f–274f
Femoral region, 7f
Femoral veins, 3f, 404f, 405, 408f, 411, 414f, 426f
Femur, 2f, 70f, 88–90, 89f–90f
Fertilization, 533–534, 534f
Fetal arteriole, 412f
Fetal circulation, 410, 412f
Fetal skull, 121, 131f
Fetal umbilicus, 412f
Fetal venule, 412f
Fibers, muscle, 46, 46f
Fibrinogen, 337
Fibroblasts, 48, 49f
Fibrocartilage, 52, 52f
Fibrocartilaginous pad, 130
Fibrocytes, 48, 49f
Fibrous capsule, 131, 133f
Fibrous joints, 129–130, 130t
Fibrous layer, of eye, 302, 302f
Fibrous pericardium, 353
Fibula, 2f, 70f, 90, 91f
Fibular artery, 381f, 397, 397f
Fibular (lateral) collateral ligament, 136, 138f
Fibular (peroneal) nerve, 273f, 274, 274f
    common, 267, 271f, 273f
    deep, 273f
    superficial, 273f
Fibular vein, 404f, 408f

Fibularis brevis muscle, 161f–162f, 190t, 193, 197f, 203, 204f
Fibularis brevis tendon, 197f
Fibularis longus muscle, 161f–162f, 190t, 193, 196f–198f, 203, 204f
Fibularis longus tendon, 197f
Fibularis muscles, 190–198
Fibularis tertius muscle, 190t, 193, 197f, 203, 204f
Field effect, 371
Field of view, of microscopes, 16–17, 16f
Filiform papillae, 466, 467f, 478
Filtration, 24, 35–36, 495
Filtration apparatus, 35f
Filtration rate, 35
Filtration solution, A–2
Filum terminale, 263, 263f
Fimbriae, 526, 530f, 533f
Fissures, 72t, 244
Fixing, of muscles, 141
Flagella, 26–27
Flat bones, 71, 71f
Flexion, 139, 139f–141f
Flexor carpi radialis muscle, 164t, 173, 173f, 182, 182f
Flexor carpi ulnaris muscle, 164t, 173, 173f, 182, 182f, 184f
Flexor digiti minimi brevis muscle, 164t, 174, 175f
Flexor digitorum longus muscle, 191t, 195, 198f, 201, 201f, 202f
Flexor digitorum profundus muscle, 164t, 173, 174f, 182, 182f, 184f
Flexor digitorum sublimis muscle, 182
Flexor digitorum superficialis muscle, 164t, 173, 174f, 182, 184f
Flexor digitorum superficialis tendon, 175f
Flexor hallucis longus muscle, 191t, 195, 198f, 201, 202f
Flexor muscles, 161f, 173, 174f
Flexor pollicis brevis muscle, 164t, 174, 175f
Flexor pollicis longus muscle, 164t, 172f, 173
Flexor pollicis longus tendon, 174f
Flexor retinaculum, 173, 175f
Floating ribs, 102, 104f
Focusing, of microscope, 14–15
Folia, 247, 248f–249f, 250
Foliate papillae, 466, 467f
Follicle(s)
    of hair, 59, 60f, 62–63, 63f–64f
    of ovaries, 331, 331f, 525–526, 527f–529f, 531f, 534f

Follicle cells
 ovarian, 331, 331f, 527f
 thyroid, 328, 328f
Follicle-stimulating hormone (FSH), 327, 528
Fontanels, 121, 130, 131f
Foot
 bones of, 90–92, 92f
 inversion and eversion of, 141, 141f
 muscles of, 193–197, 196f–198f
Foot processes, 238f
Foramen, 72t
Foramen caecum, 467f
Foramen lacerum, 117, 118f, 118t, 119f
Foramen magnum, 117, 118f–119f, 118t
Foramen ovale, 117, 118, 118f–119f, 118t, 123f, 358, 410, 412f
Foramen rotundum, 118, 118t, 119f, 123f
Foramen spinosum, 118f–119f, 118t, 123f
Foramina of Monro, 241
Forced expiratory vital capacity (FEV), 459–460
Forearm, 83–85, 85f
 muscles of, 163t–164t, 169–176, 172f–174f
Forebrain, 242f, 243, 243t, 247
Forelimb muscles, 182, 182f–184f
Foreskin, 177, 519
Formed elements, of blood, 337–344, 339f, 342t
Fornix, 251f, 526f, 528
Fossa, 72t
Fossa ovalis, 358, 360f, 410
Fourth ventricle, 243, 243f–244f, 249f, 250, 252, 257f
Fovea capitis, 89f, 136, 137f
Fovea centralis, 302, 303f, 304f
Free edge, of nail, 64, 65f
Free fatty acids, 491
Free radicals, 26
Free ribosome, 25f
Frenulum, 519
Friction pads, 177
Frog
 heart contraction in, 372, 373f
 heart rate of, 372–374, 373f
 muscle physiology in, 150–154
 nerve conduction in, 278–280, 279f
 pithing of, 151, 151f
 skin removal from, 151, 152f
Frog Ringer's solution, A-2
Frontal bone, 111, 112f, 113f, 115, 115f–116f, 119f–120f, 131f, 436f
Frontal (anterior) fontanel, 130, 131f

Frontal lobe, 244, 245f, 247f–248f, 255f–258f
Frontal plane, 5, 5f
Frontal sinus, 118, 119f, 120f, 121, 121f, 436f
Frontalis muscle, 210t, 212, 214
FSH, 327, 528
Functional layer, of uterus, 526–528, 530f
Fundus
 of eye, 302
 of stomach, 471f, 472, 480
 of uterus, 526, 530f
Fungiform papillae, 466, 467f

G
G_0 phase, 27
G_1 phase, 27, 27f
G_2 phase, 27, 27f
Galea aponeurotica, 212, 212f, 214
Gallbladder, 2f, 465f, 472, 473f, 475f, 477, 477f, 479f, 480, 481f
Gametes, 515, 525
Ganglion cells, 302
Ganglionic layer, 302, 303f
Gap junctions, 46f
Gastric arteries, left, 395, 396f, 398f, 399, 399f
Gastric glands, 472, 472f
Gastric nodes, 426f
Gastric pit, 471f
Gastric veins, 409f
Gastrocnemius muscle, 2f, 138f, 152, 161f–162f, 191t, 195, 195f–198f, 201, 201f
Gastrocnemius tendon, 198f
Gastroepiploic vein, 409f, 416f
Gastroesophageal opening, 471f
Gastroesophageal reflux disease, 469
Gastrointestinal tract, 469–473, 470f
Gastroomental vein, right, 409f
Gastrosplenic vein, 415, 416f
Gelatinous matrix, 316f
Gemellus muscles, 192, 194f
Gender determination, 88
Genicular artery, 397f
Genioglossus, 466, 466f
Geniohyoid muscle, 480f
Genital arteries, 398f
Genital region, 7f
Genitalia, external
 female, 528, 530f
 male, 519–520
Genitofemoral nerve, 271f
Genu, 113f, 116f, 247, 249f, 257f
GERD, 469
Germinal center, of lymph node, 424f
GH, 326
Gingivae, 129f, 465, 466f, 468f
Glabella, 113f

Glans clitoris, 528
Glans penis, 516f, 519, 519f, 521f
Glenohumeral joint, 134, 136f
Glenohumeral ligament, 136f
Glenoid cavity, 81, 82f
Glenoid fossa, 136f
Glenoid labrum, 134, 136f
Glial cells, 48, 48f, 235–236, 235f
Gliding joints, 133, 135f
Globin, 347
Globulins, 337
Glomerulus, 496, 497, 498, 498f, 501f
Glossopharyngeal (IX) nerve, 251, 253f, 254t–255t, 256f
Glottis, 438, 442f
Glucagon, 329, 330f
Glucocorticoids, 329
Glucose, 329, 490, 498, 509t
Gluteal artery, superior, 397f
Gluteal nerve, 271f
Gluteal region, 7f
Gluteal tuberosity, 89, 89f
Gluteus maximus muscle, 162f, 190t, 192, 194f–197f, 199, 200f
Gluteus medius muscle, 162f, 190t, 192, 194f, 199, 200f
Gluteus minimus muscle, 190t, 192, 194f
Gluteus muscle, 199, 200f
Glycocalyx, 25f
Glycolipid, 25f
Glycoproteins, 25f, 345
Glycosuria, 508
Goblet cells, 43, 43f, 44f, 436, 471f, 473, 474f, 476f
Golgi apparatus (Golgi body), 25f, 26, 27t
Gomphosis, 131, 131f
Gonad(s), 330–331
Gonadal arteries, 395, 396f, 398f, 399
Gonadal veins, 409f, 410, 410f, 414f
Gonadotropin-releasing hormone (GnRH), 528, 531f
Gonadotropins, 326
Graafian follicle, 525, 527f–529f
Gracilis muscle, 161f–162f, 189, 190t, 191f–195f, 199, 199f, 272f
Graded response, 147
Granular leukocytes, 339–340, 339f–340f, 342t
Granulocytes. See Granular leukocytes
Granulosa cells, 527f, 529f
Gray commissure, 264, 264f
Gray matter, 236, 250, 250f, 257f, 264, 264f
Great cardiac veins, 356f–358f, 358, 404f
Great saphenous vein, 198, 404f, 405, 408f, 411, 414f

Greater alar cartilage, 120f, 436f
Greater cornua, 103f, 105
Greater curvature, of stomach, 471f, 472, 480
Greater ischiadic (sciatic) notch, 87, 88f
Greater omentum, 387, 474, 480, 481f
Greater pelvis, 88, 88f
Greater trochanter, 88, 89f–90f
Greater tubercle, 83, 84f
Greater vestibular glands, 528
Greater wings, of sphenoid bone, 117, 118, 118f–119f, 123f
Green sensitive cones, 307
Groove, 72t
Gross anatomy, 1
Ground bone, 53, 53f, 75
Growth hormone (GH), 326
Growth plate, 71
Gustation, 293–295
Gustatory center, 246
Gustatory hair, 294f
Gyri, 243–244, 255f

H
Habenula, 251f
Hair, 2f, 59, 60f, 62–63, 63f
Hair bulb, 62, 63f–64f
Hair cells, of ear, 314f, 316, 316f–317f
Hair follicle, 59, 60f, 62–63, 63f–64f
Hair papilla, 62, 63f–64f
Hair root, 62, 63f–64f
Hair shaft, 62, 63f–64f
Hallux, 90–92, 92f
Hamate bone, 85, 86f
Hamstrings, 162f, 190t, 192–193, 195f
Hamulus, of hamate, 85, 86f
Hand
 bones of, 85–86, 86f
 muscles of, 164t, 169–176, 175f–176f
Handle, of malleus, 312f
Hard palate, 117, 118f, 120f, 436f, 466, 466f, 480f
Harvard step test, 461–462
Haustra, 473, 475f
Haversian canal, 75
Haversian systems, 73f–74f
HCl. See Hydrochloric acid
Head (anatomical)
 of bone, 72t
 of epididymis, 517f, 518, 521f
 of femur, 88, 89f–90f
 of fibula, 91f
 of humerus, 83
 of malleus, 312f
 of metacarpal, 85, 86f
 of pancreas, 477
 of phalanx, 85, 86f
 of radius, 83, 85f

Head (anatomical)—(contd.)
  of rib, 102, 104f
  of sperm cells, 516f, 517
  of ulna, 83, 85f
Head (of body)
  arteries of, 381f, 383–386, 385f
  endocrine glands of, 326f
  muscles of, 210t, 212f–213f, 213–214
  veins of, 403–405, 404f–406f, 411
Head piece, of ophthalmoscope, 306
Hearing, 311–316
Hearing pathway, 313–314, 315f
Hearing tests, 316–318
Heart, 3f, 6f, 353–364, 356f, 421, 421f
  blood vessels of, 356f–357f, 358, 358f, 382f
  cat, 390f
  electrical conductivity of, 365–370, 366f
  functions of, 371–378
  model of, examination of, 353–356
  sheep, dissection of, 360–362, 361f–362f, 375–376
Heart attacks, 367
Heart block, 367, 368
Heart murmur, 375
Heart rate, 372–374, 373f, 429, 429f, 460–461
Heart sounds, 374–375
Heart valves, 372, 373f. See also specific valves
Heart wall, 353, 355f
Helicotrema, 314f
Helix, of ear, 311, 312f
Hematocrit, 347–348, 347f
Hematopoiesis, 338
Hematopoietic marrow, 72
Hematuria, 508
Heme, 347
Hemiazygos vein, 410, 411f
Hemispheres, cerebral, 246–247
Hemoglobin, 347–348, 509t
Hemolysis, 35f, 346, 508
Hemolytic disease of the newborn (HDN), 347
Hemopoiesis, 338
Hemorrhoid(s), 473
Hemorrhoidal veins, 473
Henle, loop of, 498, 498f, 500, 501f–502f
Hepatic artery, 395, 396f, 398f, 399, 399f, 476–477, 477f–478f
Hepatic ducts, 477, 477f–478f
  common, 477, 479f
  left, 477, 479f
  right, 477, 479f
Hepatic flexure, 473, 475f
Hepatic nodes, 426f
Hepatic phagocytic cells, 478f

Hepatic portal system, 408, 409f, 415, 416f
Hepatic portal vein, 404f, 409f, 412f, 415, 477, 477f–478f
Hepatic triad, 477, 478f
Hepatic veins, 404f, 409f, 415, 416f
Hepatocytes, 476, 478f
Hepatopancreatic ampulla, 477, 479f
Hilum, 495, 497f, 500, 503f
Hindbrain, 242f, 243, 243t, 245f, 247–250
Hinge joints, 134, 135f
Hip bones, 70f, 86, 87f
Hip joint, 134–136, 137f
Hip, muscles of, 189–195, 194f
Hippocampus, 251f, 295f
His, bundle of, 365, 366f
Histology, 41
Homeostasis, 23
Homunculus, 244
Horizontal plane, 4, 5f
Horizontal plate, of palatine bone, 118f, 120f, 436f
Hormonal action, external influences on, 326–328, 328f
Hormones, 325–336
Horned (bipartite) uterus, 532
Humeral circumflex artery, 384f
Humeral joint, 134, 136f
Humeral ligament, transverse, 136f
Humeroradial joint, 134, 137f
Humeroscapular joint, 134, 136f
Humeroulnar joint, 134, 137f
Humerus, 2f, 70f, 83, 84f, 136f
Hyaline cartilage, 51–52, 52f, 71
Hyaline cast, in urine, 510, 511f
Hydrocephaly, 243
Hydrochloric acid, in stomach, 472
Hydrochloric acid solution, A–2
Hydrogen ion, 455
Hydrogen peroxide, 26
Hydrostatic pressure, 35
Hydroxyapatite, 71
Hydroxyapatite crystal, 53
Hymen, 528
Hyoid bone, 70f, 103f, 105, 437f, 480f
Hyperextension, 139, 139f
Hypermetropic vision, 306
Hyperopia, 306
Hyperpolarization, 371
Hyperreflexic, 281
Hypertension, 429
Hypertonic solutions, 32, 35f
Hypochondriac regions, 7, 8f
Hypodermis, 59, 60f, 64f
Hypogastric region, 7, 8f
Hypoglossal (XII) nerve, 252, 253f, 254t, 255f, 255t, 256f, 268f
Hypoglossal canal, 119f
Hyponychium, 64, 65f
Hypophyseal fossa, 118, 123f

Hypophysis, 326. See also Pituitary gland
Hyporeflexic, 281
Hypotension, 429
Hypothalamus, 3f, 247, 249f–250f, 252, 257f, 325f–326f, 326, 327, 327f
Hypothenar eminence, 174, 175f
Hypotonic solutions, 32, 35f

I
Ileal artery, 400f
Ileocecal junction, 473f
Ileocecal valve, 472, 475f
Ileocolic arteries, 395
Ileocolic veins, 416f
Ileum, 472, 473f, 474f–475f, 481f, 482
Iliac arteries, 395
  common, 381f, 395, 396f–397f, 412f
  external, 381f, 395, 397f–399f, 400, 400f
  internal, 381f, 395, 397f–399f, 400, 400f, 412f
  left, 396f
  right, 396f, 399f
Iliac blade, 88f
Iliac crest, 87, 87f–88f, 88
Iliac fossa, 87, 87f–88f
Iliac nodes, 426f
Iliac region, 7, 8f
Iliac spine
  anterior inferior, 87, 87f–88f
  anterior superior, 87, 87f–88f, 191f
  posterior inferior, 87, 87f–88f
  posterior superior, 87, 87f–88f
Iliac veins
  common, 404f, 405, 408f–410f, 411, 414f
  external, 404f, 405, 408f–410f, 411, 414f
  internal, 404f, 405, 408f–410f, 411, 414f, 426f
Iliacus muscle, 191f, 272f
Iliocostalis cervicis muscle, 224f
Iliocostalis lumborum muscle, 224f
Iliocostalis muscle, 221, 222t, 224f, 225f, 230f
Iliocostalis thoracis muscle, 224f
Iliofemoral ligaments, 137f
Iliohypogastric nerve, 271f
Ilioinguinal nerve, 271f
Iliolumbar arteries, left, 397f–398f
Iliolumbar veins, 414f, 415
Iliopsoas muscle, 189, 190t, 191f
Iliotibial tract, 191f
Ilium, 87, 87f–88f
Immune competence, 425

Immunity, 329, 341
Incisive canal, 436f
Incisive fossa, 118f
Incisors, 114f, 120f, 436f, 466, 468f
Incomplete tetanus, 148, 149f, 151
Incus, 311, 312f
Independent variables, in experiments, A–4
Indeterminate hair, 62
India ink, A–2
Inferior, 5t
Inferior angle, of scapula, 82f
Inferior colliculi, 248, 249f, 252, 252f, 257f, 315f
Inferior lobe, of lung, 438, 438f, 443f
Inferior meatus, 436f
Inferior angle, of scapula, 81
Infraglenoid tubercle, 82, 82f
Infraorbital foramen, 111, 113f, 116f, 118t
Infraorbital margin, 113f
Infrapatellar bursa, deep, 139f
Infraspinatus muscle, 162f, 163t, 165f, 166, 168f, 181f, 182, 216f
Infraspinous fossa, 81, 82f
Infundibulum, 250f, 252, 256f–259f, 326, 326f, 526, 530f, 533f
Inguinal canals, 223f, 395, 397f, 517f, 518, 521, 521f
Inguinal ligament, 223f, 409f
Inguinal lymph node, 2f, 423f, 426f
Inguinal region, 7f
Inguinal ring, superficial, 517f
Initial weight, 33
Inner cell mass, 534, 534f
Inner ear, 311–316, 312f, 313f–314f
Inner elastic membrane, 379, 380f
Innervation, of muscles, 159
Inorganic matter, of bones, 71
Insertions, of muscles, 159, 179
Inspiration, 447–448, 448f
Inspiratory capacity, 453f
Inspiratory reserve volume (IRV), 453, 453f
Insula, 246
Insulin, 329, 330f
Integral proteins, 24, 25f
Integrating center, 280
Integument, 59
Integumentary glands, 62, 62f
Integumentary system, 2f, 3, 59–68, 60f
Interatrial septum, 358, 359f–360f
Intercalated disks, 47
Intercondylar eminence, 90, 91f
Intercondylar fossa, 89f
Intercostal arteries, 382, 382f, 389
  anterior, 382f, 396f
  posterior, 382f, 396f
Intercostal muscles, 229f

external, 222, 222t, 226f–227f, 229f
internal, 222, 222t, 226f–227f, 229f
Intercostal veins, 410, 411f
Interference, in electrocardiograph, 367
Interlobar arteries, 496, 498f
Interlobar veins, 497, 498f
Interlobular arteries, 496, 498f
Interlobular veins, 497, 498f
Intermediate cuneiform bone, 90, 92f
Intermediate filaments, 24, 26f
Intermediate mass, 257f
Interneurons, 234, 235f, 280
Interphase, 27, 27f, 29f
Interspinalis muscle, 224f
Interstitial cells, 331, 331f, 515, 516f, 518f
Interstitial fluid, 421
Interstitial lamellae, 74f
Intertransversarii muscle, 224f
Intertrochanteric crest, 88, 89f–90f
Intertrochanteric line, 88, 89f–90f
Intertubercular groove, 83, 84f
Interventricular arteries
  anterior, 356f, 358, 358f
  posterior, 357f, 358, 358f
Interventricular foramina, 241, 243f–244f, 249f
Interventricular groove (sulcus), 356, 356f–357f, 358, 362f
Interventricular septum, 355f, 358, 359f, 361
Intervertebral disks, 97, 98f
Intervertebral foramen, 98, 98f
Intervertebral notch
  inferior, 99f, 100f, 101f
  superior, 99f, 101f
Intervertebral symphysis, 132f
Intestinal arteries, 395, 398f
Intestinal gland, 470f
Intestines, 387. *See also* Large intestine; Small intestine
Intraocular pressure, 299
Intrinsic muscles, 174, 175f, 437
Introduction, of lab report, A-3–A-4
Inversion, 141, 141f
Involuntary muscle, 46–47
Iodine, and starch digestion, 488–489
Iodine solution, A-2
Iris, 299, 299f, 301, 302f, 308
Irregular bones, 71, 71f
Irritability, 277
Ischiadic (sciatic) notch
  greater, 87, 88f
  lesser, 88f
Ischial ramus, 87, 88f
Ischial spine, 87, 88f–89f

Ischial tuberosity, 87, 87f–88f
Ischiocavernosus muscle, 521f
Ischium, 87–88, 87f–88f
Islets of Langerhans, 329, 330f
Isotonic solutions, 32, 35f
Isthmus
  of thyroid gland, 328, 328f
  of uterus, 526, 530f

J
Jejunal arteries, 400f
Jejunum, 472, 473f, 481f, 482
Joint(s), 129–146
  actions at, 139–141
  classification of, 130t, 135t
  composition of, 129
  movement of, 133–134, 135f
  types of, 129–133
Joint capsule, 131, 133f
Joint cavity, 133f
Jugular foramina, 117, 118, 118f–119f, 118t
Jugular notch, 104f, 105
Jugular veins, 3f
  cat, 216f
  external, 404f, 405, 406f, 411, 413f, 426f
  internal, 243, 403, 404f–407f, 411, 413f
  transverse, 411, 413f
Juxtaglomerular apparatus (JGA), 500, 502f
Juxtamedullary nephrons, 497, 502f

K
Keratinization, 62
Keratinized stratified squamous epithelium, 44, 62
Ketones, in urine, 508, 509t
Ketonuria, 508
Ketosis, 508
Kidney(s), 2f, 495, 496f, 498f–501f, 500
Kidney stones, 511
Kinetic balance, 311
Kinetic energy, 30–31
Kinetochores, 28
Kinocilium, 316f
Knee joint, 136, 138f
Korotkoff's sounds, 430, 430f
Kupffer cells, 477
Kyphosis, 97

L
Lab reports, A-3–A-4
Labia (lips), 465
Labia majora, 526f, 528, 530f, 533f
Labia minora, 526f, 528, 530f
Labial frenulum, 465, 466f
Labyrinths, 313, 313f–314f
Lacrimal apparatus, 299–300, 301f
Lacrimal bones, 112, 112f–113f, 116f, 120f

Lacrimal canaliculi, 301f
Lacrimal caruncle, 299, 299f
Lacrimal duct, 301f
Lacrimal gland, 299–300, 301f
Lacrimal sac, 301f
Lacteals, 473, 474f
Lactiferous ducts, 532
Lactiferous glands, 62
Lactiferous sinuses, 532, 532f
Lacunae, 51, 53, 53f, 74f, 75
Lag phase, of muscle, 148, 148f, 150, 153, 153f
Lambdoid suture, 112, 115f–116f, 131f
Lamellae, 74f, 75, 75f
Lamellated corpuscles, 287, 288f
Lamina propria, 380f, 470, 470f–472f, 472, 499f
Laminae, 97, 99f–101f
Langerhans, islets of, 329, 330f
Large intestine, 2f, 398f, 399, 465f, 473, 475f, 476f, 482
Laryngopharynx, 436f, 437, 466f, 469, 480f
Larynx, 2f, 356f, 435f–436f, 437–438, 437f, 441, 442f, 443f
Latent phase, of muscle, 148, 148f, 150
Lateral, 4f, 5t
Lateral border, of scapula, 81, 82f
Lateral column, 264f
Lateral commissure, 299, 299f
Lateral condyles
  femoral, 89, 89f–90f
  tibial, 91f
Lateral cuneiform bone, 90, 92f
Lateral epicondyle
  femoral, 89, 89f–90f
  humeral, 83, 84f
Lateral facet, of patella, 91f
Lateral fissure, 244, 245f, 246
Lateral geniculate nucleus, 303f
Lateral horn, of spinal cord, 264, 264f
Lateral incisor, 468f
Lateral longitudinal arch, 92f
Lateral malleolus, 90, 91f
Lateral menisci, 136, 138f
Lateral process, of breast, 532f
Lateral rotation, 140f, 141
Lateral supracondylar ridge, 83, 84f
Lateral ventricles, 241, 243f–244f, 250f, 252, 257f
Latissimus dorsi muscle, 162f, 163t, 164, 165f, 167f–168f, 171f, 179f–180f, 180, 183f, 223f, 225f, 229f
Latissimus dorsi tendon, 170f
Lead I, II, III, in electrocardiogram, 366, 367f
Left atrium, 353, 355f–362f, 356, 360

Left bundle branch, 365, 366f
Left bundle branch block, 368
Left cerebral hemisphere, 246, 247f, 256f
Left colic flexure, 473, 475f
Left lobe, of liver, 476, 477f
Left lung, 438, 443f
Left ventricle, 356, 356f–357f, 360, 362f
Leg, 6
  bones of, 90, 91f
  cat, 198–204, 199f–204f
  muscles of, 189–193, 190t–191t, 191f–192f, 193–197, 196f–198f
Length, measurement of, A–1, A–1t
Length-tension relationship, in muscle, 149
Lens, 300–302, 302f, 307, 307f
Lesser cornua, 103f, 105
Lesser curvature, of stomach, 471f, 472, 480
Lesser ischiadic (sciatic) notch, 88f
Lesser omentum, 474, 477f, 480, 481f
Lesser trochanter, 88, 89f–90f
Lesser tubercle, 83, 84f
Lesser wing, of sphenoid bone, 117–118, 119f, 123f
Leukocytes. *See* White blood cell(s)
Levator anguli oris muscle, 212f–213f
Levator labii superioris alaeque nasi muscle, 212f–213f
Levator labii superioris muscle, 210t, 212, 212f–213f
Levator palpebrae superioris muscle, 300f–301f
Levator scapulae muscle, 163t, 164, 165f, 168f, 180, 181f, 216f, 224f
Levator scapulae ventralis muscle, 180, 181f, 216f
Leydig (interstitial) cells, 515, 516f, 518f
LH, 327, 331–332, 332f, 528
Ligaments, 75, 131
Ligamentum arteriosum, 356f, 358
Ligamentum teres, 136, 137f
Ligation, of nerve, 153
Light microscopy, 13, 14f
Light-touch receptor mapping, 289
Limbic system, 250, 251f
Line, of bones, 72t
Linea alba, 223f
Linea aspera, 89f–90f, 90
Linea semilunaris, 223f
Lingual artery, 385f, 389f
Lingual frenulum, 466, 466f, 478, 480f
Lingual tonsils, 424, 425f, 436f
Lingual vein, 406f
Lip(s), 465, 480f

Lipids, digestion of, 491–492
Litmus cream, 491, A–2
Litmus solution, A–2
Liver, 2f, 399, 465f, 474–477, 477f–478f, 481f
Liver lobules, 476, 478f
Liver sinusoids, 476, 478f
Lobar bronchi, 438, 438f–439f
Lobes
  of breasts, 532, 532f
  of liver, 476, 477f, 481f
  of lungs, 438, 438f, 443f
  of thyroid gland, 328
Lobules
  of ear, 311, 312f
  of liver, 476, 478f
  of mammary glands, 532f
Long bones, 71, 71f–73f
Longissimus capitis muscle, 224f
Longissimus cervicis muscle, 224f
Longissimus muscle, 221, 222t, 224f, 225f, 230f
Longissimus thoracis muscle, 224f
Longitudinal arches, of foot, 92f
Longitudinal fissure, 246, 247f, 257f
Longitudinal layer, of muscularis, 470f–472f, 472
Loop of Henle, 498, 498f, 500, 501f–502f
Loose connective tissue, 48–49, 49f
Lordosis, 97
Lower body, arteries of, 395–402
Lower esophageal sphincter, 469, 471f, 472
Lower extremity, 6, 7f
Lower limbs
  arteries of, 381f, 395–397, 397f
  bones of, 81, 88–92
  muscles of, 189–208
  veins of, 405, 408f–410f, 411
Lower respiratory tract, 435f
Lubb/dupp sounds, 375
Lugol's iodine solution, A–2
Lumbar arteries, 396f
Lumbar curvature, 97, 98f
Lumbar enlargement, 263, 263f
Lumbar nerves, 265, 267f
Lumbar nodes, 426f
Lumbar plexus, 267, 267f
Lumbar regions, 7, 7f–8f, 98f
Lumbar veins, 410, 411f
Lumbar vertebrae, 98–101, 98f, 101f
Lumbodorsal fascia, 164
Lumbosacral plexus, 267, 267f, 271f
Lumbosacral trunk, 271f
Lumbrical muscles, 172f, 175f
Lumen, 469, 474f
Lunate bone, 85, 86f
Lunate surface, 88f
Lung(s), 2f, 356f, 435f, 438–441, 438f–440f, 443f
Lung function, 459–464

Lunula, 64, 65f
Luteinizing hormone (LH), 327, 331–332, 332f, 528
Luteotropin (LTH), 327
Lymph, 421, 421f
Lymph capillaries, 421, 421f, 423, 423f
Lymph flow, 421, 421f, 423f
Lymph nodes, 2f, 421f, 423–424, 423f, 424f, 440f
Lymph nodules, 423–424, 473, 474f
Lymph organs, 423, 423f, 424–425
Lymph tissue, 423
Lymphatic duct, 423, 423f, 426f
Lymphatic nodule, 424f, 470f
Lymphatic system, 2f, 3, 421–428, 421f, 423f, 426f
Lymphatic vessels, 2f, 421, 421f, 423, 470f
  afferent, 424, 424f
  deep, 440f
  efferent, 424, 424f
  superficial, 440f
  valves of, 422, 423f
Lymphatics, 423–424
Lymphocytes, 339f, 340–342, 341f, 342t
Lysosomes, 25f, 26, 27t

M
Macromolecules, 488
Macrophages, 442f
Macula
  of ear, 314, 316f
  of eye, 302, 303f
Macula lutea, 303f, 304f
Magnetic resonance imaging (MRI), 1, 1f
Magnification, 16–17, 16f
Main bronchi, 438, 438f–439f
Major calyces, 495, 497f, 500, 503f
Male contraception, 520–521
Male gametes, 515
Male reproductive system, 3, 515–524, 516f
Male sterilization, 520
Malleolus
  lateral, 90, 91f
  medial, 90, 91f
Malleus, 311, 312f
Maltose, 490
Maltose solution, A–2
Mammary artery, 389, 389f
Mammary glands, 2f, 178f, 531, 532f
Mammary (suspensory) ligaments, 531, 532f
Mammary plexus, 2f, 423f
Mammary vein, internal, 413f
Mammillary bodies, 247, 248f–249f, 251f, 253f, 326f, 327f

Mandible, 70f, 112, 112f–114f, 116f, 120f
Mandibular angle, 114f, 115
Mandibular condyle, 114f, 116f, 117
Mandibular foramen, 114f, 117, 118t, 120f, 121
Mandibular fossa, 117, 118f, 125f
Mandibular nodes, 426f
Mandibular notch, 114f, 116f, 117
Mandibular ramus, 114f, 116f, 117
Mandibular symphysis, 112, 113f
Manual region, 7f
Manubriosternal symphysis, 132f
Manubrium, 104f, 105
Marginal arteries, 358, 358f
Mass, measurement of, A–1, A–1t
Masseter muscle, 210t, 212, 212f–216f, 217f, 466, 476f, 479f
Mast cells, 48, 49f
Mastication, muscles of, 210t, 214, 214f, 466
Mastoid (posterolateral) fontanel, 130, 131f
Mastoid notch, 118, 118f
Mastoid process, 115, 116f, 118f, 124, 125f
Materials, in lab report, A–4
Matrix, 48
Mature (graafian) follicle, 525, 527f–529f
Maxilla, 112, 112f–113f, 116f, 117, 118f, 120f, 436f
Maxillary artery, 383, 385f
Maxillary bone, 120f
Maxillary sinus, 121, 121f
Maximal stimulus, 148
Maximum recruitment, 150, 153
Maximum recruitment voltage, 278
Measurement, A–1, A–1t
Meatus, 72t
Mechanoreception, 311
Mechanoreceptors, 287
Medial, 4f, 5t
Medial border, of scapula, 81, 82f
Medial commissure, 299, 299f
Medial condyle
  femoral, 89, 89f–90f
  tibial, 91f
Medial cuneiform bone, 90, 92f
Medial epicondyle
  femoral, 89, 89f–90f
  humeral, 83, 84f
Medial facet, of patella, 91f
Medial geniculate nucleus, 315f
Medial longitudinal arch, 92f
Medial malleolus, 90, 91f
Medial menisci, 136, 138f
Medial rotation, 140f, 141
Medial supracondylar ridge, 83, 84f
Median fissure, anterior, 264, 264f

Median nerve, 266, 268f, 271f, 274, 274f
Median plane, 5, 5f
Median sulcus, posterior, 264, 264f–265f
Mediastinum, 5, 6f, 353
Medulla
  of adrenal glands, 329, 330f
  of hair, 63, 63f
  of kidney, 495, 497f–498f, 498, 501f–502f
  of lymph nodes, 424, 424f
  of ovary, 527f
Medulla oblongata, 233, 242f, 247, 248, 248f–250f, 252–253, 252f, 255f
Medullary cavity, 72, 73f
Medullary cord, of lymph node, 424, 424f
Medullary ray, 497f
Medullary sinus, of lymph node, 424, 424f
Meibomian gland, 301f
Meissner's corpuscles, 287, 288f
Melanin, 61, 301
Melanocytes, 61, 63f
Melatonin, 326
Membranous labyrinth, 313, 313f–314f
Membranous urethra, 500f, 517f, 519, 519f, 521f
Meningeal dura, 241, 242f
Meninges (brain), 241, 242f
Meninges (spinal cord), 265, 265f
Menisci
  lateral, 136, 138f, 139f
  medial, 136, 138f, 139f
Meniscofemoral ligament, posterior, 138f
Menstrual cycles, 528–529, 531f
Menstruation, 529
Mental foramina, 112, 113f, 114f, 116f, 118t
Mentalis muscle, 210t, 212, 212f–213f
Merkel discs, 287, 288f
Merocrine sweat glands, 62
Mesencephalic aqueduct, 241, 243, 249f
Mesencephalon, 241, 242f
Mesenchyme, 48
Mesenteric arteries, 395
  anterior, 398f, 400f
  inferior, 381f, 395, 396f, 399f, 400, 400f
  posterior, 398f
  superior, 381f, 395, 396f, 399, 399f
Mesenteric nodes, 426f
Mesenteric veins
  inferior, 404f, 409f, 415, 416f
  superior, 404f, 409f, 415, 416f

Mesentery, 468f–469f, 469, 473f, 475f, 482
Mesoderm, 534, 534f
Mesometrium, 533f
Mesosalpinx, 530f, 533f
Mesothelium, 42, 42f
Mesovarium, 527f, 530f, 533f
Metabolic rate, basal, 329
Metacarpals, 70f, 83–85, 86f
Metaphase, 28, 29f, 30t
Metatarsals, 70f, 90, 91, 92f
Metencephalon, 243t
Methods, in lab report, A–4
Methylene blue, 31, A–2
Methylene blue solution, A–2
Metric system, A–1, A–1t
Microfilaments, 24, 26f
Microglia, 236, 238f
Microscope, 13–18, 14f, 15t
Microscopy, 13–22
Microtubule network, 25f
Microtubules, 24, 26f
Microvilli, 25f, 27, 498
Midbrain, 233, 241, 242f, 243t, 247–248, 250, 252f
Middle ear, 311, 312f
Middle lobe, of lung, 438, 438f, 443f
Middle meatus, 436f
Middle phalanx, 85, 86f, 90–91, 92f
Midline, 4f
Midpiece, of sperm, 516f, 517
Midsagittal plane, 5, 5f
Midstream collection, 507
Mifepristone, 531
Milk glands, 62
Milk teeth, 467
Mineralocorticoids, 329
Minor calyces, 495, 497f, 500, 503f
Minute ventilation, 454
Mitochondria, 24, 25f–26f, 27t, 147f, 516f, 517
Mitosis, 27–28, 27f, 29f–30f, 30t
Mitral (bicuspid) valve, 355f, 359, 359f, 361f, 375f–376f
Mixed nerves, 265
Modality, 287
Molars, 114f, 467, 468f
Molasses solution, A–2
Monoclonal antibody, 332
Monocytes, 339f, 341–342, 341f, 342, 342t
Monoglycerides, 491
Monosynaptic reflex arc, 280, 281
Monro, foramina of, 241
Mons pubis, 526f, 528, 530f
Morning after pill, 531
Morula, 534, 534f
Motor cortex, primary, 244, 245f–246f
Motor division, of peripheral nervous system, 233

Motor homunculus, 244
Motor (efferent) neurons, 147f, 234, 235f, 280, 281f
Motor speech area, of brain, 244, 245f
Motor unit, 147, 147f
Mouth, 465, 465f–466f
Mucosa
  of gastrointestinal tract, 470, 470f
  of large intestine, 476f
  of small intestine, 474f
  of stomach, 471f–472f, 472
Mucous cells, surface, 471f, 472
Mucous epithelium, 470
Mucous membrane, 435, 470, 470f, 472f
Mucous neck cells, 471f
Mucus, 473
Multiaxial joints, 133
Multifidus muscle, 222, 222t, 224f, 225f, 230f
Multinucleate muscle cells, 46
Multiple motor unit summation, 148, 148f, 153
Multiple wave summation, 148
Multipolar neuron, 234, 235f–236f, 265
Multistix procedure, for urinalysis, 508, 508f
Muscle(s), 161f–162f. See also specific muscles
  cardiac, 46–47, 47f
  fixing of, 141
  length-tension relationship in, 149
  nomenclature of, 159–160, 169
  origins, insertions, 159, 179
  physiology of, 147–158
  skeletal, 46, 46f
  smooth, 47, 47f
Muscle contraction, 147–158, 148f, 153
Muscle fatigue, 153–155
Muscle fibers, 46, 46f
Muscle recruitment, 153–154
Muscle twitch, 148, 148f
Muscular arteries, 379
Muscular function, of heart, 365
Muscular system, 2f, 3
Muscular tissue, 41, 46–47
Muscularis
  of gastrointestinal tract, 470, 470f
  of large intestine, 476f
  of small intestine, 474f
  of stomach, 471f–472f, 472
Muscularis externa, 470
Muscularis mucosa, 470, 470f, 472, 472f
Musculocutaneous nerve, 266, 268f, 270f, 274, 274f

Musculotendinous cuff, 163t
Myelencephalon, 243t
Myelin, 235
Myelin sheath, 235, 235f, 237f–238f
Myelinated axon, 237f
Myelinated nerve fibers, 236, 250
Myenteric plexus, 470f
Mylohyoid muscle, 209, 210t, 211f, 214, 217f, 480f
Myocardial infarcts, 367
Myocardium, 353, 355f
Myocytes, 46
Myofibrils, 147, 147f
Myogenic conduction, 371
Myometrium, 526, 530f
Myopia, 306
Myopic vision, 306
Myosin filaments, 147

N

Nail(s), 59, 64, 65f
Nail bed, 64, 65f
Nail body, 64, 65f
Nail fold, 64, 65f
Nail groove, 64, 65f
Nail matrix, 65f
Nail root, 64, 65f
Nares
  external, 435
  internal, 117, 118f, 436
Nasal bones, 111, 112f–113f, 116f, 120f, 436f
Nasal cartilage, 120f, 121, 435–436, 436f
Nasal cavity, 2f, 113f, 435–436, 435f–436f
Nasal concha
  inferior, 112f, 120f, 121, 436, 436f
  middle, 113f, 120f, 121, 122f, 436, 436f
  superior, 120f, 121, 122f, 436, 436f
Nasal region, 7f
Nasal septum, 113f, 117, 121, 435, 436f
Nasal spine, anterior, 113f, 436f
Nasal vestibule, 435, 436f
Nasolacrimal canal, 112, 116f
Nasolacrimal duct, 300, 301f
Nasopharynx, 436f, 437, 466f, 469, 480f
Nasus (nose), 435–436, 436f
Navicular, 90, 92f
Near point determination, 304
Nearsightedness, 306
Neck (anatomical)
  of bone, 72t
  of femur, 88, 89f–90f
  of humerus, 83, 84f
  of rib, 102, 104f
  of tooth, 466, 468f

Neck (of body)
  arteries of, 383, 385f
  muscles of, 209–213, 210t, 211f, 228f
  veins of, 405f–406f, 411
Neck (surgical), of humerus, 83, 84f
Negative afterimages, 308
Negative feedback inhibition, 326
Nephron(s), 497–498, 502f
Nerve(s), 2f, 59, 60f
  physiology of, 277–283
  spinal cord associated, 263–276
  structure of, 265f
Nerve cell, 233–234
Nerve cell body, 48, 48f, 233, 235f–236f
Nerve conduction, 278–279
Nerve impulse, 147, 279
Nervous layer, of eye, 302
Nervous system, 2f, 3, 233–240
  divisions of, 233, 234f
  functions of, 234
  physiology of, 277–286
Nervous tissue, 41, 47–48, 327
Neural arch, 97
Neuroglia, 48, 48f, 235f, 236–237
Neurohypophysis, 326, 327f. See also Pituitary gland
Neurolemmocyte, 235–236, 237f–238f
Neuromuscular junction, 147, 147f
Neurons, 47–48, 48f, 233–234, 235f–236f
Neuroplasm, 233
Neurotransmitters, 235, 237f
Neutrophils, 339, 339f–340f, 341, 342t
Nipple, 531, 532f
Nissl bodies, 48f
Nissl substance, 233, 235f
Nitric acid, for decalcifying bones, A–2
Nitrites, in urine, 509, 509t
Nociceptors, 287, 290
Node of Ranvier, 235–236, 235f, 237f–238f
Nonkeratinized stratified squamous epithelium, 44, 44f, 45f, 466
Nonstriated muscle, 47
Norepinephrine, 330, 371
Normal reflex, 281
Nose, 435–436, 435f–436f
Notch, 72t
Novocain (procaine hydrochloride), 280, A–2
Nuchal lines, inferior and superior, 118f
Nuchal region, 7f
Nuclear envelope, 25f, 26, 28, 30f
Nuclear pores, 25f, 26
Nucleolus, 25f, 26, 28, 30f

Nucleoplasm, 25f
Nucleus, 25f–26f, 27t, 29f, 233, 235f, 237f, 516f
Nutrient foramen, 72, 83
Nystagmus, 319

O

Oblique layer, of muscularis, 471f, 472
Oblique line, of mandible, 113f
Oblique muscle
    inferior, 300f, 300t, 301f
    superior, 300f, 300t
Obturator artery, 397f
Obturator externus muscle, 192, 194f, 272f
Obturator foramen, 87, 87f–88f
Obturator internus muscle, 192, 194f
Obturator nerve, 267, 271f, 272f
Occipital artery, 383, 385f, 389
Occipital bone, 112, 112f, 115, 115f–116f, 118f–120f, 131f
Occipital condyles, 117, 118f
Occipital (posterior) fontanel, 130, 131f
Occipital lobe, 244–246, 245f, 247f, 255f
Occipital nerve, lesser, 268f
Occipital protuberance, external, 115, 116f, 118f
Occipital sinus, 405f
Occipitalis muscle, 210t, 212, 214, 217f
Occipitofrontalis muscle, 212, 212f–213f
Oculomotor (III) nerve, 251, 253f, 254t, 255f, 255t, 256f
Odontoid process, 98, 100f
Odor, of urine, 508, 509t
Oil glands, 59, 60f, 62, 62f–63f
Oil immersion lens, 18
Olecranon bursa, 137f
Olecranon fossa, 83, 84f
Olecranon process, 83, 85f, 137f, 171f
Olfaction, 295–298, 295f
Olfactory bulb, 251f, 252, 255f–257f, 295f
Olfactory center, 246
Olfactory cortex, 251f
Olfactory discrimination, 296
Olfactory foramina, 119f, 121, 122f, 436f
Olfactory (I) nerve, 251, 253f, 254t–255t, 295, 295f
Olfactory recess, 120f
Olfactory reflex, 295
Olfactory tract, 248f, 252, 253f
Oligodendrocytes, 236, 238f
Olivary nucleus, superior, 315f
Olive, of medulla oblongata, 253f
Omenta, 469f, 474, 477f, 480, 481f

Omohyoid muscle, 209, 210t, 211f–214f
Oocytes, 330, 331f, 525, 527f–529f, 534f
Ophthalmic veins, 405f
Ophthalmoscope, 306, 306f
Opponens digiti minimi muscle, 164t, 174, 175f
Opponens pollicis muscle, 164t, 174, 175f
Opsin, 308
Optic canal, 112, 113f, 118t
Optic chiasma, 247, 248f–249f, 251, 253f, 303f, 326f
Optic disk, 303f, 308
Optic foramen, 119f, 123f
Optic (II) nerve, 251, 253f, 254t–255t, 300f, 302, 302f, 303f, 307, 307f, 308
Optic radiations, 303f
Optic tract, 250f, 303f
Optical cell counter, 348, 348f
Oral cavity, 2f, 436f, 437, 465, 465f, 466, 466f, 480f
Oral region, 7f
Orbicularis oculi muscle, 210t, 212, 212f–213f, 301f
Orbicularis oris muscle, 212, 212f–213f, 214f
Orbit(s), 112
Orbital fissure
    inferior, 112, 113f, 118f, 118t
    superior, 112, 113f, 118t, 123f
Orbital plate
    of ethmoid bone, 121, 122f
    of frontal bone, 113f
Organ of Corti, 313, 314f
Organ systems, 1–3, 2f–3f
Organelles, 24–26, 27t
Organic matter, of bones, 71
Organization, of body, 1–11
Orientation, nervous system and, 233
Origin, of muscles, 159, 179
Oropharynx, 437, 466, 466f, 469, 480f
Osmosis, 24, 30, 32–35, 33f, 35f
Osmotic potential, 32
Osmotic pressure, 32
Ossicles, 311, 312f
Osteoblasts, 75, 75f
Osteoclasts, 75, 75f
Osteocytes, 53, 74f, 75, 75f
Osteogenic progenitor cells, 75
Osteons, 53, 53f, 73f–74f, 75, 75f
Ostium, 533
Otolithic membrane, 316f
Otoliths, 314, 316f
Oval window, 312f, 313, 313f–314f
Ovarian arteries, 395, 396f
Ovarian cycle, 528–529, 531f

Ovarian epithelium, 527f
Ovarian follicles, 331f, 525, 527f–529f, 531f, 534f
Ovarian ligament, 528, 530f
Ovaries, 2f–3f, 325f, 331, 331f, 525–526, 521f–526f, 531f, 533, 533f
Oviducts, 533, 533f
Ovulation, 525, 528–529, 529f, 531f, 534f
Oxygen, diffusion of, 31
Oxygen free radicals, 26
Oxytocin, 328, 328f

P

P wave, 365, 366f
Pacemaker, 365
Pacinian corpuscles, 287, 288f
Packed cell volume (PCV), 347–348, 347f
Pain receptors, 287
Pain, referred, 290, 290f
Palate
    hard, 117, 118f, 120f, 436f, 466, 466f, 480f
    soft, 436f, 466, 466f, 480f
Palatine bone, 112f, 117, 118f, 120f, 436f
Palatine process, 117, 118f, 120f, 436f
Palatine suture, median, 117, 118f
Palatine tonsils, 424, 425f, 436f, 467f, 480f
Palmar aponeurosis, 172f, 173
Palmar arch arteries, 383
    deep, 383, 384f
    superficial, 383, 384f
Palmar arch veins, 405, 407f
Palmar interossei, 175f
Palmaris longus muscle, 164t, 173, 173f, 182, 182f
Palpebral conjunctiva, 301f
Palpebral fissure, 301f
Pancreas, 2f–3f, 325f, 329, 330f, 465f, 472, 473f, 477, 479f, 481f, 482
Pancreatic duct, 330f, 477, 479f
Pancreatic islets, 329, 330f
Pancreatin, 487, 491–492
Pancreatin solution, A–2
Pancreatoduodenal veins, 416f
Papillae, 59
    of duodenum, 479f
    of hair, 62, 63f–64f
    of kidney, 495, 497f, 500, 503f
    of tongue, 466, 467f, 478, 480f
Papillary layer of dermis, 59, 60f
Papillary muscles, 355f, 358, 359f–360f, 361
Parafollicular cells, 328f, 329
Parahippocampal gyrus, 251f
Paranasal sinuses, 121, 121f, 436f
Parasagittal plane, 5, 5f

Parasympathomimetic substance, 374
Parathormone, 329
Parathyroid glands, 3f, 325f, 329, 329f
Parathyroid hormone (PTH), 329
Parietal bone, 112, 112f, 113f, 115f–116f, 119f–120f, 131f
Parietal cells, 471f, 472
Parietal lobe, 244, 245f, 247f, 255f
Parietal pericardium, 3, 353, 356f
Parietal peritoneum, 42, 223f, 387, 469, 469f, 481f, 496f, 499f
Parietal pleura, 4, 439, 439f, 441
Parotid duct, 473, 476f, 479f
Parotid glands, 473, 476f, 478, 479f
Particle weight, 31, 32t
Patella, 70f, 90, 91f, 161f, 191f, 193f
Patellar groove, 89f
Patellar ligament, 136, 138f–139f, 191f
Patellar reflex, 281, 281f
Patellar region, 7f
Patellar tendon, 136, 138f, 191f
Pectinate muscles, 358, 360f
Pectineal line, 89, 89f
Pectineus muscle, 161f, 190t, 191, 192f–193f
Pectoantebrachialis muscle, 179, 180f
Pectoral girdle, 81
Pectoral muscles, 179, 180f
Pectoralis major muscle, 2f, 161f, 163t, 164, 166f–167f, 179, 180f, 223f
Pectoralis minor muscle, 163t, 164, 167f, 179, 180f, 227f
Pedal region, 7f
Pedicles, 97, 99f–101f
Peduncles, cerebral, 247, 249f, 252f, 255f
Pelvic arteries, 397f, 400f
Pelvic brim, 87f–89f, 88
Pelvic cavity, 5, 6f
Pelvic girdle, 81, 86–88, 87f–88f
Pelvic outlet, 89f
Pelvic veins, 405, 408f, 410
Pelvis, 2f, 86–87, 87f–89f
Pelvis, renal, 495, 497f, 500, 503f
Penile urethra, 500f, 519, 519f
Penis, 500f, 516f–517f, 519–520, 519f, 521, 521f–522f
Pepsin, 472
Pepsinogen, 472
Peptide bonds, 492
Perforating canals, 72, 73f–74f
Perforating fibers, 72
Perfume, dilute solution, A–2
Pericardial cavity, 5, 353, 356f
Pericardial fluid, serous, 353
Pericardium, 353, 356f–357f
Perichondrium, 51, 52f
Perilymph, 313, 313f–314f
Perilymphatic cells, 313f–314f

Perimetrium, 526, 530f
Perineum, 520, 520f, 530f
Perineurium, 265, 265f
Periodontal ligaments, 131, 131f, 468f
Periosteal dura, 241, 242f
Periosteum, 72, 73f–75f, 133f
Peripheral nervous system (PNS), 233, 234f
Peripheral proteins, 24, 25f
Perirenal fat, 495, 496f
Peristalsis, 469, 497
Peritoneum, 469, 469f, 470f, 481f, 482, 499f
Peritubular capillaries, 496, 498f
Permanent teeth, 467
Peroneus muscles, 193, 196f
Peroxisome, 25f, 26, 27t
Perpendicular plate, of ethmoid bone, 113f, 120f, 121, 122f, 435, 436f
Petrosal sinus, inferior and superior, 405f
Petrous portion, of temporal bone, 118, 119f, 124, 125f
Peyer's patches, 473
pH, 509, 509t
Phagocytic vesicle, 25f
Phalanges, 70f, 86f, 90–91, 92f
Pharyngeal constrictor muscles, 469
Pharyngeal tonsils, 425, 425f, 436f
Pharyngotympanic tube, 311
Pharynx, 2f, 435f, 436, 437, 465f, 469
Phasic receptors, 289
Phosphate buffer solution, A–2
Phospholipid bilayer, 24, 25f
Photoreceptive layer, 302, 303f
Photoreceptors, 287, 303f
Phrenic artery, 382f
Phrenic nerve, 265, 268f
Phrenic vein, 410f
Physical stimuli, nerve response to, 277–286
Physiology, 1
Pia mater, 241, 242f, 249f, 265, 265f
Pigmentation of skin, 61
Pineal gland (pineal body), 3f, 247, 249f, 252f, 257f, 325f, 326, 326f
Pinna, 311, 312f
Piriformis muscle, 192, 194f
Pisiform bone, 85, 86f
Pithing, in frog preparation, 151, 151f
Pituitary gland, 3f, 247, 248f–249f, 253f, 325f, 326–328, 326f–327f, 531f
  anterior, 326, 327f
  posterior, 326, 327f
Pivot joints, 134, 135f
Placenta, 410, 412f

Plane joints, 134, 135f
Planes of sectioning, 4–5, 5f
Plantar arteries, 397, 397f
Plantar flexion, 189
Plantar response, 283, 283f
Plantar veins, 408f
Plantaris muscle, 198f
Plaque, arteriosclerotic, 397–398, 398f
Plasma, 53, 54f, 337, 421f
Plasma cells, 340–341
Plasma (cell) membrane, 24, 25f–26f, 34–35, 35f
Platelets, 53, 54f, 337, 339f, 342t
Platysma muscle, 179, 209, 210t, 212f–213f, 214, 466
Pleural cavities, 5, 6f, 356f, 439, 439f, 441
Plexuses, 265–273, 266t, 267f–268f
Pneumocytes, type I and type II, 439–441, 442f
Polarization, cardiac, 365, 371
Pollex, 85, 86f
Polycythemia, 347
Polymorphonuclear leukocytes, 339
Polypeptide chains, 492
Polysynaptic reflex arcs, 280
Pons, 233, 242f, 247, 248, 248f–251f, 252f, 253f
Popliteal artery, 381f, 397, 397f–398f, 400, 429f
Popliteal ligament, oblique, 138f
Popliteal nodes, 426f
Popliteal region, 7f
Popliteal tendon, arcuate, 138f
Popliteal vein, 404f, 408f, 411, 414f
Popliteus muscle, 191t, 195, 198f, 201
Portal system, 409f, 415, 497
Portal (hepatic) triad, 477, 478f
Portal vein, 404f, 409f, 412f, 415, 477, 477f–478f
Position, anatomic, 3–4, 4f
Positioning, and blood pressure, 431
Positive afterimages, 308
Postcentral gyrus, 244, 245f–246f
Posterior, 3, 4f, 5t
Posterior arch, cervical, 100f
Posterior chamber (eye), 300, 302f, 308
Posterior compartment, of eye, 300, 302, 302f
Posterior horn, of spinal cord, 264, 264f
Posterior ligament of incus, 312f
Posterior median sulcus, 264–265, 264f
Posterior vein of left ventricle, 358f
Postsynaptic membrane, 147f, 237f
Postural muscles, 221, 222t, 224f
Postural reflex test, 318

Potassium dichromate solution, 31–32, A–2
Potassium permanganate solution, 31–32, A–2
PQ interval, 368
Precentral gyrus, 244, 245f–246f
Prefrontal area, 245f
Pregnancy prevention, 520–521, 529–531
Premolars, 114f, 467, 468f
Premotor area, 245f
Prepared slides, 15, 17–18
Prepuce, 177, 516f, 519, 528, 530f
Presynaptic terminal, 147f, 234, 235f, 237f
Primary auditory cortex, 245f, 246
Primary bronchi, 437f–438f, 438
Primary follicles, 525, 527f–529f
Primary motor cortex, 244, 245f–246f
Primary oocytes, 525, 527f–529f
Primary sensory cortex, 244, 245f
Primary somatic sensory cortex, 245f, 246
Primary spermatocytes, 515, 516f, 518f
Primary teeth, 467
Prime mover, 141
Primordial follicles, 525, 527f–529f
Procaine hydrochloride (Novocain), 280, A–2
Procerus muscle, 212f–213f
Process, 72t
Progenitor cells, osteogenic, 75
Progesterone, 331, 529, 531f
Projections, 71, 72t
Prolactin, 327
Pronation, 140f
Pronator muscles, 169, 173f
Pronator quadratus muscle, 163t, 169, 173f, 182, 184f
Pronator teres muscle, 163t, 169, 170f, 173f, 182
Prophase, 28, 29f, 30t
Proprioception, 290
Proprioceptors, 287, 290
Prosencephalon, 241, 242f
Prostate gland, 500f, 516f–517f, 518, 519f, 521, 521f
Prostate gland ducts, 519f
Prostatic urethra, 500f, 517f, 518, 519f, 521f
Proteins
  digestion of, 492
  in urine, 509
  integral, 24, 25f
  peripheral, 24, 25f
Protraction, 141, 141f
Proximal, 4, 4f, 5t
Proximal convoluted tubule, 498, 498f, 501f–502f

Proximal phalanx, 85, 86f, 90–91, 92f
Pseudostratified ciliated columnar epithelium, 436
Pseudostratified columnar epithelium, 44
Pseudounipolar neuron, 234, 236f
Psoas major muscle, 189, 191f, 272f, 496f
Psoas minor muscle, 189
Pterygoid muscles, 213, 214, 466
  lateral, 135f, 210t, 214f
  medial, 210t, 214f
Pterygoid plate
  lateral, 117, 118f, 123f
  medial, 117, 118f, 120f, 123f
Pubic crest, 87f–88f
Pubic ramus, inferior and superior, 87, 88f
Pubic region, 7f
Pubic symphysis, 86, 132f
Pubis, 87–88, 87f–88f
Pubofemoral ligament, 137f
Pudendal artery, internal, 397f
Pudendal cleft, 530f
Pudendal nerve, 271f
Pulmonary arteries, 3f, 355f, 358, 361
  left, 356f–357f, 359f
  right, 356f–357f, 359f
Pulmonary capacities, 449–454, 453f
Pulmonary capillaries, 440f
Pulmonary circulation, 421f
Pulmonary function, 459–464
Pulmonary health, 459–464
Pulmonary obstruction, 460
Pulmonary restriction, 460
Pulmonary semilunar valve, 355f, 358, 359f, 361, 374, 375, 375f–376f
Pulmonary trunk, 353, 355f–356f, 358, 358f–359f, 361, 375, 381f, 412f
Pulmonary veins, 355f, 356, 358, 361, 404f, 440f
  left, 356f–357f
  right, 356f–357f
Pulmonary ventilation, 447
Pulmonary volume, 447–458, 453f
Pulp cavity, of tooth, 466, 468f
Pulse pressure, 431
Pulse rate, 374–375, 429, 429f
Puncta, 301f
Punctate distribution, 287
Pupil, 299, 299f, 302f, 308
Pupillary reactions, 306–307
Purkinje fibers, 365, 366f
Pyloric antrum, 471f
Pyloric canal, 471f
Pyloric orifice, 471f

Pyloric part, of stomach, 471*f*, 472, 480
Pyloric sphincter, 471*f*, 472, 480
Pyuria, 509

Q
Q10, 374
QRS complex, 366*f*, 368
QT interval, 368
Quadrangular membrane, 437*f*
Quadrants, of abdomen, 7, 8*f*
Quadrate lobe, of liver, 476, 477*f*
Quadratus femoris muscle, 192, 194*f*
Quadratus lumborum muscle, 221, 222*t*, 224*f*
Quadriceps femoris muscle, 2*f*, 138*f*, 161*f*, 190*t*, 191–192, 191*f*
Quadriceps femoris tendon, 136, 138*f*, 139*f*
Quadriceps muscle, 191*f*
Quantities, measurement of, A–1

R
Radial artery, 381*f*, 383, 384*f*, 389, 389*f*, 429, 429*f*
Radial collateral ligament, 137*f*
Radial muscles, of eye, 301
Radial nerve, 266, 268*f*–269*f*, 274, 274*f*
Radial notch, 83, 85*f*
Radial tuberosity, 83, 85*f*
Radial vein, 405, 407*f*, 411, 413*f*
Radius, 2*f*, 70*f*, 83, 85*f*
Ramus, 72*t*. *See also specific types*
Ranvier, node of, 235–236, 235*f*, 237*f*–238*f*
Reabsorption, renal, 495
Receptors, 280, 287–290
Recruitment, muscle, 153–154
Rectum, 2*f*, 465*f*, 473, 481*f*, 482, 516*f*, 521*f*, 526*f*, 533*f*
Rectus abdominis muscle, 2*f*, 161*f*, 221, 222*t*, 223*f*, 228*f*
Rectus femoris muscle, 161*f*, 190*t*, 191*f*–195*f*, 192, 199, 272*f*
Rectus femoris tendon, 137*f*
Rectus muscle
 inferior, 300*f*, 300*t*, 301*f*
 lateral, 300*f*, 300*t*
 medial, 300*f*, 300*t*
 superior, 300*f*, 300*t*, 301*f*
Rectus sheath, 228*f*
Red blood cell(s), 53, 54*f*, 338–339, 339*f*, 342*t*
 crenation of, 35, 35*f*
 hemolysis of, 35, 35*f*, 346, 509
 in urine, 508, 510, 510*f*–511*f*
Red bone marrow, 51, 72
Red pulp, of spleen, 425, 425*f*
Red sensitive cones, 307
Referred pain, 287, 290, 290*f*

Reflection, of muscle, 164
Reflex arc, 280, 281, 281*f*
Reflexes, 280–283. *See also specific types*
Refractory period, 148, 278
Regional anatomy, 1
Regions, 6–7, 7*f*–8*f*
Relaxation phase, of muscle, 148, 148*f*, 150, 153, 153*f*
Renal arteries, 381*f*, 396*f*, 399, 399*f*, 495, 496*f*, 500
 left, 395, 396*f*, 397*f*–398*f*
 right, 395, 396*f*
Renal calculi, 511
Renal capsule, 495, 496*f*, 498, 500, 503*f*
Renal columns, 495, 497*f*–498*f*
Renal corpuscle, 498, 498*f*, 501*f*–502*f*
Renal cortex, 495, 497*f*–498*f*, 498, 501*f*–502*f*
Renal fat pads, 495
Renal medulla, 495, 497*f*–498*f*, 498, 501*f*–502*f*
Renal papilla, 495, 497*f*, 500, 503*f*
Renal pelvis, 495, 497*f*, 500, 503*f*
Renal pyramids, 495, 497*f*–498*f*, 500, 502*f*
Renal sinus, 495, 497*f*, 500
Renal tubules, 498*f*, 500, 501*f*
Renal veins, 409*f*, 410, 410*f*, 414*f*, 415, 495, 496*f*–498*f*, 500
Repolarization, 366, 366*f*
Reproductive system, 3
 female, 2*f*, 3, 525–536, 526*f*
 male, 3, 515–524, 516*f*
Residual volume, 453*f*, 454
Resistance, to air flow, 455
Respiratory bronchioles, 439, 440*f*
Respiratory centers, 250
Respiratory epithelium, 435–436, 439*f*
Respiratory gases, acid-base effects of, 455
Respiratory membrane, 441
Respiratory muscles, 221, 222*t*, 226*f*–227*f*
Respiratory sounds, 454, 455*f*
Respiratory system, 2*f*, 3, 435–446, 443*f*
 function of, 447–458
 structure of, 435–441, 435*f*
Respirometer, 447, 453
Resting membrane potential, 277–278, 371
Results, in lab report, A–4
Rete testis, 516*f*, 518
Reticular connective tissue, 50, 51*f*
Reticular fibers, 48, 49*f*, 50, 51*f*
Reticular layer of dermis, 59, 60*f*
Retina, 302*f*, 303*f*, 307, 307*f*, 308
Retinal artery, 302*f*

Retinal vein, 302*f*
Retraction, 141, 141*f*
Retromandibular vein, 405*f*–406*f*
Retroperitoneal, 495
Retropharyngeal lymph nodes, 426*f*
Rh antigen, 346–347
Rh antiserum (anti-D), 347
Rh factor, determination of, 346–347
Rh immune globulin, 347
Rhodopsin, 308
Rho-GAM, 347
Rhombencephalon, 241, 242*f*
Rhomboideus capitis muscle, 216*f*, 229*f*
Rhomboideus major muscle, 163*t*, 164, 165*f*, 168*f*, 181*f*, 228*f*–229*f*
Rhomboideus minor muscle, 163*t*, 164, 165*f*, 168*f*, 181*f*, 228*f*–229*f*
Rhomboideus muscles, 164, 165*f*, 168*f*, 181
Rib(s), 2*f*, 70*f*, 97, 102, 103*f*
Rib facet, 98, 99*f*, 100*f*
Ribonucleic acid (RNA), ribosomal, 26
Ribosomal RNA, 26
Ribosomes, 24, 25*f*, 26, 26*f*, 27*t*
Right atrium, 353, 356, 356*f*–357*f*, 358*f*–359*f*
Right bundle branch, 365, 366*f*
Right bundle branch block, 368
Right cerebral hemisphere, 246–247, 247*f*
Right colic flexure, 473, 475*f*
Right lobe, of liver, 476, 477*f*
Right lung, 438
Right ventricle, 356, 356*f*–357*f*, 358, 358*f*–359*f*, 361*f*–362*f*
Ring, of ophthalmoscope, 306
Ringer's solution, A–2
Rinne test, 317–318, 318*f*
Risorius muscle, 210*t*, 212, 212*f*–213*f*
Rods, 302, 303*f*, 304
Romberg test, 319
Root
 of hair, 62, 63*f*–64*f*
 of nail, 64, 65*f*
 of penis, 519*f*, 520
 of spinal nerves, 263*f*
 of tongue, 467*f*
 of tooth, 129*f*, 466, 468*f*
Root canal, 466, 468*f*
Root sheath, of hair, 62–63, 64*f*
Rootlets, of spinal nerve, 264*f*, 266*f*
Rotating disk, of ophthalmoscope, 306
Rotation, 140*f*, 141
Rotational acceleration, 316

Rotator cuff muscles, 134, 136*f*, 166, 167*f*–168*f*
Rough endoplasmic reticulum, 24–26, 25*f*, 27*t*
Round ligament, 477*f*, 528, 530*f*, 533*f*
Round window, 312*f*, 313, 313*f*–314*f*
Rugae, 471*f*, 472, 480, 500*f*, 526*f*, 528

S
S phase, of interphase, 27*f*
Saccular macula, 316*f*
Saccule, 314, 316*f*
Sacral arteries, lateral and median, 396*f*–397*f*
Sacral canal, 101, 102*f*
Sacral crest, 101, 102*f*
Sacral crest, lateral and median, 101, 102*f*
Sacral (pelvic) curvature, 97, 98*f*
Sacral foramen
 anterior, 101, 102*f*
 posterior, 101, 102*f*
Sacral hiatus, 101, 102*f*
Sacral nerves, 265, 267*f*
Sacral plexus, 267*f*, 274, 274*f*
Sacral promontory, 87*f*, 89*f*, 98*f*, 101, 102*f*
Sacral region, 7*f*, 98*f*
Sacroiliac joint, 87*f*, 101
Sacrum, 70*f*, 87, 98*f*, 101, 102*f*–103*f*
Saddle joints, 134, 135*f*
Sagittal plane, 5, 5*f*
Sagittal sinus
 inferior, 405*f*
 superior, 242*f*, 404*f*–405*f*
Sagittal suture, 112, 115*f*, 131*f*
Salivary amylase, 473
Salivary glands, 2*f*, 178, 217*f*, 465*f*, 473–474, 476*f*, 478
Salt solution, nerve response to, 280
Saltatory conduction, 236
Saltwater solution, A–3
Saphenous veins
 great, 404*f*, 405, 408*f*, 411, 414*f*
 small, 404*f*, 408*f*
Sarcolemma, 147*f*
Sarcomeres, 147
Sarcoplasm, 147*f*
Sartorius muscle, 2*f*, 161*f*, 189, 190*t*, 191*f*–195*f*, 199, 200*f*, 272*f*
Scala media (cochlear duct), 313, 314*f*–315*f*
Scala tympani, 313, 314*f*–315*f*
Scala vestibuli, 313, 314*f*–315*f*
Scalene muscles, 210*t*, 211*f*, 214, 216*f*
 anterior, 210*t*, 226*f*
 middle, 210*t*, 226*f*
 posterior, 210*t*, 226*f*
Scaphoid bone, 83–85, 86*f*

Scapula, 70f, 81, 82f
Scapular ligament, transverse, 136f
Scapular muscles, 168f, 182
Scapular nerve, dorsal, 268f
Scapular notch, 81, 82f
Scapular region, 7f
Scapular spine, 81
Scapular veins, transverse, 411, 413f
Schlemm, canal of, 301, 302f
Schwann cell, 235f, 236, 237f
Sciatic nerve, 151, 200f, 267, 271f, 274, 274f, 276f, 279
Sclera, 299, 299f, 302, 302f, 307, 307f
Scleral venous sinus, 301, 302f
Scoliosis, 97
Scrotum, 177, 515, 516f–517f, 521, 521f
Sebaceous (oil) glands, 59, 60f, 62, 62f–63f
Sebum, 62
Secondary bronchi, 437f–438f, 438
Secondary epiphysis, 73f
Secondary follicles, 525, 527f–529f
Secondary oocytes, 525, 528f, 534f
Secondary spermatocytes, 516f, 517, 518f
Secondary teeth, 467
Secretion, renal, 495
Secretory vesicles, 25f
Sectioning planes, 4–5, 5f
Sediment study, 509–511, 509f–511f
Segmental arteries, 496, 497f–498f
Segmental bronchi, 437f–438f, 438
Selectively permeable membrane, 32
Selectivity, of filtration membrane, 35
Sella turcica, 118, 119f, 123f, 326f
Semen, 519
Semicircular canals, 312f–313f, 314–316, 314f, 317f
  anterior, 316, 317f
  lateral, 316, 317f
  posterior, 316, 317f
Semilunar notch, 83, 85f
Semilunar valves, 355f, 358, 358f, 359, 359f, 375
Semimembranosus muscle, 162f, 190t, 193, 195f, 199, 200f
Semimembranosus tendon, 138f
Seminal fluid, 519
Seminal vesicle, 516f–517f, 518, 519f
Seminiferous tubules, 331, 331f, 515, 516f, 518f
Semispinalis capitis muscle, 224f, 228f
Semispinalis cervicis muscle, 224f
Semispinalis muscle, 222, 222t, 224f, 228f
Semispinalis thoracis muscle, 224f
Semitendinosus muscle, 162f, 190t, 192–193, 195f, 199, 200f
Sensorineural deafness, 317

Sensors, in electrocardiogram, 366
Sensory cortex, primary, 244, 245f
Sensory division, of peripheral nervous system, 233
Sensory homunculus, 244
Sensory nerves, 263
Sensory (afferent) neurons, 234, 235f, 280, 281f
Sensory receptors, 235f–236f, 287–290
Sensory speech area, 245f
Septa, 516f
Septal area, 251f
Septal cartilage, 435, 436f
Septal cells, 441
Septum pellucidum, 247, 249f
Serosa
  of female reproductive system, 530f
  of gastrointestinal tract, 469, 469f, 470, 470f
  of small intestine, 474f
  of stomach, 471f–472f, 472
Serous membranes, 6
Serous pericardial fluid, 353
Serous pericardium, 353
Serratus anterior muscle, 161f, 166f–167f, 170f, 221, 222t, 223f, 229f
Sertoli cells, 515, 516f, 518f
Sesamoid bone, 90
Sex hormones, 326
Shaft
  of femur, 89f
  of hair, 62, 63f–64f
  of humerus, 83
  of metacarpal, 85, 86f
  of phalanx, 85, 86f
Sharpey's fibers, 72
Sharps container, 160
Sheep
  brain of, 252, 255f–257f
  eye of, 307–308, 307f
  heart of, 360–362, 361f–362f, 375–376
  kidney of, 500, 503f
Short bones, 71, 71f
Shoulder joint, 134, 136f
Shoulder, muscles of, 160–169, 163t, 179–182, 181f
Sigmoid colon, 473, 475f
Sigmoid sinus, 405f
Simple columnar epithelium, 43, 43f, 472, 473, 474f
Simple cuboidal epithelium, 42, 43f
Simple epithelium, 42–43, 42t
Simple squamous epithelium, 42, 42f, 422, 472
Single pith, 151
Sinoatrial (SA) node, 365, 371

Sinuses, 72t, 121, 121f. See also specific types
Sinusoids, of liver, 476, 478f
Skeletal muscle, 46, 46f, 159, 161f–162f, 469
Skeletal system, 2f, 3, 69–76
Skin, 2f, 59–62, 60f
Skin receptors, 287–290, 288f
Skull, 2f, 70f, 111–128, 112f–117f, 119f–120f, 121f–124f
  anterior view, 112–113, 113f
  fetal, 121, 131f
  inferior view, 115–118, 117f
  lateral view, 116f
  midsagittal view, 118–121, 120f
  openings of, 117t, 118t
  sinuses of, 121, 121f
  superior view, 112–115, 115f
Slides, 15, 17
Sliding filament model, 147
Small intestine, 2f, 398f, 399, 465f, 472–473, 473f–474f
Smell, sense of. See Olfaction
Smooth endoplasmic reticulum, 24–26, 25f, 27t
Smooth muscle, 47, 47f, 469, 499f
Snellen test, 305
Sodium chloride solution, A–3
Sodium hydroxide solution, A–3
Sodium-potassium pump, 365, 371
Soft palate, 436f, 466, 466f, 480f
Soleus muscle, 161f–162f, 191t, 195, 196f–198f, 201, 201f
Solute perspective, of osmosis, 32
Solutes, 31
Solutions, 32, 35f
Solvent perspective, of osmosis, 32
Solvents, 31
Soma (nerve cell body), 48, 48f, 233, 235f–236f
Somatic motor nervous system, 233
Somatic nerves, 265–273, 266t
Somatic reflex, 280
Somatic sensory association area, 245f
Somatic sensory cortex, primary, 245f, 246f
Somatostatin, 329
Somatotropin, 326
Sound location test, 318
Specific gravity, of urine, 508, 509t
Spence, tail of, 531
Sperm, 516f, 517, 533–534, 534f
Sperm cells, 27, 515–517, 516f, 518f, 534f
Spermatic cord, 517f, 518–519, 521f–522f
Spermatic fascia, 517f
Spermatids, 516f, 517, 518f
Spermatocytes, 515–517, 516f, 518f
Spermatogenesis, 517
Spermatogonia, 515, 516f, 518f

Spermatozoa, 330, 515
Spermicides, 529
Sphenoid bone, 112, 112f, 113f, 115, 116f, 118, 118f–120f, 121, 123f, 436f
Sphenoidal (anterolateral) fontanels, 130, 131f
Sphenoidal sinus, 120f, 121, 121f, 436f
Sphincter pupillae, 301
Sphygmomanometer, 429, 430f
Spinal artery, anterior, 386f
Spinal cord, 2f, 233, 234f, 263–276, 263f–273f
Spinal cord central canal, 243f
Spinal cord smear, 48f
Spinal curvature, 97, 98f
Spinal nerves, 234f–235f, 263–276, 263f–266f
Spinal plexuses, 265–273, 267f–268f
Spinalis muscle, 221, 222t, 224f, 225f
Spinalis thoracis muscle, 224f
Spindle apparatus, 27
Spindle fibers, 28, 29f
Spine
  of bone, 72t
  of ilium, 86, 87f–88f, 191f
  of ischium, 87, 88f–89f
  of scapula, 81, 82f, 168f
Spinodeltoid muscle, 181, 181f
Spinotrapezius muscle, 180, 181f, 229f
Spinous process, 97, 99f–101f
Spiral lamina, 313f–314f
Spiral ligament, 313f–314f
Spiral organ, 313, 314f
Spirogram, 459–460
Spirometer, 449–452, 449f
Splanchnic nerve, 266f
Spleen, 2f, 399, 423f, 424–425, 424f–425f, 481f, 482
Splenic artery, 381f, 395, 396f, 398f, 399, 399f, 425f
Splenic flexure, 473, 475f
Splenic vein, 404f, 408, 409f, 425f
Splenium, 247, 249f
Splenius capitis muscle, 162f, 224f, 228f
Splenius cervicis muscle, 228f
Splenius muscle, 222, 222t, 224f, 228f–229f
Spongy bone, 53, 53f, 72, 73f
Spongy urethra, 500f, 517f, 519, 519f
Squamous cells, in urine, 510
Squamous epithelium, 42, 42f, 44, 44f, 45f, 62, 421, 472
Squamous portion, of temporal bone, 118, 119f, 124, 125f
Squamous suture, 115, 116f, 131f
Stapedius muscle, 312f

Stapes, 311, 312f
Starch, digestion of, 487–489
Starch solution, A–3
Static balance, 311
Stem cells, osteogenic, 75
Stereocilia, 313f–314f, 316f
Stereoscopic vision, 304
Sterilization, male, 520–521
Sternal angle, 104f–105f, 105
Sternal end
  of clavicle, 82, 83f
  of rib, 102, 104f
Sternal symphyses, 132f
Sternocleidomastoid muscle, 161f–162f, 166f, 209, 210t, 211f–212f, 214, 228f
Sternocostal synchondrosis, 132f
Sternohyoid muscle, 209, 210t, 211f–214f, 214
Sternomastoid muscle, 214, 216f–217f
Sternothyroid muscle, 209, 210t, 211f, 214, 216f
Sternum, 2f, 70f, 104f, 105, 226f
Stethoscope, 374
Stimulus, for muscle twitch, 147, 147f, 280
Stimulus-dependent force generation, 149–150
Stomach, 2f, 399, 465f, 470–472, 471f, 480, 481f
Straight arterioles, of uterus, 528
Straight sinus, 405f
Strata, of epidermis, 61–62, 61f
Stratified epithelium, 42, 42t, 44–45, 62
Stratified squamous epithelium, 44, 44f, 45f, 62, 466
Stratum basale, 61–62, 61f, 63f
Stratum corneum, 61f, 62
Stratum germinativum, 61, 61f
Stratum granulosum, 61, 61f
Stratum lucidum, 61f, 62
Stratum spinosum, 61, 61f
Stretch reflexes, 281–282
Striations, 46
Stroma, 525, 527f–528f
Stylohyoid muscle, 211f
Styloid processes, 83, 85f, 115, 116f, 117, 118f, 125f
Stylomandibular ligament, 135f
Stylomastoid, 118t
Stylomastoid foramen, 118f
Subacromial bursa, 136f
Subarachnoid space, 241, 242f, 244f, 252, 265, 265f
Subcapsular sinus, 424f
Subclavian arteries, 381f–383f, 382, 389,
  left, 382, 382f, 388, 388f, 389, 389f–390f, 396f
  right, 382, 382f, 388, 388f–390f, 389, 396f

Subclavian veins, 403, 404f, 406f–407f, 411, 413f
Subclavius muscle, 167f
Subcutaneous fat, 179
Subdural space, 241, 242f, 249f
Sublingual glands, 474, 476f, 478, 479f
Submandibular duct, 476f, 479f
Submandibular glands, 473–474, 476f, 478, 479f
Submucosa
  of gastrointestinal tract, 470, 470f
  of large intestine, 476f
  of small intestine, 474f
  of stomach, 471f–472f, 472
Submucosal plexus, 470f
Subpubic angle, 86, 87f, 88, 89f
Subscapular arteries, 384f, 389, 389f–390f
Subscapular fossa, 81, 82f
Subscapular vein, 411, 413f
Subscapularis muscle, 136f, 163t, 166, 167f–168f, 182
Substantia nigra, 248
Substrate, for enzyme, 488f, 489
Subthreshold stimulus, 148
Sucrose solution, concentrated, A–3
Sudoriferous (sweat) glands, 59, 62, 62f
Sugar, production in digestion, 489–490
Sugar solution, A–3
Sulcus, 72t, 244, 255f
Summation, 147–148, 148f, 151, 154
Superficial, 5t
Superficial fascia, of penis, 519f
Superior, 3, 4f, 5t
Superior angle, of scapula, 81, 82f
Superior border, of scapula, 81, 82f
Superior colliculi, 248, 249f, 252, 252f, 257f, 303f
Superior ligament of malleus, 312f
Superior lobe, of lung, 438, 438f, 443f
Superior meatus, 436f
Superior olivary nucleus, 315f
Supination, 140f
Supinator muscle, 163t, 169, 176f, 182, 184f, 269f
Supraclavicular nerves, 268f
Supracondylar ridges, lateral and medial, 83, 84f
Supraglenoid tubercle, 81, 82f
Supramaximal stimuli, 148
Supraorbital foramina, 112, 113f, 116f, 118t
Supraorbital margins, 112, 113f, 116f
Suprapatellar bursa, 138f, 139f
Suprarenal arteries, 395, 396f, 399f
Suprarenal veins, 410, 410f
Supraspinatus muscle, 163t, 166, 168f, 181f, 182

Supraspinatus tendon, 167f
Supraspinous fossa, 81, 82f
Suprasternal notch, 105
Surface area, and digestion, 492
Surface mucous cells, 471f, 472
Surfactant, 441
Surgical neck, of humerus, 83, 84f
Suspensory ligaments, 301, 302f, 307, 307f, 528, 530f, 531
Sustentacular (Sertoli) cells, 515, 516f, 518f
Sutural bones, 115, 115f–116f, 131f
Suture, 131
Sweat ducts, 59, 60f
Sweat glands, 59, 60f, 62, 62f
Sympathomimetic drugs, 374
Symphysis, 130, 132f, 133f. See also specific types
Symphysis pubis, 86, 87f–89f, 132f, 520f, 521, 526f, 532
Synapse, 147, 234–235, 237f
Synaptic cleft, 147f, 237f
Synaptic vesicle, 147f, 237f
Synarthrotic joints, 133
Synchondrosis, 130, 132f
Syncytial muscle cells, 46
Syndesmosis, 130, 130f
Synergists, 141
Synostosis, 131
Synovial cavity, 131
Synovial fluid, 129, 131, 133f
Synovial joints, 130t, 131–133, 133f–135f
Synovial membrane, 131, 133f
Systemic anatomy, 1
Systemic circulation, 421f
Systole, 365, 376f
Systolic pressure, 430f, 431

T
T cells, 340–341, 425
T_3, 329
T_4, 329
Tachycardia, 368
Tail, 177
  of epididymis, 517f, 518, 521f
  of pancreas, 477, 482
  of sperm, 516f, 517
Tail of Spence, 531
Talus, 90, 92f
Tapetum lucidum, 307, 307f
Target areas, 325
Tarsal bones, 70f, 90, 92f
Tarsal (meibomian) gland, 301f
Tarsal plate, 301f
Tarsal region, 7f
Taste, 293–294
Taste area, 245f
Taste buds, 293, 294f, 466, 467f
Taste cell, 294f
Taste pore, 294f
Taste receptors, 294–295, 295f
Teats, 177

Tectorial membrane, 313–314, 314f–315f
Tectum, 247, 249f
Teeth, 114f, 129f, 466–469, 468f
Telencephalon, 242f, 243t
Telophase, 28, 30f, 30t
Temperature
  and diffusion rates, 31–32
  and heart rate, 374
  judgment of, 287
  measurement of, A–1, A–1t
  nerve response to, 280
Temperature receptors, 287, 289
Temporal artery, superficial, 383, 385f, 429f
Temporal bone, 112, 112f, 113f, 115, 116f, 117f, 118f–120f, 124, 125f, 312f
Temporal lines, inferior and superior, 116f
Temporal lobe, 245f, 246, 248f, 255f–258f, 314
Temporal process, 115, 116f, 118f
Temporal vein, superficial, 406f
Temporalis muscle, 2f, 159, 210t, 212, 212f–213f, 214, 214f, 217f, 466
Temporomandibular joint, 134, 135f
Tendinous intersections, 221, 223f, 228f
Tendon(s), 72, 179
Tendon sheath, 133, 133f–134f
Teniae coli, 473, 475f
Tensor fasciae latae muscle, 161f, 190t, 191f, 192, 199, 200f
Tensor tympani muscle, 312f, 315f
Teres major muscle, 162f, 163t, 165f, 166, 167f–168f, 170f, 182
Teres minor muscle, 162f, 163t, 165f, 166, 167f–168f, 170f, 182
Terminal bronchiole, 438f, 440f
Terminal sulcus, 467f
Tertiary bronchi, 437f–438f, 438
Testes, 3f, 325f, 331, 331f, 515–519, 516f–518f, 521
Testicular arteries, 395, 396f, 517f, 518
Testicular nerves, 517f, 518
Testicular vein, 517f, 518
Testosterone, 331, 515
Tetanus, 148, 149f, 151
Thalamic nuclei
  anterior, 251f
  lateral geniculate, 303f
  medial geniculate, 315f
Thalamus, 247, 249f–250f, 252, 252f, 257f
Theca externa, 527f, 529f
Theca interna, 527f, 529f
Thenar eminence, 174, 175f

Index

Thermoreceptors, 287, 289
Thigh, muscles of, 189–193, 190t, 191f–194f, 195f, 198–204, 199f–200f
Third ventricle, 241, 243f–244f, 249f–250f, 252, 257f
Thoracic aorta, 381f, 382, 382f, 388, 396f
Thoracic arteries
    internal, 382f, 384f–385f, 396f
    lateral, 384f
    long, 389, 389f
    ventral, 389, 389f
Thoracic cage, 102, 104f
Thoracic cavity, 5, 6f, 356f
Thoracic curvature, 97, 98f
Thoracic duct, 2f, 423, 423f, 426f
Thoracic muscles, 221–223, 222t, 226f–227f
Thoracic nerves, 265–267, 267f–268f
Thoracic region, 7f, 98f
Thoracic veins, 410, 411f, 413f, 415
Thoracic vertebrae, 98, 100f
Thoracic volume, 447–458, 448f
Thoracoacromial artery, 384f
Thoracodorsal vein, 413f
Threshold, 148, 152–154
Threshold stimulus, 152–154
Threshold voltage, from sciatic nerve stimulation, 278–279
Throat. *See* Pharynx
Thumb, muscles of, 164t, 174, 177f
Thymosin, 329
Thymus, 2f, 6f, 325f, 329, 423f, 425, 426f
Thyrocervical artery, 389, 389f–390f
Thyrocervical trunk, 384f
Thyrohyoid membrane, 437f
Thyrohyoid muscle, 211f
Thyroid arteries, superior, 385f
Thyroid cartilage, 436f, 437, 437f, 441, 442f
Thyroid gland, 3f, 325f, 328–329, 328f
Thyroid notch, superior, 437f
Thyroid vein, superior, 406f
Thyroid-stimulating hormone (TSH), 326
Thyrotropin, 326
Thyroxine (T_4), 329
Tibia, 2f, 70f, 90, 91f
Tibial arteries, 398f, 400
    anterior, 381f, 397, 397f
    posterior, 381f, 397, 397f, 429f
Tibial (medial) collateral ligament, 136, 138f
Tibial condyles, 90, 91f

Tibial nerve, 267, 271f, 273f, 274, 274f
Tibial tuberosity, 90, 91f
Tibial veins, 411, 414f
    anterior, 404f, 408f, 414f
    posterior, 404f, 408f, 414f
Tibialis anterior muscle, 159, 161f, 190t, 193, 196f–198f, 201f, 203, 203f
Tibialis anterior tendon, 203f
Tibialis posterior muscle, 191t, 195, 198f, 201, 202f
Tibialis posterior tendon, 202f
Tibiofemoral joint, 136, 138, 138f–141f
Tidal volume (TV), 449, 453, 453f
Tight junctions, 46, 46f
Time, measurement of, A–1, A–1t
Tissues, 41–58
Tongue, 294–295, 294f, 436f, 466, 466f–467f, 478, 480f
Tonic receptors, 289
Tonic water, A–3
Tonsils, 2f, 423f, 424, 425f, 436f, 480f
Torso, muscles of, 221–232
Total lung capacity, 453f, 454
Touch receptors, 287–290
Trabeculae, 72, 73f, 75f
Trabeculae carneae, 355f, 358, 359f–360f, 361
Trachea, 2f, 6f, 356f, 435f–436f, 438, 438f–439f, 441, 443f, 480f
Tracheal cartilage, 437f, 438, 439f, 441
Tracheal membrane, posterior, 438, 439f, 441
Tracheobronchial tree, 438, 438f–439f
Tracking, visual, 302
Tracts, of spinal cord, 264, 265
Transection, of muscle, 179
Transfusion reaction, 345–346
Transitional cells, in urine, 510, 510f
Transitional epithelium, 44–45, 497, 499f
Transverse colon, 475f, 481f, 482
Transverse fissure, 245f, 246
Transverse foramen, 98, 98f–100f
Transverse jugular vein, 411, 413f
Transverse lines, 101, 102f
Transverse plane, 4, 5f
Transverse process, 97, 99f–101f
Transverse scapular vein, 411, 413f
Transverse sinus, 405f
Transverse tubules, 147
Transversus abdominis muscle, 160, 221, 222t, 223f, 228
Transversus thoracis muscle, 226f
Trapezium bone, 83, 85, 86f

Trapezius muscle, 160, 161f–162f, 163t, 165f, 179f–180f, 211f–212f, 225f, 228f, 268f
Trapezoid bone, 83, 86f
Treppe, 153, 153f
Triceps brachii muscle, 162f, 163t, 165f, 168f, 169, 170f–171f, 179f, 182, 182f–183f, 269f
Triceps brachii reflex, 282, 282f
Triceps brachii tendon, 136f
Tricuspid valve, 358, 359f–362f, 375, 375f
Trigeminal nerve, 251, 255t
Trigone, 497, 499f–500f
Triiodothyronine (T_3), 329
Triquetrum bone, 85
Trochlea of humerus, 83, 84f
Trochlear nerve, 251, 255t
Trophoblast, 534
True vocal cords, 437
Tubal ligation, 529
Tunica adventitia, 379, 380f
Tunica albuginea, 515, 521
Tunica intima, 379, 380f
Tunica media, 379, 380f
Tunica vaginalis. *See* Scrotum
Two-Point Discrimination Test, 289
Tympanic cavity, 311
Tympanic membrane, 311

U

Ulna, 85f
Ulnar artery, 383, 384f, 389, 389f
Ulnar nerve, 266, 270f, 274, 274f
Ulnar notch of radius, 85f
Ulnar tuberosity of radius, 85f
Ulnar vein, 405, 407f, 411
Umami solution, A–3
Umbilical cord, 412f
Umbilical vein, 410, 412f
Unipolar neuron, 234, 236f
Universal donor, 346
Universal recipient, 346
Unmyelinated nerve fibers, 236
Ureters, 2f, 495, 496f, 497, 497f–500f, 500, 503f, 516f–517f, 533f
Urethra, 2f, 495, 496f, 497–498, 500, 500f, 516f, 519f, 521f, 526f, 533f
Urethral opening, 499f, 533, 533f
Urethral orifice, external, 497, 517f, 519f, 526f, 528, 530f
Urethral sphincters, external and internal, 500f
Urinalysis, 507–514
Urinary bladder, 2f, 495, 496f, 497, 499f, 500, 503f, 516f–517f, 519f, 521f–522f, 533f
Urinary sphincter, external, 499f
Urinary system, 2f, 3, 495–506, 503f

Urine, characteristics of, 507–511
Urobilinogen, 509, 509t
Urochrome, 507
Urogenital diaphragm, 500f
Urogenital opening, 177
Urogenital orifice, female, 532, 533f
Urogenital triangle, 520, 520f, 528
Uterine cavity, 530f
Uterine epithelium, 534f
Uterine glands, 526–528, 530f
Uterine horns, 532, 533f
Uterine tube, 2f, 525, 526f, 530f, 533, 533f
Uterus, 2f, 526–528, 526f, 530f–531f, 532, 534f
Utricle, 314, 316f
Utricular macula, 316f
Uvula, 436f, 437, 466, 466f

V

Vagina, 2f, 526f, 528, 530f, 533, 533f
Vaginal canal, 528, 533
Vaginal orifice, 526f, 528, 530f
Vaginal vestibule, 533
Vagus (X) nerve, 251–252, 253f, 254t–255t, 256f, 371
Vallate papillae, 294f
Values, measurement of, A–1
Valves
    of heart, 375–376, 376f
    of lymphatic vessels, 422, 423f
    of veins, 422, 422f
Variables, in experiments, A–4
Vas deferens, 520, 520f
Vasa recta, 496, 498f
Vasa vasorum, 380f
Vascular layer, of eye, 302, 302f
Vasectomy, 520, 520f
Vasopressin, 327
Vastus intermedius muscle, 161f, 190t, 191f, 192, 193f, 199, 272f
Vastus lateralis muscle, 161f, 190t, 191f, 192, 193f, 199, 200f, 272f
Vastus medialis muscle, 161f, 190t, 191f, 192, 193f, 199, 199f, 272f
Vater, ampulla of, 477, 479f
Vein(s), 60f, 353, 379, 403–420, 404f, 421, 421f
    of abdomen, 410, 410f
    of brain, 403–405, 405f
    cat, 388, 410–416, 413f–416f
    deep, 401, 403, 407f
    of head and neck, 403–405, 404f–406f
    of heart, 356f–357f, 358f
    of kidney, 495, 498f
    of lower limb, 404f, 405, 408f
    of pelvis, 405, 408f, 410
    of thorax, 410, 411f

Vein(s),—(contd.)
  of trunk, 404f, 408
  of upper limb, 404f, 405, 407f
  structure of, 379, 380f, 403
  superficial, 403
  valves of, 422, 422f
Vena cava
  inferior, 3f, 226f, 355f, 356, 356f–359f, 358, 360, 362f, 388f, 397f, 399, 400f, 404f, 405, 406f, 408f–410f, 411, 411f, 414f, 416f, 496f
  superior, 3f, 355f, 356, 356f–359f, 358, 360, 362f, 388f, 404f, 405, 406f, 411f, 426f
Venous sinus, 244f
Ventral, 3, 4f, 5t
Ventral (anterior) column, 264f
Ventral ramus, 265, 266f
Ventral root, 235f, 265, 265f–266f
Ventricles, of brain, 241–243, 243f–244f, 249f–250f, 252, 257f
Ventricles, of heart, 353–356
  left, 356, 356f–357f, 359f, 360, 362f
  right, 356, 356f–357f, 358, 358f–359f, 360, 362f
Ventricular depolarization, 366, 366f
Ventricular repolarization, 366, 366f
Ventricular systole, 365
Venules, 379, 421, 421f, 423f
Vermiform appendix, 2f, 465f, 473f, 474, 475f–476f
Vertebra prominens, 98, 100f
Vertebrae, 97–101, 98f, 99f
Vertebral arch, 97, 99f

Vertebral arteries, 383, 384f–385f, 389, 389f, 413f
  left, 383
  right, 383, 389f, 413f
Vertebral column, 2f, 70f, 97, 98f
Vertebral foramen, 97, 99f–100f
Vertebral region, 7f
Vertebral ribs, 102
Vertebral veins, 403, 405f, 411
Vertebrochondral ribs, 102
Vertebrosternal ribs, 102
Vertical plate, of palatine bone, 120f
Vesicles, 26
Vestibular folds, 436f, 437, 437f, 442f
Vestibular glands, greater, 528
Vestibular membrane, 313, 314f–315f
Vestibular nerve, 316f
Vestibule
  ear, 312f–313f, 314, 314f, 316f
  nasal, 435, 436f
  oral, 480f
  vaginal, 528, 530f
Vestibulocochlear (VIII) nerve, 251, 253f, 254t–255t, 256f, 312f, 314, 315f
Vibrissae, 177
Villi, 473, 474f, 482
Vinegar solution, A–3
Visceral arteries, 382f
Visceral pericardium, 3, 355f
Visceral peritoneum, 387, 469, 469f, 470, 470f–472f, 482
Visceral pleura, 4, 356f, 439, 439f, 441
Visceral reflex, 280
Vision, 299–310
Visual acuity, 305
Visual area, of brain, 244

Visual association area, 245f
Visual cortex, 245f, 303f
Visual cues, in olfaction, 295
Visual field, 303f, 304, 305f
Visual pathways, 303f
Visual tracking, 302–304
Vital capacity (VC), 450, 451t–452t, 453, 453f
Vitreous chamber, 302, 302f
Vitreous humor, 302f–303f, 307f, 308
Vocal cords, 437, 437f
Vocal folds, 436f, 437, 437f, 442f
Voice box. See Larynx
Volkmann's canals, 72, 73f–74f
Volume, measurement of, A–1, A–1t
Voluntary muscles, 46, 159
Vomer, 112f–113f, 117, 118f, 120f, 435, 436f
Vulva, 528

W
Warm receptors, 289
Water, as solvent, 31
Wave summation, 148, 149f
Weber test, 317, 318f
Weight, initial, 33
Wernicke's area, 244, 245f
Wet mount preparation, 15, 15f, 17
Wet spirometer, 449
White blood cell(s), 53, 54f, 337, 339–342, 339f, 342t
  in urine, 509t, 510, 510f–511f
White blood cell count, 339, 341, 341f
White commissure, 264f

White matter, 236, 250, 250f, 257f, 264, 264f
White pulp, of spleen, 425, 425f
Willis, circle of, 386f
Windpipe. See Trachea
Wisdom teeth, 467, 468f
Working distance, of microscope, 16f
Wormian bones, 115
Wright's stain, A–3
Wrist extensors, 162f
Wrist flexors, 161f

X
Xiphihumeralis muscle, 179, 180f, 228f
Xiphisternal symphysis, 132f
Xiphoid process, 104f, 105

Y
Yellow bone marrow, 51, 72
Yolk sac, 534f

Z
Z disk, 147
Zona fasciculata, 329, 330f
Zona glomerulosa, 329, 330f
Zona pellucida, 527f, 529f
Zona reticularis, 329, 330f
Zygomatic arch, 115, 116f, 118f, 214f
Zygomatic bone, 112, 112f–113f, 116f, 118f
Zygomatic process, 115, 116f, 118f, 124, 125f
Zygomaticus major muscle, 210t, 212, 212f–213f
Zygomaticus minor muscle, 210t, 212, 212f–213f
Zygote, 533–534